ELECTRICAL PROPERTIES OF POLYMERS

Second edition

Fully revised and expanded, this new edition of Anthony Blythe's successful title on electrical properties of polymers covers both the fundamental and recent developments in this growing area. The book provides a broad and comprehensive account of the topic, describing underlying physical principles and synthesis through to emerging technologies. The second edition places particular emphasis on the new generation of conductive polymers, describing emerging uses of polymers in industrial applications and covering topics such as light emitting diodes, flexible polymers and soft electronics.

Written in an accessible style, without complicated theory, this book combines key concepts with applications. With the inclusion of further reading material at the end of each chapter for interested readers, this book is an authoritative guide to advanced-level undergraduates and graduates studying polymer materials and physical sciences. It will also be of significant interest to researchers working in this evolving field.

Anthony Blythe has been active in the area of polymer materials and processing for over 40 years. He has successfully initiated numerous ground-breaking projects in the polymer field, specifically for the wire and cable industry, and his achievements are recognised by the several patents he holds. He has also been an external lecturer to established universities in the UK and has authored more than 40 scientific papers.

David Bloor is Emeritus Professor of Physics at the University of Durham. Since his retirement in 2002 he has continued his research in the area of optical materials and conductive materials, with particular interest in the developing areas of liquid crystalline materials and display devices. He holds a number of patents and has authored over 300 scientific papers.

ELECTRICAL PROPERTIES OF POLYMERS

Second edition

TONY BLYTHE
Former Chief Scientist, BICC Cables

DAVID BLOOR
University of Durham, UK

CAMBRIDGE
UNIVERSITY PRESS

CAMBRIDGE UNIVERSITY PRESS
Cambridge, New York, Melbourne, Madrid, Cape Town, Singapore, São Paulo

CAMBRIDGE UNIVERSITY PRESS
The Edinburgh Building, Cambridge, CB2 2RU, UK

Published in the United States of America by Cambridge University Press, New York

www.cambridge.org
Information on this title: www.cambridge.org/9780521

Printed in the United Kingdom at the University Press, Cambridge

A record for this book is available from the British Library

Library of Congress in Publication data

ISBN-13 978-0-521-55219-6 hardback

ISBN-10 0-521-55219-2 hardback

Contents

Preface

The predecessor to this book was published in the Cambridge Solid State Sciences Series in 1979 following a series of lectures given to graduate students in the Polymer Physics Group at Leeds University. It was written as an introductory text on electrical aspects of polymeric materials, principally electrically insulating polymers, but with a description of work on electrically conductive organic materials up to the mid-1970s. Since that time intrinsically conductive polymers have progressed from laboratory curiosities with either poor chemical and physical stability frequently, indeterminate composition, to widely explored materials that have demonstrated new physics and use in diverse commercial applications. An account of these developments has displaced most of the content on organic conductors from the original volume.

Emphasis is mainly laid on the description and explanation, in molecular and electronic terms, of the observed phenomena, so as to give a basic understanding of the electrical behaviour of polymers. Principles of measurement methods are also stressed, since a sound framework of experimental techniques and data is so essential for proper scientific development of such a subject. The choice of subject matter was made with the aim of being educative and stimulating rather than exhaustive and exhausting. It is hoped that the book will be of most help to those venturing into research in polymer science, or to those joining the industrial sectors that utilise both insulating and conductive plastics, who want an insight into the somewhat specialised area of electrical properties. Only a general knowledge of physics and chemistry is assumed, and for this reason an introduction to polymer structure is included.

The electrical properties of polymers are a subject which is inherently interdisciplinary in nature. The development of intrinsically conductive polymers has benefited immensely from the contributions of synthetic chemists.

Electrical properties are closely allied with the mechanical properties of polymers studied by both physicists and engineers. Established models of the semiconductive properties of inorganic substances and metals provide a starting point for understanding similar properties of polymers, although subsequent critical input from across the disciplines eventually led to the view that the underlying physics is distinctly different. A primary objective, therefore, was to collate the relevant aspects of these contingent subjects to form a more unified treatment than is generally available.

Since the early days of plastics technology, when such materials were regarded electrically as simply good insulators, observations of subtleties in electrical response have shed a great deal of light on the underlying microscopic structure and molecular dynamics. This has contributed to polymer science in a general way, and has, at the same time, enabled the development of materials which meet exacting electrical engineering requirements. This process has continued in the development of intrinsically conductive polymers over the last 25 years of the twentieth century. The book encompasses both this well-established area of dielectric science and the modern frontiers concerned with non-linear optical and electrically conductive plastics. Research along these lines has demonstrated the feasibility of obtaining materials with entirely novel sets of properties, that have opened up new areas of application for polymers.

The bibliography has been considerably expanded, partly due to the inclusion of new topics, e.g. linear and non-linear optical properties, and partly reflecting the steady and sizeable flow of papers on intrinsically conductive polymers. The authors have made extensive use of the literature in the preparation of this book. The selection of papers included in the bibliography is meant to be indicative of this literature since it cannot possibly include more than a small fraction of the published output. The final choice is not intended as any indication of the quality of work that is not included.

SI units have been used throughout the book.

Special thanks are due to Professor I. M. Ward for encouraging us to update the earlier book and for reading and commenting on the whole of the draft manuscript. We should also like to thank Dr G. H. Cross and Dr G. R. Davies for constructive criticism on much of the manuscript, and Professors J. S. Dugdale and A. P Monkman for suggestions on Chapters 4 and 9 respectively. ARB is grateful for the invaluable help and advice from colleagues at BICCGeneral, and DB for similar assistance from colleagues at the University of Durham. Valuable inputs were received from Professor D. Baeriswyl, Professor A. Kaiser, Professor W. R. Salaneck, Professor G. Weiser, Dr D. de Leeuw and Dr S. Roth. Dr Mile H. Bertein, Mr W. Reddish, Dr D. J. Groves and Mr J. Billing kindly supplied photographs.

Permission to reproduce the material listed below is gratefully acknowledged.

Covion Organic Semiconductors GmbH for Fig. 10.17.

OSRAM Opto Semiconductors Inc. for Fig. 10.19.

Philips Research for Figs. 10.10 and 10.18.

The American Institute of Physics for Figs. 3.16, 4.13, 7.21, 7.22, 8.2, 8.3, 8.26, 8.28, 8.29(a), 9.15, 9.34, 10.3, 10.8, 10.9, 10.12, 10.23 and 10.28.

The American Chemical Society for Figs. 3.11, 7.19, 9.32, 10.2(b), 10.6 and 10.11(b).

The American Physical Society for Figs. 4.14, 6.3, 8.27, 8.29(b), 9.1, 9.21, 9.36, 10.1 and 10.2(a).

The Institution of Electrical Engineers for Figs. 6.4, 6.5 and 6.8.

The Institute of Electrical and Electronics Engineers, Inc. for Figs. 6.9 and 7.9.

The Institute of Physics for Figs. 6.15, 7.16, 8.13, 8.15, 8.17, 8.30(a), 9.27, 9.33, 9.37, 9.38 and 10.15.

The Japanese Physical Society for Fig. 3.20.

The Royal Society for Fig. 6.2.

The Royal Society of Chemistry for Fig. 5.1.

Dr Dietrich Steinkopff Verlag for Fig. 3.7.

Elsevier for Figs. 613, 8.12, 8.19, 8.20, 8.22(b), 8.30(b), 9.7, 9.19(a) & (c), 9.23(b), 9.24, 9.26, 9.29, 9.30, 10.4, 10.20, and 10.25.

Imperial College Press for Fig. 9.23(a).

Macmillan Magazines Ltd for the following figures from *Nature*: Figs. 8.22(a), 9.19(b), 9.28, 10.33, 10.7(a), 10.11(a) and 10.20.

Marcel Dekker, Inc. for Figs. 3.8, 3.10, 9.31 and 10.2(c).

Oxford University Press for Fig. 6.6.

John Wiley and Sons, Inc. for Figs. 3.13 and 8.16.

Wiley-VCH for Figs. 8.8, 8.9, 9.20, 9.22(b), 9.35, 10.5, 10.7(b), 10.12, 10.14, 10.22 and 10.26.

Finally, we are especially grateful to our wives for their constant moral support during this project.

1
Introduction

1.1 Overview

Although unified by direct concern with the effects produced by electric fields, the subject of the electrical properties of polymers covers a diverse range of molecular phenomena. Until the latter part of the twentieth century, known polymers generally displayed a much less striking electrical response than inorganic materials. While inorganic solids behave as semiconductors and metals (where the electrical response is overwhelmingly one of electronic conduction), or as insulators, the common polymers were distinctly confined to the latter category. From 1950 onwards this situation changed with the emergence of polymers displaying semiconducting properties and the discovery in the 1970s of polymers that even had metallic levels of conductivity. Since then there has been an explosion in fundamental and applied research and numerous conductive polymers have been produced and studied. While the macroscopic properties of these materials resemble those of their inorganic counterparts, their microscopic behaviour is markedly different and reflects fundamental differences between the rigid crystal lattices of inorganic materials and the deformable, molecular structure of polymers.

The absence of conduction in insulating polymers allows a whole set of more subtle electrical effects to be readily observed. For instance, polarisation resulting from distortion and alignment of molecules under the influence of an applied field becomes apparent. Examination of such polarisation not only gives valuable insight into the nature of the electrical response itself, but it also provides a powerful means of probing molecular dynamics. In conductive polymers the presence of charge carriers produces pronounced local deformation of the molecular framework, significantly larger than that found in inorganic semiconductors. The Coulomb interaction of charge carriers is less effectively screened in conductive polymers. Hence electron–phonon and electron–electron interactions have more profound effects in conductive polymers. It should also be noted that single crystal polymers are rare and the complex morphology of partially crystalline polymers has an impact on their electrical properties. For all these reasons electrical studies form a desirable supplement to studies of mechanical and thermal properties aimed at reaching an understanding of the behaviour of polymers on a molecular basis.

In the remainder of this chapter we outline the main features of polymer structure and its relation to electrical properties, as a basis for the subsequent more detailed explanation and discussion of electrical behaviour.

1.2 Structure of polymers

Most of the synthetic high polymers with which this book is concerned are organic compounds consisting of long, chain-like molecules where repeated molecular units are linked by covalent bonds. A single molecular chain can contain a thousand or more repeat units and reach a total length in excess of 1 μm. Such large molecules have complex shapes and form solids that may be either amorphous or crystalline, but are commonly partially crystalline. The atoms, which form the *backbone* of organic polymers, are predominantly carbon atoms, sometimes in combination with oxygen and/or nitrogen. The nature of the chemical bonding in the polymer directly influences the electrical properties.

1.2.1 Chemical structure – saturated polymers

The carbon atoms in a polymer are primarily bound together by *covalent bonds* formed by the interaction of two electrons, one from each of the atoms linked by the bond. We illustrate this process with the simpler example of bond formation between two hydrogen atoms. As the atoms are brought together their 1*s*-electron orbitals overlap forming two new σ-orbitals around the atoms, which are symmetric and have zero angular momentum with respect to the interatomic axis. One orbital, the bonding orbital, is formed by the linear superposition of the two *s*-orbitals, shown by the solid curves in Fig. 1.1(a). This gives a high electron density between the hydrogen nuclei, the dashed line in Fig. 1.1(a), which reduces their electrostatic repulsion. Thus the nuclei are drawn together and an electron in the bonding orbital has a lower energy than one in the isolated atom orbital. The other, the antibonding orbital, is formed by the superposition of *s*-orbitals with opposite sign, see solid curves in Fig. 1.1(b), giving a lower electron density between the nuclei, shown by the dashed curve in Fig. 1.1(b). The electrostatic repulsion between

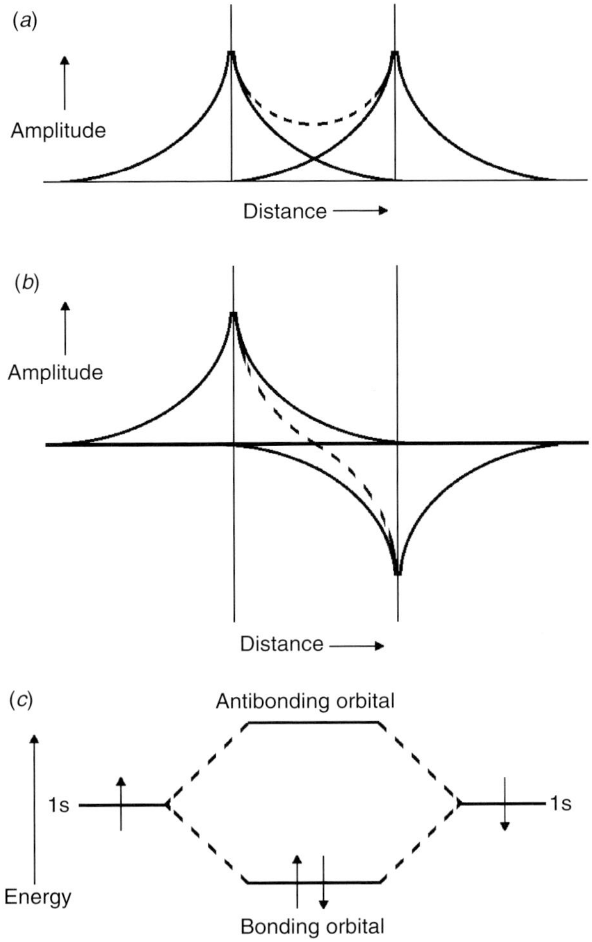

Fig. 1.1 Overlap of hydrogen atom orbitals to form (a) a bonding orbital, and (b) an antibonding orbital, with energies as shown in (c).

the nuclei is not screened and the energy of an electron in this orbital is increased, as indicated in Fig.1.1(c). The Pauli exclusion principle allows the two electrons from the hydrogen atoms to *pair* in the bonding orbital, if they are of opposite spin, to give a stable molecule with total energy less than that of the two isolated hydrogen atoms.

A neutral carbon atom has six electrons, which occupy the $1s$, $2s$ and $2p$ orbitals giving an electronic configuration of $1s^2 2s^2 2p^2$. When a carbon atom forms a bond with another atom, an s-electron is promoted to the vacant p-orbital to give an sp^3 configuration (s, p_x, p_y, p_z) in the outer valence shell. These electronic orbitals do not bond separately but hybridise, i.e. mix in linear combinations, to produce a set of orbitals oriented towards the corners of

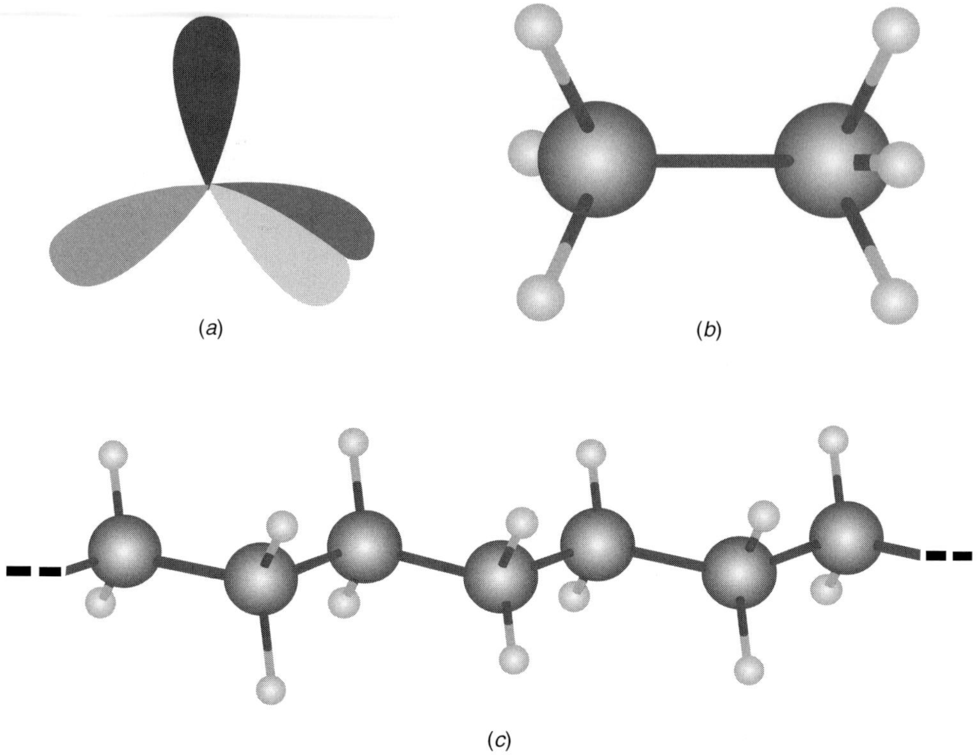

Fig. 1.2 (a) sp^3 hybrid orbitals, and the molecular structure of (b) ethane and (c) polyethylene. H atoms have light shading.

a regular tetrahedron. Figure 1.2(a) shows the form of the wavefunctions that describe the hybridised orbitals. The sp^3 hybrids allow a much greater degree of overlap in bond formation with another atom, and this, in turn, produces extra bond strength and stability in the molecules. The arrangement of bonds resulting from overlap with sp^3 hybrid orbitals on adjacent atoms gives rise to the tetrahedral structure that is found in the lattice of diamond and in molecules such as ethane, C_2H_6, as depicted in Fig. 1.2(b). In these structures all the available electrons are tied up in strong covalent bonds, *σ-bonds*. Carbon compounds containing σ-bonds formed from sp^3 hybrid orbitals are termed *saturated* molecules. Polyethylene, $-(CH_2)-_n$, is a typical saturated polymer where the σ-bonds between adjacent carbon atoms form the zigzag of the polymer backbone and the two remaining bonds are to the pendent hydrogen atoms, as in Fig. 1.2(c). The strong σ-bonding gives rise to chemically stable compounds and localises the electrons in the molecular structure rendering saturated compounds electrically insulating.

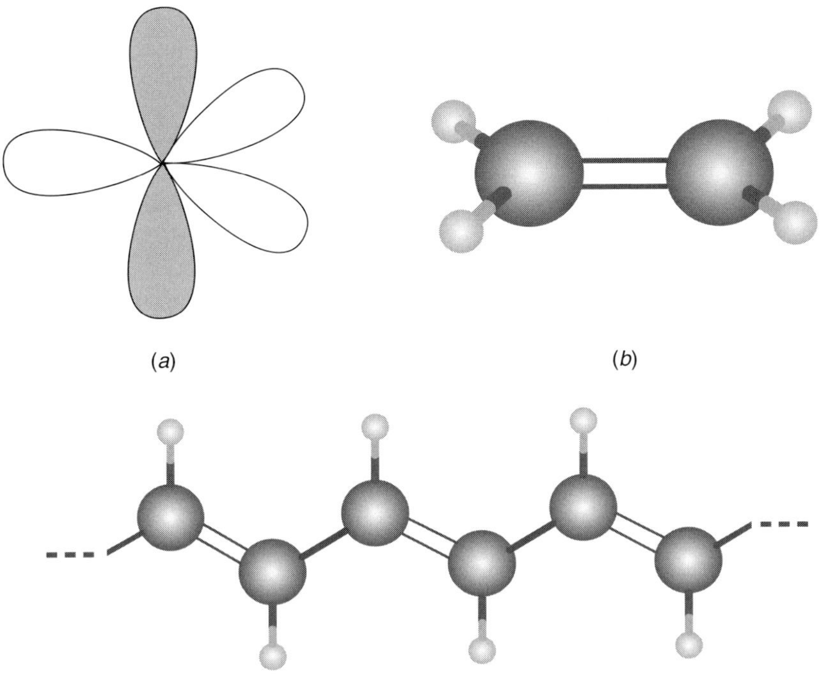

Fig. 1.3 (a) sp^2 hybrid orbitals, and the molecular structure of (b) ethylene and (c) polyacetylene.

1.2.2 Chemical structure – unsaturated polymers

While tetrahedral bonding is common in carbon compounds, it is not the only possibility. A planar σ-bond structure occurs if the hybrid orbitals are formed from one s- and two p-orbitals. sp^2 hybrid orbitals, which lie in a plane with angles of 120° between them, are shown by the unshaded lobes in Fig. 1.3(a), whilst the remaining p-orbital (p_z), which lies orthogonal to the plane of the σ-bonds, is shown by the shaded lobes. This p_z electron is then available to form additional π-orbitals, with unit angular momentum about the bond axis, with p_z electrons in adjacent atoms. The resulting *π-bond* will form parallel to the underlying σ-bond, resulting in a multiple (double) bond, as in ethylene, C_2H_4, as shown in Fig. 1.3(b). Compounds where orbitals are used in this way to form double bonds between carbon atoms, rather than to form single bonds with other atoms, e.g. hydrogen, are said to be *unsaturated*. Where successive carbon atoms in a chain engage in π-bonding, the structure is said to be *conjugated*, and may be represented as a sequence of alternating single and double bonds, as shown in Fig. 1.3(c), which illustrates the bonding in polyacetylene, $\text{-}(CH{=}CH)\text{-}_n$.

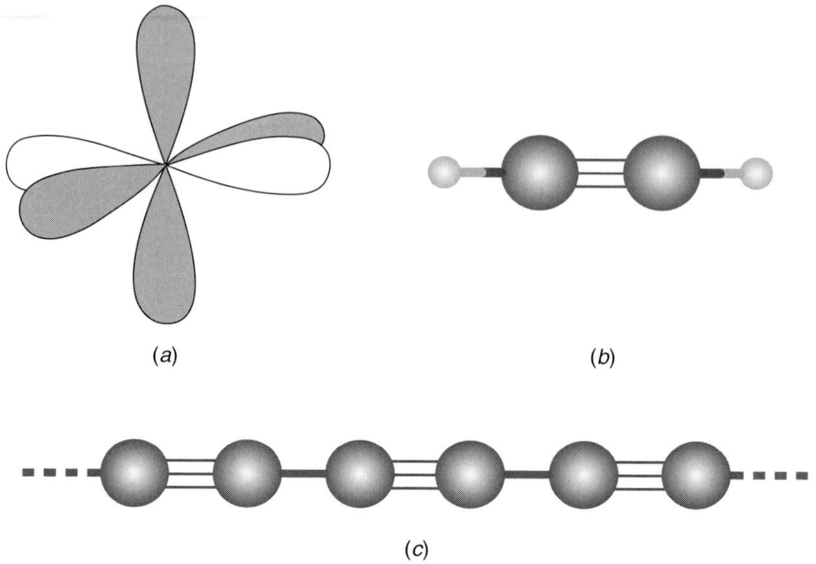

Fig. 1.4 (a) *sp* hybrid orbitals, and the molecular structure of (b) acetylene and (c) polyyne.

Furthermore, compounds containing triple bonds can be obtained from *sp* hybrid orbitals. *sp* hybridisation produces two orbitals oriented at 180° to each other, as shown by the filled lobes in Fig. 1.4(a), which can form co-linear σ-bonds. This leaves two *p*-electrons, shown by the shaded lobes in Fig. 1.4(a), which are free to form π-bonds aligned with the σ-bond, as in acetylene, C_2H_2, shown in Fig. 1.4(b).

The polymer polyyne, $\text{-(}C{\equiv}C\text{)-}_n$, also known as polycarbene, and shown in Fig. 1.4(c), is formed with *sp* hybrids in this way. Its structure may be either an alternation of single and triple bonds or a sequence of double bonds, depending on the terminal groups – see Fig. 1.5.

The formation of multiple bonds leads to a reduction in the separation of the carbon atoms involved. Thus a triple bond is shorter than a double bond, which is shorter than a single bond. This gives rise to an alternation of bond length in conjugated polymers such as polyacetylene and polyyne, Figs. 1.3(c) and 1.4(c). The electrons in the π-bonds are less strongly bound than the electrons in the σ-bonds. Hence the π-electrons can be more easily removed from the bonds than can the σ-electrons. In consequence unsaturated compounds are less chemically stable than saturated compounds. The multiple bonds are subject to chemical attack, being reduced to saturated (single) bonds in the process. The weaker binding of the π-electrons also means that

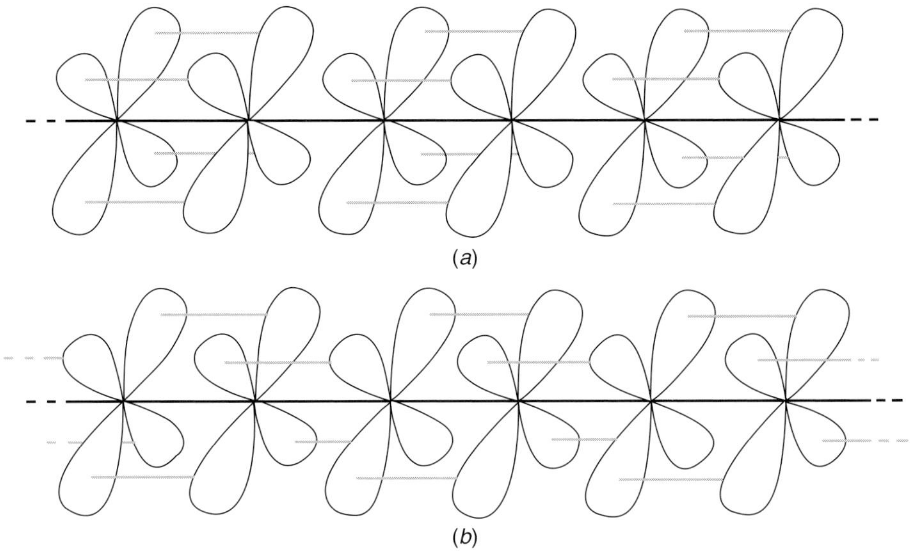

Fig. 1.5 The alternative structures of polymers formed with *sp*-hybridised orbitals, (a) polyyne and (b) polycarbene. Overlap of the π-electron orbitals, shown by the off-axis lines, gives an interatomic π-bond. The σ-bonds are shown by on-axis lines.

unsaturated compounds have the potential to display either semiconducting or metallic behaviour.

1.2.3 Preparation of polymers

The process of polymerisation in which small molecules of the starting material, the monomer, undergo chemical reaction together to form long chains, may proceed in a variety of ways. These may be divided into two principal categories, addition and condensation polymerisation, which have major structural implications for the final product.

Addition polymerisation is typified by the conversion of ethylene to polyethylene according to the following equation:

$$n\,CH_2{=}CH_2 \rightarrow \left[\!-\underset{H_2}{C}-\underset{H_2}{C}-\!\right]_n .$$

In this joining together of molecules of ethylene the necessary linking bonds are obtained by breaking the original double bonds, which are relatively unstable.

The chain reaction may be initiated with the aid of a catalyst, often a peroxide, which provides a chemically reactive species to attack a limited

Table 1.1 *Examples of vinyl and related polymers*

Monomer formula	Polymer name	Polymer structure
$CH_2{=}CH_2$	Polyethylene	$-(CH_2)-_n$
$CH_2{=}CH{-}CH_3$	Polypropylene	$-(CH_2{-}CH(CH_3))-_n$
$CH_2{=}CH{-}Cl$	Poly(vinylchloride)	$-(CH_2{-}CHCl)-_n$
$CH_2{=}CH{-}C_6H_5$	Polystyrene	$-(CH_2{-}CH(C_6H_5))-_n$
$CH_2{=}C(COOCH_3)(CH_3)$	Poly(methylmethacrylate)	$-(CH_2{-}C(CH_3)(COOCH_2))-_n$
$CF_2{=}CF_2$	Polytetrafluoroethylene	$-(CF_2)-_n$

number of double bonds. Sequential addition of ethylene molecules then proceeds until all the ethylene is used up. None of the atoms of the original monomer, ethylene in this case, is excluded from the final polymer, the addition process only involving rearrangement of bonds. The polymer chains produced are of finite length because of the presence of termination reactions, which compete with the monomer addition reaction. Similarly, polymerisation of acetylene leads to the conjugated polymer polyacetylene:

$$n\,CH{\equiv}CH \rightarrow \left[-CH{=}CH-\right]_n.$$

Addition polymers include the important class derived from vinyl monomers, general formula $CH_2{=}CHX$, having a backbone consisting entirely of carbon atoms. Some examples of vinyl polymers and closely related ones are given in Table 1.1.

A second important class of addition polymers is that derived from monomers containing triple bonds and aromatic groups, when the polymers possess conjugated backbones containing multiple bonds. This group includes polyacetylene, and some further examples are shown in Table 1.2. The conjugated backbones are inflexible and the polymers are insoluble in the solvents normally used for polymerisation reactions, which limits the length of the polymer chains that can be attained. This problem can be overcome by either the inclusion of long flexible pendent groups, which render the polymer soluble,

Table 1.2 *Examples of conjugated polymers*

Monomer formula	Polymer name	Polymer structure
HC≡CH	Polyacetylene	$\left[\underset{H}{C}{=}\underset{H}{C} \right]_n$
	Poly(*p*-phenylene)	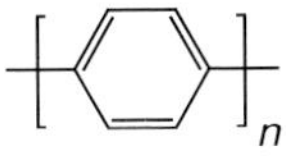
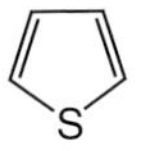	Polythiophene	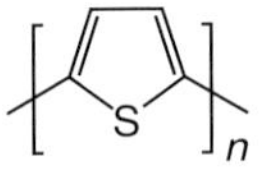
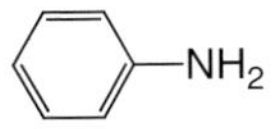	Polyaniline	See Fig. 9.6 [a]
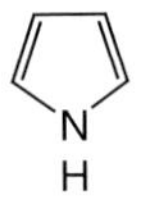	Polypyrrole	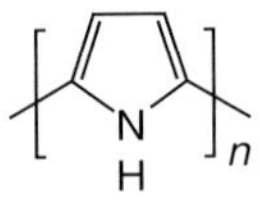

[a] Polyaniline has a complex structure dependent on the degree of oxidation and protonation.

or the preparation of a non-conjugated, soluble precursor polymer, which is subsequently converted into the desired, conjugated product.

Polyethers, which contain oxygen as well as carbon atoms in their backbones, are examples of addition polymers that are generally made by a ring-opening reaction. Thus the cyclic compound ethylene oxide polymerises to a polyether as follows:

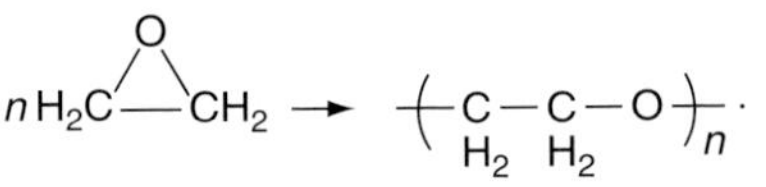

Condensation polymerisation occurs when multifunctional monomers, which possess more than one chemically reactive group per molecule, react together with the elimination of small molecules, typically either water or HCl. With a bifunctional monomer the product is a linear polymer, e.g. the polyamide Nylon-6 is derived from ε-amino caproic acid as follows:

$$NH_2(CH_2)_5CO\boxed{OH + H}NH(CH_2)_5COOH \rightarrow$$
$$NH_2(CH_2)_5CO{\cdot}NH(CH_2)_5COOH + H_2O \text{ etc.}$$

Quite often polymerisation proceeds by interaction of pairs of complementary monomers. Thus Nylon-6,6 is formed by reaction of adipic acid with hexamethylene diamine:

$$HOOC(CH_2)_4COOH + NH_2(CH_2)_6NH_2 \rightarrow$$
$$HOOC(CH_2)CONH(CH_2)_6NH_2 + H_2O \text{ etc.}$$

The formation of poly(ethyleneterephthalate) from ethylene glycol and terephthalic acid is another very well-known example:

$$n HO(CH_2)_2OH + n HOOC{\cdot}C_6H_4{\cdot}COOH \rightarrow$$
$$HO[(CH_2)_2OCO{\cdot}C_6H_4{\cdot}COO]_nH + nH_2O.$$

Conjugated polymers can also be prepared by condensation polymerisation, usually from a dihalide with the elimination of the related acid. The dehalogenation reaction for the preparation of poly(*p*-phenylenevinylene) is a typical example:

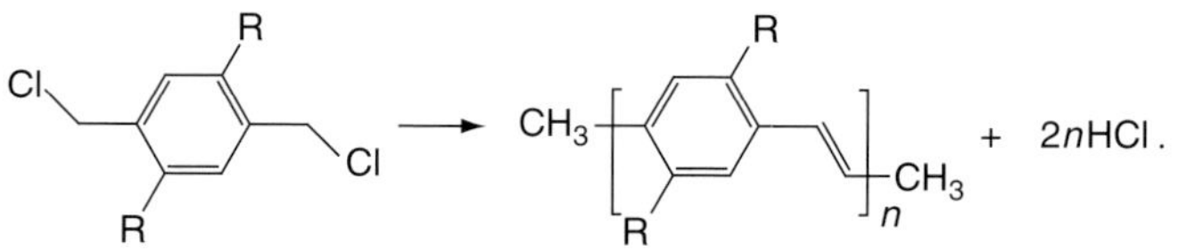

Some common condensation polymers are given in Table 1.3.

The final product of polymerisation, be it of an addition or condensation type, will contain chains covering a range of lengths, and it is only possible to specify an average molar mass.[1] The number-average molar mass $(M_n)_{av}$ is defined by

$$(M_n)_{av} = \frac{\sum M_x n_x}{\sum n_x},$$

where M_x and n_x are the molar mass and relative number, respectively, of the species containing x repeat units.

Electrochemical polymerisation offers an alternative synthetic route in some cases. Initiation here proceeds by electrochemical oxidation or reduction of monomer molecules at an electrode to generate reactive ions, i.e. species with both an unsatisfied valency and a net charge, which are termed radical ions. This method is of limited applicability in saturated polymers that are electrically non-conducting, as an insulating barrier quickly builds up on the electrode where the radical ions are generated and the polymerisation is terminated. This does not occur with conjugated polymers, which generally possess some degree of electrical conductivity, and thick layers of conjugated

[1] The molar mass M of a molecular compound is the mass per mole of the molecules (kg/mol). 1 mole is the amount of the compound that contains as many molecules as there are atoms in 12 g of C^{12}.

Table 1.3 *Examples of condensation polymers*

Polymer name	Chemical formula
Polyester (Terylene, Dacron)	$\{(CH_2)_2O{\cdot}CO{\cdot}C_6H_4{\cdot}CO{\cdot}O\}_n$
Polyamide (Nylon-6,6)	$\{NH(CH_2)_6NH{\cdot}CO(CH_2)_4CO\}_n$
Polycarbonate	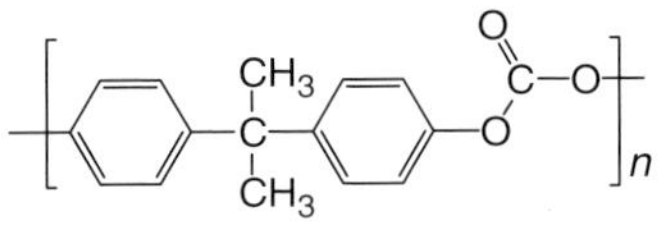
Polyethersulphone	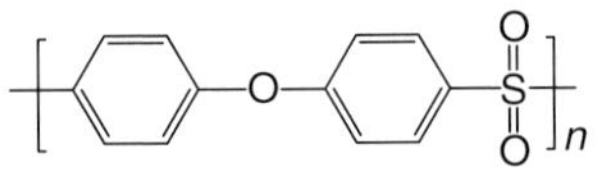
Polyphenylenevinylene	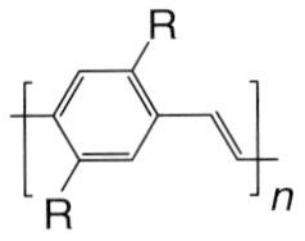

polymers can be deposited electrochemically. The process proceeds through the generation of monomer radicals that can either add to a growing chain or form oligomers, which are subsequently oxidised and form progressively larger chain segments. For simple monomers there are often several alternative sites at which reaction can occur, so that the product is not a linear chain but contains branches and crosslinks (such variants are discussed further in the next section). An example of the reactions that can occur in electrochemical polymerisation of pyrrole is shown in Fig. 1.6.

Solid-state polymerisation is quite rare, although a number of reactions that lead to highly regular, crystalline polymers are known. Reaction between molecules in a crystal lattice can follow the generation of radicals by ionising radiation, ultra-violet light and occasionally heat. Such reactions are termed *topo-chemical* reactions. The product is generally a dimer, but the reaction can propagate in an appropriately aligned stack of molecules to form a polymer. Such a regular reaction between identically oriented molecules is described as *topo-tactic*. If the heat of reaction is large, the energy released can melt the monomer crystal and the product is a disordered, amorphous polymer. In cases where the heat of reaction is modest the crystal structure is retained,

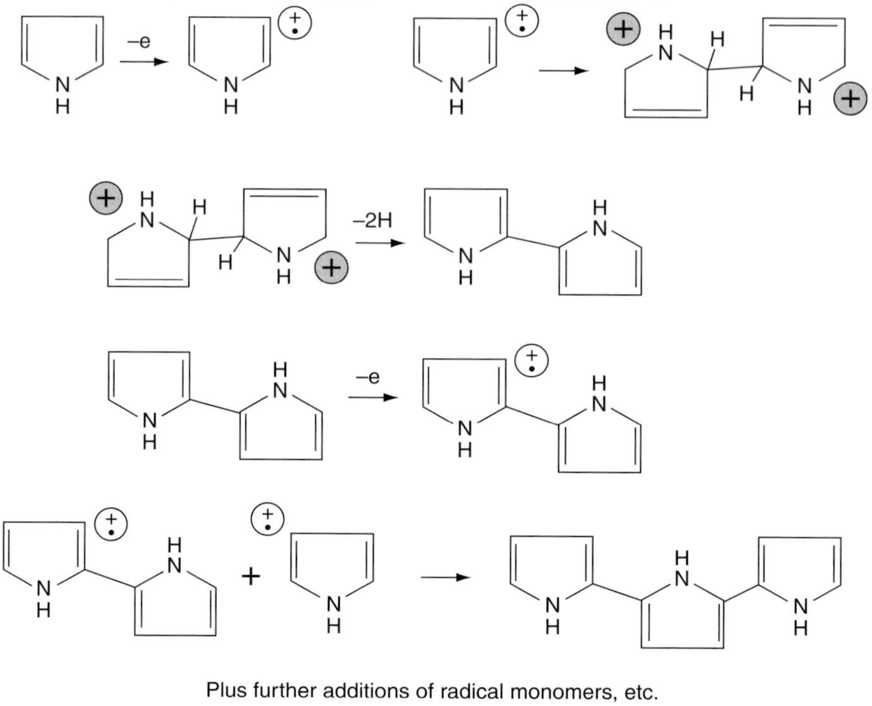

Fig. 1.6 Postulated mechanism for the anodic step-wise polymerisation of pyrrole.

provided that the internal strain due to the mismatch between the monomer lattice spacing and the polymer repeat distance is modest. Where the mismatch is large, the crystal lattice is disrupted and a fibrous crystalline product is obtained, but where it is small the product consists of extended polymer chains in a regular crystal lattice.

Methyl methacrylate polymerises in the solid state to give an amorphous poly(methylmethacrylate), as do a number of other methacrylate monomers. Poly(oxymethylene) is a fibrous, oriented, crystalline polymer that is obtained by the ring opening polymerisation of cyclic oxymethylenes, $\text{(}CH_2O\text{)}_m$, where m can be between 3 and 6.

A notable case of solid-state polymerisation is that of diacetylenes giving rise to a special class of highly ordered conjugated polymers as follows:

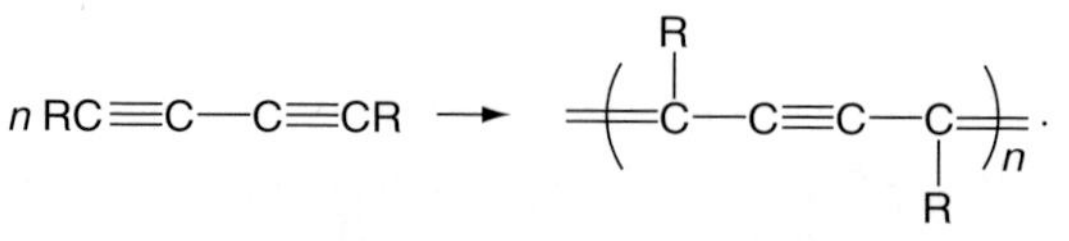

The pendent groups determine the packing of the monomers in the crystal lattice and crucially affect whether polymerisation is possible and the perfection of the product. In several instances the products retain the perfection of the initial monomer crystal (Bloor, 1982).

1.2.4 Chemical and physical variants

Many variations from linear, regular chain formation may occur during polymerisation. The first set of variations arises from the presence of alternative reactive centres either on the monomer or on the already formed polymer chains. Thus the uniformity of *head-to-tail* monomer assembly:

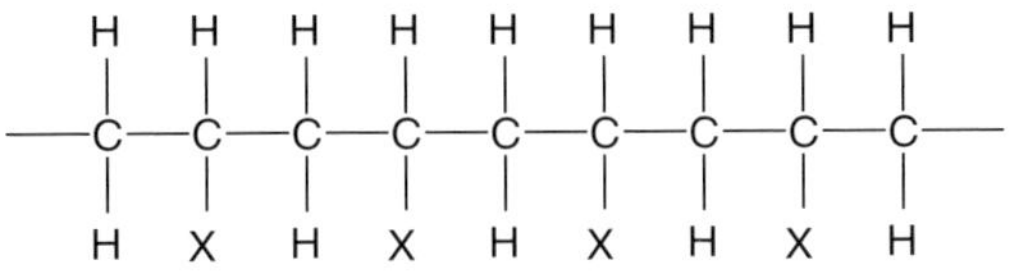

may be broken, and some *head-to-head* and *tail-to-tail* structures incorporated in the chain:

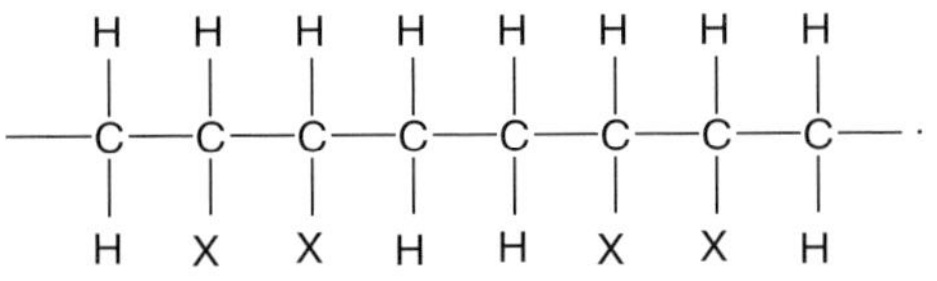

Similar structures will occur in conjugated polymers such as polypyrroles and polythiophenes prepared from monomers with one or more different substituents at the positions on the aromatic ring not involved in the polymerisation, see Fig. 1.5. The reactivity of one particular site in the monomer usually predominates, so that the effect is relatively small. A 1% head-to-head content may, however, seriously upset crystallisation.

Some polymerisations do not continue in a linear fashion, and *branching* occurs, as shown in Fig. 1.7(a). When branching is prevalent, it can have a serious effect on properties. Thus the polymerisation of ethylene under high-pressure conditions gives a product that has so many side chains on each main chain that crystallisation is appreciably suppressed. This material is softer than the highly crystalline linear form from a catalytic low-pressure process. The two forms of polyethylene may be distinguished by a difference in density between them, the more crystalline material being denser.

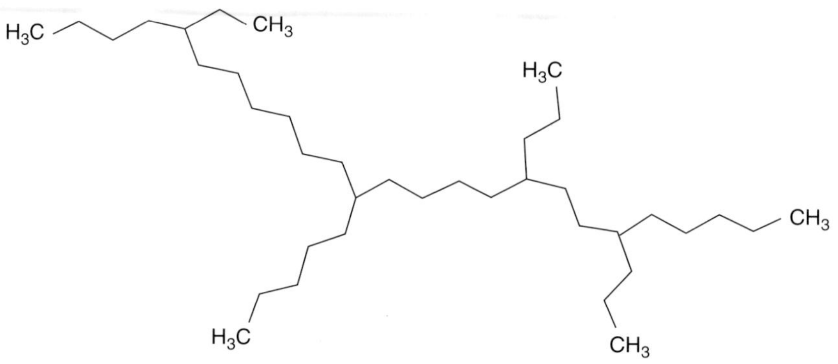

(*a*)

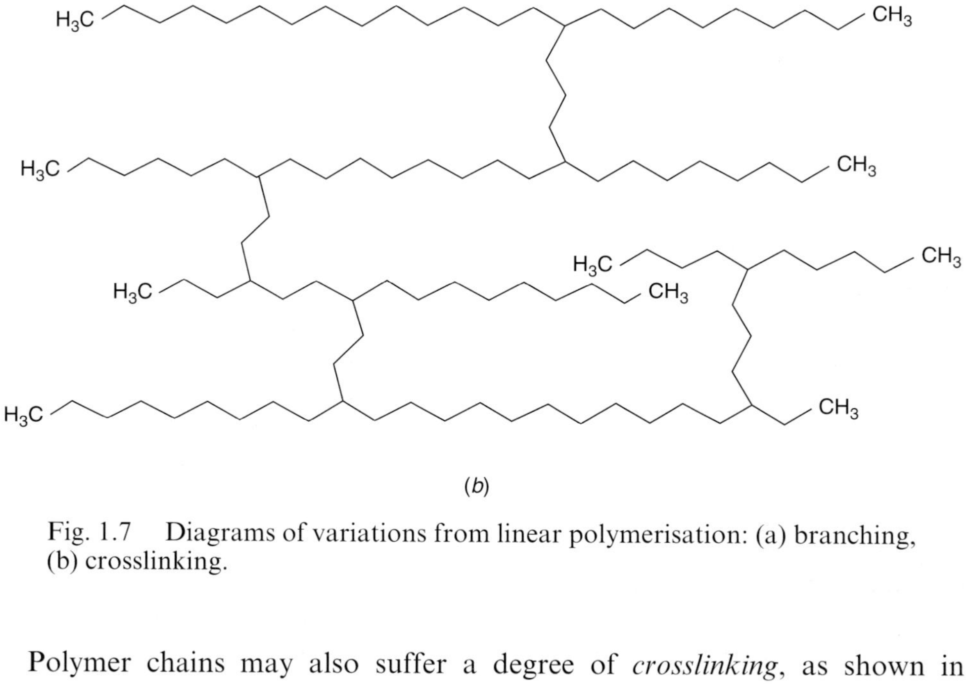

(*b*)

Fig. 1.7 Diagrams of variations from linear polymerisation: (a) branching, (b) crosslinking.

Polymer chains may also suffer a degree of *crosslinking*, as shown in Fig. 1.7(b), either during the initial polymerisation, or at some subsequent stage. In a condensation polymerisation where the monomer molecules each possess three reactive groups, i.e. are trifunctional, crosslinking is inevitably a major feature throughout the polymerisation. Furthermore, it may be formally shown in such a system that once two-thirds of the total number of reactive groups have reacted, the polymer will be rapidly combined into one giant, crosslinked molecule or *gel*. The onset of this 3-dimensional network formation will be reflected in a sudden increase in molar mass to a very large value over a small extent of reaction. When extensive crosslinking is promoted, as in *thermosetting* materials, the product is typically a hard, brittle resin at

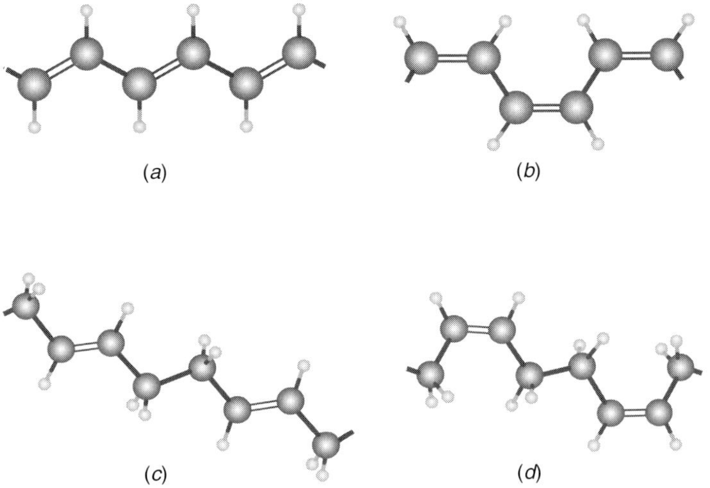

Fig. 1.8 *trans*- and *cis*-isomers in polyacetylene, (a) and (b), and polybutadiene, (c) and (d), respectively.

room temperature, and it is unable to flow at high temperatures. For some purposes crosslinks are often deliberately introduced into a linear polymer by a post-curing process, e.g. by heating with peroxides. In the rubber industry light crosslinking with sulphur (vulcanisation) is used to convert the starting polymers (natural or synthetic latex) into a tough product with a high extensional elasticity. As noted above, the electro-polymerisation of conjugated polymers from multifunctional monomers also leads to a crosslinked polymer. In contrast, uncrosslinked saturated polymers, generally called *thermoplastics*, soften and flow easily at high temperatures. This will also occur for conjugated polymers with large pendent groups, where the disordering of the pendent groups outweighs the stiffness of the polymer chain. However, in the absence of large pendent groups conjugated polymers often decompose before they melt.

As in any other chemical compound, different geometrical arrangements of substituent groups are possible in a polymer where rigid molecular units are involved. This gives rise to *trans*- and *cis-configurational isomerism* in polymers containing double bonds in their repeat units, as in polyacetylene and natural and synthetic rubbers. The structures of the *trans*- and *cis*-isomers of polyacetylene and polybutadiene are illustrated in Fig. 1.8.

Such differences in structure can have a profound effect on the physical properties of a polymer. Thus natural rubber, which comprises *cis*-1,4-poly(isoprene), is a soft rubbery material at room temperature, whereas gutta-percha, which comprises the corresponding *trans*-isomer, is semi-crystalline and hard. The method of polymerisation determines the isomeric form of the polymer.

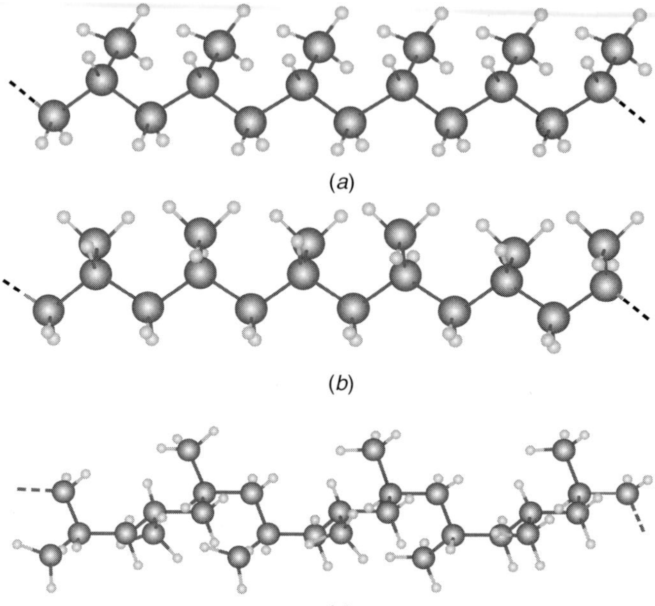

(a)

(b)

(c)

Fig. 1.9 Stereoisomers of polypropylene: (a) isotactic, (b) syndiotactic (schematic) and (c) syndiotactic (helical).

When the repeat units of a polymer chain are themselves asymmetric, on account of their containing a carbon atom with four different substituents, the large number of possible permutations of right-handed (*d*) and left-handed (*l*) units represents a large number of *steric*-isomers. This kind of isomerism in polymers is called *tacticity*. When the arrangement along the chain is completely ordered, the polymer is said to be *stereoregular*. In a substituted vinyl polymer like polypropylene, $-(CH_2-CH(CH_3))-_n$, there are three main types of steric isomer:

a. *Isotactic* In this isomer all repeat units are identical, either right-handed (*d*) or left-handed (*l*) forms, and arranged in a linear, head-to-tail fashion. In the fully extended (planar zigzag) chain all the methyl substituents lie on the same side of the chain (Fig. 1.9(a)).
b. *Syndiotactic* Here successive repeat units alternate in configuration, *dldldl*. . ., and successive methyl groups along the fully extended chain lie on alternate sides (shown schematically in Fig. 1.9(b)). More elaborate ordering does not often occur in practice.
c. *Atactic* This label is applied to completely random arrangements of *d* and *l* units along the chain.

Similar structures are found in conjugated polymers with asymmetric repeat units.

A high degree of stereoregularity usually increases crystallinity, leading to enhanced rigidity. For this reason certain catalysts that favour stereoregularity, e.g. Natta–Ziegler catalysts for propylene polymerisation, are of great commercial importance.

1.2.5 Conformation and hindered rotation

Rotation about a single, covalent bond between two carbon atoms is only restricted by the spatial interaction of the other atoms or groups attached to the two carbons. The distinct molecular arrangements, which are thereby stabilised when the interaction is strong, are called *conformational* isomers. The principles and nomenclature of conformation can be most easily described in the case of a simple organic molecule like *n*-butane, and the situation is illustrated in Fig. 1.10. From the potential-energy curve with respect to the angle φ of rotation around the central C–C bond we can readily appreciate that the highest energy occurs in the *fully eclipsed* state ($\varphi = 180°$), where the terminal methyl groups most strongly interact with each other. Conversely, the most stable state corresponds to the *trans*-conformation ($\varphi = 0°$) whose energy is about 15 kJ mol^{-1} lower than that of the eclipsed form. In effect, this imposes a barrier against rotation about the C–C bond. Two other staggered states, called *gauche*-conformations ($\varphi = \pm 120°$), are also relatively stable. In a saturated carbon-chain polymer the rotational arrangement about the whole sequence of bonds along the backbone determines the shape or conformation of the molecule. For example, the steric interaction of the CH_3 units in syndiotactic polypropylene results in a sequence of *trans*- and *gauche*-conformers along the backbone, leading to the helical structure with three repeat units per turn shown in Fig. 1.9(c).

In solution, the flexibility of the polymer chain allows the polymer to collapse into a compact, randomly coiled structure rather than to retain an extended form. A transition between these two forms can be observed depending on solvent quality and temperature and is best known as the 'rod-to-coil' transition of biomacromolecules such as DNA.

Since the energy barriers to internal rotation are rather low, typically not much larger than thermal energies at room temperature, polymer chains may usually reorganise themselves rapidly under an applied stress at modest temperatures. This, in turn, endows an uncrosslinked or *thermoplastic* polymer with pronounced flexibility and toughness. Natural or synthetic rubbers owe their extreme pliability partly to extremely low rotational barriers about single bonds; although the double bonds that are present along the chain are

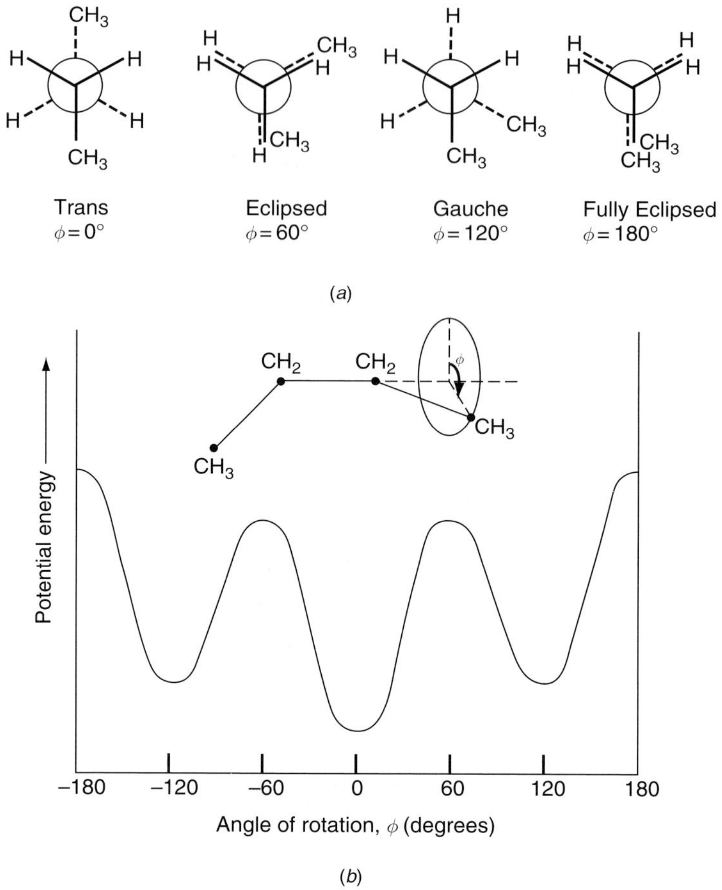

Fig. 1.10 Hindered rotation in *n*-butane: (a) schematic view along the central C–C bond, showing various conformations, (b) potential-energy curve for rotation about the central C–C bond.

themselves rigid, they impose reduced steric hindrance upon rotation about adjacent single bonds. At elevated temperatures thermal energies are quite sufficient for most rotational barriers to be rapidly overcome, and this is the basis for one of the most useful features of thermoplastics – they may be softened and re-formed by the application of heat.

In the case of a conjugated polymer like polyacetylene, as shown in Figs. 1.8(a) and 1.8(b), the situation with respect to rotation about the nominally single bonds is very different; it is controlled by the electronic structure of the backbone bonds rather than by steric hindrance. As will become clear later, when this type of polymer is discussed in relation to conduction, the partial double bond character acts to keep the whole molecule planar, so that the molecule resembles a ribbon that can only undergo small deformations from its extended, flat conformation. In effect, the polymer has a worm-like form with gradual changes in orientation along its length, and it can only collapse into a random coil under extreme conditions. Polyyne has an axially symmetric, but very stiff, backbone and behaves as a rigid rod.

1.2.6 Copolymers

Copolymers, in which individual molecular chains contain a mixture of two or more different repeat units, are often encountered in practice. The compositional pattern of copolymers depends on the conditions of polymerisation and may be either random,

–A–B–B–A–B–A–A–,

or divided into *blocks* containing long sequences of like units,

$-[A]_l-[B]_m-[A]_n-$.

By special methods it is also possible to produce *graft* copolymers where branches of a different polymer are tagged on to a polymer backbone:

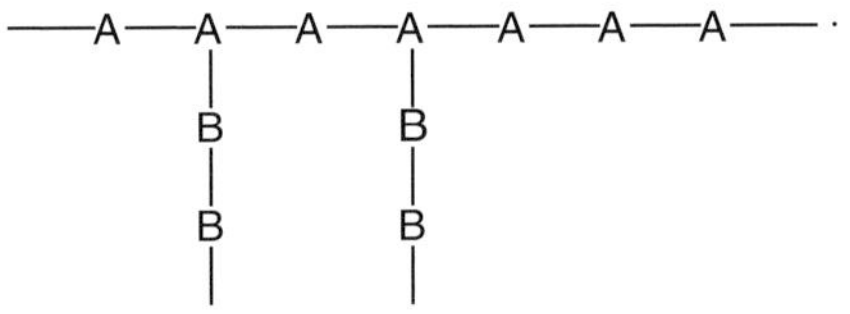

Also, chains may be subsequently *extended* with a different polymer:

B–B–A–A–A–A–A–A–A–B–B.

In random copolymers chain flexibility and crystallinity can be quite different from those in either of the component *homo*polymers. In consequence, the copolymer may well present an entirely new set of physical properties. On the other hand, the different chain segments in block or graft copolymers often segregate into effectively separate phases, when the properties of the composite resemble those of a mixture of the individual homopolymers.

1.2.7 Crystallisation and orientation

Long, flexible molecules naturally become highly contorted and entangled with each other in the molten or dissolved state, and reorganisation into a regular, crystalline array is then an awkward process. If a molten polymer is rapidly cooled or quenched there is insufficient time for any crystallisation to occur, and a transparent, amorphous solid is obtained. This apart, at least some crystallisation will usually take place in most polymers, and the typical milky appearance of a polymer is indicative of a semi-crystalline state consisting of a mixture of small crystalline regions (crystallites) and amorphous matter. The exact degree of crystallinity, most often in the range 5 to 50%, depends on the inherent ease of crystallisation of the polymer molecules (as mentioned previously, factors like branching, crosslinking, stereo-irregularity and copolymerisation can suppress crystallisation) and on the thermo-mechanical history of the particular sample under observation. 100% crystallinity can only be achieved by solid-state polymerisation.

The amorphous or disordered phase of a polymer normally shows a transition from a rubbery to a glassy condition at some characteristic transition temperature T_g. Crystallisation can only proceed above the *glass transition temperature* where sufficient molecular mobility is present for reorganisation to occur. The glass transition is a difficult phenomenon to explain in detail and the subject is developed further in Section 3.3.

When a polymer is stretched or drawn, the internal shearing action tends to align the long molecules preferentially in the stretch direction, and this orientation can be detected by optical methods. Not surprisingly such 1-dimensional ordering tends to induce crystallisation. In distinctly non-crystallisable polymers like commercial polystyrene and poly(methyl methacrylate), made by free-radical initiation at high temperature, 3-dimensional order is still absent after stretching, although the resulting structure must be regarded as a somewhat elongated, tangled skein in these cases.

The extra strength that orientation of chain molecules imparts to a material in the direction of orientation (at the expense of strength in the other directions) is widely exploited in the commercial production of films and fibres. Typically, these materials are made by subjecting the polymer to a high degree of drawing. Crystallising the polymer while it is held under tension above T_g *sets* the orientation produced by drawing.

The structure of truly crystalline regions of a polymer may be determined from wide-angle X-ray diffraction patterns. In this way the unit cell dimensions and molecular configurations have been obtained for all the common polymers. In general, the molecules of the crystalline phase adopt an extended zigzag or helical conformation and lie parallel to each other.

By contrast, total molecular architecture, involving the conjunction of crystalline and amorphous parts, has proved to be much less amenable to investigation. For a long time structural interpretations were based on the *fringed-micelle* model in which molecules are supposed to wander through

(*a*)

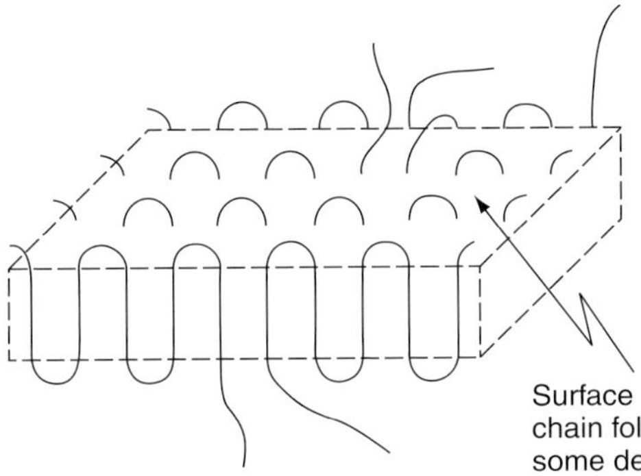

(*b*)

Fig. 1.11 Diagrams of polymer crystallinity models: (a) fringed micelles, (b) a chain-folded lamella.

ordered and disordered regions, as shown in Fig. 1.11(a). While this structure is appropriate for very stiff polymers, it has been found not to be appropriate for flexible polymers. The latter adopt a specific mode of crystallisation of single crystals from solution in which the principal crystal habit, well exemplified by linear polyethylene, is a lamella, with lateral dimensions of the order of 10–20 μm and with a thickness of only about 10 nm. Electron-diffraction studies show, surprisingly, that the chain axis (usually denoted the c-axis) of the polymer molecules lies approximately perpendicular to the plane of the lamella, even though the lamella thickness is much shorter than the chain. This is a result of the chains folding back on themselves in order to fit into the crystalline region, as shown in Fig. 1.11(b). The exact arrangement of the folds is difficult to characterise unambiguously, but in some cases it is believed to be tight, with chains re-entering the crystal at adjacent sites. It is now generally held that chainfolding is also a common feature in the crystallisation of flexible polymers from the melt.

When a quiescent melt is cooled the nucleation process governs the gross features of crystallisation. Microcrystalline structures, called spherulites, grow radially from nucleation centres, and the ultimate size of the spherulite will be determined by the original concentration of nuclei. In a very clean and pure polymer where suitable nuclei are scarce, the spherulites may grow large enough to be seen with the naked eye. The main structure of a spherulite is a radial array of twisted lamella-like blades called fibrils. Although some secondary crystallisation may take place later between the fibrils, considerable amounts of non-crystalline material are left there, and between adjacent spherulites. The orientation produced by any mechanical shearing in a melt will modify the course of a crystallisation process, and the final structure will possess a degree of anisotropy. While this morphology is typical of saturated polymers, it is quite uncommon for conjugated polymers, where the formation of the tight chain folds necessary for the formation of lamellae is energetically unfavourable because of the chain stiffness. However, it should be emphasised that the overall morphology of conjugated polymers, with the exception of the rare examples of those obtained by solid-state polymerisation, remains a mixture of ordered crystalline and disordered amorphous regions just as for saturated polymers.

The interplay of orientation and crystallisation leads to a wide range of super-molecular structures or *morphologies*. Each different morphology represents to the user a different compromise in physical properties, so that characterisation and control of morphology becomes very important for the efficient application of polymeric materials.

1.3 Polymers as non-conductive materials

In polymers with chemically saturated structures the electrons are tightly bound in the σ-bonds between atoms. Since all the available electrons are fixed in the σ-bonds, there are no electrons free to carry an electrical current. Hence, solid, saturated polymers are characteristically electrical insulators. Because of this, any electrostatic charges that they acquire are retained for a long time. Since charges may be deposited by mere contact with a different material, the charged condition is frequently encountered with articles made from polymers. It is worth pointing out here that, although contact charges represent only a slight imbalance of charge compared with the total amounts of positive and negative charge present in matter, they can nevertheless give rise to electric fields which are high enough to cause sparking in air. Thus a surface charge density resulting from just one or two extra electronic charges per million surface atoms is sufficient to generate a field that exceeds the air breakdown value. The occurrence of sparking means that an insulating specimen will often present a complex and confusing distribution of charge on its surface, reflecting a whole history of charging and discharging events. In recent years there has been a resurgence of interest in electrostatics, and new experimental techniques have vastly improved the scope for scientific study. The stimulus has come partly from recognition of the possible benefits to be gained commercially from control and exploitation of static charges on polymeric materials. As no known insulator is completely free of conduction processes, charges are expected to diffuse away eventually under the influence of their own field, even though this may take many years in extreme cases. This is particularly true in polymers where the structure is disordered, creating a low density of defect sites, which can be occupied by charge carriers. Low-level conduction in essentially insulating materials can take a variety of forms. It may very often be attributed to impurities that provide a small concentration of charge carriers in the form of electrons or ions. At high fields the electrodes may also inject new carriers into the polymer, causing the current to increase more rapidly with voltage than in accord with Ohm's law. At very high fields these and other processes, often involving conduction over a surface, inevitably lead to complete failure of the material as an insulator.

1.4 Polymers as conductive materials

Initially one might have expected that π-bonding in a conjugated chain, like that of polyacetylene, would produce bonds of equal length, with the p_z orbital of each carbon overlapping equally with those of both of its neighbours along the chain. Such uniform orbital overlap would form a π-electron wavefunction extending over the whole polymer backbone. The electrons would then be *delocalised* and free to move along the polymer backbone. In this idealised case the polymer can be described as a one-dimensional metal with a half filled conduction band. However, *localisation* of electrons in the double (or triple) bonds, which characterise conjugated polymers, lowers the overall energy of the electron system. This introduces an alternation in bond lengths (an example of Jahn–Teller distortion) and opens up an energy gap in the electron density of states. There is a concomitant increase in the elastic energy of the backbone bonds, but this is less than the energy gained by the electron system. This has a profound effect on the properties of the polymer, with the presence of the energy gap turning it into a semiconductor. The transformation from a metallic to a semiconducting form, the Peierls transition, will be discussed at greater length later along with other, more realistic, models for conjugated polymers.

It is possible to excite π-electrons out of the localised π-bonds by the absorption of visible photons to give rise to electrical conduction, i.e. photoconductivity. For saturated polymers much more energetic ultra-violet photons are needed to achieve the same result for the electrons in the σ-bonds. Interaction of conjugated polymers with electron donating or accepting chemicals can also result in the addition or removal of electrons from the polymer. This can produce sufficient free charge carriers in the polymer that it becomes metallic. It was the discovery of this effect in polyacetylene in the late 1970s that sparked a major increase in the study of conjugated polymers, which continues to this day. The significance of this discovery was recognised by the award of a Nobel Prize in 2000. There will undoubtedly be continuing scientific interest in intrinsically conductive polymers. However, it is the emerging commercial use of conjugated polymers, described in the following section, that will provide the impetus for further research.

1.5 Electrical applications of polymers

The electrical insulating quality inherent in saturated polymers has long been exploited to constrain and protect currents flowing along chosen paths in conductors and to sustain high electric fields without breaking down. Insulating polymeric materials for early electrical equipment were made from naturally occurring products. For example, the first trans-Atlantic telephone cables, laid in the 1860s, were insulated with gutta-percha, which is one of the polymers extracted from rubber trees. As synthetic high polymers became available in the twentieth century, the range of insulators was continually improved. The great virtue of these new materials, such as polystyrene, was their combination of high quality of insulation with ease of fabrication by moulding. Polyethylene, which combines superb insulating properties with mouldability and a high degree of toughness and flexibility, arrived on the scene just in time for the more demanding applications of insulation in coaxial cables for radar apparatus and television. More extreme requirements for very low-conductivity materials, e.g. as used in electret microphones, have been met by fluorinated polymers. High-performance thin films have also been developed for various types of capacitors.

The choice of material for a particular application naturally depends on being able to reach a good compromise amongst a whole range of considerations, including mechanical properties, ease of fabrication into a final product, and cost. The basic insulating properties of polymers are more than adequate for many purposes, and any development effort may then be primarily concentrated on improving other aspects of the material's performance. High on the list will be a need for good chemical and physical stability in the working environment, which might involve exposure to strong sunlight, organic solvents and high temperatures. Commercial success for a polymeric material for certain highly specialised electrical applications will, on the other hand, depend on very careful grooming with respect to the most demanding electrical quality expected of it. Thus, some applications will require that every precaution be taken to eliminate vestigial DC conduction effects or to reduce variation of polarisation effects with AC frequency to an absolute minimum. In order to meet the specifications for high-voltage use, priority must be given to the need for a high breakdown strength that can be maintained in the face of practical environmental conditions. Only on the basis of detailed knowledge and understanding of the molecular structure and behaviour of the basic polymers can one hope to approach the optimum in performance. This has motivated the investigation in depth of the electrical behaviour of many polymeric systems.

The wide use of polymers has largely sprung from the ease with which their physical form can be manipulated from melts and solutions. The availability of a conducting plastic with similar ease of fabrication is obviously desirable! A lightweight, readily mouldable, highly conductive material was long recognised as a worthwhile goal to strive for, prior to the discovery of metallic conduction in polyacetylene in the 1970s. This discovery spawned a huge research effort, which has led to the development of useful products. Initially, applications were in areas requiring intermediate conductive properties, e.g. for the discharge of static electricity and graded cable insulations. Static charges are generally an embarrassment on manufactured articles and materials – they attract dust, cause film to cling to metal surfaces and can give electric shocks to people. There is real benefit to be gained from a slight degree of conduction, which allows charges to leak away to earth. This application can be met to some extent by modifications to existing polymers. Thus conductive composites made by mixing graphitic powder into the bulk of a polymer are commonly used. Surface treatments with so-called antistatic agents are also widely used on ordinary plastics to give sufficient conduction over the surface without destroying other valuable properties of the material. Unfortunately, this kind of solution is not entirely satisfactory, because the mechanical properties of composites are usually inferior to those of the host polymer and surface treatments tend to wear off. An inherent, bulk conduction is superior, and intrinsically conductive polymers have been used to discharge the imaging drum in Xerox copiers and as antistatic coating on textiles. Conductive polymers have also proved to be competitive with composites as absorbers of electromagnetic radiation.

The electrochemical properties of conductive polymers make them suitable for use as electrodes in batteries. Although specific conductivities, i.e. conductivity per unit mass, higher than those of copper have been achieved, such polymers lack the long-term stability required for power and signal transmission. Despite this the use of conductive polymers in soft electronics is being actively pursued. Soft electronics are devices on flexible substrates of relatively low level of sophistication designed for short-term applications, which can be prepared inexpensively, e.g. by printing. A significant breakthrough to application has followed from the discovery of electro-luminescent light emission from semiconducting polymers. In a relatively short period this has gone from a laboratory curiosity to prototype display devices. Polymeric electro-luminescent displays are competing in the market place with existing technologies such as liquid crystal displays. The need to achieve a balance across a broad spectrum of properties in order to meet applications requirements noted above also applies to conductive polymers. Given the much shorter history of these polymers relative to their insulating counterparts, the extent to which this difficult requirement has been met is encouraging for the emergence of further technological applications.

1.6 Further reading

Many basic texts on polymer science are available. The books by Billmeyer (1971), Brydson (1999) and Boyd and Phillips (1993) provide a wide coverage of the subject. Nicholson (1997) provides a useful elementary introduction to polymer chemistry. The text by Grosberg and Khokhlov (1997) provides an excellent introduction to the basics of polymer properties and behaviour in relation to their underlying molecular structure. The proceedings of the 81st Nobel Symposium, Salaneck, Lundström and Rånby (1993), gives an overview of the science of conjugated polymers.

2

Dielectrics in static fields

2.1 Electrostatic relations

The degree to which a material responds to an applied electric field can be most easily appreciated in the case of a parallel-plate capacitor. Suppose first that a fixed voltage V is connected across such a capacitor where the plates are separated by a distance d in a vacuum, as shown in Fig. 2.1(a). Neglecting edge effects, the electric field E thereby produced in the region between the plates will be uniform, having a magnitude

$$E = \frac{V}{d}. \tag{2.1}$$

We should note that, since an electric field has direction as well as magnitude, it is more fully represented as a vector quantity denoted by $\boldsymbol{E}$, with components E_x, E_y, E_z. In the present case, where the field is necessarily perpendicular to the plates, it follows from Coulomb's law[1] that the charges $+Q$ and $-Q$ per unit area stored on the plates are directly proportional to the magnitude of the field, i.e.

$$Q = \varepsilon_0 E. \tag{2.2}$$

The constant of proportionality ε_0 is called the permittivity of free space and has the value 8.85×10^{-12} $\mathrm{Fm^{-1}}$. The vacuum capacitance per unit area of

[1] In SI units Coulomb's fundamental inverse-square law governing the force $\boldsymbol{F}$ between two point charges q_1 and q_2 separated by the vector distance $\boldsymbol{r}$ in a vacuum takes the form:

$$\boldsymbol{F} = \frac{q_1 q_2 \mathbf{r}}{4\pi\varepsilon_0 r^3},$$

where ε_0 is the permittivity of free space. The factor 4π is present in the denominator of the above formula, because permittivity (along with permeability) is rationalised, in order to avoid the continual appearance of the factor 4π in Maxwell's electromagnetic field equations.

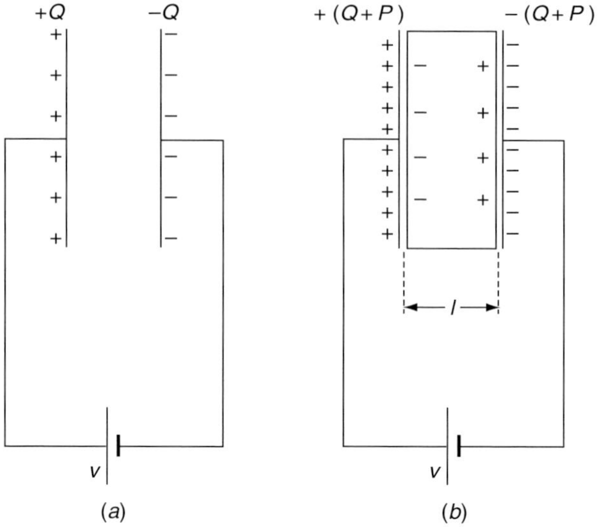

Fig. 2.1 Charges on a parallel-plate capacitor with (a) a vacuum between the plates, and (b) a dielectric between the plates.

electrode C_0 is defined as the ratio of the stored charge per unit electrode area to the applied voltage, i.e.

$$C_0 = \frac{Q}{V}. \tag{2.3}$$

Now consider the capacitor with the material of interest between its plates, as shown in Fig. 2.1(b). The material will respond to the applied electric field by redistributing its component charges (electrons and protons) to some extent, positive charges being attracted towards the negative electrode and vice versa. This effect is called polarisation of the material. If the material is isotropic, the effect in the uniform field of the parallel plate capacitor will amount to the production of a small dipole moment $\boldsymbol{P}\mathrm{d}\nu$ aligned in the field direction in each elemental volume $\mathrm{d}\nu$. The polarisation $\boldsymbol{P}$ is thus defined as a vector quantity expressing the magnitude and direction of the electric moment per unit volume induced in the material by the applied field.[2] Each volume element of a polarised material will behave electrically as though it consisted of charges $+q$ and $-q$ separated by a distance $\boldsymbol{l}$ in the field direction, giving a dipole moment $Q\boldsymbol{l} = \boldsymbol{P}\mathrm{d}\nu$. These dipoles will combine in an analogous manner to the head-to-tail addition of magnets, to produce charges $+P$ and $-P$ per unit area on the surfaces adjacent to the electrodes. Green's theorem

[2] Unless the field is very high, the magnitude of the polarisation is directly proportional to the field. In the case of an anisotropic material its direction is not necessarily parallel to the field.

may be used to show rigorously that the field due to a uniform polarisation P in a sample is equivalent to that from a distribution of charge P_n on the bounding surface, where P_n is the normal component of the polarisation at the surface. The presence of these polarisation or *bound* charges means that more charge can be stored on the capacitor electrodes for the same applied voltage and the capacitance of the system is thereby increased. The ratio ε of the increased capacitance to the vacuum capacitance,

$$\varepsilon = \frac{C}{C_0} = \frac{Q + P}{Q}, \tag{2.4}$$

varies from material to material depending on the amount of polarisation which occurs in the material.

This characteristic ratio is essentially independent of the applied voltage and is therefore also independent of the electric field; it is commonly called the dielectric constant of the material. Substituting for Q from Equation (2.2) in Equation (2.4), and adopting the more general vector notation,

$$\begin{aligned} \varepsilon &= \frac{(\varepsilon_0 \boldsymbol{E} + \boldsymbol{P})}{\varepsilon_0 \mathbf{E}} \\ &= 1 + \frac{\boldsymbol{P}}{\varepsilon_0 \boldsymbol{E}} \\ &= 1 + \chi, \end{aligned} \tag{2.5}$$

where $\chi = \boldsymbol{P}/\varepsilon_0 \boldsymbol{E}$ is the electric susceptibility of the material.

The quantity $\varepsilon_0 \varepsilon \boldsymbol{E}$, called the electric displacement $\boldsymbol{D}$ in the material, may be obtained by rearranging Equation (2.5):

$$\boldsymbol{D} = \varepsilon_0 \varepsilon \boldsymbol{E} = \varepsilon_0 \boldsymbol{E} + \boldsymbol{P}. \tag{2.6}$$

This is the fundamental electric field equation that applies at any point in an isotropic medium. In this context the quantity $\varepsilon_0 \varepsilon$ is the absolute permittivity of the material, and the ratio ε, which we have called the dielectric constant of the material, is more properly termed the relative permittivity (with respect to the absolute permittivity of free space ε_0) and we shall use this term. The flux of dielectric displacement begins and ends on free charge and otherwise is continuous, even at an interface between two media. Electric field, on the other hand, is discontinuous at an interface between two different materials as a result of the different degrees of polarisation.

If we think of polarisation on the molecular level, we can say that the effect of the applied electric field is to induce an electric dipole $\boldsymbol{m}$ on each individual molecule, the magnitude depending on the local electric field strength $\boldsymbol{E}^{\mathrm{L}}$ at the molecule:

$$\boldsymbol{m} = \alpha \boldsymbol{E}^{\mathrm{L}}. \tag{2.7}$$

The constant of proportionality α is called the polarisability of the molecule. For isotropic molecules, which include the monatomic molecules of the inert gases, the average direction of the induced molecular dipoles is in the direction of the applied field. (Since the local field is, as we shall see later, proportional to the overall applied field, we have $\boldsymbol{m} = \text{const.}\boldsymbol{E}$.) The total dipole moment per unit volume developed in this way, the polarisation $\boldsymbol{P}$, is then related to the number of molecules per unit volume N_0:

$$\boldsymbol{P} = N_0 \alpha \boldsymbol{E}^{\mathrm{L}}. \tag{2.8}$$

In most cases, molecules are anisotropic, with their polarisability varying with the direction of the applied field, and any induced molecular dipole will in general *not* lie in the same direction as that of the applied field. More properly, polarisability is expressed as a symmetric, second-rank tensor quantity, relating the induced dipole vector to the applied field vector:

$$m_{ij} = \alpha_{ij} E_j, \tag{2.9}$$

and has principal components α_1, α_2 and α_3, when transformed to its principal axes:

$$\tilde{\alpha} = \begin{vmatrix} \alpha_1 & 0 & 0 \\ 0 & \alpha_2 & 0 \\ 0 & 0 & \alpha_3 \end{vmatrix}. \tag{2.10}$$

We can then treat ε as a scalar quantity, which is independent of the orientation of the electric field vector, as in Equation (2.6). This will not be true if the polymer has a preferred orientation conferred either by mechanical deformation or solid-state polymerisation. In this situation the relative permittivity will be a tensor so that

$$\boldsymbol{D} = \varepsilon_0 \tilde{\varepsilon} \boldsymbol{E}. \tag{2.11}$$

The number of components in the relative permittivity tensor is limited by the sample symmetry. In axially oriented samples two independent components are found, parallel and perpendicular to the alignment direction ($\varepsilon_{\parallel}$ and $\varepsilon_{\perp}$ or ε_1 and ε_2), since the sample is isotropic in the plane normal to the direction of orientation, i.e. $\varepsilon_1 = \varepsilon_2$.

In this discussion we have dealt with a linear dielectric where the polarisation is proportional to the applied field. As with most physical phenomena, the dielectric response of materials is non-linear. Fortunately the higher order terms, in $\boldsymbol{E}^2$, $\boldsymbol{E}^3$, etc., are small and can usually be neglected. They can, however, become important in certain circumstances, as will be seen later.

2.2 Molecular polarisability

We must now look in more detail at polarisation at the molecular or microscopic level. There are three components of molecular polarisation.

a. *Electronic polarisation* An electric field will cause a slight displacement of the electrons of any atom with respect to the positive nucleus. The shift is quite small because the applied electric field is usually quite weak relative to the intra-atomic field at an electron due to the nucleus. Thus, taking the charge on a proton to be 1.6×10^{-19} C and a typical atomic radius to be 10^{-10} m, the electric field on an electron will be of the order of 10^{11} Vm^{-1}, whereas externally applied fields seldom exceed 10^8 Vm^{-1}. Electronic polarisation can, however, react to very high frequencies and is responsible for the refraction of light.

The electronic polarisability of a spherical atom may be calculated in a number of simplified ways. In the oldest approximation, an atom is regarded as a conductive sphere of radius R, when the polarisability may be shown to be $4\pi\varepsilon_0 R^3$, a quantity that is closely related to the actual volume of a molecule. In the more realistic semi-classical Bohr model of a hydrogen atom, the application of a field normal to the plane of the electron orbit, radius R, will produce a small shift, $-x$, in the orbit, as shown in Fig. 2.2. To a first approximation the distance of the orbit from the nucleus will still be R and the dipole moment μ induced in the atom will have magnitude ex. At equilibrium, the external field acting on the electron is balanced by the component of the Coulombic field from the positive nucleus in the field direction:

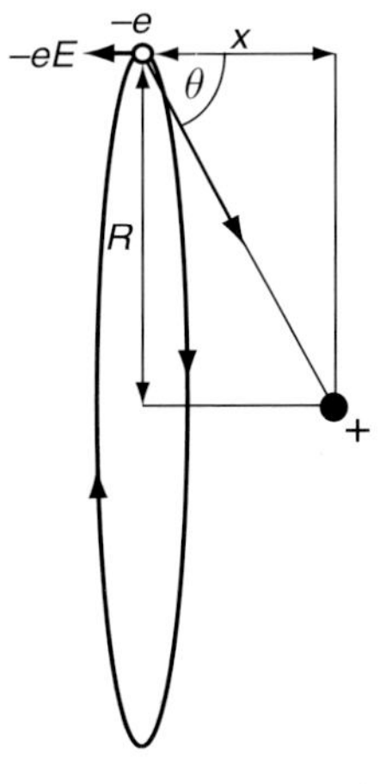

Fig. 2.2 Dipole induction by displacement of an electron orbit in an atom.

$$E = \frac{e \cos\theta}{4\pi\varepsilon_0 R^2}, \tag{2.12}$$

where $\cos\theta \approx x/R$.

The electronic polarisability α of an atom is then given by:

$$\alpha = \frac{\mu}{E} = 4\pi\varepsilon_0 R^3. \tag{2.13}$$

The more correct result from a quantum mechanical treatment is larger than this by a factor 9/2.

We can understand how anisotropy of electronic polarisation arises in molecules in terms of a simple diatomic molecule, consisting of two similar atoms of radius R at separation L, in an applied field E.

Each atom will respond to both the applied field itself and to the field from the dipole induced in its partner. In the case of parallel alignment with the field (Fig. 2.3(a)), the fields are additive and the dipole $\mu_{||}$ induced on each atom, polarisability α, is determined by the equation:

$$\mu_{||} = \alpha\left(E + \frac{\mu_{||}}{2\pi\varepsilon_0 L^3}\right). \tag{2.14}$$

Solving this equation for $\mu_{||}$:

$$\mu_{||} = \frac{\alpha}{1 - \frac{\alpha}{2\pi\varepsilon_0 L^3}} E, \tag{2.15}$$

and the corresponding polarisability $\alpha_{||}$ of the whole molecule becomes:

$$\alpha_{||} = 2\frac{\mu_{||}}{E} = \frac{2\alpha}{1 - \frac{\alpha}{2\pi\varepsilon_0 L^3}}. \tag{2.16}$$

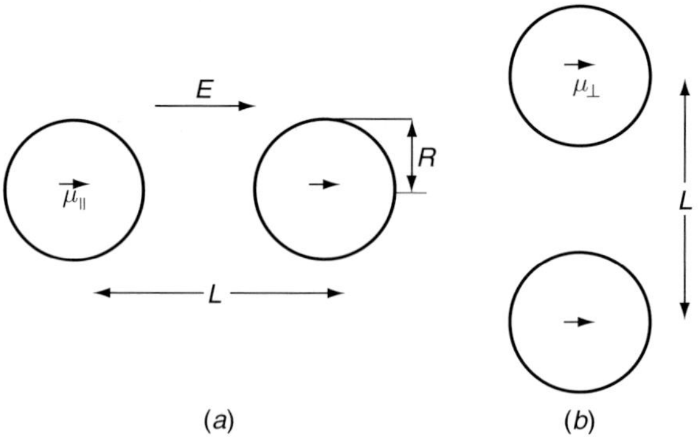

Fig. 2.3 Polarisation of a diatomic molecule: (a) aligned parallel to the applied field, and (b) aligned perpendicular.

In the case of the perpendicular alignment (Fig. 2.3(b)), the fields from the induced dipoles oppose the applied fields at their partner atoms, and the induced dipoles are determined by the equation:

$$\mu_\perp = \alpha(E - \frac{\mu_\perp}{4\pi\varepsilon_0 L^3}). \tag{2.17}$$

The perpendicular polarisability of the whole molecule then becomes:

$$\alpha_\perp = \frac{2\alpha}{1 + \frac{\alpha}{4\pi\varepsilon_0 L^3}}. \tag{2.18}$$

If the polarisability of the component atoms is taken to be $4\pi\varepsilon_0 R^3$, we obtain:

$$\alpha_{\parallel} = \frac{8\pi\varepsilon_0 R^3}{1 - 2\left(\frac{R}{L}\right)^3} \quad \text{and} \quad \alpha_\perp = \frac{8\pi\varepsilon_0 R^3}{1 + \left(\frac{R}{L}\right)^3}. \tag{2.19}$$

Hence, to a first degree of approximation the degree of anisotropy $(\alpha_{\parallel} - \alpha_\perp)$ is given by:

$$\Delta\alpha \approx 24\pi\varepsilon_0 \frac{R^6}{L^3},$$

$$\text{or} \quad \Delta\alpha \approx 6\alpha\left(\frac{R}{L}\right)^3. \tag{2.20}$$

b. *Atomic polarisation* An electric field can also distort the arrangement of atomic nuclei in a molecule or lattice. The movement of heavy nuclei is more sluggish than that of electrons, so that atomic polarisation cannot occur at such high frequencies as electronic polarisation, and it is not observed above infra-red frequencies. Now, in the case of molecular solids we know from vibrational spectroscopy that the force constants for bending or twisting of molecules, involving changes in angles between bonds, are generally much lower than those for bond stretching, so we may expect bending modes to make the major contribution to atomic polarisation. The magnitude of atomic polarisation is usually quite small, often only one-tenth of that of electronic polarisation, although there are exceptions where a particular mode of bending produces relatively large departures from the normally symmetric arrangements of positive and negative centres within the molecule. In ionic compounds the effect can sometimes be very large. For instance, there is a special contribution in the case of crystalline sodium chloride from a relative shift of all the positive ions with respect to the negative ones.

c. *Orientational polarisation* If the molecules already possess permanent dipole moments, there is a tendency for these to be aligned by the applied field to give a net polarisation in that direction. This effect is discussed in the next section, and the rate of dipolar orientation, which is highly dependent on molecule–molecule interaction, is the subject of the next chapter. Suffice it to say at this stage that orientation of molecular dipoles can make a contribution which is large, but which may be slow to develop, to the total polarisation of a material in an applied field.

Figure 2.4 depicts the characteristic stepwise fall in polarisation of a material as the measurement frequency is raised, rendering it impossible for preceding components of molecular polarisation to make their contribution. The relative permittivity follows a similar pattern.

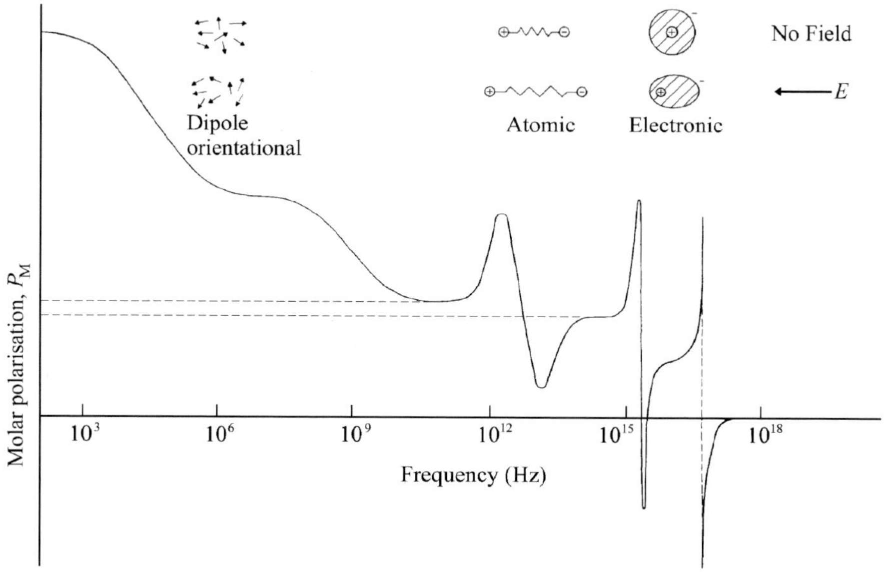

Fig. 2.4 Dispersion of molar polarisation in a dielectric (schematic).

2.3 The local field

As soon as we consider the molecular nature of a material, we realise that the internal electric field will vary from point to point as a consequence of the interaction of fields from the dipoles which are induced on each molecule by the applied field, although the space-average electric field over a volume large in comparison with molecular size (this is equivalent to the classical electric field based on a continuum model) may still be uniform. The field acting on an individual polarisable entity like an atom or molecule is called the local field $\boldsymbol{E}^L$, and it is an important concept in linking observable bulk behaviour of a material with the properties of its constituent atoms or molecules.

In calculating a local field it is convenient to divide it into two principal components:

$$\boldsymbol{E}^{\mathrm{L}} = \boldsymbol{E}^{\mathrm{C}} + \Sigma \boldsymbol{E}^{\mathrm{D}}, \tag{2.21}$$

where $\boldsymbol{E}^{\mathrm{C}}$ is the field due to the real charges on the electrodes which sustain the externally applied field, and $\Sigma\boldsymbol{E}^{\mathrm{D}}$ is the sum of fields from all the molecular dipoles apart from that on the molecule at the reference point within the sample. A general evaluation is difficult because the sum $\Sigma\boldsymbol{E}^{\mathrm{D}}$ does not converge for an infinitely large sample and it depends on the shape of the sample. An approach that was first adopted by Lorentz and which has been of great value in the development of dielectric theory is based on the following simple model. Consider a sample of material contained between the electrodes of a parallel-plate capacitor, as shown in Fig. 2.5, and suppose that an imaginary sphere is drawn round a particular molecule at point A. The local field at A may then be written as

$$\boldsymbol{E}^{\mathrm{L}} = \boldsymbol{E}^{\mathrm{C}} + \boldsymbol{E}^{\mathrm{P}} + \boldsymbol{E}^{\mathrm{M}}, \tag{2.22}$$

where $\Sigma\boldsymbol{E}^{\mathrm{D}}$ has been subdivided into a component $\boldsymbol{E}^{\mathrm{P}}$ from all the sample which lies outside the sphere and a component $\boldsymbol{E}^{\mathrm{M}}$ from that inside the sphere. If the radius r of the sphere is chosen to be small in comparison with the size of capacitor, but large in comparison with molecular dimensions, it is permissible to calculate $\boldsymbol{E}^{\mathrm{P}}$ on the basis of classical electrostatic theory, assuming that the material outside the sphere is a continuum with a relative permittivity ε. Discrete molecular structure need only be taken into account in calculating $\boldsymbol{E}^{\mathrm{M}}$. Let us now consider the separate components of the local field.

Since the total charge per unit area on the electrodes as shown previously is $(Q+P)$, we have directly for the magnitude of the field from this source

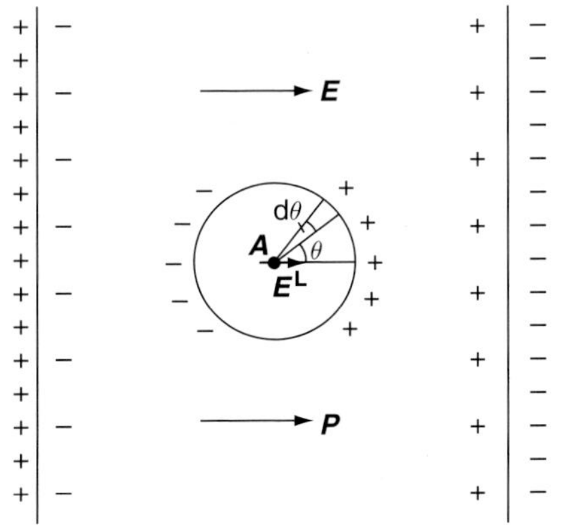

Fig. 2.5 Model for the Lorentz local field.

$$E^{\mathrm{C}} = \frac{(Q + P)}{\varepsilon_0}. \tag{2.23}$$

Substituting for Q from Equation (2.2) we obtain

$$\boldsymbol{E}^{\mathrm{C}} = \boldsymbol{E} + \frac{\boldsymbol{P}}{\varepsilon_0}, \tag{2.24}$$

where $\boldsymbol{E}$ is the applied field ($E = V/d$).

The field at A from the polarisation of the material outside the sphere may be further resolved into two parts in terms of the apparent surface charges which it produces at its boundaries. The first part is due to the apparent charges $-P$ and $+P$ per unit area at the surface adjacent to the electrodes and has the value $-\boldsymbol{P}/\varepsilon_0$. The second part is due to the apparent charges P $\cos\theta$ per unit area on the surface of the sphere, where θ is the angle shown in Fig. 2.5, and it may be evaluated by integrating over surface elements confined between angles θ and $(\theta + \mathrm{d}\theta)$. Such an element has an area $2\pi r^2 \sin\theta \mathrm{d}\theta$. By symmetry the fields perpendicular to $\boldsymbol{P}$ at the centre of the sphere cancel out and the net field in the direction parallel to $\boldsymbol{P}$ takes the value

$$\int_0^{\pi} \frac{2\pi r^2 \sin\theta \,\mathrm{d}\theta \, P \cos\theta \cos\theta}{4\pi\varepsilon_0 r^2} = \frac{P}{3\varepsilon_0}. \tag{2.25}$$

Hence we have

$$\boldsymbol{E}^{\mathrm{P}} = -\frac{\boldsymbol{P}}{\varepsilon_0} + \frac{\boldsymbol{P}}{3\varepsilon_0}. \tag{2.26}$$

$\boldsymbol{E}^{\mathrm{M}}$, the field at A due to the molecules inside the sphere, will depend on the detailed arrangement of the molecules. For certain special cases, including simple cubic lattices and completely random arrays,

$$\boldsymbol{E}^{\mathrm{M}} = 0. \tag{2.27}$$

In these cases we finally obtain for the local field at A:

$$\boldsymbol{E}^{\mathrm{L}} = \boldsymbol{E} + \frac{\boldsymbol{P}}{\varepsilon_0} - \frac{\boldsymbol{P}}{\varepsilon_0} + \frac{\boldsymbol{P}}{3\varepsilon_0} + 0, \tag{2.28}$$

$$\text{i.e.} \qquad \boldsymbol{E}^{\mathrm{L}} = \boldsymbol{E} + \frac{\boldsymbol{P}}{3\varepsilon_0}.$$

Substituting for $\boldsymbol{P}$ from Equation (2.5),

$$\boldsymbol{E}^{\mathrm{L}} = \frac{(\varepsilon + 2)}{3}\boldsymbol{E}. \tag{2.29}$$

2.4 The Clausius–Mosotti relation

Knowing the local field operating at each molecule from Equation (2.29) above, we may now calculate the individual contributions to the polarisation from Equation (2.7):

$$\boldsymbol{m} = \alpha\frac{\varepsilon + 2}{3}\boldsymbol{E}. \tag{2.30}$$

The total polarisation is given by substitution in Equation (2.8):

$$\boldsymbol{P} = N_0\alpha\frac{\varepsilon + 2}{3}\boldsymbol{E}. \tag{2.31}$$

Again substituting for $\boldsymbol{P}$ from Equation (2.5), we obtain the Clausius–Mosotti relation:

$$\frac{\varepsilon - 1}{\varepsilon + 2} = \frac{N_0\alpha}{3\varepsilon_0}. \tag{2.32}$$

If M_{W} is the molar mass of the material and ρ is its density, this equation may be written:

$$\frac{\varepsilon - 1}{\varepsilon + 2}\frac{M_{\mathrm{W}}}{\rho} = \frac{N_{\mathrm{A}}\alpha}{3\varepsilon_0}, \tag{2.33}$$

where N_{A} is Avogadro's number. The quantity $N_{\mathrm{A}}\alpha/3\varepsilon_0$ is called the molar polarisation P_{M}, and it has the dimensions of a volume. So long as the foregoing theory remains valid, the molar polarisation will be constant for a particular material irrespective of the temperature or pressure. We might expect this to be true for a gas or vapour where the density is low, so that intermolecular interaction plays only a minor role; the approximation $\boldsymbol{E}^{\mathrm{M}} = 0$ is then valid. We must also remember that we have not yet taken into account the effect of any orientation of molecular dipoles in the applied field. Fortunately, this latter complication may be circumvented altogether by using the high-frequency value of the relative permittivity which is obtainable from Maxwell's identity between relative permittivity and the square of the refractive index of light n:

$$\varepsilon = n^2. \tag{2.34}$$

Orientation of molecules is much too slow to contribute to polarisation at these high frequencies. Then substituting (2.34) in (2.33) gives a quantity that is usually called the molar refraction of the material:

$$R_M = \frac{n^2 - 1}{n^2 + 2} \frac{M_W}{\rho} = \frac{N_A \alpha}{3\varepsilon_0}. \tag{2.35}$$

Equation (2.35), known as the Lorenz–Lorentz relation, provides a method of calculating the molecular polarisability from a macroscopic, observable quantity, the refractive index. We must make the proviso that we stay away from any resonant absorption frequency, where the refractive index is anomalously high. If the refractive index refers to optical frequencies, the polarisability α will be purely electronic in origin. In practice, electronic polarisabilities derived in this way are remarkably insensitive to temperature and pressure, even for highly condensed phases in which intermolecular forces must be large. This is illustrated for the particular case of xenon in Table 2.1.

Table 2.1 *Molecular polarisability of xenon*

State	Polarisability(10^{-40} Fm^2)
Gas at room temperature and pressure	4.45
Liquid at the b.p.	4.27
Liquid at the m.p.	4.26
Solid at the m.p.	4.32

A useful property of molar refraction is its additivity. Thus, to a first approximation the molar refraction of a given molecule may be predicted by taking the sum of contributions from its various parts. By applying this principle to a large number of compounds, a self-consistent set of polarisabilities may be derived for a wide range of atoms, ions and molecules. For organic compounds a particularly successful approach has been to regard the molar refraction of a symmetrical hydrocarbon like methane as the sum of four equal C–H bond refractions. On this kind of basis a self-consistent set of bond refractions has been established, and a short list (Vogel *et al.*, 1952) is given in Table 2.2.

It is then a simple matter to calculate the molar refraction of any given compound by adding up the values for all its component bonds. Thus the molar refraction for butan-2-one, CH_3–CO–CH_2–CH_3, would be calculated as 20.66 cm^3. The observed value is 20.79 cm^3 which is in line with the 1% agreement that is usually obtained. The method works well because it inherently takes into account the different polarisabilities of bonding electrons. In the case

Table 2.2 *Bond refractions for the sodium D-line*

Bond	Refraction(10^{-6} m^3)
C–H	1.676
C–C	1.296
C=C	4.17
C≡C (terminal)	5.87
C≡C (non-terminal)	6.24
C–C (aromatic)	2.688
C–F	1.44
C–Cl	6.51
C–Br	9.39
C–I	14.61
C–O (ether)	1.54
C=O	3.32
C–S	4.61
C–N	1.54
C=N	3.76
C≡N	4.82
O–H (alcohol)	1.66
O–H (acid)	1.80
N–H	1.76

of a high polymer, e.g. polyethylene, CH_3–$(CH_2)_n$–CH_3, one can ignore the chain ends and simply calculate a molar refraction of the repeat unit –(CH_2)–. The system fails for molecules which contain conjugated double bonds where electrons are delocalised over many atoms. Exceptionally high molar refractions are observed for these molecules; the effect is called optical exaltation.

2.5 Polar molecules

In contrast to molar polarisation calculated from optical refractivities, that calculated from relative permittivities observed at lower frequencies is by no means always independent of temperature. Actually, materials tend to fall into one of two classes. Those in one class show a relatively constant molar polarisation in accord with the simple Clausius–Mosotti relation, whilst the members of the other class, which contains materials with high relative permittivities, show a molar polarisation that decreases with increase in temperature. Debye recognised that permanent molecular dipole moments were responsible for the anomalous behaviour. From theories of chemical bonding we know that certain molecules which combine atoms of different electronegativity are partially ionic and consequently have a permanent dipole moment. Thus chlorine is highly electronegative and the carbon–chlorine bond consequently entails an excess negative charge near the chlorine atom, leaving the carbon with a net positive charge:

$$\overset{+}{\mathrm{C}}-\overset{-}{\mathrm{Cl}}.$$

The order of magnitude of such a dipole will be equal to the product of the charge on an electron (1.6×10^{-19} C) and a typical bond length (10^{-10} m) giving a value of $\sim 10^{-29}$ Cm. The old electrostatic Debye unit of molecular dipoles was in fact equivalent to 3.335×10^{-30} Cm. Orientation of such molecular dipoles will clearly produce an extra contribution to the molar polarisation in addition to dipoles *induced* by the applied field, and it is not unreasonable to expect that the equilibrium degree of orientation in a given field will depend on temperature.

First consider the application of an external electric field to an assembly of rigid molecular dipoles, i.e. let there be no deformational polarisation. If the permanent moment of each dipole is μ, then the potential energy u of a dipole in the applied field will be:

$$u = -\boldsymbol{\mu}.\boldsymbol{E}^{\mathrm{L}} = -\mu E^{\mathrm{L}}\cos\theta, \tag{2.36}$$

where θ is the angle between the axis of the dipole and the direction of the local electric field $\boldsymbol{E}^{\mathrm{L}}$ (see Fig. 2.6).

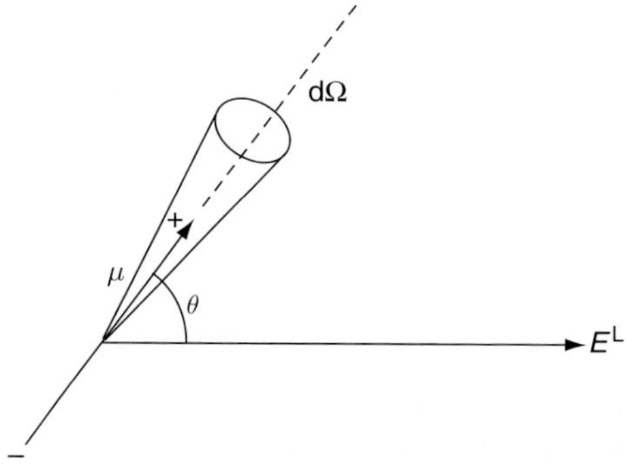

Fig. 2.6 Orientation of a molecular dipole in an applied electric field.

If the electric field is the only orientational force acting on the molecules, the distribution of dipole orientations will be given by Boltzmann's law. Thus the number of dipoles pointing in the directions confined within the element of solid angle dΩ will be

$$A\mathrm{e}^{-u/kT}\mathrm{d}\Omega,$$

where A is a constant which is determined by the total number of dipoles per unit volume. Hence the average molecular moment $\overline{m}$ in the direction of the electric field will be given by

$$\overline{m} = \frac{\int A e^{\mu E^{L} \cos\theta/kT} \mu \cos\theta \, d\Omega}{\sum A e^{\mu E^{L} \cos\theta/kT} d\Omega}, \tag{2.37}$$

where the integration is taken over all possible directions.

Now the element of solid angle $d\Omega$, which includes all the directions in which the dipole lies at an angle in the range θ to $\theta + d\theta$ with respect to the direction of the applied field, is $d\Omega = 2\pi \sin\theta d\theta = 2\pi \, d(\cos\theta)$. Making this substitution and introducing the variable $x = \mu E^{L}/kT$, we obtain

$$\overline{m} = \frac{2\pi\mu \int_0^{\pi} e^{x\cos\theta} \cos\theta \, d(\cos\theta)}{2\pi \int_0^{\pi} e^{x\cos\theta} d(\cos\theta)}. \tag{2.38}$$

Integrating by parts:

$$\begin{aligned} \frac{\overline{m}}{\mu} &= \frac{\left[\frac{\cos\theta}{x} e^{x\cos\theta} - \frac{1}{x^2} e^{x\cos\theta}\right]_0^{\pi}}{\left[\frac{1}{x} e^{x\cos\theta}\right]_0^{\pi}} \\ &= \frac{\frac{1}{x}(e^{x} + e^{-x}) - \frac{1}{x^2}(e^{x} - e^{-x})}{\frac{1}{x}(e^{x} - e^{-x})} \\ &= \frac{(e^{x} + e^{-x})}{(e^{x} - e^{-x})} - \frac{1}{x} = L(x), \end{aligned} \tag{2.39}$$

where $L(x)$ is the Langevin function known from magnetic theory and may be expanded as a power series:

$$L(x) = \frac{x}{3} - \frac{x^3}{45} + \cdots \tag{2.40}$$

Figure 2.7 shows a graph of the Langevin function. It is clear that there will be a saturation effect at very high fields. In practice such fields are unapproachable and it is usually not necessary to take into account more than the first term of the Langevin function, so that:

$$\overline{m} \approx \frac{\mu^2 E^{L}}{3kT}. \tag{2.41}$$

Consequently, polarisation due to dipolar orientation is directly proportional to the local field strength and inversely proportional to temperature. We can think of the quantity $\mu^2/3kT$ as an orientational polarisability and simply add

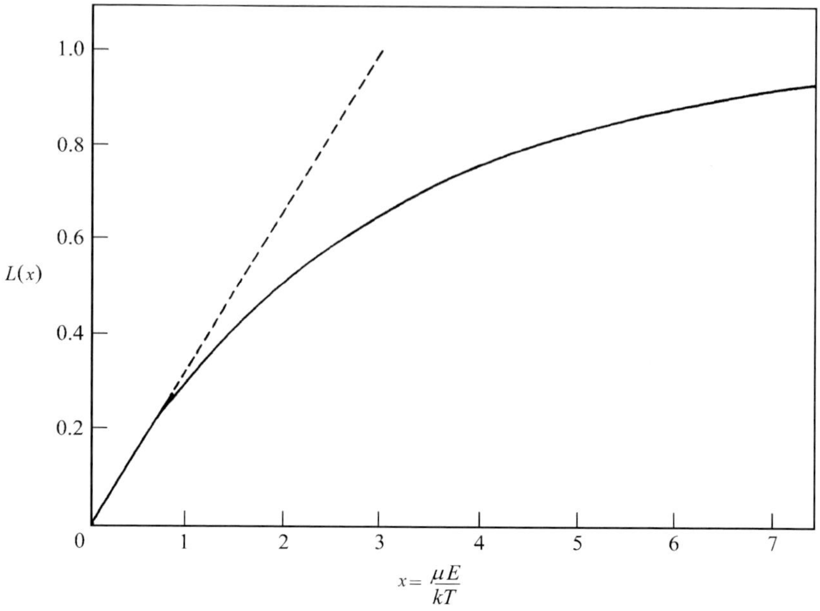

Fig. 2.7 Graph of the Langevin function, $L(x)$.

it to the normal deformational polarisability α_0 to give a total effective polarisability:

$$\alpha = \alpha_0 + \frac{\mu^2}{3kT}. \tag{2.42}$$

Taking into account the possible anisotropy of the deformational polarisability of molecules requires that

$$\alpha_0 = \frac{\alpha_1 + \alpha_2 + \alpha_3}{3}, \tag{2.43}$$

where α_1, α_2 and α_3 define the ellipsoid of the deformation–polarisability tensor.

The complete expression (Debye equation) for the molar polarisation is then:

$$P_{\mathrm{M}} = \frac{\varepsilon - 1}{\varepsilon + 2}\frac{M_{\mathrm{W}}}{\rho} = \frac{N_{\mathrm{A}}}{3\varepsilon_0}\left(\frac{\alpha_1 + \alpha_2 + \alpha_3}{3} + \frac{\mu^2}{3kT}\right). \tag{2.44}$$

This formula provides a simple way of determining molecular dipole moments from experimental measurements. The method is to plot molar polarisation or effective polarisability against $1/T$, as in Fig. 2.8; the slope of

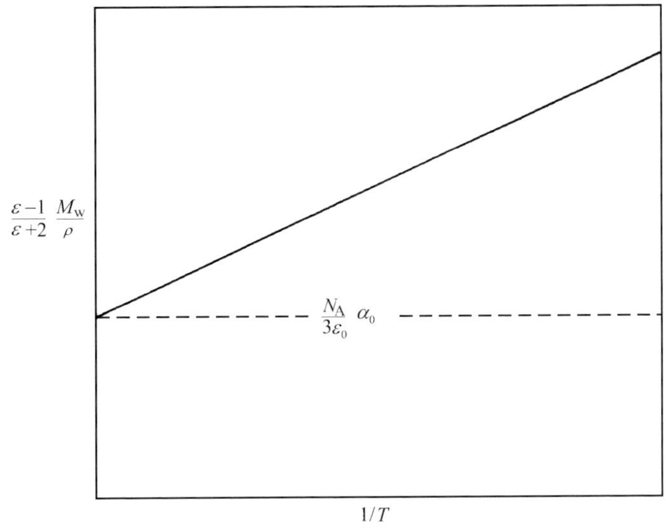

Fig. 2.8 Temperature dependence of molar polarisation.

the straight line is then simply related to the square of the dipole moment. The method is most aptly applied to gases where one might expect the simple theory, which essentially neglects intermolecular effects, to be most accurate. The method may also be extended to dilute solutions of polar molecules in non-polar solvents. For this purpose, simple additivity of solute and solvent polarisations is assumed, each component making a contribution proportional to its molar fraction f in the solution. The polarisation of the solution, experimentally observed through its relative permittivity, then complies with the following equation:

$$\frac{\varepsilon - 1}{\varepsilon + 2}\frac{M_{W1}f_1 + M_{W2}f_2}{\rho} = P_{M1}f_1 + P_{M2}f_2, \qquad (2.45)$$

where subscripts 1, 2 denote solute and solvent respectively. Knowing the molar polarisation of the pure solvent P_{M2}, the molar polarisation of the solute P_{M1} can be evaluated. In practice the latter varies with the concentration, usually decreasing at high concentrations (high values of f_1). This effect is attributed to association between solute molecules, and is eliminated by extrapolating the results to infinite dilution, where each solute molecule must be entirely surrounded by solvent molecules. This approach has been worked extensively, and a long list of molecular dipole moments has been compiled (Wesson, 1948). An illustrative sample of results is given in Table 2.3. It is reasonable to suppose that various atomic groupings will have characteristic moments, and that when several groups are combined in one complicated

Table 2.3 *Molecular dipole moments*

Compound	Dipole moment(10^{-30} Cm)
H_2O	6.1
HF	6.4
HCl	3.6
HBr	2.6
CH_4	0
CCl_4	0
CO_2	0
NH_3	4.9
CH_3Cl	5.3
C_2H_5OH	5.7
$(C_2H_5)_2O$	3.8
C_6H_5Cl	5.7
C_6H_5Br	5.8
$C_6H_5NO_2$	14.0
$C_6H_5CH_3$	1.2

molecule their moments will add vectorially. This principle has proved to be a very valuable test both for assessing theories about molecular configuration and for predicting values of dipole moments. Group moments depend considerably on the character of their bonding to the rest of the molecule in which they are present, however, and quantitative discrepancies are bound to occur. Thus the dipole moment of a substituent chlorine atom will be different in aliphatic and aromatic compounds, and two groups, which are close to each other in a molecule, may affect each other. Nevertheless, a useful assignment of group moments has been established and this enables one to make a good estimate of the dipole moment of most molecular configurations. For example, from the list of group moments given in Table 2.4, the predicted molecular moment for *p*-chloronitrobenzene is 8.3×10^{-30} Cm, which agrees well with the observed value 9.3×10^{-30} Cm.

The directionality of group moments is usually clear from chemical evidence or by a study of the resultant dipole moment of a compound following the introduction of other polar substituents.

The situation is much more complicated in solids because the intermolecular effects can no longer be ignored, i.e. the approximation $\boldsymbol{E}^{\mathrm{M}} = 0$ inherent in the simple formula for the local field (2.29) is not generally true. Consequently, although we can predict the molecular dipole moment from known group moments, it is not possible to calculate the molar polarisation and thereby the relative permittivity, without further elaboration of the dielectric model. In the case of a polymer there are further complications which arise from the flexibility of the long chains.

Table 2.4 *Group dipole moments*

Group	Aliphatic compounds, moment (10^{-10} Cm)	Angle[a] (degrees)	Aromatic compounds, moment (10^{-30} Cm)	Angle (degrees)
$-CH_3$	0	0	1.3	0
–F	6.3		5.3	
–Br	6.7		5.7	
–I	6.3		5.7	
–OH	5.7	60	4.7	60
$-NH_2$	4.0	100	5.0	142
–COOH	5.7	74	5.3	74
$-NO_2$	12.3	0	14.0	0
–CN	13.4	0	14.7	0
$-COOCH_3$	6.0	70	6.0	70
$-OCH_3$	4.0	55	4.3	55

[a]The angle denotes the direction of the moment with respect to the bond attaching the group to the rest of the molecule.

The weakness of the simple theory is highlighted if we solve the Debye version of the Clausius–Mosotti relation (2.44) for ε, neglecting the contribution of α_0:

$$\varepsilon = \frac{1 + [(2N_A\mu^2)/(9\varepsilon_0 kT)]}{1 - [(N_A\mu^2)/(9\varepsilon_0 kT)]}. \quad (2.46)$$

This predicts that as the temperature is decreased, ε becomes infinite at a temperature T_c given by

$$T_c = \frac{N_A\mu^2}{9k\varepsilon_0}, \quad (2.47)$$

and we may rewrite the relation in terms of this critical temperature:

$$\varepsilon - 1 = \frac{3T_c}{T - T_c}. \quad (2.48)$$

The implication is that lowering the temperature eventually allows the polarisation to increase to such an extent that the molecular dipoles may *spontaneously* align themselves parallel to each other, even in the absence of an applied field.

This is called the ferroelectric effect, by analogy with ferromagnetism; Equation (2.48) is analogous to the Curie–Weiss law. A few crystalline materials do indeed display ferroelectric behaviour, but the phenomenon is rare.

This means that the Debye equation fails for most materials in the regime of high local fields and it is necessary to improve the theoretical treatment to cope with this situation.

Onsager derived an improved formula by adopting a better model for the calculation of the local field at a molecule. His model consists of a spherical cavity which is excised in the dielectric material and which is just large enough to accommodate one molecule. The molecular dipole is supposed to be a point dipole μ at the centre of the sphere, radius a. Onsager then said that the local field operating on the dipole at the centre of the cavity could be resolved into two components, a cavity field $\boldsymbol{G}$ and a reaction field $\boldsymbol{R}$:

$$\boldsymbol{E}^{\mathrm{L}} = \boldsymbol{G} + \boldsymbol{R} . \tag{2.49}$$

The cavity field is defined as that present at the centre of an empty cavity in the presence of the applied field $\boldsymbol{E}$, i.e.

$$\boldsymbol{G} = \frac{3\varepsilon}{2\varepsilon + 1}\boldsymbol{E} . \tag{2.50}$$

The reaction field is defined as the extra field at the centre of the cavity due to the polarisation of the material external to the cavity caused by the presence of the point dipole at the centre of the cavity, i.e.

$$\boldsymbol{R} = \frac{2(\varepsilon - 1)}{\varepsilon_0(2\varepsilon + 1)a^3}\boldsymbol{\mu} . \tag{2.51}$$

Using this local field one finally obtains the relation

$$\frac{(\varepsilon - n^2)(2\varepsilon + n^2)}{\varepsilon(n^2 + 2)^2}\frac{M_{\mathrm{W}}}{\rho} = \frac{N_{\mathrm{A}}\mu^2}{9\varepsilon_0 kT}, \tag{2.52}$$

where the deformational polarisability of the molecules has been taken into account using the equation for molar refraction (2.35). This formula very successfully describes the behaviour of many fluids. It is an improvement on the Clausius–Mosotti relation in that it avoids artificially predicting a ferroelectric effect (often referred to as the ferroelectric catastrophe). Strong local forces have, however, still been neglected, and it is therefore not surprising that it fails to explain the behaviour of associated liquids, like water, and many solids.

Kirkwood took a more rigorous statistical-mechanical approach in an attempt to incorporate the effect of local ordering. His theory is only valid for rigid dipoles, and it was left to Fröhlich to extend the treatment properly to a system of polarisable dipolar molecules. The work is well described in Fröhlich's classic text (1949). The final outcome was the following formula, which connects the macroscopic relative permittivity with the molecular dipole moment:

$$\frac{(\varepsilon - n^2)(2\varepsilon + n^2)}{\varepsilon(n^2+2)^2}\frac{M_W}{\rho} = \frac{N_A g \mu^2}{9\varepsilon_0 kT} \ . \tag{2.53}$$

The parameter g is the correlation factor first introduced by Kirkwood (1939) to take account of local order:

$$g = 1 + z\overline{\cos\gamma} \ , \tag{2.54}$$

where z is the number of nearest neighbours of a molecule in the system and γ is the angle of the reference molecule with respect to a nearest neighbour. The Fröhlich equation reduces to the Onsager equation where $g = 1$.

The usefulness of Fröhlich's formula (2.53) is mainly restricted by our ignorance of the correlation factor g, which necessarily depends on the shapes of molecules and the disposition of the permanent dipoles within them, the anisotropy of polarisability and the presence of charge distributions of higher orders of symmetry. The theory gives us a good general understanding of the behaviour of polar materials, but deviations from the simple Debye equation (2.44) can often only be discussed in qualitative terms.

Local ordering effects have long been recognised experimentally in measurements of dipole moments of polar solutes in non-polar solvents, where the value obtained on the basis of the simple model differs from the value obtained for the pure solute in the gas phase, even when the results are extrapolated to infinite dilution. This so-called solvent effect is due to the Onsager reaction field. If there is no strong local ordering, Onsager's formula (2.52) is valid and the apparent solution moment is related to the isolated molecule or gas moment by

$$\boldsymbol{\mu}_{\mathrm{app}} = \frac{(2\varepsilon_0 + 1)(n^2+2)}{3(2\varepsilon_0 + n^2)}\boldsymbol{\mu} \ . \tag{2.55}$$

Usually, though, the solvent molecules do organise themselves in a special way around the polar species for the reason cited above. The effect depends principally on the anisotropy of the solute molecules, and many semi-empirical equations have been devised on this basis to explain the apparent change of dipole moments in solutions. An alternative approach has been to quote gas- and solution-determined dipole moments separately and to use the appropriate value in any particular situation.

In the case of polymers the principal ordering effect is an *intra*molecular one, as we shall see in the next section.

2.6 Relative permittivity of polymers

The foregoing theoretical development provides a general basis for our understanding and interpretation of the relative permittivity of polymers, which are our primary concern here. A convenient way of thinking about the dielectric behaviour of polymers is to consider not the complete long-chain molecules as the polarisable entities, but rather the component repeat units. For most polymeric materials, the degree of polymerisation is greater than a hundred, so that the effect of chain ends can be neglected for most purposes. The predominant constraint on each segment is due to its chemical linkage to the rest of the molecular chain on either side, and the strong intramolecular ordering, or correlation, of the segments along the chain can be formally taken into account through an intramolecular *g*-factor. We shall summarise and illustrate the results that have been obtained by this general approach, initially discussing a few overriding, simple principles, which give physical insight into dielectric properties of polymers.

2.6.1 Non-polar polymers

Let us consider an essentially non-polar polymer, e.g. polyethylene, $CH_3–(CH_2)_n–CH_3$. The density of solid polyethylene covers a range from 0.92 to 0.99 Mgm^{-3} depending on the extent of chain branching which determines its crystallinity. We may therefore test the validity of the Clausius–Mosotti relation. From published tables of bond polarisabilities, the Clausius–Mosotti relation for an assembly of $–CH_2–$ units becomes

$$\frac{\varepsilon - 1}{\varepsilon + 2} = K\rho, \quad \text{where } K = 0.327 \times 10^{-3}, \tag{2.56}$$

$$\text{i.e.} \qquad \varepsilon = 2.276 + 2.01(\rho - 920)10^{-3}.$$

Experimental measurements confirm the linear dependence of relative permittivity on density. Furthermore, the agreement on the value of K (mean experimental value 0.326 compared with 0.327 theoretical) is very satisfactory.

Conjugated polymers, which contain multiple carbon–carbon bonds, can be expected to have somewhat higher relative permittivities, since the bond polarisabilities of multiple bonds are higher than those of single C–C bonds, see Table 2.2. Most measurements of relative permittivities have been made on the conductive forms of these polymers, but these are outside the scope of this chapter and will be discussed later. The intrinsic energy gap in the semiconducting forms is large, typically > 2 eV. In consequence, in samples with low impurity content, the electrical conductivity at room temperature, where $kT \sim 0.025$ eV, is low, as the number of thermally excited carriers is small. The samples are essentially insulating and can be treated as dielectrics. Measured values of the mean relative permittivity of lightly rolled samples of polyacetylene perpendicular and parallel to the rolling direction are 4.0 ± 0.2 and 5.7 ± 0.2 for *trans*-polyacetylene (Devreux *et al.*, 1981). Values of 3.6 and 4.8 were obtained for similar samples of *cis*-polyacetylene, although rolling converts some of the *cis*-polyacetylene to the *trans*-form, so that these values are less accurate. (Although these measurements were made at 9.1 GHz, in the absence of any relaxation due to dipole reorientation, see Fig. 2.4, the relative permittivities will be close to the low frequency values.) A number of determinations have been made of the dielectric properties of the single crystal poly(diacetylene *p*-toluenesulphonate).

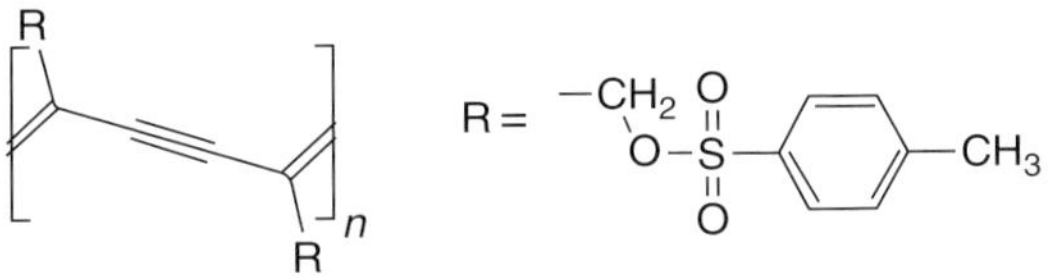

The principal components of the relative permittivity tensor, measured parallel to the crystal axes, *a*, *b* and *c*, at 1 kHz and room temperature, are (Orczyk, 1990):

$$\varepsilon_a = 3.90 \pm 0.02, \ \varepsilon_b = 6.80 \pm 0.03, \ \varepsilon_c = 4.35 \pm 0.01. \qquad (2.57)$$

The conjugated polymer chain lies parallel to the crystallographic *b*-axis. All the values are quite large, but this is to be expected given the structure of the bulky pendent groups in this polymer. However, the value parallel to the conjugated polymer chain is distinctively the largest.

2.6.2 Polar polymers

When permanent dipole moments are present in a polymeric chain we can distinguish two different situations: one where the whole polymer backbone, together with its polar groups, is rigidly fixed in a single conformation, and one where the backbone is flexible and pendent side groups can rotate freely. The former is more characteristic of the crystalline state than of amorphous or fluid states, although occasionally a particular configuration may be especially favoured and survive even in non-crystalline environments.

If a polymer is held in a fixed conformation, the resulting moment of a complete molecule will depend very much on whether the moments of the

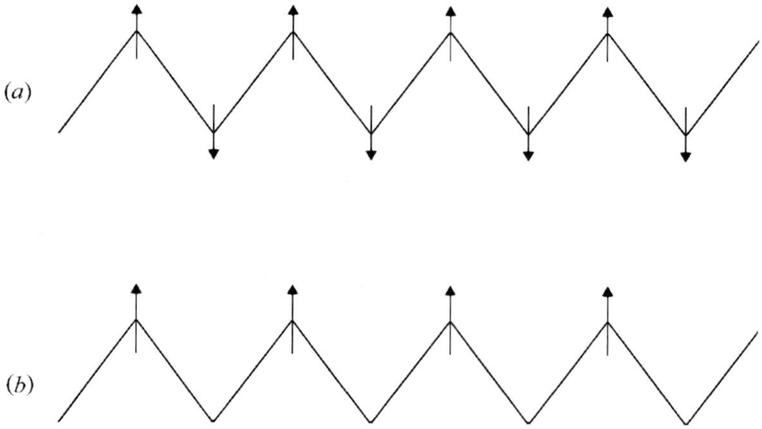

Fig. 2.9 Schematic arrangement of polar groups in a polymer chain: (a) balanced dipoles, (b) additive dipoles.

individual segments reinforce or compensate each other. In extended configurations of poly(tetrafluoroethylene), the high dipole moments of alternate $-CF_2-$ groups balance each other exactly (Fig. 2.9(a)) and for this reason the relative permittivity of this polymer is low, like that of a non-polar material. However, some defects are always present in the conformations of the molecules (helical conformations, which also give dipole balancing, are typical of the crystalline phase of this polymer), and these are responsible for the small dipolar orientation effects that can just be detected. In contrast, the C–Cl dipole moments in poly(vinylchloride) are additive in the preferred planar zigzag conformation (Fig.2.9(b)), so that the relative permittivity of this polymer is high.

A particularly striking additive effect is observed in the case of certain synthetic polypeptides, e.g. poly(γ -benzyl-L-glutamate), which has a formula $(-CO-CHR-NH-)_n$, $R = -CH_2-CH_2-COO-CH_2-C_6H_5$. In solution this substance readily forms an α-helix, maintained by hydrogen bonding, with an axial dipole moment per repeat unit of about 4.14×10^{-30} Cm. Since a typical molar mass is about 500 000, the total axial dipole moment for a molecule is about $10\,000 \times 10^{-30}$ Cm.

2.6.3 Mean-square moment

Most usually, polymer molecules will not be in a single, fixed conformation, and the experimentally observable quantity, the mean-square dipole moment, is an average over many different conformations. At any given instant the resultant moment $\boldsymbol{M}$ of a whole molecule will be given by the vector sum of all the segment moments, $\boldsymbol{m}_k$:

$$\boldsymbol{M} = \sum_{k=1}^{n} \boldsymbol{m}_k. \tag{2.58}$$

The mean-square moment for an assembly of such molecules will be defined theoretically by

$$\begin{aligned}\overline{M^2} &= \overline{\sum_{i=1}^{n} \boldsymbol{m}_i \cdot \sum_{j=1}^{n} \boldsymbol{m}_j} \\ &= \sum_{i=1}^{n} \sum_{j=1}^{n} \overline{\boldsymbol{m}_i \cdot \boldsymbol{m}_j} \\ &= m^2 \left(n + \sum_{i=1}^{n} \sum_{j=1(i \neq j)}^{n} \overline{\cos\theta_{ij}} \right),\end{aligned} \tag{2.59}$$

where $\overline{(\cos\theta_{ij})}$ is the cosine of the angle θ_{ij} made between the dipole of repeat unit *i* and that of unit *j* averaged over all polymer molecules. In a real situation this depends principally on the constraints of chemical bonding, steric hindrance between neighbouring parts of the chain and dipole–dipole interaction along the chain, as well as interactions with neighbouring molecules. From Equation (2.59) we may obtain an expression for the effective mean-square moment per repeat unit of the polymeric chains, and thereby define a segmental correlation factor g_r for the polymer:

$$\begin{aligned}\frac{\overline{M^2}}{n} &= m^2 \left(1 + \frac{1}{n} \sum_{i=1}^{n} \sum_{j=1(i \neq j)}^{n} \overline{\cos\theta_{ij}} \right) \\ &= g_r m^2.\end{aligned} \tag{2.60}$$

Intramolecular forces must dominate the correlation, and this notion provides the basis for a theoretical discussion of g_r-values. Intermolecular forces may be taken into account as a secondary perturbation. An important case concerns the arrangement of segmental dipoles, aligned perpendicularly with respect to the chain contour, along a polymeric molecule which has a backbone of carbon atoms. Part of the dipolar correlation is fixed, of course, by the tetrahedral nature of the carbon valence, but part depends on the possible rotation about the C–C bonds of the chain. For completely free rotation it may be shown that $g_r = 11/12$.

Only rarely is this situation approached in real molecules, and in the *n*-paraffins, for example, the *trans*-isomer is about 3 kJ mol^{-1} lower in energy than the *gauche*-isomer, so that at room temperature the *trans*-isomer is appreciably favoured, and the g_r-value for the CH_2 dipoles is reduced.

The series of polyethers of general formula $-((CH_2)_x-O)-_n$ has been treated theoretically in detail (Read, 1965) and illustrates the principles of segmental dipole correlation very well. Confining ourselves here to a qualitative discussion, we first note that the main dipole moment, essentially the same in all of these molecules, is that of the oxygen atom in the chain ($\mu = 3.8 \times 10^{-30}$ Cm). In the simplest member of the series, poly(methylene oxide), the adjacent dipoles along the chain are very close to one another and the dipole–dipole repulsion is strong. For this reason a helical, *gauche*-conformation, which gives a low $\overline{M^2}/n$ value by dipole balancing, is preferred energetically. The low relative permittivity ($\varepsilon \approx$ 3.5) is observed to increase with temperature and this may be explained by an increase in the higher energy *trans*-conformations, which produce parallel alignment of adjacent dipoles, as the thermal energy increases.

In the next member of the series, poly(ethylene oxide), the dipoles are further apart. An intermediate $\overline{M^2}/n$ value with negligible temperature dependence is observed, suggesting that the *trans*- and *gauche*-isomers have approximately equal energies. At room temperature $\varepsilon \approx 4.5$.

By the time the third member of the series, poly(trimethylene oxide), is reached, dipole–dipole interaction along the chain is very weak indeed, and the *trans*-conformation is slightly more favoured energetically (by about 4 kJ mol^{-1}). Since the *trans*-conformation involves parallel alignment of the oxygen dipoles, the observed decrease of relative permittivity with temperature is expected.

Finally, in poly(tetramethylene oxide), the *trans*-conformation is again the most favoured, but here the *trans*-conformation incurs anti-parallel alignment of the oxygen dipoles. The relative permittivity therefore increases with temperature.

Strictly speaking, the foregoing discussion of segmental correlation has assumed that the polymers are *iso*tactic. When the chemical repeat units along the polymeric chains are not all stereo-identical, and *syndio*tactic or even *a*tactic isomers are present, the simple molecular concept of a segmental correlation factor no longer applies. A more elaborate treatment is then required (Volkenstein, 1963). Riande and Saiz (1992) discuss examples in more detail.

2.7 Low relative permittivity polymers

The development of polymers with low relative permittivity provides an informative and instructive exercise in polymer design using the principles outlined above. The context is the electronics industry, which makes extensive use of high-performance polymers as insulating layers to encapsulate electronic components, to separate them with interlayers and to provide isolated pathways to connect them on multilayer printed circuit boards. Now the performance and speed of computers is limited by the coupling capacitance between circuit paths and the integrated circuit components, and the situation becomes more severe as microprocessors become smaller, demanding thinner and thinner insulations. Any decrease that can be made in the relative permittivity of the insulating material reduces the deleterious stray and coupling capacitances, thereby alleviating the problem, and this has driven a search for low relative permittivity polymers.

For advanced electronics applications polyimides have been the materials of choice for high-performance insulation. The simplest member of the polyimide series has the structural formula:

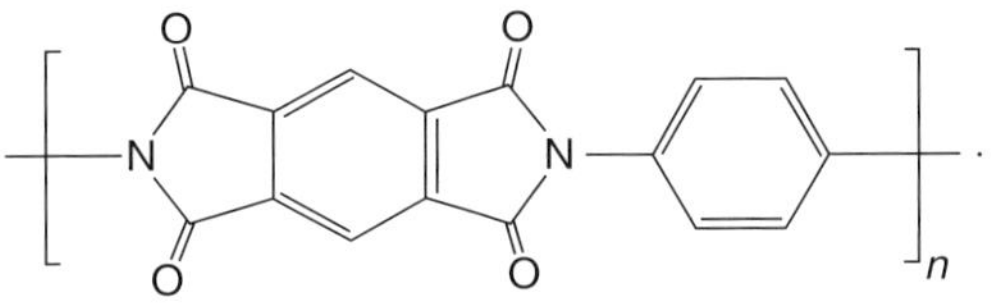

These polymers exhibit exceptional thermal and mechanical properties, are easily processable as thin films from soluble precursors and have good dielectric properties. Clearly, they might be made even better by reducing their relative permittivity, and it was early recognised that one way of doing this was to substitute fluorine for some of their hydrogen atoms. The basis for this is the lower polarisability of the C–F unit compared with that of the C–H unit, as indicated in Table 2.2. The disadvantage of fluorine substitution is the highly polar character of C–F, as indicated in Table 2.4. Although this will not affect the relative permittivity at high frequencies, an increase in the relative permittivity from dipolar orientation could intrude at lower frequencies. Polyimides are, however, generally used below their glass transition temperature (see Section 3.3), so that dipolar orientation is normally too slow to cause an appreciable effect at operational frequencies. Furthermore, using a balanced substitution will avoid introducing any net dipole effect at all, e.g.

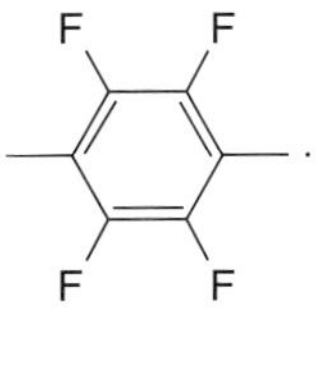

An additional benefit of fluorination is an increase in hydrophobicity through the effect of the highly polar C–F groups. This means that the level of absorbed water in the polymer at typical ambient humidity is much reduced. Water has a very high relative permittivity, so that its presence can significantly increase the relative permittivity of a material, and the level will also be liable to vary with ambient conditions, bringing uncertainty into design considerations for electronic systems.

Adopting the fluorination approach in polyimides, relative permittivities can be reduced from ~3.4 to ~2.8, a common structure incorporating the hexafluoro*iso*propylydene moiety (6F):

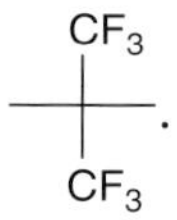

Early indications were that the fluorination effect on relative permittivity was larger than that to be expected from changes in bond polarisabilities alone. More detailed work by Hougham, Tesoro & Viehbeck (1996) on a series of polyimides with differing degrees of fluorine substitution clearly shows that approximately 50 % of the reduction in relative permittivity is attributable to a change in *free volume* brought about by the fluorine substitution. Free volume in a polymer is volume in addition to that *occupied* by the atoms on the basis of their van der Waals radii. The free volume associated with 1 mol of repeat units of the polymer may be estimated by subtracting the occupied molar volume of a repeat unit (using group additivity tables), V_0, from the total molar volume, M/ρ, where M is the molar mass of the repeat unit. The fractional free volume v_{ff} is then given by:

$$v_{\mathrm{ff}} = \frac{M/\rho - V_0}{M/\rho}. \tag{2.61}$$

An increase in fractional free volume will reduce the number of polarisable groups per unit volume, and thereby reduce the relative permittivity of the polymer. Quantitatively, the effect may be estimated by means of the Clausius–Mossotti/Lorenz–Lorentz model for dielectric mixing (Böttcher, 1978):

$$\frac{\varepsilon'_{\mathrm{mix}} - 1}{\varepsilon'_{\mathrm{mix}} + 2} = \sum_i \theta_i \frac{\varepsilon'_i - 1}{\varepsilon'_i + 2}. \tag{2.62}$$

Here θ_i is the volume fraction of the ith component, and ε'_i is its relative permittivity in pure form. For the case of a two-component mixture of free space and volume occupied by the polymer, we may obtain, taking the relative permittivity of free space to be 1:

$$\frac{\varepsilon'_{\text{mix}} - 1}{\varepsilon'_{\text{mix}} + 2} = 0 + (1 - V_{\text{ff}})\frac{\varepsilon'_{\text{VO}} - 1}{\varepsilon'_{\text{VO}} + 2}, \tag{2.63}$$

where ε'_{VO} is the relative permittivity of the pure polymer with no free space within it. Knowing the relative permittivity of one particular mixture – that comprising the experimental polymer – together with an estimate of its fractional free volume, ε'_{VO} may be determined for the polymer concerned. Hence the relative permittivity of a polymer may be hypothetically normalised to the same fractional free volume as that of a fluorinated version, enabling a comparison to be made of the relative permittivities at equal fractional free volumes. This immediately gives an estimate of the contribution that the reduced polarisability of the C–F bond alone makes to the reduction in the relative permittivity of the fluorinated polymer. Subtracting this from the experimentally observed reduction in total relative permittivity leaves that attributable to the increase in free volume brought about by fluorination.

In this way Hougham, Tesoro & Viehbeck (1996) found that the free volume effect was generally large, but varied with the exact form of the structures undergoing substitution. Results for three cases are given in Table 2.5, with the structures of the repeat units of the polymers shown in Fig. 2.10. The large free volume-related reductions that are observed in relative permittivity from fluorine substitutions in $-CH_3$ groups suggests that these substitutions upset packing density efficiencies much more than substitutions around planar aromatic rings. Direct measurements of fractional free volume using the positron annihilation lifetime spectroscopy technique confirm the relationship between fluorination and increase in free volume in these kinds of polymeric structures (Hougham, Zhang & Jean, 1996).

Table 2.5 *Change of relative permittivity attributable to change in fractional free volume*

Polymer pair	$\%\Delta\varepsilon'$ due to $\Delta\nu_{\text{ff}}$
(6FDA–PDA) → (6FDA–TFPDA)	25 ± 20
(6FDA–DAT) → (6FDA–TFMPDA)	49 ± 31
(6FDA–2DAT) → (6FDA–2TFMPDA)	94 ± 12

Generally, we must conclude that control of free volume is just as important as selection of groups with low polarisability, in order to achieve polymer molecular structures with low relative permittivities.

There are now alternative polymeric film-forming materials that achieve low ε values without the need for fluorine substitution, e.g. crosslinked

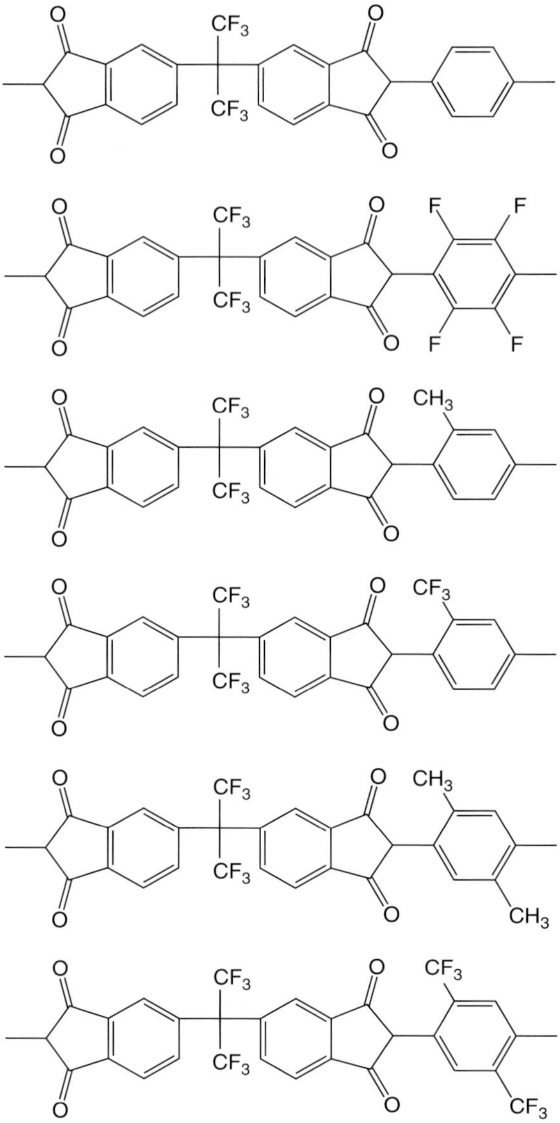

Fig. 2.10 Molecular structures of fluorine-containing polyimides.

polyphenylenes (Martin *et al.*, 2000). These avoid problems of compatibility of some metal components with fluorine-containing polymers.

2.8 Further reading

The original monographs by Debye (1929) and Fröhlich (1949) still provide an excellent introduction to molecular dielectric theory. Hinchcliffe and Munn

(1985) is a more recent text that presents electromagnetism in the context of molecules and molecular properties. Smith (1955) gives detailed information about measurements of dipole moments, and data on polymers are reviewed by McCrum, Read & Williams (1967).

3

Dielectric relaxation

3.1 General theory

3.1.1 Complex relative permittivity and dielectric loss

Orientation of molecular dipoles is a relatively slow process if comparison is made with electronic transitions or molecular vibrations, which have frequencies generally above 10^{12} Hz. Furthermore, it does not consist of a uniform switch in the arrangement of all molecules; it is more in the nature of a slight adjustment of their average orientations in the face of continued thermal agitation. Only when sufficient time is allowed after the application of an electric field for the orientation to attain equilibrium will the maximum polarisation, corresponding to the highest observable relative permittivity, be realised in a material. If ample time is allowed, then the observed relative permittivity is called the static relative permittivity (or static dielectric constant) ε_s. On the other hand, if the polarisation is measured immediately after the field is applied, allowing no time for dipole orientation to take place, then the observed instantaneous relative permittivity, denoted ε_∞, will be low and due to deformational effects alone. Somewhere in between these extremes of timescale there must therefore be a dispersion from a high to a low relative permittivity. Let us start our examination of this rate effect by considering the application of an alternating electric field $\boldsymbol{E}$, amplitude $\boldsymbol{E}_0$ and angular frequency ω, across a dielectric material:

$$\boldsymbol{E} = \boldsymbol{E}_0 \cos \omega t. \tag{3.1}$$

This will produce polarisation, which alternates in direction, and, if the frequency is high enough, the orientation of any dipoles which are present will inevitably lag behind the applied field. Mathematically, we can express this as a phase lag δ in the electric displacement:

$$\boldsymbol{D} = \boldsymbol{D}_0 \cos(\omega t - \delta), \tag{3.2}$$

which may be written

$$\boldsymbol{D} = \boldsymbol{D}_1 \cos \omega t + \boldsymbol{D}_2 \sin \omega t, \tag{3.3}$$

where

$$\boldsymbol{D}_1 = \boldsymbol{D}_0 \cos \delta \text{ and } \boldsymbol{D}_2 = \boldsymbol{D}_0 \sin \delta.$$

This leads us to define two relative permittivities

$$\varepsilon' = \frac{\boldsymbol{D}_1}{\varepsilon_0 \boldsymbol{E}_0} \quad \text{and} \quad \varepsilon'' = \frac{\boldsymbol{D}_2}{\varepsilon_0 \boldsymbol{E}_0}, \tag{3.4}$$

linked by the relation

$$\frac{\varepsilon''}{\varepsilon'} = \tan \delta. \tag{3.5}$$

It is convenient to combine these two quantities into a complex relative permittivity (or complex dielectric constant):

$$\varepsilon^* = \varepsilon' - \mathrm{i}\varepsilon''. \tag{3.6}$$

The meaning of the real and imaginary parts may be readily appreciated by considering the material in a capacitor (capacitance C_0 when empty), as shown in Fig. 3.1. The current I which flows in the external circuit after application of

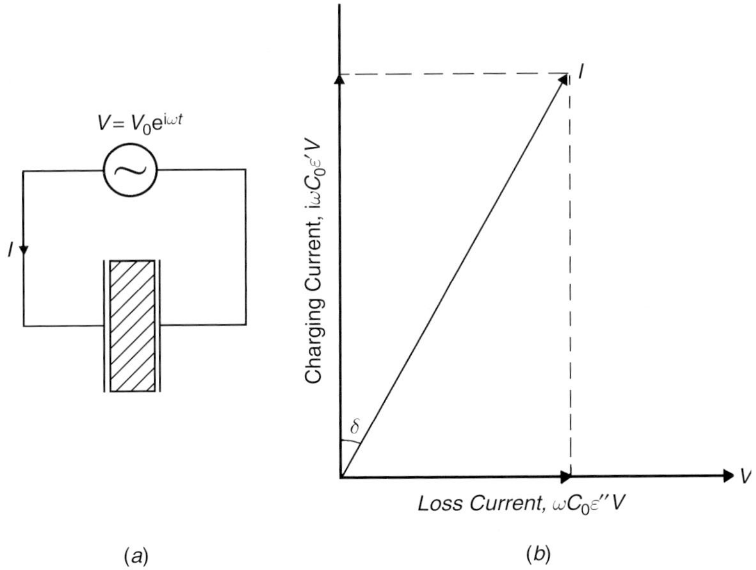

Fig. 3.1 AC losses in a dielectric: (a) circuit diagram, (b) Argand diagram of complex current–voltage relationship.

an alternating voltage given by the real part of $V = V_0 e^{i\omega t}$ may be calculated as follows:

$$\begin{aligned} I &= \varepsilon^* C_0 \frac{dV}{dt} \\ &= i\omega\varepsilon^* C_0 V \\ &= \omega C_0(\varepsilon'' + i\varepsilon')V. \end{aligned} \tag{3.7}$$

This implies that we have a capacitive component of the current,

$$I_C = i\omega C_0 \varepsilon' V, \tag{3.8}$$

which leads the voltage by 90°, and a resistive component,

$$I_R = \omega C_0 \varepsilon'' V, \tag{3.9}$$

which is in phase with the voltage. Work can only be done by the latter component, and the physical meaning of the useful quantity tan δ, previously defined by Equation (3.5), becomes apparent:

$$\tan\delta = \frac{\varepsilon'}{\varepsilon''} \propto \frac{\text{energy dissipated/cycle}}{\text{energy/cycle}}. \tag{3.10}$$

ε'' is called the dielectric loss factor and tan δ is usually called the dielectric loss tangent or dissipation factor.

ε' and ε'' are experimentally observable quantities which may be used to characterise the dielectric dispersion over a range of frequencies. In order to be able to interpret any such dispersion behaviour, it is first necessary to forge a link between these macroscopic, observable quantities and molecular properties, by postulating a reasonable model that describes the way the molecules respond to the applied field.

3.1.2 The dielectric relaxation process

The basic theory of dielectric relaxation behaviour, pioneered by Debye, begins with a macroscopic treatment of frequency dependence. This treatment rests on two essential premises: exponential approach to equilibrium and the applicability of the superposition principle. In outline, the argument is as follows.

Consider the application at time $t = 0$ of a steady field $\boldsymbol{E}_0$ across a dielectric. The resulting electric displacement $\boldsymbol{D}(t)$ at subsequent times t will then comply with the equation:

$$\boldsymbol{D}(t) = \varepsilon_0[\varepsilon_\infty + (\varepsilon_s - \varepsilon_\infty)\Psi(t)]\boldsymbol{E}_0. \tag{3.11}$$

The first term on the right-hand side, $\varepsilon_0\varepsilon_\infty \boldsymbol{E}_0$, represents the instantaneous response of the material to the field. The second term, $\varepsilon_0(\varepsilon_s - \varepsilon_\infty)\Psi(t)\boldsymbol{E}_0$, represents the slower contribution from polarisation of dipoles, with the factor $\Psi(t)$ describing the time development of the underlying orientation process. By this definition $\Psi(0) = 0$ and $\Psi(\infty) = 1$. We further assume that the rates at which dipolar polarisation ${}^{\mathrm{D}}\boldsymbol{P}(t)$ progresses towards its equilibrium value ${}^{\mathrm{D}}\boldsymbol{P}(\infty) = \varepsilon_0(\varepsilon_s - \varepsilon_\infty)\boldsymbol{E}_0$ is proportional to its degree of departure from equilibrium, i.e.

$$ {}^{\mathrm{D}}\boldsymbol{P}(t) = -[{}^{\mathrm{D}}\boldsymbol{P}(\infty) - {}^{\mathrm{D}}\boldsymbol{P}(t)]/\tau. \tag{3.12} $$

Here τ is a characteristic time constant, usually called the dielectric relaxation time. To conform to analogous theories of visco-elastic behaviour we should really use the term dielectric *retardation* time, because it refers to a gradual change in a strain (the polarisation or resulting electric displacement) following an abrupt change in stress (the applied field). Dielectric *relaxation* time is, however, still most commonly used in spite of this inconsistency. By integration of Equation (3.12):

$$ {}^{\mathrm{D}}\boldsymbol{P}(t) = {}^{\mathrm{D}}\boldsymbol{P}(\infty)(1 - \mathrm{e}^{t/\tau}), \tag{3.13} $$

so that

$$ \Psi(t) = 1 - \mathrm{e}^{-t/\tau}. \tag{3.14} $$

If polarisation of dipoles is a linear function of applied field, then a higher field $\boldsymbol{E}_0 + \boldsymbol{E}_1$, applied at time $t = 0$, will produce a proportional increase in electric displacement at all subsequent times:

$$ \boldsymbol{D}(t) = \varepsilon_0[\varepsilon_\infty + (\varepsilon_s - \varepsilon_\infty)\Psi(t)](\boldsymbol{E}_0 + \boldsymbol{E}_1). \tag{3.15} $$

According to Boltzmann's superposition principle for such linear systems, if the extra field is added at time t_1, the total displacement at times $t > t_1$ will be

$$ \begin{aligned} \boldsymbol{D}(t) = {} & \varepsilon_0[\varepsilon_\infty + (\varepsilon_s - \varepsilon_\infty)\Psi(t)]\boldsymbol{E}_0 \\ & + \varepsilon_0[\varepsilon_\infty + (\varepsilon_s - \varepsilon_\infty)\Psi(t - t_1)]\boldsymbol{E}_1, \end{aligned} \tag{3.16} $$

i.e. the sum of the two electric displacements at their respective elapsed times. In general, for the series of all field increments up to time t,

$$ \boldsymbol{D}(t) = \sum_{t_i = -\infty}^{t} \varepsilon_0[\varepsilon_\infty + (\varepsilon_s - \varepsilon_\infty)\Psi(t - t_i)]\boldsymbol{E}_i, \tag{3.17} $$

whence for a continuously varying field

$$\boldsymbol{D}(t) = \varepsilon_0\varepsilon_\infty \boldsymbol{E}(t) + \int_{-\infty}^{t} \varepsilon_0(\varepsilon_s - \varepsilon_\infty)\Psi(t-s)\frac{\mathrm{d}\boldsymbol{E}_i}{\mathrm{d}s}\mathrm{d}s, \tag{3.18}$$

where s is a variable time. Equation (3.18) expresses the value of the electric displacement actually present in the material at time t in terms of the entire past history of subjection to applied field. Integrating by parts:

$$\boldsymbol{D}(t) = \varepsilon_0\varepsilon_\infty \boldsymbol{E}(t) + \varepsilon_0(\varepsilon_s - \varepsilon_\infty) \int_{-\infty}^{t} \Psi'(t-s)\boldsymbol{E}(s)\mathrm{d}s. \tag{3.19}$$

From Equation (3.14) the function $\Psi'(t)$, called the *dielectric decay function*, is given by

$$\Psi'(t) = \frac{1}{\tau}\,\mathrm{e}^{-t/\tau}. \tag{3.20}$$

Now, differentiation of Equation (3.19) with respect to t gives

$$\frac{\mathrm{d}\boldsymbol{D}(t)}{\mathrm{d}t} = \varepsilon_0\varepsilon_\infty\frac{\mathrm{d}\boldsymbol{E}(t)}{\mathrm{d}t} + \varepsilon_0(\varepsilon_s - \varepsilon_\infty)\frac{\mathrm{d}}{\mathrm{d}t}\left[\int_{-\infty}^{t} \Psi'(t-s)\boldsymbol{E}(s)\mathrm{d}s\right], \tag{3.21}$$

and from Equation (3.20) it follows that

$$\frac{\mathrm{d}}{\mathrm{d}t}\left[\int_{-\infty}^{t} \Psi'(t-s)\boldsymbol{E}(s)\mathrm{d}s\right] = \Psi'(0)\boldsymbol{E}(0) - \int_{-\infty}^{t} \frac{1}{\tau}\Psi'(t-s)\boldsymbol{E}(s)\mathrm{d}s. \tag{3.22}$$

Since, also from Equation (3.20) $\Psi'(0) = 1/\tau$, combining Equations (3.19), (3.21) and (3.22) finally gives the following differential equation for the electric displacement in the material:

$$\tau\frac{\mathrm{d}\boldsymbol{D}(t)}{\mathrm{d}t} + \boldsymbol{D}(t) = \tau\varepsilon_0\varepsilon_\infty\frac{\mathrm{d}\boldsymbol{E}(t)}{\mathrm{d}t} + \varepsilon_0\varepsilon_s\boldsymbol{E}(t). \tag{3.23}$$

For the particular case of an alternating field represented by the real part of $\boldsymbol{E}_t = \boldsymbol{E}_0\mathrm{e}^{\mathrm{i}\omega t}$, which produces an alternating displacement $\boldsymbol{D}(t) = \boldsymbol{D}_0\mathrm{e}^{\mathrm{i}\omega(t-\delta)}$, the solution to Equation (3.23) may be written

$$\begin{aligned}\varepsilon^* &= \frac{\boldsymbol{D}(t)}{\varepsilon_0\boldsymbol{E}(t)} \\ &= \varepsilon_\infty + \frac{\varepsilon_s - \varepsilon_\infty}{1 + \mathrm{i}\omega\tau},\end{aligned} \tag{3.24}$$

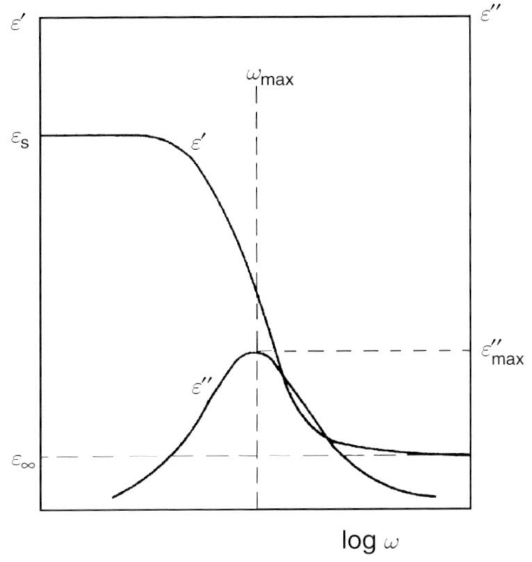

Fig. 3.2 Debye dielectric dispersion curves.

which is called the Debye dispersion equation. Equating real and imaginary parts of the two sides,

$$\varepsilon'(\omega) = \varepsilon_\infty + \frac{\varepsilon_s - \varepsilon_\infty}{1 + \omega^2\tau^2}, \tag{3.25}$$

$$\varepsilon''(\omega) = \frac{\varepsilon_s - \varepsilon_\infty}{1 + \omega^2\tau^2}\omega\tau. \tag{3.26}$$

The graphs of ε' and ε'' against frequency of the applied field (logarithmic scale) through the dispersion region are shown in Fig. 3.2. The dielectric loss peak has a half-height width of 1.14 decades. The maximum loss value occurs when $\omega\tau = 1$, corresponding to a critical frequency $\omega_{max} = 1/\tau$, and location of this peak provides the easiest way of obtaining the relaxation time from experimental results. The difference in relative permittivity measured at low and high frequencies is called the strength of the relaxation and it is related to the area under the absorption curve:

$$\Delta\varepsilon = (\varepsilon_s - \varepsilon_\infty) = \int_{-\infty}^{+\infty} \varepsilon''(\omega)\mathrm{d}(\ln\omega). \tag{3.27}$$

(This is a particular case of the Kramers–Krönig relation that expresses the general interdependence of quantities like ε' and ε'' for any linear system.)

3.1.3 Departures from the Debye model

The dependence of ε'' on ε' may be used to test how well the Debye model fits a real case. If we eliminate the parameter $\omega\tau$ between Equations (3.25) and (3.26) we obtain

$$\left(\varepsilon' - \frac{\varepsilon_s + \varepsilon_\infty}{2}\right)^2 + \varepsilon''^2 = \left(\frac{\varepsilon_s - \varepsilon_\infty}{2}\right)^2. \quad (3.28)$$

This is the equation of a circle, centre $[(\varepsilon_s + \varepsilon_\infty)/2, 0]$, radius $(\varepsilon_s - \varepsilon_\infty)/2$, so that a plot of ε'' against ε' should give a semicircle, as shown in Fig. 3.3. Experimental results for many polar liquids give excellent agreement with this theoretical curve, their relaxation times being of the order of 10^{-11} s.

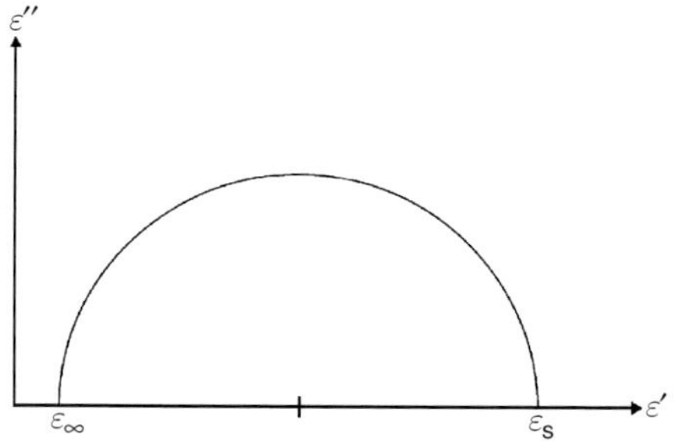

Fig. 3.3 Semicircular Cole–Cole plot.

Relaxations observed in polymers show broader dispersion curves and lower loss maxima than those predicted by the Debye model, and the $(\varepsilon'' - \varepsilon')$ curve falls inside the semicircle. This led Cole and Cole (1941) to suggest the following semi-empirical equation for dielectric relaxations in polymers:

$$\varepsilon^* = \varepsilon_\infty + \frac{\varepsilon_0 - \varepsilon_\infty}{1 + (\mathrm{i}\omega\tau)^\alpha}, \quad (3.29)$$

where α is a parameter, $0 < \alpha \leq 1$.

Equation (3.29) better describes a broad dispersion, and gives an $\varepsilon'' - \varepsilon'$ plot where the centre of the semicircle is depressed below the abscissa. It corresponds to a superposition of a group of Debye-like relaxation processes with a range of relaxation times that are symmetrically distributed about τ. Such a spread of relaxation times was seen as a very likely explanation of the broadness of the relaxation in a system of entangled, long-chain molecules where the forces restraining the orientation of segmental dipoles might be expected to cover a very wide range. Although the exact form of the distribution of relaxation times in the Cole–Cole equation is complicated and is not based on any special model, the parameter α is a convenient one for specifying the broadness of experimental relaxation peaks, and it has been used extensively for this purpose. Davidson and Cole (1950) improved the fit with experiment by using a slightly different semi-empirical equation:

$$\varepsilon^* = \varepsilon_\infty + \frac{\varepsilon_0 - \varepsilon_\infty}{1 + (\mathrm{i}\omega\tau)^\beta}, \tag{3.30}$$

where β is a parameter, $0 < \beta \leq 1$.

Equation (3.30) corresponds to a skewed distribution of relaxation times about τ, but it again has no particular theoretical foundation apart from the improved agreement with experiment for certain materials.

More recently, it has been suggested that a departure from the assumed exponential form of the approach to equilibrium may be responsible for the breadth of dipolar relaxation in polymers. A decay function that describes a wide range of slow relaxation processes is the so-called *stretched exponential*:

$$\Psi'(t) = \mathrm{e}^{-\left({}^{t}/_{\tau_{\mathrm{WW}}}\right)^\beta}, \tag{3.31}$$

where τ_{WW} is again a characteristic relaxation time and $0 < \beta \leq 1$. Although Kohlrausch originally introduced this decay function as long ago as 1863 to describe mechanical creep in glassy fibres, it was not until 1970 that Williams and Watts showed that it could also account very successfully for dielectric relaxation in polymers. Mathematically, the function may be expressed as a superposition of uncoupled Debye-like processes:

$$\mathrm{e}^{-\left({}^{t}/_{\tau_{\mathrm{WW}}}\right)^\beta} = \int_0^\infty \mathrm{e}^{-t/\tau} \rho(\tau)\mathrm{d}\tau, \tag{3.32}$$

where $\rho(\tau)$ is the distribution of Debye relaxation times. The physical meaning of $\rho(\tau)$ remains, however, unclear, although the decay function may be derived from various models of complex correlated processes.

Another way of describing experimentally observed dielectric relaxation in polymers is based on an empirical function formulated by Havriliak and Negami (1967) for the frequency domain:

$$\chi^*(\omega) = \chi_\infty + (\chi_s - \chi_\infty)\frac{1}{[1 + (\mathrm{i}\omega\tau_{\mathrm{HN}})^\alpha]^\gamma}, \tag{3.33}$$

where χ_∞, χ_{s} represent the high and low frequency limits of the susceptibility (as defined in Equation (2.5)), τ_{HN} is again a characteristic relaxation time and α, γ are parameters ranging between 0 and 1. Although the function may be represented mathematically as a distribution of elementary Debye relaxations, it is especially good in fitting the asymptotic dielectric response at both low and high frequencies, predicting that the imaginary component of the susceptibility, χ'', varies as $\omega^{-\alpha\gamma}$ at very high frequencies, and as $\omega^{-\alpha}$ at low frequencies. This turns out to be typical of many systems, and has led to the concept of a *universal* dielectric response attributed to a *many-body* interaction in solid materials (Jonscher, 1977), rather than to a superposition of elementary Debye processes.

The Kohlrausch–Williams–Watts and Havriliak–Negami formalisms are equally capable of representing real experimental data, and this is their main value, rather than an ability to *explain* the underlying relaxation processes. They are rooted in the time and frequency domains, respectively, and there is no analytical way of transforming from one to the other, but their effective equivalence has been convincingly demonstrated by numerical methods (Alvarez, Alegría and Colmenero, 1991).

3.2 Thermal activation of dipolar relaxation

In order to progress further in the interpretation of dipolar relaxation behaviour we must develop a molecular model in still more detail. Many different models, depending on the type of material concerned, have been proposed and these have led to a large number of theoretical treatments. We shall confine ourselves to two theories, which are particularly relevant to polymers, and provide a suitable basis for general discussion of the main features encountered. First, in this section, we will consider the theory which has been central to the understanding of the temperature dependence of nearly all reaction rate processes: thermal activation over a potential-energy barrier.

The model (Fig. 3.4) simply consists of two molecular states, A and B, separated by a potential-energy barrier of height ΔU^*, and the reaction whose rate is calculated is the change A → B. A and B may be taken to represent two orientations of a dipolar group about a valence bond linking the group to the rest of the molecule, the *reaction coordinate* being the angle of rotation about the bond. The reaction will follow the usual monomolecular rate equation:

$$-\frac{\mathrm{d}c_\mathrm{A}}{\mathrm{d}t} = k_\mathrm{A} c_\mathrm{A}, \tag{3.34}$$

where C_A is the concentration of dipoles in state A and k_A is the rate constant. It is clear that molecules must acquire an extra amount of energy ΔU^* in order to surmount the energy barrier prior to passing from state A to state B. Now it is well known from Boltzmann statistics that the probability that a given

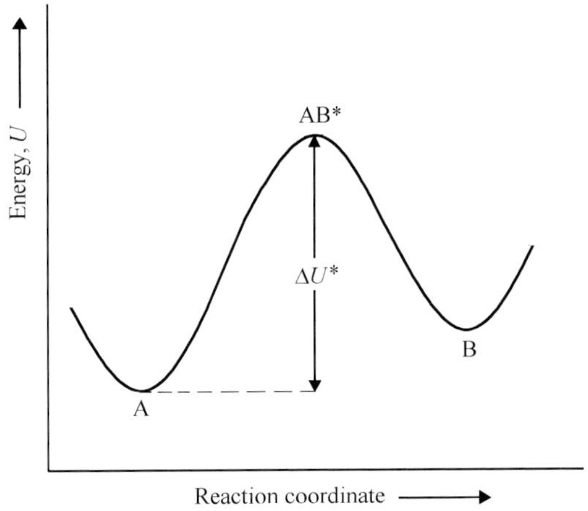

Fig. 3.4 Potential-energy diagram for a thermally activated process.

molecule possesses energy in excess of an amount ΔU^* is proportional to $e^{-\Delta U^*/kT}$, where k is the Boltzmann constant, so that we may expect the rate constant k_A to be given by

$$k_A = Ae^{-\Delta U^*/kT}, \tag{3.35}$$

where A is a constant or a function which only varies slowly with temperature, and $\Delta U*$ is called the activation energy. The relaxation time τ may be roughly identified with $1/k_A$, so that we shall have

$$\ln\tau = \frac{\Delta U^*}{kT} + \text{constant}. \tag{3.36}$$

This equation, an example of the well-known Arrhenius law, means that a plot of $\ln\tau$ against $1/T$ should give a straight line whose slope is directly related to the activation energy.

In order to calculate an absolute reaction rate we may use the transition-state theory (Glasstone, Laidler and Eyring, 1941). In this theory the intermediate stage in the reaction, corresponding to the peak in the potential-energy curve and called the transition state AB*, is treated as a pseudo-stable state whose equilibrium concentration can be calculated in terms of statistical mechanical partition functions. A partition function summarises the distribution of molecules over all their possible energy states and it determines the free energy of a system. It may be factorised with respect to each independent degree of freedom (Ubbelohde, 1952). The partition functions are somewhat arbitrary for the transition state, because there is one special degree of freedom – the *vibration* over the potential hill or pass along the reaction coordinate. It is supposed that the frequency ν of this slow (negative force constant) vibration gives the velocity of the change from the transition state to the final state. On this basis,

$$k_{\mathrm{A}} = K^{*}\nu, \tag{3.37}$$

where K^{*} is the chemical equilibrium constant for the transition state:

$$K^{*} = \frac{c_{\mathrm{AB}^{*}}}{c_{\mathrm{A}}}, \tag{3.38}$$

where $c_{\mathrm{AB}^{*}}$ is the concentration of dipoles in the transition state. Applying the usual principle of minimisation of free energy to the equilibrium, we shall also have

$$K^{*} = \mathrm{e}^{-\Delta G^{*}/RT}, \tag{3.39}$$

where ΔG^{*} is the change in molar free energy in going to the transition state and R is the gas constant. Eyring suggests, however, that it is appropriate to separate out the partition function $kT/h\nu$ for the degree of freedom belonging to the vibration in the reaction coordinate, because this degree of freedom is only accessible to activated molecules, so that

$$K^{*} = \frac{kT}{h\nu}\mathrm{e}^{-\Delta G^{*}/RT}, \tag{3.40}$$

where h is the Planck constant. Hence, by substituting Equation (3.40) in Equation (3.37),

$$k_{\mathrm{A}} = \frac{kT}{h}\mathrm{e}^{-\Delta G^{*}/RT}, \tag{3.41}$$

and therefore:

$$\tau = \frac{1}{k_{\mathrm{A}}} = \frac{h}{kT}\mathrm{e}^{\Delta G^{*}/RT} = \frac{h}{kT}\mathrm{e}^{-\Delta S^{*}/R}\mathrm{e}^{\Delta H^{*}/RT}, \tag{3.42}$$

where ΔS^{*} is the molar entropy of activation and ΔH^{*} is the molar enthalpy of activation. We see that the pre-exponential factor is dependent on temperature, although the temperature dependence of the relaxation time will be dominated by the exponential factor. Neglecting the entropy factor, we calculate that, at room temperature,

$$\tau \approx 10^{-12}\mathrm{e}^{\Delta H^{*}/RT}. \tag{3.43}$$

3.3 Cooperative dipolar relaxation in polymers

A major feature of the behaviour of amorphous or partially crystalline polymeric materials is the glass transition. At low temperatures most plastics become hard and brittle, whereas at high temperatures they are rubbery or leathery and have great flexibility and toughness. The change from one form to the other is found to take place over a comparatively narrow range of temperature. It can be demonstrated by measuring the force needed to push a needle into a polymer at a given rate, or observed more qualitatively as a change in the ease of folding a sheet of the material. The change is not a first-order transition – no latent heat is given out or taken in and there are no discontinuities in density and other properties – but sharp changes in temperature dependence of various properties are seen, suggesting a second-order transition. The most usual way of determining the glass transition temperature T_g is to follow the volume contraction of a sample in a dilatometer and to note the temperature at which the volume–temperature graph changes in slope (Fig. 3.5). The T_g value found in this way depends on the rate of cooling, e.g. polystyrene gives the following results:

$$T_g = 105\ \mathrm{C\ at\ 1\ deg\ min^{-1}},$$
$$T_g = 100\ \mathrm{C\ at\ 1\ deg\ day^{-1}}.$$

We may attribute T_g to a major change in the segmental mobility of the polymer chains. Above T_g there is sufficient mobility, a sort of micro-Brownian

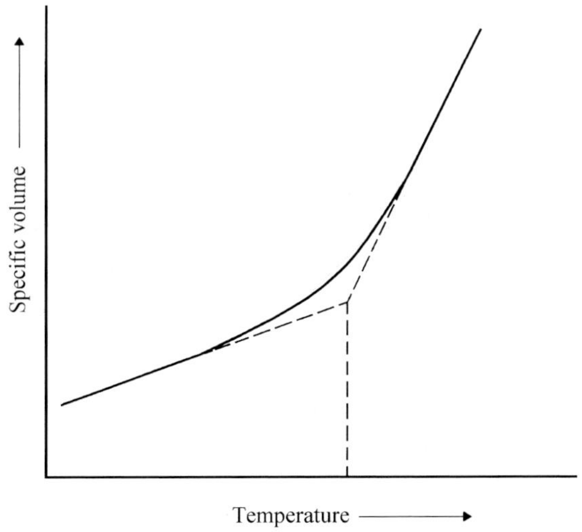

Fig. 3.5. Volumetric determination of the glass transition.

motion, which enables large-scale reorganisation of the chains to occur in response to an applied stress (e.g. a change of temperature), whereas below T_g the chains are *frozen* in position. In other words, an observed T_g is that temperature at which the time constant for a molecular rearrangement process becomes comparable with the timescale of the experiment used to measure it. As may be expected, at the onset of the molecular mobility above T_g, permanent dipoles attached rigidly to the polymer backbone become free to orient in an electric field, and the glass transition is accompanied by a major dielectric dispersion.

The temperature dependence of the dielectric relaxation time of the molecular process associated with T_g does not conform to the simple Arrhenius law. The plot of $\ln\tau$ against $1/T$ is curved, as if the activation energy were increasing towards lower temperatures. This effect is seen in other properties too, and the general inference is that a large-scale rearrangement of a long-chain molecule involves a cooperative mechanism, i.e. the movement of one molecule is not independent of its neighbours. One way of looking at this is in terms of *free volume*: for a molecular segment to move or twist there must be a vacant site for it to move into. The availability of vacant sites may be expressed as an average free volume per molecular segment, ν_f, defined by

$$\nu_f = \nu - \nu_0, \tag{3.44}$$

where ν is the actual volume occupied by a segment, and ν_0 is the close-packed-sphere volume, approximately equal to the volume per segment at 0 K. Free volume increases with temperature, as evidenced by the relatively high coefficient of expansion of a rubber. This would be predicted on the basis that an extra amount of energy is associated with a hole, so that the probability of a hole existing in a system will increase with temperature in accord with the Boltzmann distribution law. At low temperatures, where free volume is scarce, the occurrence of a hole of the right size at the right place may become the controlling factor in molecular movement. By considering the probability of holes joining together to provide the critical free volume ν_f^* which is needed for a segmental jump to take place, it may be shown (Bueche, 1962) that the dependence of the rate r of segmental motion on free volume takes the form

$$r = \text{const.}\ \exp\left(\frac{-\nu_f^*}{\nu}\right). \tag{3.45}$$

(This explains the success of the Doolittle empirical equation for the viscosity of low molar mass hydrocarbon liquids:

$$\eta = a.\exp\left(\frac{b}{\nu_{ff}}\right), \tag{3.46}$$

where a and b are constants, and ν_{ff} is the fractional free volume, ν_{f}/ν.) Now consider the ratio of the rates of segmental movement for a polymer at two different temperatures, T_1 and T_2, where the segmental free volumes are ν_{f1} and ν_{f2}, respectively:

$$\ln\frac{r_2}{r_1} = \nu_{\mathrm{f}}^*\left(\frac{1}{\nu_{\mathrm{f1}}} - \frac{1}{\nu_{\mathrm{f2}}}\right). \tag{3.47}$$

If we assume that the excess rate of expansion of the rubber over that of the glass is entirely attributable to an increasing free volume, we may write

$$\nu_{\mathrm{f2}} = \nu_{\mathrm{f1}} + \alpha\nu_1(T_2 - T_1), \tag{3.48}$$

where α is the difference between the cubical expansion coefficients above and below the glass transition temperature, and ν_1 is the actual segmental volume at temperature T_1. Then, substituting for ν_{f2} in Equation (3.47)

$$\ln\frac{r_2}{r_1} = \frac{(\nu_{\mathrm{f}}^*/\nu_{\mathrm{f1}})(T_2 - T_1)}{(\nu_{\mathrm{f1}}/\alpha\nu_1) + T_2 - T_1}. \tag{3.49}$$

This equation has the same form as the well-known WLF equation (Williams, Landel and Ferry, 1955) that correlates the mechanical behaviour of all polymers near their T_{g} provided we set $T_{\mathrm{g}} = T_1$ (T_{g} measured by the same method for each polymer). From experimental results one finds that

$$\frac{\nu_{\mathrm{f}}^*}{\nu_{\mathrm{fg}}} \approx 40 \quad \text{and} \quad \frac{\nu_{\mathrm{fg}}}{\nu_1\alpha} = 52, \tag{3.50}$$

where ν_{fg} is the segmental free volume at the glass transition, which imply that $\nu_{\mathrm{f}}^* \approx$ (size of a molecular segment). The WLF equation fits dielectric data if we set

$$\frac{r_2}{r_1} = \frac{\tau_1}{\tau_2}, \tag{3.51}$$

i.e. if we take the dipolar relaxation time as a measure of segmental mobility. The temperature dependence of relaxation may then be written in the form

$$\log a_T = \log\frac{\tau_T}{\tau_{T_{\mathrm{g}}}} = -\frac{C_1(T - T_{\mathrm{g}})}{C_2 + T - T_{\mathrm{g}}}, \tag{3.52}$$

where C_1, C_2 are universal constants.

3.4 Dielectric relaxation in solid polymers

Several distinct relaxation processes are usually present in a solid polymeric material, and these are dielectrically active if they incur significant orientation of molecular dipoles. The multiplicity of relaxation processes is seen most easily in a scan of dielectric loss at constant frequency as a function of temperature (Fig. 3.6). As the temperature is raised, molecular mobilities of

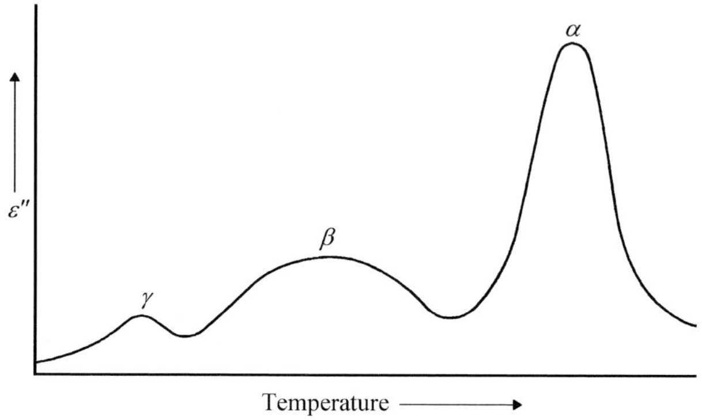

Fig. 3.6 A schematic dielectric loss curve.

various types become successively energised and available for dipolar orientation. By convention the dielectric relaxation processes are labelled α, β, and so on, beginning at the high-temperature end. The same relaxation processes are generally responsible for dispersions in mechanical properties too, although a particular molecular rearrangement process may produce a stronger dielectric than mechanical effect, or vice versa.

Some polymers are wholly amorphous and there is only one phase present in the solid material. In such cases there is always a high-temperature α-relaxation associated with the micro-Brownian motion of the whole chains and, in addition, at least one low-temperature (β, γ, etc.) subsidiary relaxation. The relative strength of the α- and β-dielectric relaxations depends on how much orientation of the dipolar groups can occur through the limited mobility allowed by the β-process before the more difficult but more extensive mobility of the α-process comes into play: there is a partitioning of the total dipolar alignment amongst the molecular rearrangement processes.

Detailed examination of the relaxations requires isothermal scans of relative permittivity and dielectric loss factor as a function of frequency f so that effective dipole movements and activation energies of relaxation times may be obtained. A typical pair of plots of ε' and ε'' values against $\log f$ is shown in Fig. 3.7. Graphs of dielectric data of this kind are sometimes called, rather

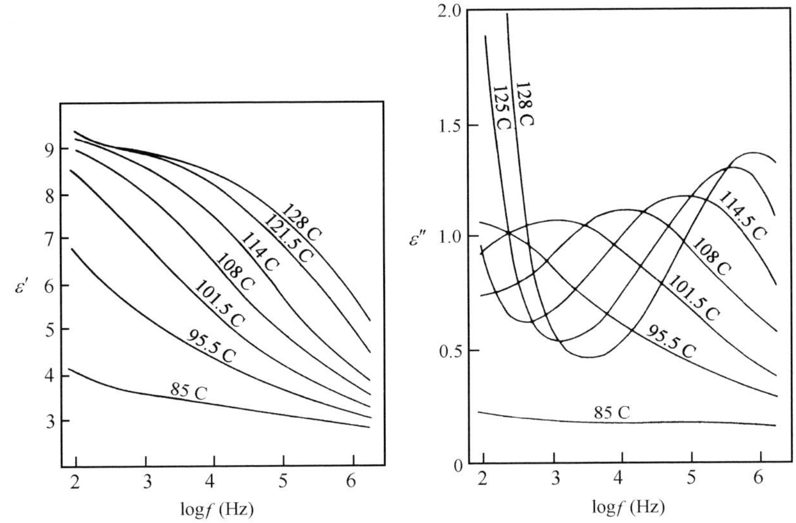

Fig. 3.7 Dielectric relaxation curves for poly(vinylchloride) in the α-relaxation region. From Ishida (1960) with permission from Dr Dietrich Steinkopff Vertag.

loosely, dielectric spectra. From a series of such plots the relaxation times can be obtained for the individual relaxation processes as a function of temperature.

The high-frequency β, γ, ... subsidiary peaks in amorphous polymers are characteristically very broad with a half-height width of several decades (compared with 1.14 decades for a single Debye relaxation process), although a good, linear Arrhenius plot is usually obtained, suggesting a non-co-operative mechanism. Figure 3.8 shows the Arrhenius plot for the β-relaxation of poly(epichlorhydrin),

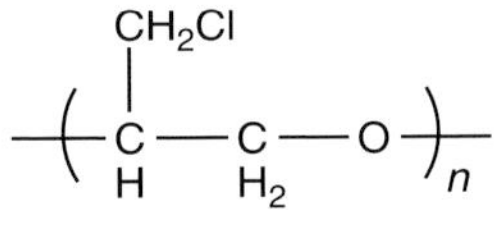

where the β-process is attributable to rotation of the polar $-CH_2Cl$ group about its C–C linkage to the main polymer chain (Blythe and Jeffs, 1969). Relatively small activation energies are found for β-type processes, and values for representative polymers are listed in Table 3.1. The mechanism of a β-type process may be one of several different types depending on the nature of the dipole group concerned and its position on the polymer chain. Among the most important mechanisms are the following.

a. *Rotation of a side group about a C–C bond.* This is probably the simplest type conceptually. It may be a small group, e.g. $-CH_2Cl$, or a more complicated side chain, e.g. $-CO{\cdot}OC_2H_5$.

Table 3.1 *Activation energies for α-relaxation in solid polymers*

Polymer	ΔH^*(kJmol^{-1})
Poly(methylmethacrylate)	84
Poly(epichlorhydrin)	35
Poly(vinylchloride)	63
Poly(vinylacetate)	42
Poly(4-chlorocyclohexylmethacrylate)	48

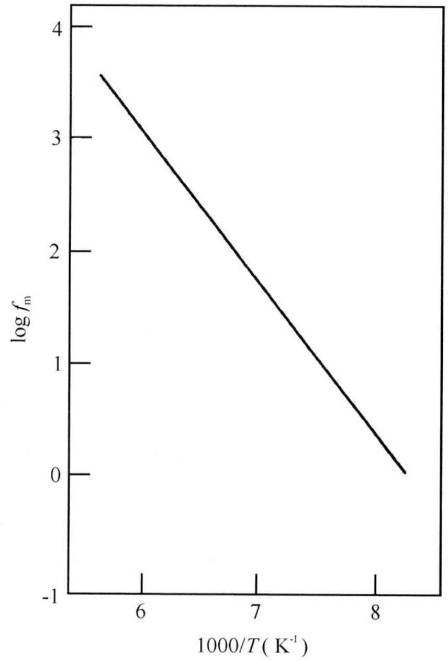

Fig. 3.8 Temperature dependence of the β-relaxation process in poly(epichlor-chlorhydrin). From Blythe and Jeffs (1969), by courtesy of Marcel Dekkes, Inc.

b. *Conformational flip of a cyclic unit*. The best-known examples involve the *cyclo*-hexyl side group. The transition from one chair-form to another alters the orientation of a polar substituent (Fig. 3.9(a)).

c. *Local motion of a segment of the main chain*. Some limited, localised movement of the main chain is a common feature. This has to be the explanation of the β-process in poly(vinylchloride), for example, where the dipolar group is directly attached to the main chain and cannot move independently of the polymer backbone. This kind of movement also arises when there are runs of four or more CH_2 units (Willbourn, 1958) and can be explained as a *crankshaft* rotation (Schatzki, 1962) about two collinear C–C bonds. The smallest segment of a $(CH_2)_n$ chain which can allow such rotation, whilst not involving the rest of the chain, is a $(CH_2)_4$ group (Fig. 3.9(b)).

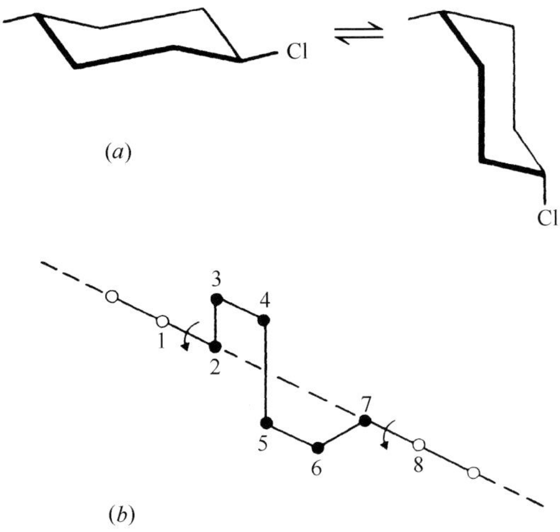

Fig. 3.9 Diagrams of molecular relaxation mechanisms: (a) conformational 'flip' of chlorohexane, (b) 'crankshaft' rotation in polyethylene.

The $\alpha(T_g)$-relaxation peak of an amorphous polymer is typically much narrower than a β-peak, although still considerably broader than that for a simple Debye process. Also, the temperature dependence of the α-process is generally much steeper than that of a β-process, signifying the greater thermal activation energy required for the larger scale of motion involved. The distinctive curvature in the Arrhenius plot of the $\alpha(T_g)$-process is exemplified in Fig. 3.10(a) by the case of poly(epichlorhydrin). At the higher temperature end the graph is roughly linear, corresponding to an activation energy (ΔH^*) of 190 kJmol^{-1} (0 to 6 C), whereas at lower temperatures, towards T_g, ΔH^* appears to increase rapidly (430 kJmol^{-1} at –20 C). Figure 3.10(b) shows that a good straight line is obtained in a WLF plot, though, confirming that the α-relaxation process near T_g is largely dependent on free volume. In some polymers the α-relaxation process catches up with the β-process at high temperatures on account of its steeper dependence on temperature, and the α- and β-peaks then tend to merge. However, they may be separated again on a temperature scan by applying high pressures, which greatly suppress the α-process, reflecting its requirement for free volume.

Molecular structure greatly affects glass transition temperatures and their associated dielectric relaxation times. Thus a bulky side group can decrease T_g by preventing the chains from packing together tightly and vice versa. T_g can also be artificially reduced by adding a plasticiser; Fig. 3.11 shows the effect on the α-relaxation of adding diphenyl to poly(vinylchloride) (Fuoss, 1941).

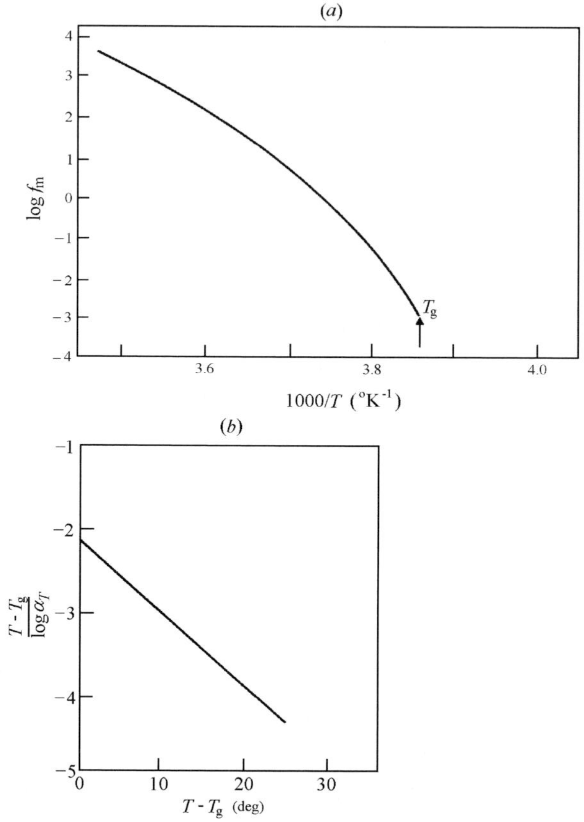

Fig. 3.10 Temperature dependence of the α-relaxation process in poly(epichlorhydrin): (a) Arrhenius plot, (b) WLF plot. From Blythe and Jeffs (1969), by courtesy of Marcel Dekker, Inc.

When a polymer possesses no very polar groups, dielectric relaxations may be very weak effects, which are scarcely observable, although the underlying molecular rearrangement processes will still be there. In these circumstances it is often possible to enhance the associated dielectric effect artificially by adding just a few polar groups, which in other ways do not disturb the system too much. This technique has been most notably applied in the case of polyethylene, which is of great technological interest. Being non-polar, its dielectric relaxations are especially weak, but by slightly oxidising the material, e.g. by milling in the presence of air, some of the $>CH_2$ groups are converted to polar $>C=O$ groups, which are readily accommodated even in the polyethylene crystal lattice. The molecular relaxation mechanisms can then be examined using dielectric methods; a valuable achievement, because the accessible frequency range of dielectric measurement is wider than that of any other kind. Various molecular processes in non-polar polymers may also show up accidentally as a

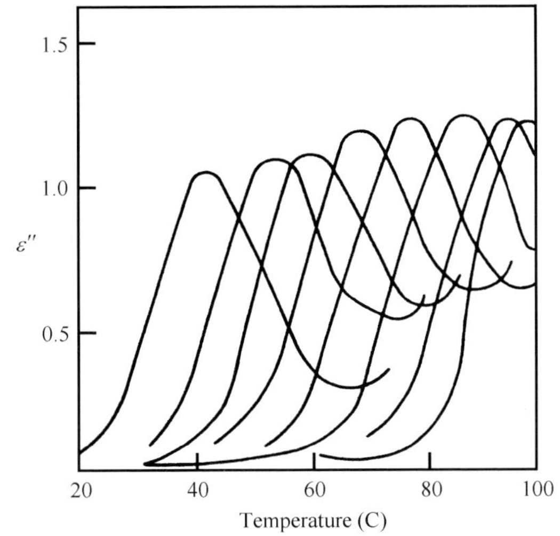

Fig. 3.11 The plasticising effect of diphenyl in poly(vinylchloride): dielectric loss curves at 60 Hz for (right to left) 0, 1, 3, 6, 9, 12, 15, 20% diphenyl. Reproduced from Fuoss (1941). Copyright of the American Chemical Society.

consequence of a coupling of the motion of the polymer chains with the displacement of some small impurity molecules, which may be polar or even ionic. By the same token a low concentration of a polar impurity may ruin the low dielectric loss of a non-polar polymer in a particular frequency region.

In partially crystalline polymers where crystalline and amorphous phases coexist in the solid, the relaxation spectrum becomes more complicated. Apart from the orientational processes taking part entirely in the amorphous regions, there are different mechanisms operating inside the crystals and at their boundaries. It is usually possible to decide whether a given loss peak belongs to the amorphous phase or is connected with crystals by varying the crystallinity. Thus, reducing the crystallinity, by rapid quenching from the melt, enhances the strength of any relaxation process originating in the amorphous phase.

In crystals of polymers like polyethylene where the molecules are in their extended *zigzag* conformation, the molecular chains lie in a staggered arrangement with respect to their neighbours. If a chain has a polar group attached to it one can envisage that the application of an electric field will in general exert a couple tending to turn the chain about its axis, the magnitude of the couple depending on the angle between the dipole and the field. The repulsive forces from the surrounding chains in the crystal lattice will constrain chain rotation to jumps between two equilibrium positions, involving a 180° turn combined with a translational shift, as shown in Fig. 3.12. On this model the molecules

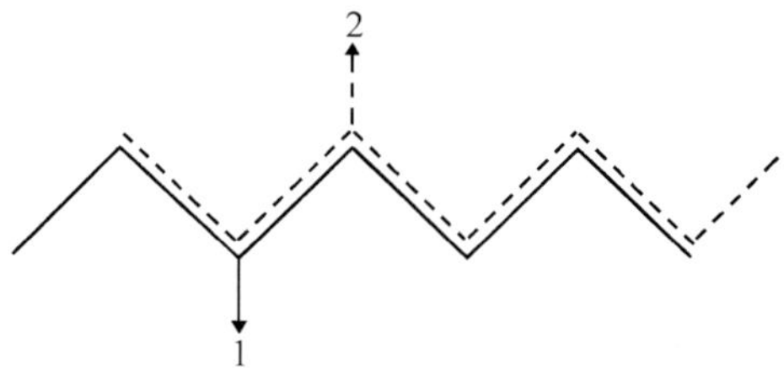

Fig. 3.12 Diagram of chain rotation in a polyethylene crystal.

turn as rigid rods, so that we may expect the activation energy for the orientation process to be directly proportional to the length of chain, which turns in the crystal. As a test of the above idea, dipolar relaxations of ketones in paraffin waxes have been examined in cases where the ketone chain length is shorter than the thickness of the paraffin crystals, so that constraints on the ends of the rotating chains may be neglected. Experiment confirms the dependence of activation energy on the length of the dipolar chain (Meakins, 1962).

In the case of long polymer chains, some direct evidence of the rotation of chains in crystals has been obtained from observations of the dependence of the relaxation strength on orientation of the chain axes with respect to the applied field. It is clear that rotation of chains which lie parallel to the field direction cannot contribute to the relaxation, and only a weak relaxation peak is found for crystals oriented in this way (Davies and Ward, 1969). If the chain-folds at the crystal boundaries were loose, a chain in the crystal would suffer little constraint on its rotation from outside the crystal, so that one might expect the activation energy for the relaxation to be directly proportional to crystal thickness (in the *c*-axis direction). For long chains, however, we must take into account the possibility of chain twisting, i.e. departure from rigid-rod behaviour. More elaborate theories suggest that twisting will develop as chains exceed 60–120 carbon atoms in length, when the activation energy should flatten out, causing the temperature of maximum dielectric loss (at a fixed frequency) to *saturate* with respect to chain length. Experimental data, though scanty, seem to suggest this view, as shown in Fig. 3.13 (Hoffman, Williams and Passaglia, 1966).

Other complications may arise in dielectric relaxation spectra of polymers from chain branching, which may introduce a distinct relaxation process connected with molecular motion at a branch point, and from crosslinking, which greatly restricts certain kinds of molecular movement.

Relaxation processes in solid polymers are summarised diagrammatically in Fig. 3.14.

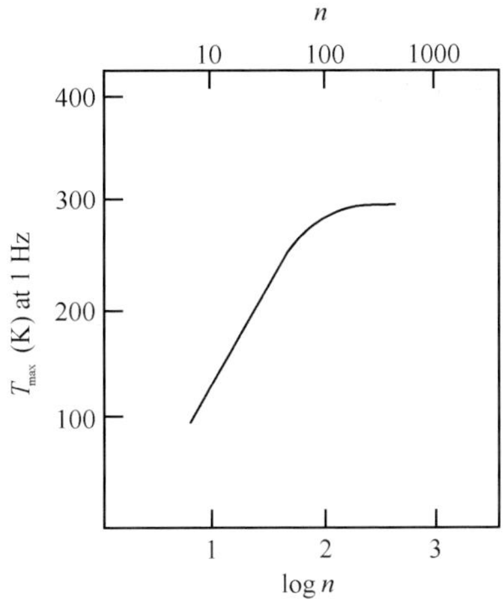

Fig. 3.13 The temperature of maximum dielectric loss at 1 Hz for crystalline long carbon-chain compounds, T_{max}, as a function of number, n, of CH_2 units in the chain. After Hoffman, Williams and Passaglia (1966). Reproduced by permission of John Wiley & Sons, Inc.

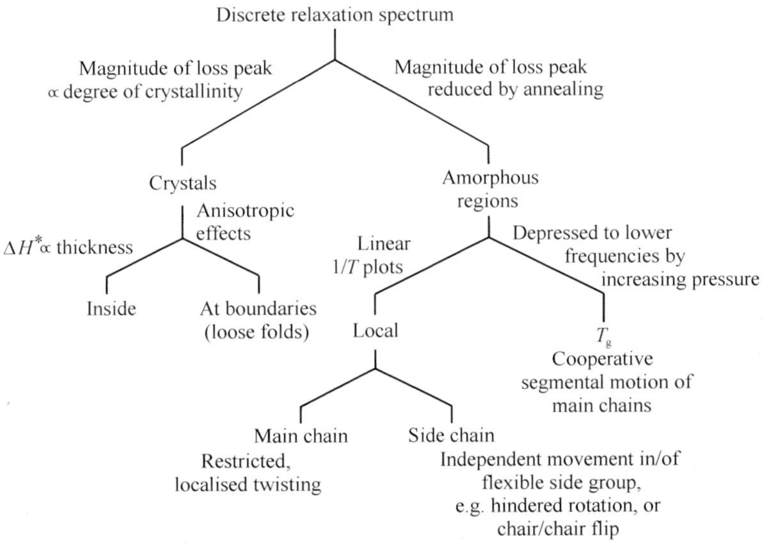

Fig. 3.14 Generalised pattern of dielectric relaxation processes in polymers.

3.5 Polymer liquids

3.5.1 Dilute solutions

End-over-end rotation of whole polymer molecules becomes a much more accessible mechanism for orientation of molecular dipoles when the material is in the liquid state, especially when it is dissolved in a solvent of low molar mass.

With this additional feature in mind, there are three aspects of polymeric molecules which must be considered in order to reach an understanding of their dielectric behaviour in the liquid state: the type of dipole present on each repeat unit of the polymer, the equilibrium conformation of the individual molecules, and the flexibility of the molecular chains.

Polar groups in a polymer molecule may be classified according to the relative geometry of their dipole moment with respect to the contour of the polymer backbone. There are three main types:

a. dipole rigidly attached to the backbone and perpendicular to it;
b. dipole rigidly attached to the backbone and parallel to it;
c. dipole attached in a side chain, which can move independently of the main chain.

Combinations of these types may also occur.

Perpendicular dipole units of type (a), which commonly occur in vinyl polymers, e.g. poly(vinylchloride), can orientate individually by segmental motion, provided that rotations about the C–C bonds of the polymer chain are sufficiently free. We know that only in cases of exceptional steric hindrance or intramolecular hydrogen bonding do intramolecular barriers to such rotations lie above 20 $kJmol^{-1}$, so that in liquids and solutions, where there is ample free volume available, we should expect dipolar relaxation to be very rapid and to be insensitive to the total length of the chain. This is borne out by experiment, and a single relaxation process, which has a low activation energy and a relaxation time independent of the degree of polymerisation, is observed (North, 1972).

Parallel dipole units of type (b) add vectorially along the length of the polymer chain to give a cumulative moment that is proportional to the end-to-end distance of the molecule (Fig. 3.15). This moment can only orient by end-over-end rotation of the whole molecule and remains unaffected by rotation of individual segments about the main chain axis. On this basis we predict a strong dependence of the associated relaxation process on molar mass (the bigger the molecule the more difficult will it be for it to turn round), although we expect that the exact form of the dependence will be related to shape as well as to length of polymer chains (see below). This behaviour has been observed (Baur and Stockmayer, 1965) in liquid poly(propylene oxide). In this polyether the $-CH_3$ side groups disturb the normal geometry of the oxygen dipole and the dipoles on each monomer unit no longer bisect the C–O–C bond angle, so that each such dipole has a small component directed parallel to the contour of the main chain. The relaxation of the cumulative moment of these parallel components is seen as a subsidiary peak on the low-frequency side of the main loss peak, which is associated with relaxation of the perpendicular components

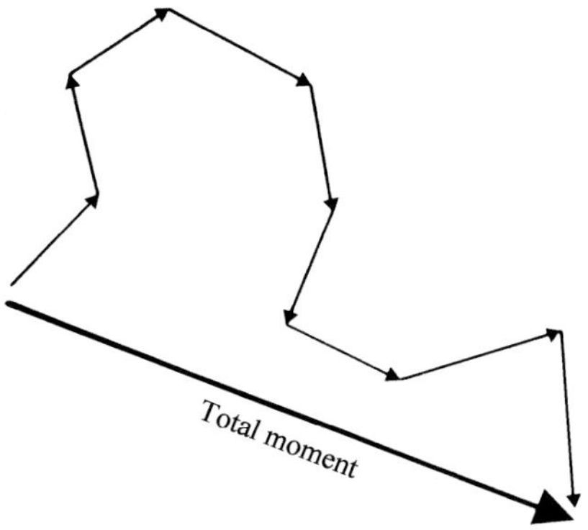

Fig. 3.15 Vectorial addition of unit dipoles along a polymer chain.

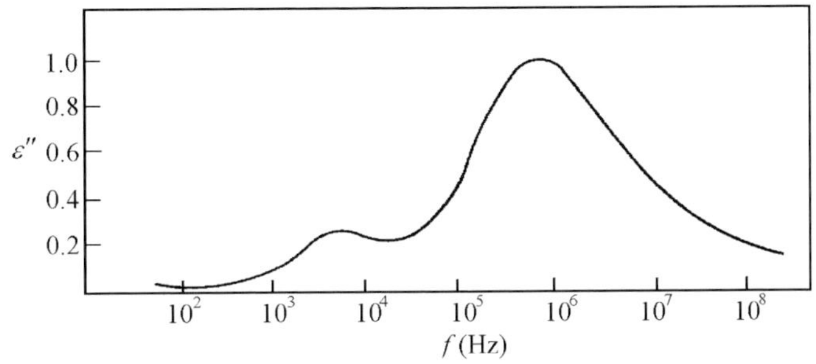

Fig. 3.16 Relaxation spectrum of liquid poly(propylene oxide) at −30 C. Reproduced with permission from the American Institute of physics from Bauer and Stockmayer (1965).

by segmental motion (Fig. 3.16). Increasing the molar mass of the polymer shifts the subsidiary peak to a lower frequency.

A dipole moment of type (c) in a flexible side chain, e.g. as in poly(methyl methacrylate), imposes a dipole moment which can be resolved into a rigid, perpendicular component and a rotatable component. One might have supposed that two separate relaxation processes might then have occurred, but in practice only one, high-frequency, relaxation is observed and one concludes that in solution the side-group rotation and segmental motion in the main chain are combined in a single, fast process (North, 1972).

Because of the profound influence that the overall shape of a polymer chain undoubtedly has on any orientation process, which involves rotation of the whole molecule, it is worthwhile digressing on the subject of equilibrium conformation. Flexible, linear polymers tend to be somewhat coiled-up in the liquid phase or in a solution. In a *good* solvent, polymer–solvent contacts are preferred energetically to polymer–polymer contacts and so the coils expand and the root-mean-square end-to-end distance, $(\overline{r^2})^{1/2}$, increases. On the other hand, in a *bad* solvent the polymer will contract and $(\overline{r^2})^{1/2}$ will decrease – ultimately phase separation will occur if the solvent is bad enough. At a certain temperature in a relatively poor solvent the effect of the intermolecular interactions will just offset the increase in end-to-end distance of the polymer due to excluded volume effects (occupation of space by a chain of finite thickness) and the polymer will then behave just like an unperturbed molecular chain of negligible thickness. The solvent in this ideal situation is called a *theta-solvent* and the chains adopt their *random-flight* or Gaussian conformation. It may be shown that the root-mean-square end-to-end distance $(\overline{r^2})^{1/2}$ is then proportional to the number of segments in the chain, and therefore to the molar mass.

The frictional motion of deformable, random-coil molecules in solution has been examined theoretically in a general way. There are two basic models: the *free-draining* model (Rouse, 1953) which neglects hydrodynamic interaction between neighbouring parts of the same chain, and the *non-free-draining* model (Zimm, 1956) in which viscous drag between different segments assumes a dominant role. The latter model may be expected to give a better account of what happens inside the coils of a long molecule, and indeed, it gives good agreement with experimental viscosity data for dilute solutions of high polymers. The simpler Rouse model gives better agreement at higher concentrations, however, where the polymer molecules might be expected to intermingle. Presumably the hydrodynamic interaction between neighbouring segments of one molecule is nullified by the invasion of foreign segments.

The relaxation time τ for orientation of the total dipole moment of a chain, substantially by end-over-end rotation, but accompanied by some distortion, may be related to the viscosity of the solvent η_s and solution η on the basis of the above models:

$$\tau(\text{free-draining}) = 1.21\frac{(\eta - \eta_s)}{cRT}M_W, \tag{3.53}$$

$$\tau(\text{non-free-draining}) = 0.85\frac{(\eta - \eta_s)}{cRT}M_W, \tag{3.54}$$

where c is the concentration of the polymer solute and M_W is the molar mass of the polymer. Although these equations do not give a great deal of physical insight into the orientation process they do predict fairly accurate values and give the correct molar mass dependence.

The backbones of some polymer molecules are of intermediate stiffness; they adopt a coil-like configuration in a solution, but their segmental mobility is so low that dipolar orientation is faster by whole molecule rotation than by twisting of individual dipole units. This behaviour has been observed in sulphones, e.g. poly(hex-1-enesulphone) and poly(2-methylpent-l-enesulphone) in benzene and toluene (Bates, Ivin and Williams, 1967). Since the dipoles are rigidly attached to the main chain in these instances only one relaxation process is seen, and the relaxation time is molar mass dependent. Cellulose esters and ethers also fall into the stiff, coil-like class, but they generally show an additional high-frequency relaxation process, because they contain dipoles in rotatable side chains as well as dipoles fixed in the main chain. The high-frequency process in cellulose acetate is independent of molar mass, having a relaxation time of about 10^{-7} s at room temperature (Kuhn and Moser, 1963). This is appreciably longer than that in vinyl acetate, implying that in the cellulose acetate derivatives the acetate side groups are subject to considerable steric hindrance and do not benefit from cooperative backbone movement. The loss peak is no wider than that for a single Debye-like process, suggesting that each acetate group undergoes orientation independently.

Certain polymer molecules have a tendency to form rigid rods in solution. The best-known examples are of biochemical origin, e.g. proteins, where the rigidity is brought about by intramolecular hydrogen bonding. A few examples of polymers that adopt the rod form simply through excessive steric hindrance are also known. The π-bonding in conjugated polymers is maximised when the chains are extended and planar. Thus, conjugated polymers are inherently stiff with a preference for the rod form. Theory indicates that the end-over-end rotational relaxation time of a rigid rod should vary as the cube of the rod length, so that the dielectric relaxation time for the axial component of the dipole moment should vary as the cube of the molar mass. Such a dependence has been observed for dilute solutions of poly(*n*-butyl*iso*cyanate) with molar masses below 10^5 (Bur and Roberts, 1969). As a polymer chain becomes longer and longer, it becomes more and more difficult to sustain the rigid-rod configuration, and some curvature is inevitable for very long chains. This will eventually lead, for sufficiently long chains, to random-coil configurations. The rotational relaxation time will then depend on the cube of the radius of gyration that varies as the square root of the chain length. We therefore expect a transition to a dependence of the dielectric relaxation time on the three-halves power of the molar mass. This occurs in the case of poly(*n*-butyl*iso*cyanate) when the molar mass is above 10^5. At high molar mass we might expect the rate of end-over-end rotational orientation to be overtaken by segmental orientation, implying a change from a stiff coil to a flexible coil. This has been observed in the case of poly(*N*-vinylcarbazole) (North and Phillips, 1968). In principle, we expect all polymers to be rod-like at very low molar masses and to progress via stiff coils to flexible coils as the molar mass is increased; this should be reflected in the dependence of dielectric relaxation time on molar mass, as indicated schematically in Fig. 3.17.

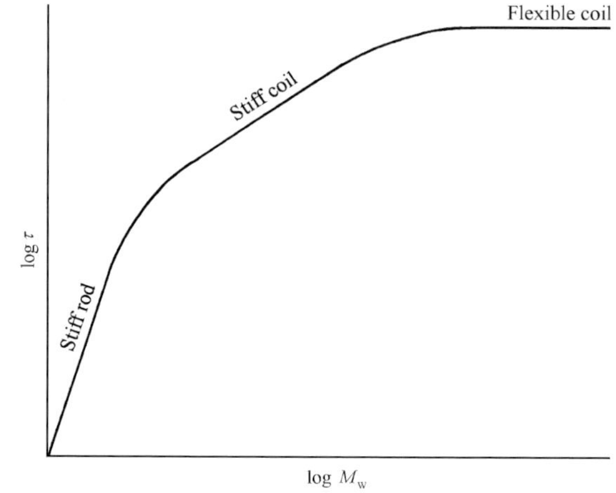

Fig. 3.17 Schematic molar mass dependence of the relaxation time for whole-molecule rotation.

Many biochemical polymer systems show polarisation effects attributable to migration of ion atmospheres surrounding the polymer chains. The mechanisms are then more akin to the behaviour of electrolytes than to dielectric materials and therefore will not be discussed further here.

3.5.2 Concentrated solutions and melts

In very concentrated solutions, and more especially in molten polymers, motion of polymer chains is highly constrained by the presence of neighbouring chains. From their theoretical work, Doi and Edwards (1986) were the first to realise that any particular chain becomes, in effect, confined within a tube, the centre line of which defines the overall path of the chain in the assembly of molecules, as depicted in Fig. 3.18. The chain cannot move through the wall of the tube, because it is blocked by other chains. Large-scale movement of the chain now takes a fundamentally different form: chains move by slithering along the corridor or tube in a snake-like fashion. De Gennes (1971) called this new mode of polymer motion *reptation,* from the Latin *reptare*, 'to crawl'. Small-scale motion can still occur by wriggling of the chain within the tube on a shorter time scale, but flow of the polymer can only occur by the slower process of

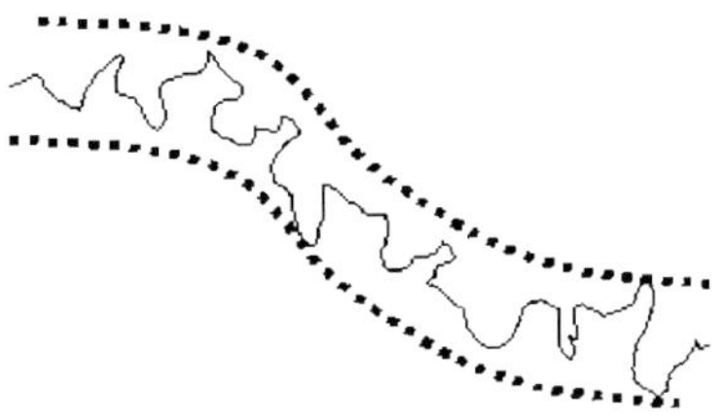

Fig. 3.18 Polymer chain in a tube.

reptation, by which chains vacate their original tubes and enter new ones. The time scale of this process is prescribed by the longest relaxation time τ^*, which is roughly the time it takes for the chain to migrate along its tube by a distance equal to the length of the tube axis, whence it may be shown that $\tau^* \propto N^3$, where N is the number of monomer units in the chain. This strong dependence on N explains why flow is so much slower in polymers compared with that in normal liquids. At times less than τ^* the material will behave elastically and only flow at times in excess of τ^*. The reptation model has been very successful in describing this behaviour of polymers (Grosberg and Khokhlov, 1997).

3.6 Interfacial polarisation

So far in this chapter we have concerned ourselves with ideal insulating specimens, which are entirely homogeneous, and connect perfectly with the electrodes, which apply the electric field. In practice a material is always likely to have regions of non-uniformity, and impurities may be present as a second phase. Effects on dielectric properties attributable to material discontinuities are usually called Maxwell–Wagner effects, since Maxwell and Wagner were the first to consider them theoretically. Complications also often arise at electrodes where contact with the specimen may be incomplete and where entities like discharged ions may form spurious boundary layers. We shall see in this section how important it is to be aware of such anomalous effects; they can give totally misleading results if they are not recognised or avoided.

3.6.1 Maxwell–Wagner effects

A common form of discontinuity occurring in a solid dielectric is one where cracks or voids are present. The relative permittivity will then simply be reduced, depending on the amount and distribution of the enclosed space or air. A much more serious type of heterogeneity, however, is one where a relatively conductive component is mixed in an insulator. Common instances of this occur where a material is contaminated with metallic particles or droplets of water. We can expect such materials to behave like 3-dimensional *RC*-networks. On this basis we would then predict an extra frequency-dependent contribution to relative permittivity and loss on account of the limited flow of current that is possible within the isolated conductive regions.

Although Maxwell (1892) considered a rather special case when he first examined the effect of discontinuities in dielectrics theoretically, he obtained a very important result. He examined the effect of a field applied across a specimen consisting of layers of two different materials with relative permittivities ε_1', ε_2' and conductivities σ_1, σ_2 respectively. His results showed that charges will accumulate in time at the interfaces between the layers whenever $\varepsilon_1'\sigma_2 \neq \varepsilon_2'\sigma_1$.

Wagner (1914) gave an approximate treatment of the important practical case where a very highly insulating dielectric suffers from inclusions of conductive impurities. Taking the model where the impurity (relative permittivity ε_2', conductivity σ_2) exists as a sparse distribution of small spheres (volume fraction f) in the dielectric matrix (relative permittivity ε_1', negligible conductivity), he derived equations for the components of the complex relative permittivity of the composite:

$$\varepsilon' = \varepsilon_\infty'\left(1 + \frac{k}{1+\omega^2\tau^2}\right), \tag{3.55}$$

$$\varepsilon'' = \frac{\varepsilon_\infty' k\omega\tau}{1+\omega^2\tau^2}, \tag{3.56}$$

where

$$\varepsilon_\infty' = \varepsilon_1'\left[1 + \frac{3f(\varepsilon_2' - \varepsilon_1')}{2\varepsilon_1' + \varepsilon_2'}\right], \tag{3.57}$$

$$k = \frac{9f\varepsilon_1'}{2\varepsilon_1' + \varepsilon_2'}, \tag{3.58}$$

and

$$\tau = \frac{\varepsilon_0(2\varepsilon_1' + \varepsilon_2')}{\sigma_2}. \tag{3.59}$$

Comparing Equations (3.55) and (3.56) with the Debye Equations (3.25) and (3.26), we can appreciate that the composite displays a dielectric relaxation that is indistinguishable in form from that due to orientation of dipoles. Equation (3.59) shows that the relaxation time decreases as the conductivity of the material of the spheres increases, and the peak in tan δ may shift right up to radio-frequencies. For example, taking $\varepsilon_1' = \varepsilon_2' = 4$, and $\sigma_2 = 10^{-4}\,\Omega^{-1}\mathrm{m}^{-1}$, we calculate a relaxation time of about 1 μs. Consequently, the effect can be very easily mistaken for one of dipole orientation, and great care must always be exercised in the interpretation of dielectric data when the presence of heterogeneities is suspected. Sillars (1937) developed the subject further and he demonstrated the importance of the shape of conductive inclusions. The dielectric loss peak is enlarged and shifted to lower frequencies by any elongation in the direction of the applied field. Experiments by Sillars (1937) and by Hamon (1953) using deliberately engineered heterogeneous systems have verified the theoretical predictions.

3.6.2 Electrode polarisation

Polarisation effects at electrodes become most prominent when the material of a specimen shows some appreciable bulk conductivity. Characteristically, there is an apparent increase in the relative permittivity at low frequencies. The anomaly originates in a high-impedance layer on the electrode surface. This may be caused by imperfect contact between the metal electrode and the specimen, aggravated by the accumulation of the products of electrolysis, etc. At low frequencies there is sufficient time for any slight conduction through the specimen to transfer the entire applied field across the very thin electrode layers, and the result is an enormous increase in the measured capacitance. For a purely capacitive impedance C_e at the electrodes, in series with the specimen proper (geometrical capacitance C_0), Johnson and Cole (1951) showed that the apparent relative permittivity ε_{app}' takes the approximate form:

$$\varepsilon_{app}' = \varepsilon' + \frac{\sigma^2 C_0}{\omega^2 \varepsilon_0^2 C_e}, \tag{3.60}$$

where $\varepsilon'$and σ are the true (frequency-independent) relative permittivity and conductivity of the material of the specimen. This formula, which fairly accurately describes the behaviour observed with many liquids, can then be used to correct low-frequency measurements for electrode effects. For solids the electrode impedance is usually more complex, and since little detail is ever known about it, the problem of extracting meaningful results from low-frequency dielectric data is often rendered intractable by conduction.

These effects are prominent in electrically conductive polymers and will be considered at greater length in Chapters 8 and 9.

3.7 Electronic effects

The contributions of atomic and electronic polarisation to the dispersion of molar polarisation have been illustrated in Fig. 2.4 and described briefly in Section 2.2. The former arises from the interaction of radiation with molecular vibrations that are associated with changes in the molecular dipole, e.g. the stretching vibration of the HCl molecule, where, as seen from Equation (2.15), changes in the atomic separation directly affect the dipole moment. The dispersion of atomic polarisation, which occurs typically in the infra-red region, is usually somewhat smaller than that from molecular orientation and electronic excitation, which occur at lower and higher frequencies, respectively. It is electronic excitation that is responsible, then, for the dispersion of optical refraction of materials, as the frequencies involved generally lie in the optical and ultraviolet spectral regions. The analysis of electronic excitation differs in a number of respects from that of dipole relaxation considered in Section 3.1. The radiation will propagate through the material and its wavelength, λ, is generally less than the dimensions of the sample. The excitations of electrons in atoms and molecules are characterised by quantised energy levels. The interaction of radiation and the electron systems of atoms and molecules is usually described in terms of the refractive index of the medium rather than its relative permittivity.

The electric vector of the radiation will couple strongly with the electron systems of atoms and molecules via an electric dipole interaction. This interaction is much larger than those with higher order electric multipoles, and those between the magnetic field vector and a magnetic dipole, which are usually neglected. The propagation of radiation in the medium is described by Maxwell's equations. Since the polymers of interest are insulating and nonmagnetic, the wave equation for the radiation can be written in terms of the electric field vector as:

$$\nabla^2 \boldsymbol{E} = \left[\frac{\partial^2}{{}^2} + \frac{\partial^2}{\partial y^2} + \frac{\partial^2}{\partial z^2}\right]\boldsymbol{E} = -\frac{1}{\nu^2}\frac{\partial^2 \boldsymbol{E}}{\partial t^2}, \tag{3.61}$$

where ν is the velocity of the radiation in the medium, which is given by c/n (the velocity of light in vacuum divided by the refractive index of the medium). In general this equation is simplified by considering a plane wave propagating in the z-direction when $\boldsymbol{E}$ can be polarised along either the x- or y-directions. Then for plane-wave light polarised in the x-direction:

$$E_x(z,t) = E_0 \cos(\omega t - kz) = \frac{1}{2}\left(E_0 \mathrm{e}^{\mathrm{i}(\omega t - kz)} + \text{complex conjugate}\right), \tag{3.62}$$

where k is the wave-number given by $2\pi/\lambda$.

The interaction of a light wave and electrons in atoms in a solid was first analysed by H. A. Lorentz using a classical model of a damped harmonic oscillator subject to a force determined by the local electric field in the medium, see Equation (2.28). Since an atom is small compared with the wavelength of the radiation, the electric field can be regarded as constant across the atom, when the equation of motion becomes:

$$\frac{\mathrm{d}^2\boldsymbol{x}}{\mathrm{d}t^2} + \gamma\frac{\mathrm{d}\boldsymbol{x}}{\mathrm{d}t} + \omega_0^2\boldsymbol{x} = \left(\boldsymbol{E} + \frac{\boldsymbol{P}}{3\varepsilon_0}\right)\frac{e}{m}, \tag{3.63}$$

where $\boldsymbol{E}$ and $\boldsymbol{P}$ are taken to act along the x-axis, γ is the damping constant and ω_0 is the resonant frequency of the undamped harmonic oscillator. The polarisation is equal to the dipole moment per unit volume, so that for N electrons displaced through a distance x from their equilibrium positions:

$$\boldsymbol{P} = Ne\boldsymbol{x}. \tag{3.64}$$

Equation (3.63) can then be written in terms of $\boldsymbol{P}$ as:

$$\frac{d^2\boldsymbol{P}}{\mathrm{d}t^2} + \gamma\frac{\mathrm{d}\boldsymbol{P}}{\mathrm{d}t} + \omega_0^2\boldsymbol{P} = \left(\boldsymbol{E} + \frac{\boldsymbol{P}}{3\varepsilon_0}\right)\frac{Ne^2}{m}. \tag{3.65}$$

$\boldsymbol{E}$ and $\boldsymbol{P}$ are produced by a light wave oscillating at frequency ω but will differ in phase, and, using Equation (2.5), can be written in the form:

$$\boldsymbol{E} = \boldsymbol{E}_0\mathrm{e}^{\mathrm{i}\omega t}, \quad \boldsymbol{P} = \chi\varepsilon_0\boldsymbol{E}_0\mathrm{e}^{\mathrm{i}\omega t}, \tag{3.66}$$

where χ is the complex electric susceptibility. Substituting into Equation (3.65) and rearranging leads to:

$$-\omega^2 + \mathrm{i}\gamma\omega + \omega_0^2 = \left(\frac{1}{\chi\varepsilon_0} + \frac{1}{3\varepsilon_0}\right)\frac{Ne^2}{m}. \tag{3.67}$$

From Equations (2.5) and (2.34) it follows that:

$$n^2 - 1 = \frac{\boldsymbol{P}}{\varepsilon_0\boldsymbol{E}} = \chi. \tag{3.68}$$

Hence, rearrange from Equation (3.67):

$$n^2 - 1 = \frac{Ne^2}{\varepsilon_0 m}\frac{1}{\omega_0^2 - \omega^2 + \mathrm{i}\gamma\omega - Ne^2/3\varepsilon_0 m} \tag{3.69}$$

or
$$\frac{n^2-1}{n^2+2}=\frac{Ne^2}{3\varepsilon_0 m}\frac{1}{\omega_0^2-\omega^2+\mathrm{i}\gamma\omega}. \tag{3.70}$$

Just as the dipole relaxation was described in terms of the real and imaginary components of the relative permittivity, the electronic response can be expressed in terms of the real and imaginary parts of the refractive index, i.e. n is written as $n-ik$, where k is the extinction coefficient. A propagating light wave is then described by:

$$E=E_0\mathrm{e}^{\mathrm{i}(\omega t-kz)}=E_0\mathrm{e}^{\mathrm{i}\omega(t-z/v)}=E_0\mathrm{e}^{\mathrm{i}\omega(t-zn/c)}\mathrm{e}^{-\omega zk/c}=E_0\mathrm{e}^{\mathrm{i}\omega(t-zn/c)}\mathrm{e}^{-z\alpha/2}, \tag{3.71}$$

where α is the absorption coefficient given by Lambert's law for the intensity of light of initial intensity I_0 transmitted through a distance z in an absorbing medium, i.e.

$$I=I_0\mathrm{e}^{-\alpha z}. \tag{3.72}$$

Hence k is related to α by:

$$k=\frac{\alpha c}{2\omega}=\frac{\alpha\lambda}{4\pi}. \tag{3.73}$$

The factor $Ne^2/3\varepsilon_0 m$ in Equation (3.69) is the solvation or solvatochromic shift, which is a measure of the displacement of the resonance of an isolated atom, or molecule, when it is placed in a condensed medium. Thus $\left(\omega_0^2-Ne^2/3\varepsilon_0 m\right)$ can be treated as a new resonant frequency $\omega_0'^2$. Introducing the complex refractive index then leads to the results:

$$n^2+k^2=1+\frac{Ne^2}{\varepsilon_0 m}\frac{\omega_0'^2-\omega^2}{\left(\omega_0'^2-\omega^2\right)^2+\gamma^2\omega^2} \tag{3.74}$$

and

$$nk=\frac{Ne^2}{\varepsilon_0 m}\frac{\gamma\omega}{\left(\omega_0'^2-\omega^2\right)^2+\gamma^2\omega^2}. \tag{3.75}$$

Provided that $n-1$ and k are both small, approximate expressions for n and k are:

$$n=1+\frac{Ne^2}{2\varepsilon_0 m}\frac{\omega_0'^2-\omega^2}{\left(\omega_0'^2-\omega^2\right)^2+\gamma^2\omega^2} \tag{3.76}$$

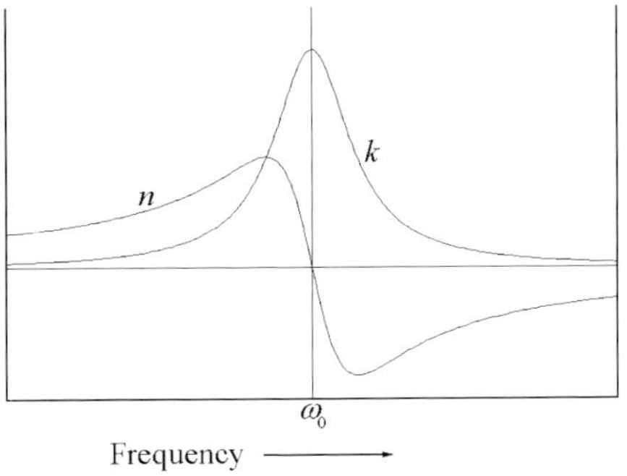

Fig. 3.19 Variation in the real and imaginary parts of the refractive index near resonance for a Lorentz oscillator.

$$\text{and} \qquad k = \frac{Ne^2}{\varepsilon_0 m} \frac{\gamma\omega}{\left(\omega_0'^2 - \omega^2\right)^2 + \gamma^2\omega^2}. \tag{3.77}$$

These functions describe an absorption with a Lorentzian line shape for which n and k vary, as shown in Fig. 3.19. This is identical to the dispersion illustrated earlier in Fig. 2.4. Quite clearly this model will also apply to molecular vibrations that produce an oscillating dipole. Hence the dispersion associated with vibrations that give rise to infra-red absorption will be of the same form, see Fig. 2.4.

The effects of electronic polarisation in insulating polymers are difficult to observe directly, since the electronic levels of saturated polymers lie in the extreme, or vacuum, ultra-violet. However, pendent groups containing –C=O and unsaturated groups give rise to absorption at longer wavelengths that can fall within the range of commercial UV-visible spectrophotometers. Examples of UV absorption spectra are shown in Fig. 3.20. Electron energy loss (EELs) measurements provide an alternative to spectroscopy for the observation of electronic states. An electron beam, which can have energies from a few eV up to 100 keV or more, is incident on the sample, and the energy of some of the electrons is reduced in inelastic collisions. Analysis of the energy spectrum of the transmitted beam enables the energy of excited states in the sample to be determined. In addition the beam can be focussed to a very small area in an electron microscope, with values between 100 and 0.7 nm being possible, so that high spatial resolution can be achieved. At electron beam energies below 5 eV energy is lost by excitation of molecular vibrations, allowing observation of vibration spectra at interfaces and in very thin films. Above 5 eV energy is lost by electronic excitation, first of electrons in the molecular orbitals and second, at higher energy, by excitation of electrons from the atomic core

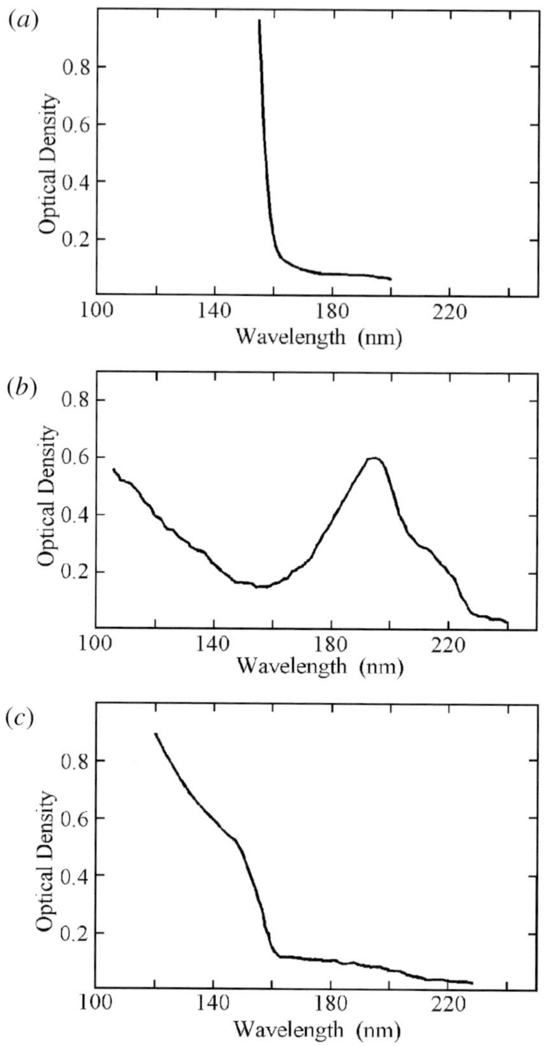

Fig. 3.20 UV absorption spectra of (a) paraffin, (b) atactic polystyrene and (c) polyvinylchloride. Reproduced with permission of the Japanese Physical Society from Onari (1969).

electron levels. The latter enables the technique to be used for chemical analysis of the sample. The core level excitations have fine structure on the low energy edge, which is caused by simultaneous excitation of the molecular electronic states.

One limitation of the technique is the energy resolution obtainable. Early instrumentation only had a resolution of the order of 1 eV, although modern instruments achieve a resolution of ~0.1 eV. These are much lower than the

Table 3.2 *Excited electronic states of polymers measured by UV spectroscopy and EELs*

Polymer	Transitions	Spectroscopic data[a] (eV)	EELs data[a] (eV)
Poly(methylmethacrylate)	$n \to \pi^*$	5.8 w	6.2 w
	$n \to \pi^*$, $\pi \to \pi^*$, $\sigma \to \sigma^*$	6.7 w	7–12 w
	Plasmon		20 s
Polystyrene	$\pi \to \pi^*$	5.6 w	5.6 w
	$\pi \to \pi^*$	6.4 m	6.9 s
	$\pi \to \pi$, $\pi \to \sigma^*$, $\sigma \to \sigma^*$		7–12 w
	Plasmon		22.1 s
Poly(2-vinylpyridine)	$\pi \to \sigma^*$		4.7 m
	$\pi \to \pi^*$		7 s
	Plasmon		21 s
Poly(ethyleneterephthalate)	$\pi \to \pi^*$		5 w
	$\pi \to \pi^*$		6.5 m
	Plasmon		22 s

[a] Strength of features; s, strong; m, medium; w, weak

resolution that can be achieved by spectroscopic methods. However, this lower resolution is offset by the additional data that can be obtained and the high spatial resolution. Results obtained by UV spectroscopy and EELs for insulating polymers by Varlot *et al.* (1997, 1998, 2001), Ritsko *et al.* (1978) and Ritsko and Bigelow (1978) are summarised in Table 3.2. In polystyrene, poly(2-vinylpyridine) and poly(ethyleneterephthalate) the lowest energy transitions result from the excitation of the π-electrons in the conjugated rings in the pendent groups, $\pi \to \pi^*$ transitions, while in poly(methylmethacrylate) (PMMA) the transitions are from non-bonding to π-orbitals, $n \to \pi^*$ transitions. The highest energy peaks at ~22 eV are collective excitations of the σ-electrons and are termed plasmons, by analogy with the collective excitations of electrons in metals. Note that the EELs data have to be taken at short exposure times as the electron beam degrades the sample, giving rise to further conjugated structures that give peaks at energies below ~6 eV. In the case of PMMA, which is used as an electron beam resist in electro-lithography, EELs has been used to study the effects of electron beam exposure *in situ*.

EELs has also been used to study the higher lying electronic states of conjugated polymers where the energy levels of the π-electrons lie at energies below 5 eV and give rise to strong absorption in the optical region. This will be discussed further in Chapter 9.

3.7.1 Non-linear effects

The optical dispersion relations discussed above have been derived for a linear medium in which the polarisation is directly related to the electric field, see Equation (2.8). This approximation will fail in all media when the electric fields become very large, for example in an intense laser pulse. Small changes in refractive index that occur at even modest field strengths can be detected either with an interferometer or by changes in the state of polarisation of a light beam. These methods made possible the discovery of the Pockels and Kerr effects, in which the refractive index of a medium depends either linearly or quadratically on an applied static electric field. These effects were discovered well before the advent of lasers, although the related harmonic generation with high-power laser beams is a more remarkable effect and stimulated an upsurge of interest in non-linear optical effects and materials. More recently, the increasing use of optical fibre to replace copper cable in telecommunications created a need for high-speed switches and modulators based on the electro-optical effect, and for new, highly responsive materials for their fabrication. This has stimulated interest in molecules where the electron cloud extends over the whole molecule and the non-linear response to both optical radiation and static fields can be large.

In its most general form the polarisation of a molecule, P^{M}, can be written as:

$$P_i^{\mathrm{M}} = \sum_j \alpha_{ij}(-\omega;\omega_1)E_j^{\mathrm{L}}(\omega_1) + \sum_j\sum_k \beta_{ijk}(-\omega;\omega_1,\omega_2)E_j^{\mathrm{L}}(\omega_1)E_k^{\mathrm{L}}(\omega_2)$$
$$+\sum_j\sum_k\sum_l \gamma_{ijkl}(-\omega;\omega_1,\omega_2,\omega_3)E_j^{\mathrm{L}}(\omega_1)E_k^{\mathrm{L}}(\omega_2)E_l^{\mathrm{L}}(\omega_3) + \ldots, \qquad (3.78)$$

where $P_{\mathrm{i}}^{\mathrm{M}}$ is the ith component of the molecular polarisation ($\mathbf{P}^{\mathrm{M}}$), $E_{\mathrm{j}}^{\mathrm{L}}(\omega_1)$ is the jth component of the local field at frequency $\omega_1(\boldsymbol{E}^{\mathbf{L}}(\omega_1))$ acting on the molecule, etc. Note that $\boldsymbol{E}^{\mathbf{L}}(\omega_1)$, $\boldsymbol{E}^{\mathbf{L}}(\omega_2)$ and $\boldsymbol{E}^{\mathbf{L}}(\omega_3)$ may have different orientations, and that α_{ij}, β_{ijk} and γ_{ijkl} are components of the polarisability and the first and second hyperpolarisability tensors ($\underline{\underline{\alpha}}$, $\underline{\underline{\beta}}, \underline{\underline{\gamma}}$) of the molecule, respectively. Energy and momentum are conserved in the non-linear interaction, so that $-\omega$ is the sum of the components to the right of the semicolon in the brackets. The response of a bulk sample is given by:

$$P_i = \sum_j \chi_{ij}^{(1)}(-\omega;\omega_1)E_j(\omega_1) + \sum_j\sum_k \chi_{ijk}^{(2)}(-\omega;\omega_1,\omega_2)E_j(\omega_1)E_k(\omega_2)$$
$$+\sum_j\sum_k\sum_l \chi_{ijkl}^{(3)}(-\omega;\omega_1,\omega_2,\omega_3)E_j(\omega_1)E_k(\omega_2)E_l(\omega_3) + \cdots, \qquad (3.79)$$

where $\chi^{(1)}$ is $(n^2–1)$, c.f. Equation (3.68), $\chi^{(2)}$ and $\chi^{(3)}$ are the non-linear susceptibilities and the $\boldsymbol{E}$s are the externally applied fields. If a single field is being considered, Equation (3.79) is often written in the much simpler form:

$$P = \chi^{(1)}E + \chi^{(2)}E^2 + \chi^{(3)}E^3 + \cdots \tag{3.80}$$

At the molecular level, if the molecule is symmetric, then P^{M} must change in sign but not in magnitude under reversal of the fields. This requires that the term in $E^{\mathrm{L}}.E^{\mathrm{L}}$ must be zero, since it does not change sign under reversal of the fields and would produce a change in magnitude of P^{M} under field reversal. It follows that all terms even in powers of E^{L} must be zero for symmetric molecules. It also follows that these terms will be non-zero in asymmetric molecules where P^{M} can change in magnitude under field reversal. A similar argument applies to the bulk. Thus for isotropic and centrosymmetric materials, all terms in even powers of E vanish, but these terms will be finite for non-centrosymmetric media. This means that amorphous polymers, which are isotropic, can only display odd power hyperpolarisabilities. In order for them to have even power hyperpolarisabilities they must posses some polar order.

When a single, polarised radiation field is propagating in the material, Equation (3.80) will take the form:

$$\begin{aligned}P &= \chi^{(1)}E_0\cos(\omega t - kz) + \chi^{(2)}E_0^2\cos^2(\omega t - kz) + \chi^{(3)}E_0^3\cos^3(\omega t - kz) + \cdots\\ &= \chi^{(1)}E_0\cos(\omega t - kz) + \chi^{(2)}E_0^2[1 + \cos(2\omega t - kz)] +\\ &\quad \chi^{(3)}E_0^3\left[\frac{3}{4}\cos(\omega t - kz) + \frac{1}{4}\cos(3\omega t - kz)\right] + \cdots\end{aligned} \tag{3.81}$$

The first order hyperpolarisability gives rise to a second harmonic of the input radiation, while the second order hyperpolarisability gives a third harmonic. The generation of harmonics has been used to screen materials thought to have large hyperpolarisabilities. Precise measurement is made difficult by the phase difference that builds up between the beams at ω and 2ω as they propagate through a sample, due to the dispersion of the refractive index, see Figs. 2.4 and 3.18. Use of fine powders, as in the Kurtz powder test (Kurtz and Perry, 1968), reduces this effect, as the optical path in the particles is short and the effect of dispersion is minimised. However, the random paths taken by the light scattered in the powder render the method only qualitative. Phase matching of the fundamental and harmonic is possible in birefringent crystals where the refractive index for the ordinary (O) and extra-ordinary (E) polarisations at ω and 2ω can be equal, and the non-linearity couples the orthogonal polarisations. In the case of third harmonic generation there is a contribution from the medium surrounding the sample and this is difficult to eliminate.

When a static field, $E_{\rm DC}$, aligned parallel to the electric vector of the radiation field, is applied, we can replace E in Equation (3.80) by the sum of $E_{\rm DC}$ and $E_0\cos(\omega t - kz)$. Then, after expanding the squared and cubed terms, the polarisation at the radiation frequency becomes:

$$
\begin{aligned}
P(\omega) &= \chi^{(1)} E_0 \cos(\omega t - kz) + 2\chi^{(2)} E_{\rm DC} E_0 \cos(\omega t - kz) \\
&\quad + 3\chi^{(3)} E_{\rm DC}^2 E_0 \cos(\omega t - kz) + \tfrac{3}{4}\chi^{(3)} E_0^3 \cos(\omega t - kz) \\
&= \chi_{\rm eff} E_0 \cos(\omega t - kz).
\end{aligned}
\tag{3.82}
$$

This may also be obtained directly from Equation (3.79), where the first term gives a contribution $\chi^{(1)}E(\omega)$, the second term two contributions of $\chi^{(2)}E_{\rm DC}E(\omega)$ and the third term three contributions of $\chi^{(3)}E_{\rm DC}^2E(\omega)$ and one of $\chi^{(3)}E^3(\omega)$, with $\chi^{(3)}(-\omega;\, \omega,\, -\omega,\, \omega)$. From Equations (3.68) and (3.82) the refractive index at ω can be written in terms of $\chi_{\rm eff}$ as:

$$
\begin{aligned}
n^2(\omega) &= 1 + \chi_{\rm eff} = 1 + \chi^{(1)} + 2\chi^{(2)} E_{\rm DC} + 3\chi^{(3)} E_{\rm DC}^2 + \tfrac{3}{4}\chi^{(3)} E_0^2 \\
\text{and} \qquad n^2(\omega) - n_0^2(\omega) &= 2\chi^{(2)} E_{\rm DC} + 3\chi^{(3)} E_{\rm DC}^2 + \tfrac{3}{4}\chi^{(3)} E_0^2,
\end{aligned}
\tag{3.83}
$$

where $n_0(\omega) = (1 + \chi^{(1)})$ is the linear refractive index. If the non-linear terms are small we can make the approximation that $((n(\omega) + n_0(\omega)) \approx 2n_0(\omega)$ and

$$
\begin{aligned}
n(\omega) &= n_0(\omega) + \frac{\chi^{(2)}}{n_0(\omega)} E_{\rm DC} + \frac{3\chi^{(3)}}{2n_0(\omega)} E_{\rm DC}^2 + \frac{3\chi^{(3)}}{8n_0(\omega)} E_0^2 \\
&= n_0(\omega) + n_1 E_{\rm DC} + n_2(0) E_{\rm DC}^2 + n_2(\omega) I(\omega).
\end{aligned}
\tag{3.84}
$$

Thus the non-linear refractive index comprises a term linearly dependent on the static field, the Pockels effect, a term dependent on the square of the applied field, the Kerr effect, and a term dependent on the radiation intensity, the optical Kerr effect.

Just as the linear response of the medium displays a resonance in the vicinity of an electronic absorption, the non-linear properties also show resonant behaviour. These can be understood by extending the Lorentz model to the case of an anharmonic oscillator where the restoring force contains terms proportional to the square, cube, etc. of the displacement. Hence Equation (3.63) becomes:

$$
\frac{{\rm d}^2x}{{\rm d}t^2} + \gamma\frac{{\rm d}x}{{\rm d}t} + \omega_0^2 x + ax^2 + bx^3 + \cdots = \xi\left(E + \frac{P}{3\varepsilon_0}\right)\frac{e}{m},
\tag{3.85}
$$

where ξ is a parameter between 0 and 1. This equation can be solved by means of a perturbation expansion in ξ, by expanding x as:

$$x = \xi x_1 + \xi^2 x_2 + \xi^3 x_3 + \cdots \tag{3.86}$$

This will be a solution of Equation (3.85) if the terms in ξ, ξ^2, ξ^3, etc. satisfy it separately. The terms in ξ give Equation (3.63) and the previous results, Equations (3.76) and (3.77). The ξ^2 terms give the equation:

$$\frac{d^2x_2}{dt^2} + \gamma\frac{dx_2}{dt} + \omega_0^2 x_2 + a x_1^2 = 0. \tag{3.87}$$

This equation can be solved using the result of the linear Lorentz model for x_1. In the most general case of two radiation fields at frequencies ω_1 and ω_2 this gives terms at $2\omega_1$, $2\omega_2$, $(\omega_1 - \omega_2)$, $(\omega_1 + \omega_2)$ and 0, i.e. second harmonic generation, frequency mixing and optical rectification. The resonant behaviour of these effects are given by (Boyd, 2003):

$$\begin{aligned}
&\chi^{(2)}(-2\omega;\omega,\omega) = \chi_0^{(2)}(2\omega)\frac{1}{(\omega_0^2-(2\omega)^2-2i\gamma\omega)(\omega_0^2-\omega^2-i\gamma\omega)^2},\\
&\chi^{(2)}(-(\omega_1\pm\omega_2);\omega_1,\omega_2) = \chi_0^{(2)}(\omega_1\pm\omega_2)\times\\
&\frac{1}{(\omega_0^2-\omega_1^2-i\gamma\omega_1)(\omega_0^2-\omega_2^2-2i\gamma\omega_2)(\omega_0^2-\omega_1^2\mp\omega_2-i\gamma(\omega_1\pm\omega_2))},\\
&\chi^{(2)}(0;\omega,-\omega) = \chi_0^{(2)}(0)\frac{1}{\omega_0^2(\omega_0^2-\omega^2-i\gamma\omega)(\omega_0^2-\omega^2+i\gamma\omega)}.
\end{aligned} \tag{3.88}$$

Here $\chi_0^{(2)}(2\omega)$, $\chi_0^{(2)}(\omega_1 \pm \omega_2)$ and $\chi_0^{(2)}(0)$ are the non-resonant values of the hyperpolarisabilities. Thus second harmonic generation is resonantly enhanced at both the fundamental and the harmonic of the optical transition, sum and difference frequency generation at the fundamentals and the sum and difference frequencies, and the rarely observed optical rectification only at the fundamental frequency. The term ξ^3 in the expansion gives rise to effects such as third harmonic generation, $\chi^{(3)}$ $(-3\omega;\omega, \omega,\omega)$, electric field induced second harmonic generation, $\chi^{(3)}$ $(-2\omega; 0,\omega, \omega)$, the optical Kerr effect, $\chi^{(3)}$ $(-\omega; \omega, \omega, -\omega)$, etc. that will display resonances at ω, 2ω and 3ω.

This discussion has covered briefly only a few of the many non-linear effects that occur in materials. A full discussion is beyond the scope of this book and the reader is referred elsewhere for more extensive accounts (Butcher and Cotter, 1990; Boyd, 2003; Sutherland, 2003).

3.7.2 Molecular non-linearity

The non-linear polarisability, Equations (3.78) to (3.80), can be expressed as a Taylor series. This approach provides a basis for the understanding of the non-linear properties of molecules provided we make some simplifying assumptions. First, that the molecular deformability and non-linearity are much larger along one axis, the z-axis, than for other directions in the molecule, so that $\underline{\underline{\alpha}}$, $\underline{\underline{\beta}}$ and $\underline{\underline{\gamma}}$ can be represented by single tensor components $\alpha_{zz}, \beta_{zzz}, \gamma_{zzzz}$. This is reasonable for elongated molecules where the electron distribution is more spatially extended along the z-direction. Second, that the local fields are aligned along this axis. Then Equation (3.78) can be written in a simplified form, similar to that of Equation (3.80). We will consider the case of an asymmetric molecule. In an asymmetric molecule of this type, the electron distribution in the molecule will also be asymmetric and the molecule will possess a permanent dipole moment (μ) also aligned along the z-axis. The molecular polarisability in the presence of a field E^{L} can then be written as:

$$P^{\mathrm{M}}(E) = \mu + \alpha_{zz}E^{\mathrm{L}} + \beta_{zzz}E^{\mathrm{L}^2} + \gamma_{zzz}E^{\mathrm{L}^3} + \cdots, \tag{3.89}$$

or, using a Taylor series expansion:

$$P^{\mathrm{M}}(E^{\mathrm{L}}) = \mu + \left|\frac{\partial^2 P^{\mathrm{M}}}{\partial E^{\mathrm{L}}}\right|_{E^{\mathrm{L}}=0} \cdot E^{\mathrm{L}} + \frac{1}{2}\left|\frac{\partial^2 P^{\mathrm{M}}}{\partial E^{\mathrm{L}^2}}\right|_{E^{\mathrm{L}}=0} \cdot E^{\mathrm{L}^2} + \frac{1}{6}\left|\frac{\partial^3 P^{\mathrm{M}}}{\partial E^{\mathrm{L}^3}}\right|_{E^{\mathrm{L}}=0} \cdot E^{\mathrm{L}^3} + \cdots, \tag{3.90}$$

Thus the non-linear terms are determined by the evolution of the molecular polarisation as a function of the local field. The forms for α_{zz}, β_{zzz} and γ_{zzzz} deduced from Equations (3.89) and (3.90) have been used by a number of authors to describe the non-linearity of a model molecule, shown in Fig. 3.21, consisting of electron accepting and donating groups connected by a conjugated bridge (Lu *et al.*, 1994 and Meyers *et al.*, 1994). In such a molecule the z-axis will coincide with the backbone of the polyene chain. The polyene chain will allow charge to be transferred from the electron donating to the electron accepting moiety, so that an isolated molecule will possess a finite dipole moment. However, this charge transfer will in general not be complete and the actual structure will be intermediate between that of the limiting valence bond (VB, zero charge transfer) and charge transfer (CT, full charge transfer) forms shown in Fig. 3.21.

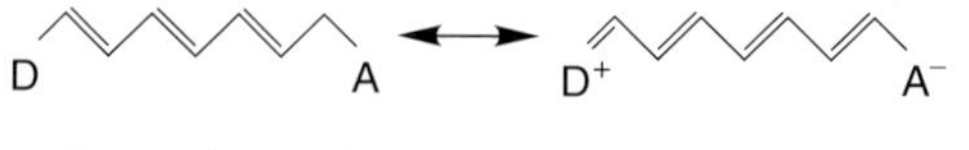

Fig. 3.21 Non-polar valence bond (at left) and 'zwitterionic' full charge transfer (at right) forms of a donor–acceptor substituted polyene.

The ground state wavefunction can be written as a linear combination of the VB and CT wavefunctions:

$$\Psi_{gr} = (1 - x)^{1/2}\Psi_{VB} + x^{1/2}\Psi_{CT}. \quad (3.91)$$

Applying perturbation theory, the value of the ground state energy, relative to the energy of the VB state, is found to be:

$$E_{gr} = \frac{\Delta E}{2} - \frac{(\Delta E^2 + 4t^2)^{1/2}}{2}, \quad (3.92)$$

where ΔE is the difference in energy of VB and CT states, $(E_{VB} - E_{CT})$, and t is the matrix element describing the interaction of the VB and CT states. The factor x in Equation (3.91) can be expressed in terms of ΔE and t as:

$$x = \frac{1}{2} - \frac{\Delta E}{2(\Delta E^2 + 4t^2)^{1/2}}. \quad (3.93)$$

The alternation of the single and double bonds in the polyene chain is inverted in the VB and CT forms, see Fig. 3.21, so that the model molecule can also be described in terms of the bond length alternation (BLA). The BLA is defined as:

$$\mathrm{BLA} = (l_{12} - l_{23})/\Delta l, \quad (3.94)$$

where l_{12} and l_{23} are the bond lengths between the atoms in Fig. 3.21 and Δl is the difference in bond length of a carbon–carbon single bond (0.145 nm) and a carbon–carbon double bond (0.133 nm). If the VB state is lower in energy than the CT state, then $x < 1/2$; if the CT state is lower in energy than the VB state, then $x > 1/2$; and if VB and CT states are equal in energy then $x = 1/2$. These situations give negative, positive and zero BLA respectively. The latter result can be appreciated intuitively, since combination of equal contributions of the extreme structures shown in Fig. 3.21 must lead to equal bond length in the polyene chain.

The degree of charge transfer will also be affected by the local electric field, which can either enhance or decrease the intrinsic degree of charge transfer in the molecule. The field will have no effect on the energy of the non-polar VB state but will shift the energy of the CT state by an amount $\mu_{CT}E^{L}$. Equations (3.92) and (3.93) remain valid, but with ΔE replaced by $\Delta E_{E^L} = \Delta E - \mu_{CT} E^{L}$. Since ΔE_{E^L} is a linear function of E^{L}, we can make the following substitution:

$$\frac{\partial}{\partial E^{L}} = \frac{\partial}{\partial \Delta E_{E^L}} \frac{\partial \Delta E_{E^L}}{\partial E^{L}} = -\mu_{CT} \frac{\partial}{\partial \Delta E_{E^L}}. \quad (3.95)$$

Then the dependence of degree of charge transfer on the field is:

$$\frac{\partial x}{\partial E^{\mathrm{L}}} = \frac{\partial x}{\partial \Delta E_{E^{\mathrm{L}}}} \frac{\partial \Delta E_{E^{\mathrm{L}}}}{\partial E^{\mathrm{L}}} = \frac{2t^2 \mu_{\mathrm{CT}}}{\left(\Delta E_{E^{\mathrm{L}}}^2 + 4t^2\right)^{3/2}}. \tag{3.96}$$

The ground state energy given by the modified version of Equation (3.92) is also dependent on the local field, i.e. $E_{\mathrm{gr}} = E_{\mathrm{gr}}(E^{\mathrm{L}} = 0) - P^{\mathrm{M}}(E^{\mathrm{L}}).E^{\mathrm{L}}$, and it follows that:

$$P^{\mathrm{M}}(E^{\mathrm{L}}) = -\left|\frac{\partial E_{\mathrm{gr}}}{\partial E^{\mathrm{L}}}\right|_{E^{\mathrm{L}}}. \tag{3.97}$$

These results can be used to calculate the variation in α_{zzz}, β_{zzz} and γ_{zzzz} as a function of either x, the BLA, E^{L} or $\Delta E_{E^{\mathrm{L}}}$. In this context E^{L} is taken to comprise the cavity field and the reaction field, as discussed in Section 2.5. The reaction field arises even in the absence of an external field because of the response of molecules in the surrounding medium to the presence of the dipolar molecule. The reaction field will in turn modify the molecular dipole, which affects the reaction field, and so on until a minimum energy state is reached. Thus the molecular dipole moment is not fixed, but will depend on the medium. This effect is particularly strong in solution, as the solvent molecules reorient in the vicinity of the solute molecule. In more polar media the reaction field is larger and so is the dipole moment. The reaction field of the host medium will also affect other properties of the solute molecule, e.g. ground and excited state energies and non-linear optical parameters. One consequence is the occurrence of marked solvatochromism, i.e. the solvent dependence of the optical absorption energy, for dipolar molecules in polar hosts as a result of the shifts in the energies of the ground and first excited states of the molecule. The dependence of the non-linear optical parameters on the polarity of the medium provides, in principle, a means of maximising non-linear effects by an appropriate choice of the medium.

The dependence of P^{M} on $\Delta E_{E^{\mathrm{L}}}$ can be deduced from the modified form of Equation (3.92) and Equations (3.95) and (3.97):

$$P^{\mathrm{M}}(\Delta E_{E^{\mathrm{L}}}) = \mu_{\mathrm{CT}}\left[\frac{1}{2} - \frac{\Delta E_{E^{\mathrm{L}}}}{2\left(\Delta E_{E^{\mathrm{L}}}^2 + 4t^2\right)^{1/2}}\right] = x\mu_{\mathrm{CT}}. \tag{3.98}$$

Thus the dipole moment is zero when $\Delta E_{E^{\mathrm{L}}}$ is $+\infty$, corresponding to the VB state with $x = 0$ and BLA $= -1$, and is μ_{CT} when $\Delta E_{E^{\mathrm{L}}}$ is $-\infty$, corresponding to

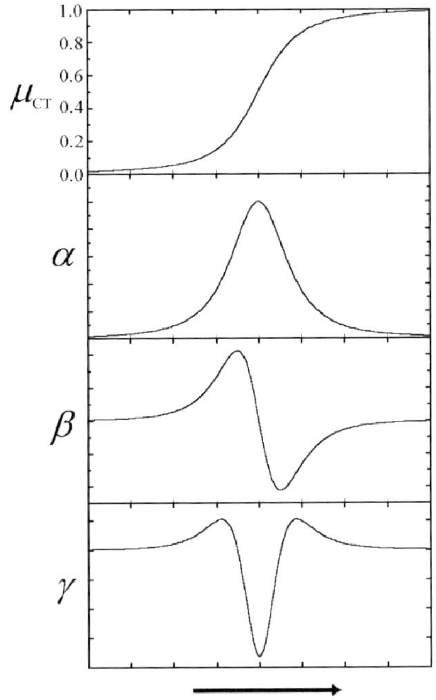

Fig. 3.22 Evolution of the dipole moment, the polarisability and the first and second hyperpolarisabilities of the model molecule of Fig. 3.21 as a function of increasing reaction field and BLA. The units of the dipole moment are μ_{CT}. α, β and γ are plotted in arbitrary units.

the CT state with $x = 1$ and BLA $= 1$. Plots of P^{M} and α_{zz}, β_{zzz} and γ_{zzzz} obtained from Equations (3.90) and (3.98) are shown in Fig. 3.22.

The plots in Fig. 3.22 show the expected increase in ground state dipole moment as the contribution of the CT state increases. The polarisability is lowest in the pure VB and CT states and reaches a maximum for zero BLA. In the limiting cases the π-electrons are localised in the double bonds, see Fig. 3.21, and, as discussed in Chapter 2, the overall polarisability is determined additively from the bond contributions. In the intermediate state of zero BLA, the π-electrons are distributed along the whole length of the conjugated bridge, and application of an external field affects this extended electron system, and thus the response is larger than that obtained by simple additivity. The first hyperpolarisability has two maxima, a positive maximum on the left-hand side and a negative maximum on the right-hand side of the diagram. Materials with large second order non-linear properties can be obtained by tailoring the structure of the molecules so that they lie close to either of these maxima. The second hyperpolarisability has three maxima, but that at zero BLA is by far the largest, since the extended electron system has a larger non-linear response to an external field than electrons localised in double bonds. Consequently, maximising third order non-linear effects requires a molecule with an extended, delocalised, polarisable electron system, such as that found in conjugated polymers.

While the classical model of an anharmonic oscillator describes the effects of non-linearity, it cannot provide information on molecular properties. Calculation of molecular properties requires a quantum mechanical model. Application of perturbation theory (Boyd, 2003) leads to the following expression:

$$\chi^{(2)}_{ijk}(-(\omega_1+\omega_2);\omega_1,\omega_2) = \frac{N}{\hbar^2}\wp_1\sum_{mn}\left(\frac{\varphi^i_{gn}\varphi^j_{nm}\varphi^k_{mg}}{(\omega_{ng}-\omega_1-\omega_2)(\omega_{mg}-\omega_2)} + \frac{\varphi^j_{gn}\varphi^i_{nm}\varphi^k_{mg}}{(\omega^*_{ng}+\omega_1)(\omega_{mg}-\omega_2)} + \frac{\varphi^j_{gn}\varphi^k_{nm}\varphi^i_{mg}}{(\omega^*_{ng}+\omega_1)(\omega^*_{mg}+\omega_1+\omega_2)}\right) \tag{3.99}$$

where $\omega_{n(m)g} = \omega^0_{n(m)g} - \mathrm{i}\gamma_{n(m)}$, the φs are the electric dipole transition moments between the levels designated by the subscripts, g denotes the electronic ground state, i, j and k are the Cartesian indices and $\wp$ is a permutation operator, which acts on ω_1, ω_2 and i, j and k. $\chi^{(2)}_{ijk}(-(\omega_1+\omega_2);\omega_1,\omega_2)$ is a complex quantity. Since many measurements are made some way from resonance, it is usual to neglect the damping term so that the hyperpolarisability is a real quantity. A number of different quantum-chemical approximations have been made to simplify the calculation of hyperpolarisabilities for molecules; Prasad and Williams (1991) present a summary of these. The simplest model involves only the ground and first excited states, which gives rise to optical absorption via an electric dipole interaction, in the determination of β. This two-level model leads to the result for second harmonic generation in the absence of damping of the electronic state:

$$\beta(-2\omega;\omega,\omega) = \frac{3}{2\varepsilon_0\hbar^2}\frac{f^2(\mu_e-\mu_g)\omega^2_{eg}}{(\omega^2_{eg}-\omega^2)(\omega^2_{eg}-(2\omega)^2)} = \beta_0\frac{f^2(\mu_e-\mu_g)\omega^4_{eg}}{(\omega^2_{eg}-\omega^2)(\omega^2_{eg}-(2\omega)^2)}, \tag{3.100}$$

where f is the oscillator strength of the absorption at frequency ω_{eg}, μ_g and μ_e are the dipole moments of the ground and excited states and β_0 is the non-resonant hyperpolarisability. It can be seen from Equation (3.100) that positive values of β, i.e. for molecules situated on the left-hand side of Fig. 3.22,

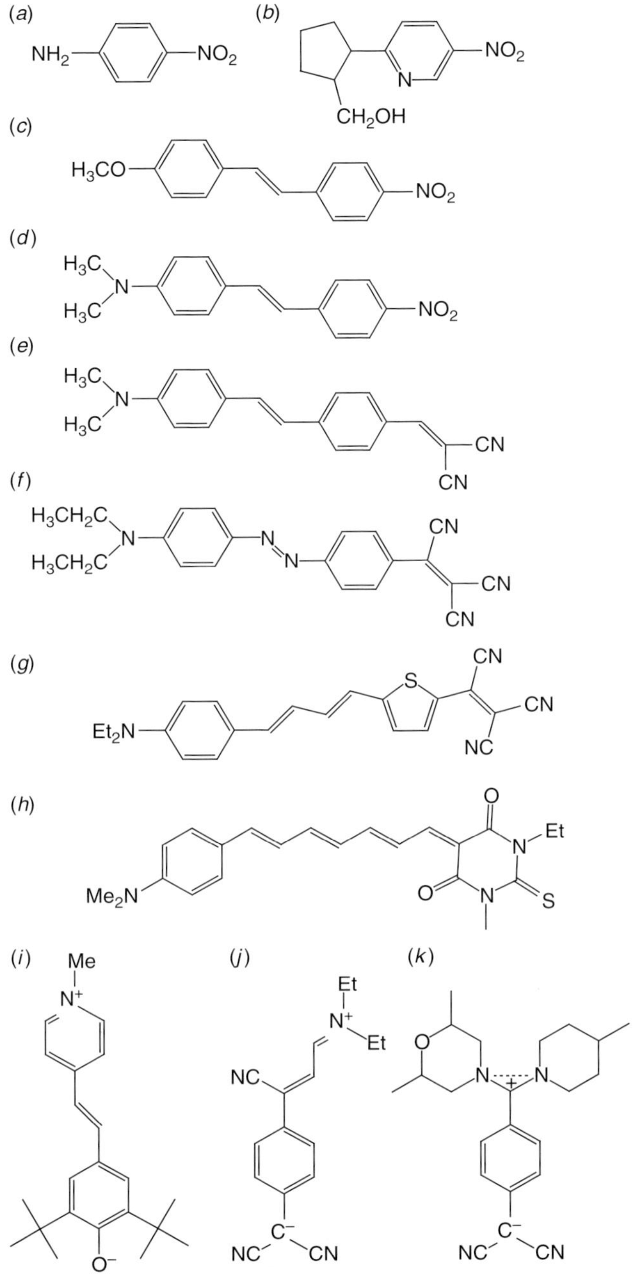

Fig. 3.23 Molecules with large β_{zzz}. (a)–(h) molecules with $\mu_g > \mu_e$ and (i)–(k) molecules with $\mu_g < \mu_e$. See the text and Table 3.3 for a description of the molecules.

Table 3.3 *Chemical names and non-linear optical properties of the molecules shown in Fig. 3.23*

Figure label	Chemical name and abbreviation	λ_{max} nm (solvent)	β_0 $10^{-40}m^4V^{-1}$	$\mu\beta_0$ $10^{-64}Cm^5V^{-1}$
a	para-nitroaniline (*p*-NA)	358(1,4-dioxane)	35	
b	2-(N-prolonol)-5-nitropyridine (PNP)	376(1,4-dioxane)	38	
c	4-methoxy-4′-nitrostilbene (MONS)	364(1,4-dioxane)	99	
d	4-dimethylamino-4′-nitrostilbene (DANS)	430($CHCl_3$)	230	508
e	4-(N,N-diethylamino)-4′-(2,2-dicyanovinyl)-stilbene (DADS)	492(DMSO)	559	
f	4-(N,N-diethylamino)-4′-tri-cyanovinylazobenzene (DADAB)	582(DMSO)	648	5740
g	(E)-benzeneamine-N, N-diethyl-4-[2-(2-tricyanovinylthiophene) butadienyl]	662		5800
h	(E,E,E) 5-[7-[4-(diethylamino)phenyl]-2, 4,6-heptatrienylidene]-1, 3-diethyldihydro-2-thioxo-4, 6-pyrimidone	624(DMSO)	3230	7130
i	4-(4-N-methylpyridino)-2,6-di-tertbutylphenoxide	524($CHCl_3$)		−300
j	(4-[1-cyano-3-(diethylamino)-2-propylidene]-2,5-cyclohexadiene-1-ylidenepropanedinitrile (DEMI)	717($CHCl_3$)	−1470	−13230
k	7-(2,6-dimethylmorphilino)-7-(4-methypiperidino)-8, 8-dicyano-quinodimethane (MORPIP)	524($CHCl_3$)	−250	−1238

correspond to $\mu_e > \mu_g$, while negative values of β, i.e. for molecules situated on the right-hand side of Fig. 3.22, correspond to $\mu_e > \mu_g$. The molecular examples in Fig. 3.22 are designed to have a large first order hyperpolarisability, and they are identified in Table 3.3, where their molecular non-linearity is also listed.

Obtaining a large molecular hyperpolarisability is not in itself sufficient to ensure a large macroscopic hyperpolarisability. As noted earlier, $\chi^{(3)}$ is always finite, but for axially elongated molecules will be a maximum only if the individual molecular contributions γ^{zzzz} are aligned. A material must be non-centrosymmetric for a finite macroscopic value of $\chi^{(2)}$. This requires crystallisation into a polar lattice. In fact, an anti-parallel alignment of pairs of molecules that crystallise in centrosymmetric lattices is more common. Furthermore, many applications require thin films, and the growth of thin films of non-centrosymmetric organic crystals suitable for application has proved to be difficult. An alternative approach is to incorporate the non-linear moiety into a polymer chain and process this to produce thin films with polar order. The non-linear moieties may be incorporated as pendent side groups, be partially integrated with the polymer chain, or be wholly incorporated in the polymer chain, as shown in Fig. 3.24. Since the non-linear moieties possess electric dipole moments, they can be oriented with an applied *poling* field. The alignment process will be easier at or above the glass transition of the polymer, where the polymer molecules are more mobile. Subsequent cooling of the polymer well below its glass transition will *freeze in* the order induced by poling. Cross-linking the polymer may also be used to lock in the poled alignment to improve the long-term stability. This process will generally be most effective for pendent species attached to the polymer chain via flexible linkages. It will be somewhat less efficient for the partially integrated species since orientation will involve some movement of the polymer backbone, and least effective for the wholly incorporated species where large-scale orientation of the polymer chains is required, despite the fact that the head-to-tail arrangement of the chromophores leads to a larger total dipole moment and stronger interaction with an applied field.

In the absence of a field the orientation of the non-linear species in an isotropic polymer will be random. An applied field will tend to orient the non-linear species, but this is opposed by their thermal motion. The macroscopic non-linearity of poled films is determined by the orientation of the non-linear species, which can be calculated if the ground state dipole and the principal component of the first hyperpolarisability are assumed to be parallel, a reasonable approximation for axially elongated molecules. The probability distribution function of the molecular orientation can be written as:

$$\mathsf{P}(\theta)\mathrm{d}\theta = e^{-U(\theta)/kT}\sin\theta\cdot\mathrm{d}\theta, \tag{3.101}$$

where θ is the angle between the applied field and the axis of the molecule, $U(\theta)$ is the potential energy of the molecular dipole in the poling field, k is the Boltzmann constant and T the absolute temperature. $U(\theta)$ contains contributions from the local anisotropy, e.g. attachment to the polymer chain, and the

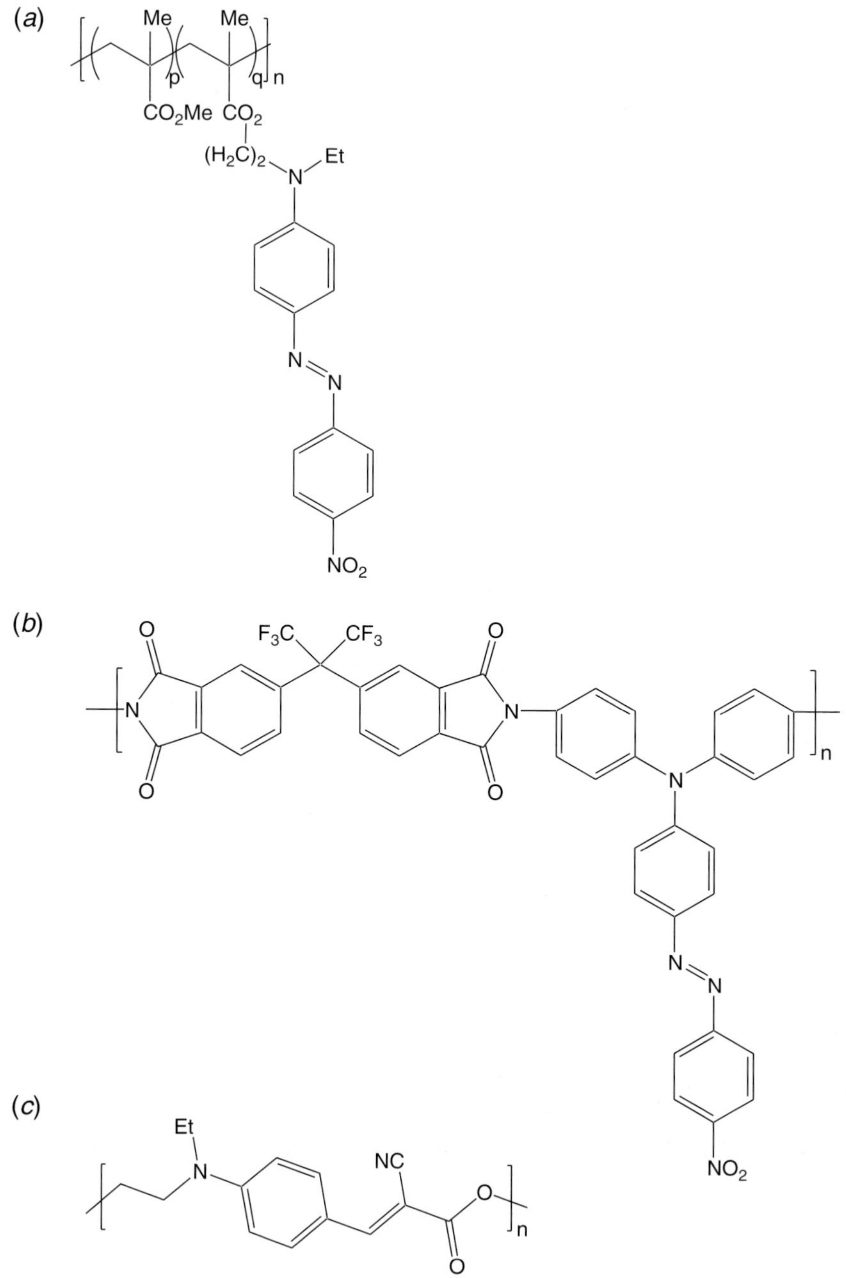

Fig. 3.24 Examples of non-linear optical polymers with (a) pendent, (b) semi-buried and (c) backbone chromophores

local field, which is taken to be the external field multiplied by a local field correction factor. Despite the fact that the polar, non-linear moieties are attached to the polymer chain, the local environment is usually taken to be isotropic and weakly interacting. Then the average orientation can be calculated using the Boltzman distribution law in the presence of the local field as:

$$\langle\cos\theta\rangle = \frac{\int_0^\pi \cos\theta e^{-\mu E^L/kT}\sin\theta d\theta}{\int_0^\pi e^{-\mu E^L/kT}\sin\theta d\theta}. \tag{3.102}$$

The solution to this equation is the Langevin function, as previously described in Section 2.5:

$$\langle\cos\theta\rangle = \coth(E_\mu) - \frac{1}{E_\mu} = L_1(E_\mu) = \frac{1}{3}E_\mu - \frac{1}{45}E_\mu^3 + \frac{2}{945}E_\mu^5 - \cdots, \tag{3.103}$$

where $E_\mu = \mu E^L/kT$. The molecular orientation induced by the field has a cylindrical symmetry with two independent coefficients in the $\chi^{(2)}$ tensor, viz. $\chi^{(2)}_{ZZZ}$ and $\chi^{(2)}_{ZXX}$ with all permutations of the indices. Using the Boltzmann distribution function, the relationships between the microscopic and macroscopic non-linearities for second harmonic generation are:

$$\chi^{(2)}_{ZZZ} = Nf_Z^2(\omega)f_Z(2\omega)\beta_{zzz}\langle\cos^3\theta\rangle = Nf_Z^2(\omega)f_Z(2\omega)\beta_{zzz}L_3(E_\mu), \tag{3.104}$$

$$\begin{aligned}\chi^{(2)}_{ZXX} &= Nf_X^2(\omega)f_Z(2\omega)\beta_{zzz}\langle\cos\theta\sin^2\theta\cos^2\delta\rangle \\ &= Nf_Z^2(\omega)f_Z(2\omega)\beta_{zzz}\tfrac{1}{2}\left[L_1(E_\mu) - L_3(E_\mu)\right],\end{aligned} \tag{3.105}$$

where the *f*s are local field factors and $L_3(\mathrm{E}_\mu)$ is the third order Langevin function. Note also that subscripts *x*,*y*,*z* are in the molecular frame of reference and *X*,*Y*,*Z* are in the laboratory frame. In the low field limit the Langevin functions $L_1(E_\mu)$ and $L_3(E_\mu)$ approximate to $E_\mu/5$ and $E_\mu/15$ respectively. Assuming that the molecular hyperpolarisability is dominated by the component β_{zzz}, the macroscopic non-linearity is:

$$\chi^{(2)}_{ZZZ} = \frac{Nf_Z^2(\omega)f_Z(2\omega)\beta_{zzz}\mu E^L}{5kT}, \quad \chi^{(2)}_{ZXX} = \frac{Nf_X^2(\omega)f_Z(2\omega)\beta_{zzz}\mu E^L}{15kT}. \tag{3.106}$$

Thus at low fields the ratio of the two components of the hyperpolarisability tensor is 3:1. At high fields $\chi^{(2)}_{ZZZ}$ saturates and $\chi^{(2)}_{ZXX}$ tends to zero, as shown in

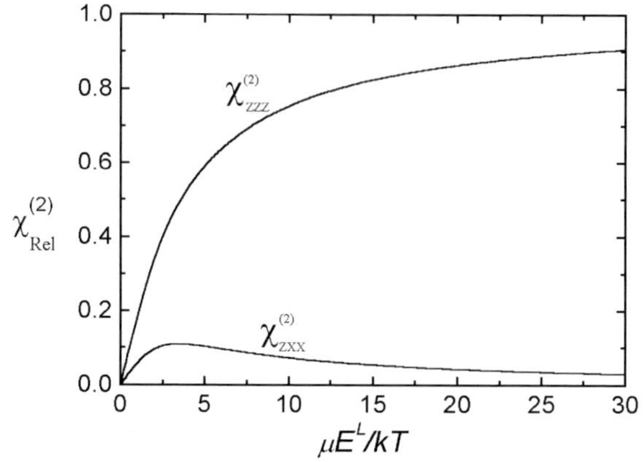

Fig. 3.25 Evolution of the first hyperpolarisability of a poled polymer film as a function of poling field.

Fig. 3.25. The degree of order induced by poling can be determined using the results above to analyse the variation in second harmonic intensity as the poled sample is rotated about an axis normal to the fundamental beam. Generally, the low field limit, given by Equation (3.106), is found to apply and the degree of orientation is low. However, for molecules with large dipole moments the divergence from the low field limit shown in Fig. 3.25 can be observed, and much higher orientation is obtained. The experimentally determined non-linearities are usually the electro-optical *(r)* and the second harmonic generation *(d)* coefficients. These are related to the hyperpolarisabilities by:

$$r_{33} = -\frac{2\chi^{(2)}_{ZZZ}(-\omega;\omega,0)}{n^4} \text{ and } d_{33} = \frac{\chi^{(2)}_{ZZZ}(-2\omega;\omega,\omega)}{2}, \tag{3.107}$$

where r_{ijk} and d_{ijk} are written in a truncated notation since they are invariant under permutation of the indices *i* and *j*. The form of the local field factors makes a detailed analysis complicated. In most analyses the elementary Onsager model is employed, even though the non-linear species are far from spherical. Böttcher (1973 & 1978) has presented a more realistic analysis using ellipsoidal cavities, although this introduces additional parameters that are not readily accessible experimentally and is rarely used. Properties of the polymers illustrated in Fig. 3.24 are given in Table 3.4.

Polymers incorporating high $\mu\beta_0$ value chromophores, such as those illustrated in Fig. 3.23, as non-bonded dopants (guest–host systems) have been reported to have r_{33} values as high as 100 pmV^{-1}. The poled polymers are, however, thermodynamically unstable. The induced non-linearity decays more rapidly for guest–host systems than for systems with covalently attached

Table 3.4 *Macroscopic non-linear optical properties of poled polymers. Measurement wavelengths are shown in square brackets.*

Polymer	T_g (°C)	r_{33} [λ] (pmV^{-1})	d_{33} [λ] (pmV^{-1})
(a)	127	18 [1.58 μm]	>50 [799 nm]
(b)	350	5 [1.305 μm]	
(c)	103		7 [1.06 μm]

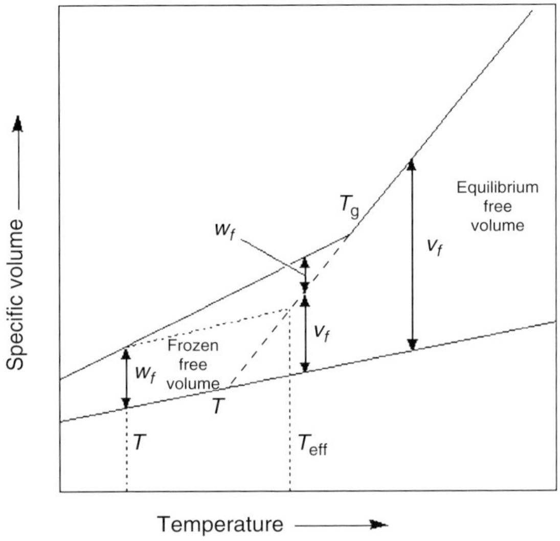

Fig. 3.26 Equilibrium and frozen free volumes of a polymer in the vicinity of the glass transition temperature.

chromophores. A modified form of the WLF Equation (3.52) can be used to describe the relaxation of the poling-induced non-linearity (Man and Yoon, 1992). In this modified equation T is replaced by T_{eff}, T_{eff} is determined by the free volume frozen in the polymer below the glass transition temperature. This is illustrated in Fig. 3.26. The upper line is that illustrated in Fig. 3.5 and the lower line is the theoretical specific volume of the polymer with no free volume, i.e. perfectly packed polymer chains. The diagonal dashed line is the volume of a polymer sample cooled from above T_g infinitely slowly.

The total free volume can be written as:

$$\nu(T) = \nu_f(T) + w_f(T_{eff}) \tag{3.108}$$

and the modified WLF equation is:

$$\log a_T = \log \frac{\tau_T}{\tau_{T_g}} = \frac{-C_1(T_{\text{eff}} - T_g)}{C_2 + (T_{\text{eff}} - T_g)}, \tag{3.109}$$

where $T_{\text{eff}} = T$ for $T \geq T_g$, $T_{\text{eff}} = T + \omega_f(T)/\Delta\alpha$ for $T_g \geq T_\infty$, $T_{\text{eff}} = T_\infty + \omega_f(T)/\Delta\alpha$ for $T_\infty \geq T$. A high T_g results in better long-term stability, c.f. polymer Fig. 3.24(b) and Table 3.4, but at the expense of lower non-linearity due to the greater difficulty in poling polymers with stiff backbones, such as polyimides. The application of these materials in devices is considered in Chapter 10, Section 10.2.3.

A full quantum mechanical description of the third order non-linearity requires forty-eight terms, which are more complex than those describing second order effects, see for example Boyd (2003). The simple model of molecular non-linearity considered above (Fig. 3.22) indicates that the delocalised π-electron systems of conjugated polymers will have large third order non-linearity. Other factors governing the third order non-linearity of conjugated polymers will be discussed in Chapter 9, Section 9.4.2.

3.8 Further reading

The original monographs by Debye (1929) and Fröhlich (1949) present the basic theory of dielectric relaxation. The books by Smyth (1955) and Hill *et al.* (1969) both give clear, critical reviews of theoretical and experimental aspects. The texts by Böttcher (1973) and (1978) present a comprehensive account of the theory of dielectrics. Results for solid polymers are most fully collected and discussed by McCrum, Read and Williams (1967), by Hedvig (1977) and by Havriliak and Havriliak (1997). The latter gives detailed analysis of experimental results using rigorous (unbiased) statistical methods. North (1972) has reviewed dielectric relaxation in polymer solutions. The principles of dielectric spectroscopy are comprehensively reviewed, together with experimental results, in the book edited by Kremer and Schönhals (2003).

Butcher and Cotter (1990), Boyd (2003) and Sutherland (2003) provide overviews of non-linear optics. Prasad and Williams (1991) have discussed the non-linear optics of molecules and polymers. Detailed accounts on this topic can be found in Bosshard *et al.* (1995), Kajzar and Swalen (1996) and Kuzyk and Dirk (1998). More specialised reviews are those of Yesodha *et al.* (2004), on non-linear polymers with azobenzene pendent groups, and Sioncke *et al.* (2003), who describe the second order non-linear optical properties of chiral molecules and polymers.

4

Electronic conduction in polymers

4.1 Introduction

Electrical conductivity of materials is a property which spans a very wide range, as may be judged from the conductivity chart (Fig. 4.1). The conductivity of insulators is typically less than $10^{-12}\,\Omega^{-1}\mathrm{m}^{-1}$, that of semi-insulating or semiconductive materials covers the range 10^{-12} to $\sim 10\,\Omega^{-1}\mathrm{m}^{-1}$, and for semi-metals and metals it is greater than $10\,\Omega^{-1}\mathrm{m}^{-1}$, with a maximum value of $\sim 10^{8}\,\Omega^{-1}\mathrm{m}^{-1}$. In contrast, superconductors have immeasurably high conductivities in their low-temperature, superconducting regimes. Organic compounds have just as wide a range of conductivities as inorganic materials, i.e. from insulating, the conductivity of pure anthracene is $\sim 10^{-20}\,\Omega^{-1}\mathrm{m}^{-1}$, to metallic and even superconductive, in a few cases, for organic charge-transfer salts. The advances made in the development of conductive polymers over the last thirty years of the twentieth century means that this is also true for this subgroup of organic materials. Polyethylene, polytetrafluoroethylene and polystyrene are amongst the best insulators known, while polyacetylene can be obtained with a higher conductivity per unit mass than copper.

Electrical conduction may occur through the movement of either electrons or ions. In each case a suitable starting point for discussion of the conduction process is, however, the basic equation

$$\sigma = qn\mu, \tag{4.1}$$

where the conductivity σ is resolved into three factors: the charge q, concentration n and drift mobility μ of the carriers. The latter parameter characterises the ease with which the charged species will move under the influence of the applied electric field E and is usually expressed as a velocity per unit field ($\mathrm{m^2V^{-1}s^{-1}}$). The mobility is the drift velocity of the carriers $\langle \nu \rangle$ per unit field, where $\langle \nu \rangle$ is defined as the average carrier velocity. In the absence of an electric field, carriers are as likely to be travelling in any one direction as any

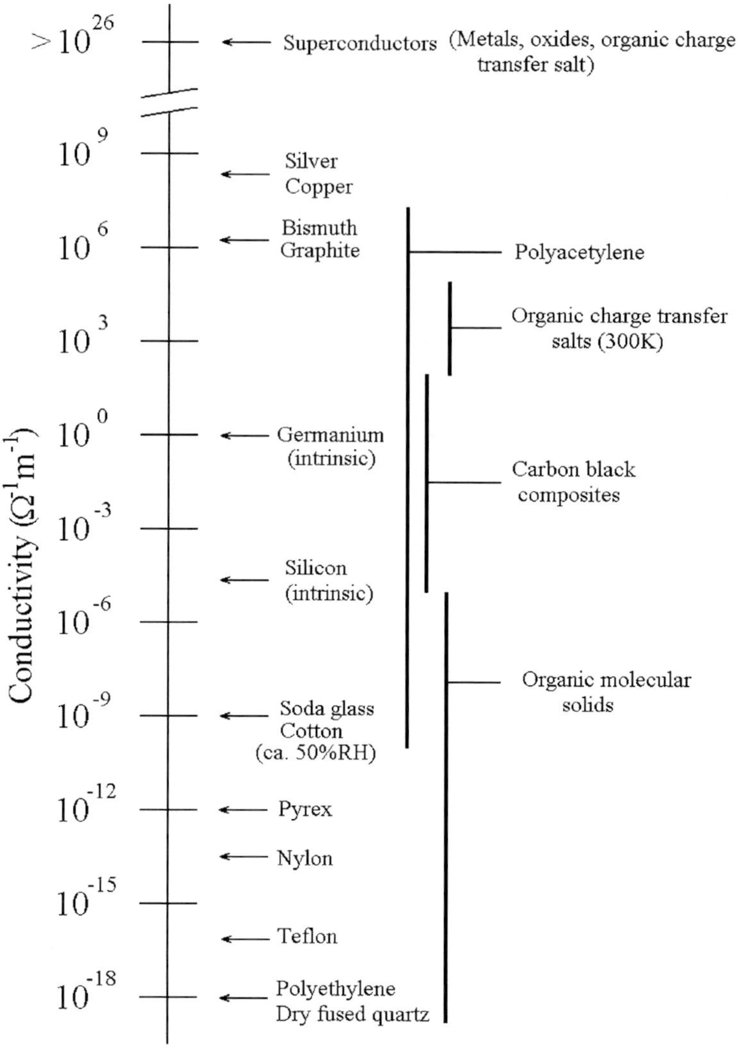

Fig. 4.1 Chart of typical conductivities.

other and ν averages to zero. In the presence of a field there will be a mean carrier velocity, either in the direction of the field or in the opposite direction, depending on whether the carrier has positive or negative charge on it, respectively. A dynamic equilibrium is attained, where the acceleration of the carriers by the field is balanced by the scattering of the carriers by collision with the host lattice (phonon interaction) or by impurities and defects scattering randomising the velocities. In terms of the simple Drude model, if the average time between scattering events is τ the drift velocity may be calculated from the electric force acting on a carrier, of mass m, over this time interval, i.e.

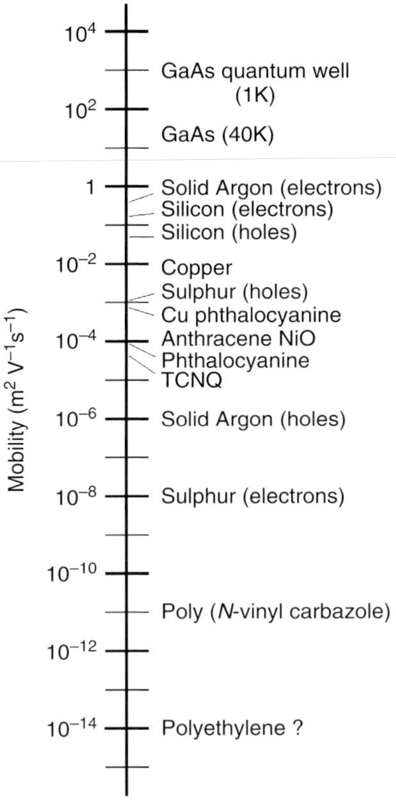

Fig. 4.2 Chart of charge-carrier mobilities.

$$\langle \nu \rangle = \frac{qE}{m} . \tau,$$

$$\text{whence} \quad \mu = \frac{\langle \nu \rangle}{E} = \frac{q\tau}{m}. \tag{4.2}$$

There may be contributions to the conductivity from several different types of carrier, notably electrons and holes (a hole is an electron vacancy carrying an equivalent positive charge) in electronic conductors, and cation and anion pairs in ionic conductors. Theories of conduction aim to explain how n and μ are determined by molecular structure and how they depend on such factors as temperature and applied field. In addition, in polymers the mobility will be affected by the sample morphology. Just as a large range of conductivity is observed for different materials so there is a large range of mobility values. Data for a selection of systems are displayed on the mobility chart (Fig. 4.2).

In polymers with chemically saturated structures, i.e. σ-bonded backbones, it is very difficult to observe any electronic conductivity at all, and what conductivity there is usually depends on the movement of adventitious ions. Consequently, any improvements in the quality of insulation are generally won by careful preparation and purification, so as to avoid as much as possible the presence of ionic impurities, including catalyst residues, products of oxidation and dissociable end groups. Conversely, useful levels of conductivity can be achieved by making composite materials in which a conductive powder is dispersed in an insulating polymer. Metal and carbon black powders have been utilised to make composite materials suitable for use in a number of practical applications. The highly conductive additive is dominant in the conduction process with the insulating polymer limiting the overall conductivity by forming barriers between the particles.

Electronic conduction is possible in polymers with chemically unsaturated (conjugated) structures, i.e. π-bonded backbones. The enthusiasm for the search and discovery of the underlying organic structures that make electronic conduction possible owes much to the biologist Szent-Györgyi (1941) who speculated about the role of semiconduction in living systems. Interestingly enough, semiconduction eventually failed, as a general foundation, to explain complex electron transfer in living cells. Nevertheless, in the meantime, chemists had been led by this work to conjecture that in certain organic compounds, where chemical reactivity had already shown that electrons are extensively delocalised in the molecules, there might be enhanced conductivity, and in 1948 Eley demonstrated conduction in one such case, the phthalocyanines. This stimulated the investigation of organic crystalline solids, leading to the discovery of highly conductive and even superconducting organic salts. This topic has grown to be beyond the scope of this book and will not be discussed in detail here. However, these materials have been incorporated into conductive composites, as discussed in Chapter 8. The same approach also laid the foundation for the great advances that have been made in recent years in developing inherently conductive polymers, which also exhibit the necessary mechanical properties to make a truly conductive plastic material. To some extent the rigidity of the conjugated polymer backbones means that the two sets of properties are mutually exclusive. Nevertheless, as we shall see, a number of approaches succeeded in overcoming this problem.

We can see that electronic conduction, ionic conduction and conduction in heterogeneous composites are all important in their own way in polymers, but they form three quite distinct subjects, and the latter two forms of conduction will therefore be treated separately in Chapter 8. Here we will discuss the basic models of electronic conduction in solid polymers. Materials will be discussed in the later chapters.

4.2 Theories of electronic conduction

Electronic conduction in organic, molecular compounds differs in several important ways from the more familiar kind in metals and inorganic semiconductors like silicon and germanium. That is not to say that they are separate subjects and, indeed, the well-known band theory of atomic lattices has provided the essential basis of concepts and language for the discussion of conduction in crystalline molecular solids. It has been used as a basis for the understanding of conduction in conjugated polymers. However, there are clearly limitations in the applicability of a model that is based on a perfect, extended-chain structure to materials with the complex morphology described in Chapter 1. Even in the idealised model there are significant differences between inorganic and polymer conductors and semiconductors. Most notably, the diamond-like lattice found in silicon and germanium is a strong structure, which suffers little distortion due to the electric field of an added electron. Hence the band structure can be calculated as though the crystal lattice were rigid. This will not be true in polymers where the softer bonds will deform in response to an added charge and this must be taken into account. In polymers the interaction between charge carriers is generally less well screened than in inorganic materials. Thus electron–electron and electron–hole interactions become more important and tend to localise the electronic states to a much larger extent than in inorganic materials. Recent studies of these effects suggest that the band model may not be the best model even for ideal polymer chains. The situation is further complicated by the inhomogeneous, partially crystalline morphology encountered in most polymers. This lack of macroscopic order means that the band model of conduction is inappropriate for any detailed discussion of electronic conduction in bulk polymeric materials, although, with that reservation, it remains a good starting point for a discussion of the conduction process. A detailed treatment of band theory would be out of place here, but we include the following descriptive review of those elements which are of most relevance to the development of our subject.

4.2.1 Band theory of conduction

We will start by looking at how energy bands arise in inorganic semiconductors. Silicon has a set of hybridised orbitals that resemble those of carbon. In the extended, crystalline solid the hybridised orbitals overlap to form a tetrahedral network of σ-orbitals. Since the individual bonding orbitals are indistinguishable, the Pauli exclusion principle prevents them having the same energy and being degenerate. The lifting of this degeneracy results in the formation of two

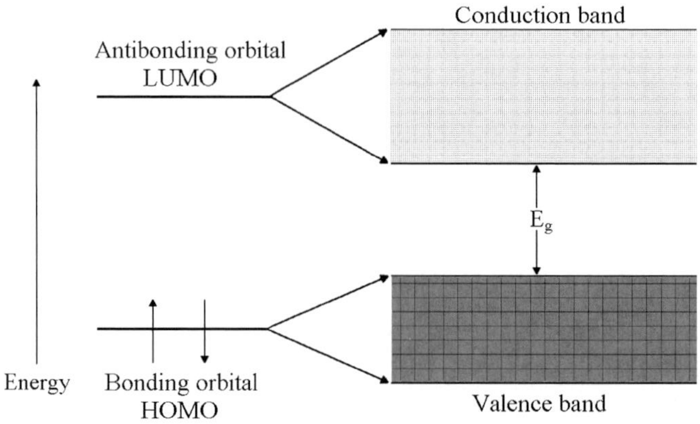

Fig. 4.3 Schematic of formation of energy bands in crystalline molecular solids.

continuous energy bands, called the valence band and the conduction band. The energy gap between these bands is a forbidden zone for electrons. In principle this model will also apply to molecules. If hydrogen molecules, see Section 1.2.1, are brought into close contact, so that orbitals on adjacent molecules interact, the bonding and antibonding orbitals will spread to form bands, as illustrated in Fig. 4.3. A similar picture will apply to carbon-based molecules including polymers. While molecules can have several molecular orbitals, generally we need to consider only the behaviour of the highest occupied molecular orbital (HOMO) and the lowest unoccupied molecular orbital (LUMO).

An important feature of the band system is that electrons are delocalised or spread over the lattice. Some delocalisation is naturally expected when an atomic orbital of any atom overlaps appreciably with those of more than one of its neighbours, but delocalisation reaches an extreme form in the case of a regular, 3-dimensional lattice. We can understand this best if we choose to think of the wave nature of electrons, and from that point of view we can formulate band theory as follows.

A free electron propagates through space as a wave[1] characterised by a wave vector k ($k = 2\pi/\lambda$, where λ is the wavelength), whose magnitude is related to the momentum p of the electron by the fundamental relation

$$k = 2\pi p/h, \tag{4.3}$$

[1] Free electron wave functions are of the form:

$$\psi_k(r) = (1/V)^{1/2}\exp(ik \cdot r),$$

where r is the position vector and V is the volume of the containing space.

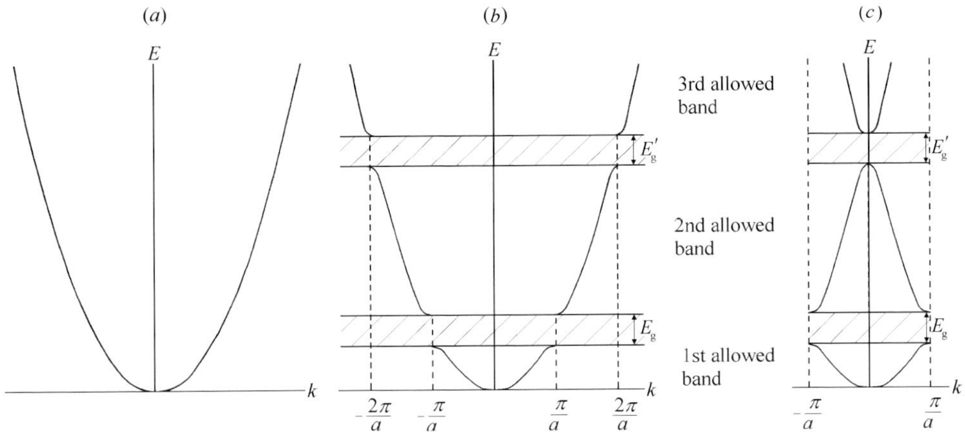

Fig. 4.4 Relation between energy E and wavevector k for (a) a free electron, (b) a nearly free electron in a monatomic linear lattice (spacing a) and (c) as (b) but plotted in the first Brillouin zone.

h being Planck's constant. In a lattice of strongly interacting atoms where the orbitals of the valence electrons overlap well, the valence electrons are nearly free to roam anywhere within the confines of the lattice. (The inner cores of electrons that are not overlapping remain, of course, localised on the atoms.) The regular array of atomic centres does impose a certain restriction on the valence electrons: in the same way that X-rays suffer Bragg reflections by the lattices of crystals (the basis of X-ray crystallography), electron waves which satisfy the Bragg reflection condition will be unable to pass through the lattice. For a simple linear lattice with a spacing a, the Bragg condition is

$$k = \pm n\pi/a, \tag{4.4}$$

where n is an integer. The excluded waves correspond to gaps in the allowed values of electron energy, and the theoretical result for a simplified model of a linear monatomic array is shown in Fig. 4.4, set alongside that for the completely free electron case. The situation is similar for 3-dimensional lattices, the allowed values of the wavevector forming regions, called Brillouin zones, in k-space, separated by forbidden energy gaps. Typically, the widths of the allowed and forbidden energy bands in solids are in the range 0.1–10 eV.

Analytical solutions for the band structure are possible under simplifying approximations. In the adiabatic approximation the electronic states are calculated for a static crystal lattice and the problem of the vibrations of the lattice is treated independently. Finally, the interaction between the electrons and the vibrations is introduced to combine these results. In discussing the electronic states we will assume a static lattice. Limiting the problem to a linear lattice, while not appropriate for crystalline solids, is a reasonable approximation to an ideal polymer chain. For a free electron the potential in the system is taken to be a constant, V_0, and the one-dimensional Schrödinger equation is then:

$$\frac{\partial^2 \varphi}{\partial x^2} + \frac{2m}{\hbar^2}(E - V_0)\,\varphi = 0. \tag{4.5}$$

Substitution of the wavefunction $\exp(\pm ikx)$ leads to the solution for the kinetic energy of the electron:

$$E_k = E - V_0 = \frac{\hbar^2 k^2}{2m} = \frac{p^2}{2m}. \tag{4.6}$$

This is the parabolic curve shown in Fig. 4.4(a). A linear atomic lattice will provide a periodic rather than a constant potential, i.e. $V(x) = V(x+a)$, where a is the repeat distance of the array. Bloch's theorem, also known as Floquet's theorem, states that possible solutions of the Schrödinger equation with a periodic potential are:

$$\varphi(x) = e^{\pm ikx} u_k(x), \tag{4.7}$$

where $u_k(x)$ is a cyclic function with a period a.

The properties of electrons in a periodic potential are demonstrated with the use of a simple square well potential, the Krönig–Penney model. The potential is zero for $0 < x < a$ and V_0 for $-b < x < 0$, i.e. it has a period $(a + b)$. The Schrödinger equations for the two regions follow directly and lead on substitution of a Bloch function to:

$$\begin{aligned}
&\frac{\partial^2 u}{\partial x^2} + 2ik\frac{\partial u}{\partial x} + \left(\alpha^2 - k^2\right) u = 0, \quad 0<x<a, \\
&\frac{\partial^2 u}{\partial x^2} + 2ik\frac{\partial u}{\partial x} - \left(\beta^2 + k^2\right) u = 0, \quad -b<x<0,
\end{aligned} \tag{4.8}$$

where $\alpha^2 = 2mE/\hbar^2$ and $\beta^2 = 2m(V_0 - E)/\hbar^2$. This leads to solutions:

$$\begin{aligned}
u_1 &= Ae^{i(\alpha-k)x} + Be^{-i(\alpha+k)x}, \qquad 0<x<a, \\
u_2 &= Ce^{i(\beta-k)x} + De^{-i(\beta+k)x}, \qquad -b<x<0.
\end{aligned} \tag{4.9}$$

These equations must be solved subject to the boundary conditions that the u-functions and their derivatives are continuous at $x = 0$ and a (or $-b$). This leads to four linear equations in the constants A, B, C and D. A solution of these equations exists if the determinant of the coefficients of A, B, C and D is zero. This leads to general solution that is further simplified if the potential barriers

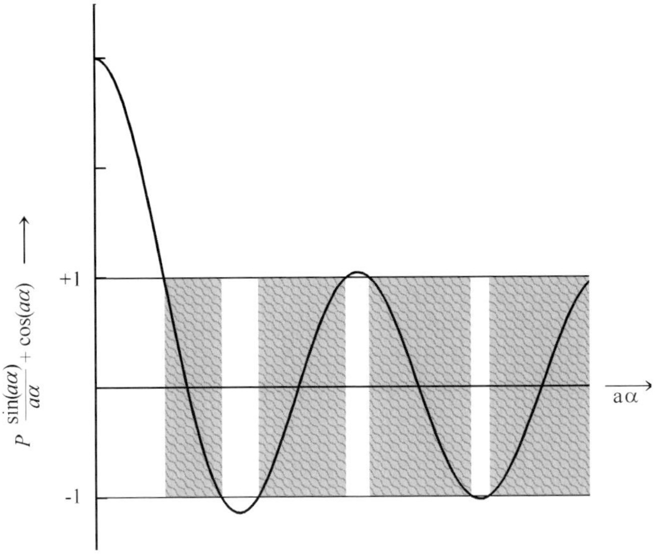

Fig. 4.5 Plot of the solution to the Krönig–Penney model for δ function barriers; the shaded regions correspond to allowed values of αa (see text).

are approximated by *delta* functions with the *area* under the barrier kept equal to $V_0 b$. This results in the equation:

$$P\frac{\sin \alpha a}{\alpha a} + \cos \alpha a = \cos ka, \tag{4.10}$$

where $P = mV_0 ba/\hbar^2$ is a measure of the barrier *area*. Since the right-hand side of Equation (4.10) must have values between -1 and $+1$ it follows that the only permissible values of αa are those for which the left-hand side lies between these limits. The left-hand side of Equation (4.10) is plotted as a function of αa in Fig. 4.5. The regions that have allowed values of αa are shown shaded. From the definition of α the horizontal axis is dependent on energy, hence the shaded areas correspond to the allowed energy bands and the clear areas to forbidden energy gaps shown in Fig. 4.4(b).

The requirement that $-1 \leq \cos ka \leq 1$ used to determine the allowed bands means that the value for k for each band is not unique in the way implied by the extended band scheme shown in Fig. 4.4(b). Instead, the band structure is more properly represented in a reduced zone scheme with $-\pi/a \leq k \leq \pi/a$, which corresponds to the first Brillouin zone, as shown in Fig. 4.4(c).

The potential utilised in the Krönig–Penney model is not a very realistic representation of the potential function of an atomic lattice. A model that can incorporate more realistic potentials is the *tight-binding model*. The lattice potential is taken to be the sum of the atomic potentials located at lattice sites s:

$$V(x) = \sum_{s=-\infty}^{\infty} U_0(x - sa). \tag{4.11}$$

The tails of the localised wavefunctions will overlap, giving rise to a small interaction between neighbouring sites. When $|x| < a/2$ the difference $W(x) = V(x) - U_0(x)$, which is a measure of the inter-site interaction, will be small. A possible solution to the Schrödinger equation is the Bloch-like function:

$$\varphi_k(x) = A \sum_s \mathrm{e}^{iksa} \varphi_0(x - sa), \tag{4.12}$$

where A is a constant and φ_0 is the wavefunction of an isolated atom centred at $x = 0$, when the Schrödinger equation becomes:

$$E\varphi_k(x) = -\frac{\hbar^2}{2m}\nabla^2\varphi_k(x) + V(x)\varphi_k(x). \tag{4.13}$$

For an isolated atom one can write:

$$E_0\varphi_0(x) = -\frac{\hbar^2}{2m}\nabla^2\varphi_0(x) + U_0(x)\varphi_0(x). \tag{4.14}$$

It follows from Equations (4.12) and (4.14) that:

$$E_0\varphi_k(x) = -\frac{\hbar^2}{2m}\nabla^2 A \sum_s \mathrm{e}^{iksa}\varphi_0(x - sa) + U_0(x) A \sum_s \mathrm{e}^{iksa}\varphi_0(x - sa). \tag{4.15}$$

The difference in energy $(E - E_0)$ is then given by:

$$(E - E_0)\varphi_k(x) = A \sum_s \mathrm{e}^{iksa} W(x - sa)\varphi_0(x - sa). \tag{4.16}$$

Multiplying by the complex conjugate of $\varphi_k\,(x)$, integrating over the linear array and rearranging gives:

$$E = E_0 + \frac{A \sum_s \mathrm{e}^{iksa} \int \varphi_k^*(x) W(x - sa)\varphi_0(x - sa)\mathrm{d}x}{\int \varphi_k^*(x)\varphi_k(x)\mathrm{d}x}. \tag{4.17}$$

Since the overlap of the localised wavefunctions is assumed to be small, the denominator in the second term in Equation (4.17) has the approximate value

A^2N, where N is the total number of atoms in the linear lattice. Substituting for the complex conjugate from Equation (4.12), with lattice site label t, gives:

$$E = E_0 + \frac{1}{N}\sum_t \sum_s e^{ik(s-t)a} \int \varphi_0^*(x - ta)W(x - sa)\varphi_0(x - sa)\mathrm{d}x. \quad (4.18)$$

This expression can be simplified by substituting $r = (t-s)$:

$$E = E_0 + \frac{1}{N}\sum_r e^{ikra} \int_{-\infty}^{+\infty} \varphi_0^*(x - ra)W(x)\varphi_0(x)\mathrm{d}x. \quad (4.19)$$

Since $\varphi_0(x)$ is a localised wavefunction the use of limits of integration of $-\infty$ to $+\infty$ will not introduce significant error. Finally, considering only the on-site and nearest neighbour interactions and taking $\varphi_0(x)$ as an even function, i.e. it does not change sign when x is replaced by $-x$:

$$\int_{-\infty}^{+\infty} \varphi_0^*(x)W(x)\varphi_0(x)\mathrm{d}x = -\alpha\,,$$

$$\int_{-\infty}^{+\infty} \varphi_0^*(x - a)W(x)\varphi_0(x)\mathrm{d}x = -\beta\,,$$

$$\int_{-\infty}^{+\infty} \varphi_0^*(x + a)W(x)\varphi_0(x)\mathrm{d}x = -\beta\,. \quad (4.20)$$

Then the energy as a function of k is:

$$E(k) = E_0 - \alpha - \beta\sum_r e^{ikra}, \quad (4.21)$$

where r takes the values ± 1. Hence:

$$E(k) = E_0 - \alpha - 2\beta\cos ka, \quad (4.22)$$

and the dispersion within the allowed energy band has the form of a cosine with a minimum value at $k = 0$, as shown in Fig. 4.6. Note that if $\varphi_0(x)$ is an odd function, i.e. $\varphi(x) = -\varphi(-x)$, the second two expressions in Equation (4.20) are positive and

$$E(k) = E_0 - \alpha + 2\beta\cos ka, \quad (4.23)$$

so that the band has a maximum at $k = 0$, see Fig. 4.6. The symmetry of the atomic function alternates between even and odd symmetry on ascending through the allowed bands, see Figs. 4.4 and 4.6.

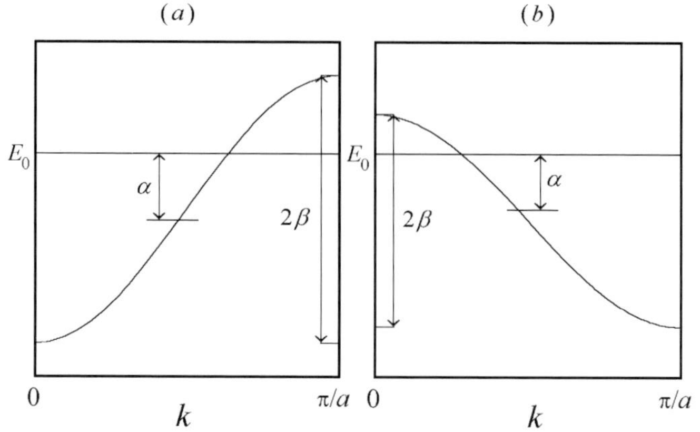

Fig. 4.6 Band dispersion in the tight binding model (a) for a symmetric (even) atomic wavefunction and (b) for an antisymmetric (odd) atomic wavefunction.

Taking into account the full detail of actual valence orbitals of component atoms, valence bands become a good deal more complicated than indicated above. For instance, it is common for different sets of valence electrons to interact and form composite bands. In crystalline solids the overall band structure will also reflect the symmetry of the lattice. The starting point for more detailed theoretical calculations is the *Hartree–Fock self-consistent field* or mean field approximation. This was first introduced to obtain solutions for the electronic states of many-electron atoms by reducing the problem to that of a single electron moving in an average potential of the nucleus and all the other electrons. The potential is calculated from an initial set of electron wavefunctions, e.g. hydrogen atom functions, and the single electron wavefunctions in this potential are determined. The process is then repeated using the new set of wavefunctions until a self-consistent set of wavefunctions is found. The approach has been extended to the study of the electronic states of molecules and of crystalline solids. Mathematically, it reduces to the solution of a series of coupled *Hartree–Fock equations*, which are single-electron equations of the form of Equation (4.13). These are not genuine Schrödinger equations since the value of the potential is dependent on the complete set of single electron functions and can only be solved by the iterative process described above. It should be noted that it is common for these calculations to be described in the literature using the Hamiltonian and Dirac notations rather than the Schrödinger and integral forms.[2]

[2] In the Hamiltonian notation, the Schrödinger equation, see Equation (4.13), is written in terms of the Hamiltonian operator as:

$$H\psi_i(x) = E\psi_i(x), \ H = -\frac{\hbar^2}{2m}\nabla + V.$$

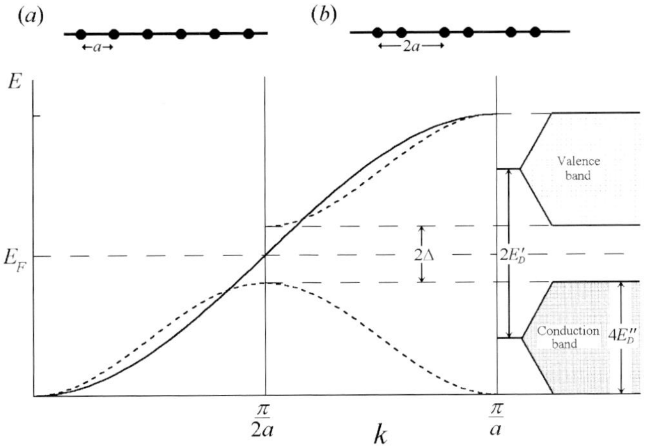

Fig. 4.7 Band structure of a 1-dimensional lattice (a) with one atom per unit cell (continuous curve) and (b) with a Peierls distortion (dotted curves).

Whether a material is a metal, a semiconductor or an insulator will depend on the band structure and the filling of the energy states within the bands. For a partially filled band, either a single band or a pair of overlapping bands, the application of an electric field is in itself sufficient to promote electrons from filled states to adjacent empty states within the same band, thereby allowing a net flow of charge. Hence materials with partially filled bands are metals. In a full band there can be no net flow of electronic charge under an external electric field. The reason is that for every wave state in which an electron is travelling in one direction there is another in which the electron is travelling in the opposite direction and there are no spare states (equal numbers of $+k$ and $-k$ values).

Models of the band structure of polymers are usually based on the simple model of an isolated linear structure, i.e. as a 1-dimensional lattice. Consider a linear lattice with one atom per repeat unit of length a, as shown in Fig. 4.7(a), and containing N atoms. Applying periodic boundary conditions to the Bloch function of Equation (4.7), i.e. setting $\varphi(x) = \varphi(x + Na)$, gives the condition $e^{ikNA} = 1$. Then taking the Nth root gives:

In the Dirac notation, quantum mechanical integrals are represented by a shorthand notation using brackets, for instance:

$$\int \psi_1^*(x)\psi_2(x)\mathrm{d}x \equiv \langle\psi_1 \mid \psi_2\rangle, \quad \int \psi_1^*(x)A\psi_2(x)\mathrm{d}x = \langle\psi_1|A|\psi_2\rangle,$$

where A is a quantum mechanical operator and the second integral returns the expectation value of the related observable quantity.

$$e^{ika} = (1)^{1/N} = e^{2\pi in/N} \quad (n = 0, 1, 2, \ldots, N-1), \tag{4.24}$$

where the expression at the left is the Nth root of unity. Taking logarithms and rearranging gives:

$$k_n = \frac{2\pi n}{Na} \qquad (n = 0, 1, 2, \ldots, N). \tag{4.25}$$

This result tells us that the total number of electron wave states in the lowest energy band in the Brillouin zone is exactly N. This means that each cell of the lattice contributes one independent k-value to the energy band. This result applies to every energy band of the system and also carries over to 3-dimensional lattices. Application of the tight-binding model will result in a valence band of the form given by Equation (4.22), as illustrated by the continuous curve in Fig. 4.7. Allowing for the two spin orientations of an electron, the Pauli exclusion principle then tells us that there will be room for two electrons per state in an energy band. Hence, if each atom contributes one bonding electron, the valence band will be only half-filled with a Fermi energy equal to $(E_0 - \alpha)$ and a Fermi momentum $\pi/2a$, see Fig. 4.7, and it follows from the argument set out above that the system should be metallic. However, Peierls (1955) showed that any 1-dimensional metal is unstable against a static lattice distortion that introduces an energy gap at the position of the Fermi level in the metal. In this instance this will be the case if the lattice distorts so that the separation of the atoms is no longer equal but consists of pairs of closely spaced atoms, see Fig. 4.7(b). The unit cell now has a size $2a$ and the Brillouin zone edge shifts from π/a to $\pi/2a$, i.e. becomes coincident with the location of the Fermi level in the equal bond length structure. The system can now be regarded as a linear array of interacting dimers and the band structure will resemble that of Fig. 4.3. The dimer *molecule* energy levels will be split into a bonding state (LUMO) and an antibonding state (HOMO) separated by twice the intra-dimer interaction energy $(2E'_D)$, which then spread into bands containing $N/2$ states of width four times the inter-dimer interaction energy $(4E''_D)$, c.f. Equation (4.22). Thus an energy gap, $2\Delta = 2E'_D - (4E''_D)$, opens at the Fermi level, see Fig. 4.7. This process also lowers the energy of the system since the electrons at E_F have their energy reduced by Δ. There is an increase in mechanical energy in the system associated with the displacement of the atoms to form the dimers, although this is less than the lowering in energy of the electron system. Peierls showed that at low temperatures where the thermal energy is negligible the energy difference between the undistorted and dimerised structures diverges as $N.\log N$ as N tends to infinity. Thus, the distorted structure is always favoured and any 1-dimensional system must be a semiconductor, or insulator, in the absence of other influences. This change in structure is referred to as the *Peierls transition*. This result has particular significance for the simplest conjugated polymer polyacetylene, as discussed below.

4.2.2 Properties of semiconductors

In pure silicon in its ground state at the absolute zero of temperature the bonding electrons provided by each silicon atom just fill the valence band. Intrinsic conduction can therefore only occur when electrons are promoted across the band gap into the conduction band by some means. Then both this electron and the hole (a vacancy with an effective positive charge) in the valence band can contribute to net charge flow. Promotional energy can be obtained by direct photon absorption, or by raising the temperature, when an electron may occasionally receive sufficient thermal energy from the lattice (electron–phonon collision). Intrinsic semiconductors are, therefore, materials with energy gaps between valence and conduction bands that are not much larger than thermal energies. Materials with large energy gaps are insulators as the chances of obtaining thermally excited carriers then become extremely small.

As electrons obey Fermi–Dirac statistics, the distribution function giving the occupation probability $f(E)$ of a state of energy E at equilibrium is

$$f(E) = \frac{1}{1 + e^{(E - E_F)/kT}}. \tag{4.26}$$

The parameter E_F, called the Fermi energy, can be regarded as the chemical potential of the electrons and corresponds to an energy level where the probability of a state being occupied is 1/2. At the absolute zero of temperature all electrons fall into the set of states with the lowest energy, and the Fermi energy of a metal then becomes identical with the energy of the highest occupied state in the partially filled valence band. At higher temperatures some states above E_F become filled at the expense of states below E_F. The position of the Fermi energy is referred to as the Fermi level.

Detailed statistical analysis (e.g. see Kittel, 1996) shows that for intrinsic band conduction in a semiconductor, where the concentration of conduction electrons n always equals the concentration of holes p, and where the energy band gap E_g is wide (i.e. $E_g \gg kT$),

$$n = p = 2\left(\frac{2\pi m^* kT}{h^2}\right)^{3/2} e^{-E_g/2kT}, \tag{4.27}$$

where m^* is the effective mass of the electron or hole. Thus, as the temperature is increased, the charge-carrier concentration increases strongly with temperature. This dominates the temperature dependence of the conductivity, giving it an Arrhenius-like character with an effective activation energy of $1/2E_g$. The factor in the activation energy reflects the dynamic equilibrium of electron and hole concentrations by continual thermal generation and recombination in accord with the Law of Mass Action (compare with Equation (8.5)). The equal concentration of electrons and holes means that the Fermi level lies at the middle of the band gap.

We have introduced an effective mass in Equation (4.27) and this reflects the fact that the electron and hole are not moving in free space but in a lattice potential. In a lattice it is the group velocity of the associated wavepacket that determines the motion of an electron. In the presence of an applied electric field F the acceleration of the electron is then:

$$\frac{\mathrm{d}v_g}{\mathrm{d}t} = \frac{\mathrm{d}}{\mathrm{d}t}\left(\frac{\mathrm{d}\omega}{\mathrm{d}k}\right) = \frac{\mathrm{d}}{\mathrm{d}t}\left(\frac{1}{\hbar}\frac{\mathrm{d}E}{\mathrm{d}k}\right) = \frac{1}{\hbar}\frac{\mathrm{d}^2E}{\mathrm{d}k^2}\frac{\mathrm{d}k}{\mathrm{d}t} = -\frac{eF}{m^*}. \tag{4.28}$$

The energy gained by the electron from the field is:

$$\mathrm{d}E = \frac{\mathrm{d}E}{\mathrm{d}k}\mathrm{d}k = -eF\mathrm{d}x = -eFv_g\mathrm{d}t = -eF\frac{1}{\hbar}\frac{\mathrm{d}E}{\mathrm{d}k}\mathrm{d}t. \tag{4.29}$$

Substituting for $\mathrm{d}k/\mathrm{d}t$ from Equation (4.29) in Equation (4.28), it follows that:

$$\frac{-eF}{\hbar^2}\frac{\mathrm{d}^2E}{\mathrm{d}k^2} = -\frac{eF}{m^*}, \text{ and } m^* = \frac{\hbar^2}{\left(\mathrm{d}^2E/\mathrm{d}k^2\right)}. \tag{4.30}$$

For free electrons Equations (4.6) and (4.30) give $m^* = m$ as expected. For the tight-binding model close to $k = 0$ using Equation (4.22), expanding $\cos(ka)$ and neglecting terms beyond $(ka)^2$ gives:

$$m^* = \frac{\hbar^2}{2\beta a}. \tag{4.31}$$

Hence, the larger β the smaller the effective mass, since, as can be seen from Fig. 4.6, the gradient and the second derivative of the dispersion curve near the origin will increase as the band gets wider. This reflects the fact that the width of an energy band reflects the strength of the interactions amongst a particular set of atomic orbitals, the interaction depending on the degree of overlap of the orbitals. In the limit of $\beta = 0$ the band becomes flat, the electrons are localised on the atomic sites and the effective mass is infinite. Hence, for large interatomic separation, the overlap is necessarily poor, the band is narrow and, as one may expect, now that the perturbations from the lattice vibrations are relatively large, the carrier mobility is low. It may be shown that when the overlap is so poor that the mobility falls below $10^{-4} m^2 V^{-1} s^{-1}$, it is no longer meaningful to use band theory – the mean free path of an electron between consecutive scattering collisions is less than both the free-electron wavelength and the interatomic spacing of the lattice.

One other feature of the energy scheme, which is especially important in semiconductors, is the occurrence of levels in the normally forbidden energy zone due to the presence of impurities. Deliberate doping with elements, which give donor states near to the conduction-band edge and electron acceptor states near to the valence-band edge, can produce enormous increases in the population of conduction electrons or holes at a given temperature. This is called extrinsic conduction. Elements from group III of the periodic table (B, Al, Ga, In) have one fewer valence electron than silicon, act as electron acceptors and give rise to hole conduction, forming a *p*-type material, since the charge carriers are positive. Elements from group V of the periodic table (P, As, Sb, Bi) have one more valence electron than silicon and act as electron donors, forming an *n*-type material with electron conduction. The quantity of dopant required is small since the acceptor and donor impurity levels lie close to the respective bands, releasing charge carriers even at low temperature. Essentially all the donor and acceptor impurities are ionised at room temperature. The Fermi level will now lie between either the donor states and the conduction band or the acceptor states and the valence band for *n*- and *p*-type material respectively. For *n*-type dopants the conductivity is $10^1 \Omega^{-1} m^{-1}$ at a concentration of about $10^{21} m^{-3}$ dopant atoms (10^{-6} atomic%); it rises further to $10^3 \Omega^{-1} m^{-1}$ at about $10^{23} m^{-3}$ dopant atoms (10^{-4} atomic%) and $10^4 \Omega^{-1} m^{-1}$ at about $10^{25} m^{-3}$ dopant atoms (10^{-2} atomic%). The conductivities are slightly lower for similar levels of *p*-type dopants. Thus quite small concentrations of dopants are required to produce material suitable for incorporation in electronic devices.

In metals, where the valence bands are only partially filled even in the ground state, the effective concentration of carriers is scarcely affected by temperature. Actually, the conductivity decreases slightly with increasing temperature, because at higher temperatures the lattice vibrations scatter electrons and their mobility goes down. For the same reason mobility also decreases with temperature in the case of silicon, but the conductivity still increases with temperature, because the steep rise of carrier concentration swamps the more gradual decline in mobility.

Dopants and other impurities and lattice defects will also affect the mobility by acting as scattering centres for electrons and holes. In high purity materials scattering by lattice vibrations (carrier–phonon interactions) is the principal scattering mechanism at room temperature, but scattering by impurities becomes dominant at low temperature as the lattice vibrations are frozen out. If the localised levels associated with impurities and lattice defects are well removed from either the valence or conduction bands then they act as traps for charge carriers. Once in a deep trap the carrier is no longer able to participate in conduction. In shallower trap levels it is possible that the carrier can be thermally excited out of the trap. The effect is to slow down the movement of the carrier and reduce the carrier mobility.

Photoconduction

Exposure of a semiconductor to light or other electromagnetic radiation may produce a temporary increase in the population of free charge carriers, and the resulting extra flow of current under the influence of an applied electric field is called *photoconduction.* The photons of the radiation can interact with the semiconductor in a variety of ways to generate carriers. The simplest process is one of absorption of single photons to promote an electron directly from the valence band to the conduction band to give an electron–hole pair, thereby enhancing the concentration of intrinsic carriers. For this to be possible the photon energy must exceed the band gap and there is consequently a threshold wavelength for this type of photoconduction. The threshold is also apparent in the absorption spectrum itself, where an absorption *edge* occurs at the frequency corresponding to the onset of electronic transitions across the band gap. The generation of free carriers requires that on excitation the electron and hole have sufficient energy to overcome their Coulomb attraction; otherwise *geminate recombination* is inevitable. Once their separation is such that the electrostatic interaction is less than the thermal energy, they can easily move apart. The distance at which the Coulomb and thermal energies are equal is termed the Coulomb capture radius. If this is small, direct generation of mobile charge carriers is easy; if it is large it becomes a much less likely process. The electrostatic interaction is affected by both the relative permittivity and the carrier concentration of the material. If both quantities are large, the Coulomb interaction is screened and the Coulomb capture radius will be small. This is the situation in extrinsic semiconductors and to a lesser extent in intrinsic semiconductors, where, although the permittivity can be high, the carrier concentration will be low if the energy gap is large. Conversely, where both are low, the Coulomb capture radius will be large. However, if direct generation of carriers by photon absorption is unfavourable, generation may still proceed, but in a less direct manner. If the photoexcited electron and hole do not escape their mutual Coulomb attraction, the process of geminate recombination can

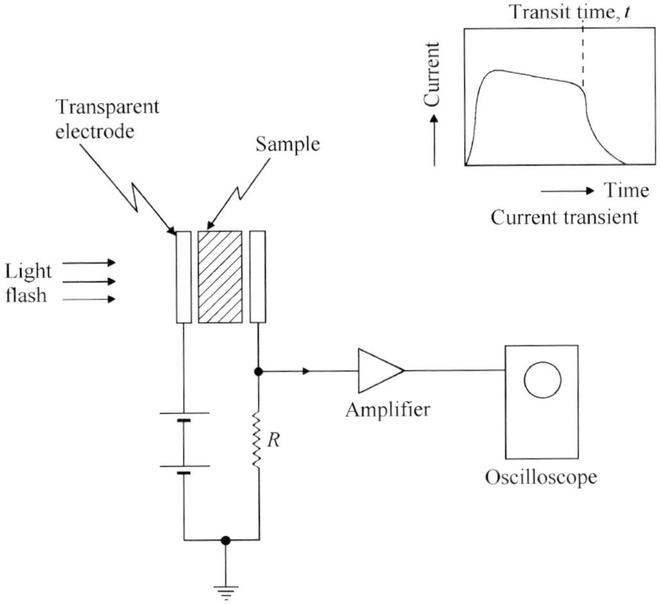

Fig. 4.8 Diagram of a pulsed photoconductivity experiment. The inset diagram shows the form of the oscilloscope trace.

lead to the production of an *exciton*. Excitons are localised, but mobile, excited electronic states. Since they consist of a bound electron and hole, they cannot by themselves transport charge, but excitons can generate unbound carriers by several mechanisms. Thus two excitons may subsequently collide to produce an electron–hole pair, the exciton fusion process, or a single, highly energetic, exciton may interact with a defect and separate into free charge carriers, exciton fission. A particular case of fission occurs when an exciton migrates to a surface and reacts with a surface state to inject a carrier of one sign into the bulk. The latter process is referred to as *photoinjection* of carriers. Such indirect processes are important in organic materials where the Coulomb capture radius can be large and the probability of direct generation of carriers is very small.

When the light is switched off, the photoconduction will decay as the carrier population gradually returns to equilibrium. By studying photoconduction kinetics it is often possible to determine the dominant mechanism of carrier loss: neutralisation at electrodes, recombination of electrons with holes, or trapping at defects or impurity centres.

Pulsed photoconductivity provides a powerful means of measuring carrier mobilities. The usual experimental arrangement shown in Fig. 4.8 (Kepler, 1960) uses a sandwich cell with a transparent front electrode. If the coefficient of absorption of the light by the photoconductor is high, a flash of light effectively generates a sheet of carriers near to the front electrode. Depending on the direction of the electric field applied across the sample, positive or negative carriers then drift towards the opposite electrode, producing a current in the external circuit; the current pulse terminates as the drifting carriers reach the electrode. The inset in Fig. 4.8 shows a typical trace of a current transient obtained on an oscilloscope on which the transit time t may be measured. Knowing the thickness of the sample d and the applied voltage V, the mobility of the carrier is then given by

$$\mu = \frac{d^2}{Vt}. \tag{4.32}$$

In some cases the drift of carriers may be seriously interrupted by capture at trapping sites in the solid. If the traps are energetically shallow, i.e. the depth of the potential well is comparable with thermal energies (kT), the carriers will soon escape again, and the only effect will be to reduce the apparent mobility. The trap-limited mobility μ_T will be given by

$$\mu_T = \mu \frac{\tau_c}{\tau_c + \tau_T}, \tag{4.33}$$

where μ is the mobility of carriers between traps, τ_c is the mean lifetime of carriers between traps, and τ_T is the mean time spent in a trap. In this case the apparent mobility will increase with temperature on account of the more rapid thermal release from the traps at higher temperatures:

$$\tau_T \propto e^{U_T/kT}, \tag{4.34}$$

where U_T is the depth in energy of the trap.

Massive photoinjection of carriers into a material where traps greatly reduce the carrier mobility gives rise to a space-charge effect analogous to that in a thermionic valve. As shown by Mott and Gurney (1948) the current density j at low fields, where the space-charge field is important, varies with applied voltage V (sample thickness d) as follows:

$$j \propto \frac{V^2}{d^3}, \tag{4.35}$$

where the current increases as the square of the voltage. In this regime the interpretation of results of pulse experiments becomes more complicated, although transit times can still be obtained – a *knee* may be discerned in the current transient as the carrier front reaches the opposite electrode, see Section 8.4.

Polarons

The presence of mobile charge species has been assumed to have no effect on the crystal lattice, the rigid lattice model. In reality there will always be some distortion of the lattice in the vicinity of an uncompensated charge. In metals it is the other free electrons that are polarised by an added charge, thus the field of the charge is screened and the effect on the crystal lattice is minimal. In intrinsic semiconductors there are many fewer free charges available to compensate the effect of an added charge, so that there will be some polarisation of the lattice. The effects will be small if the carrier is in a wide band, i.e. has a small effective mass and high velocity. However, the lattice polarisation will not be negligible if the carrier is either in a narrow band, i.e. has a large effective mass and low velocity, or if the lattice is polar, i.e. ionic, or if the carrier is localised in a trap. In these situations the lattice around the carrier becomes polarised and the carrier and the distortion move through the lattice together. This new particle is termed a *polaron*.

In a simple model of a polaron, Mott (1987) considers an electron interacting with two atoms in a simple cubic lattice. In terms of an internal co-ordinate q, e.g. the displacement of the two atoms from their equilibrium positions, the elastic energy for a displacement q is just Aq^2. The presence of the electron (or hole) will introduce an additional energy term $-Bq$. The total energy will then have a minimum when:

$$\frac{\mathrm{d}(Aq^2 - Bq)}{\mathrm{d}q} = 0, \text{ or } q_{\min} = \frac{B}{2A}. \tag{4.36}$$

At this minimum the elastic energy is increased by $Aq_{\min}^2 = \frac{1}{2}Bq_{\min}$, and the electron energy is lowered by an amount $-Bq_{\min}$, and the overall reduction in energy is $\frac{1}{2}Bq_{\min}$. Thus the electron produces a potential well, which will, in principle, always contain one localised state. However, if the depth of the well is small, the tails of the electron wavefunction extend to large distances outside the well and the electron is no longer really localised spatially in the lattice. This is the situation encountered in semiconductors with broad bands such as silicon and germanium, where polaron formation is not observed. A bound state will exist if the well is deep enough to limit the extent of the wavefunction outside the well. This will be the case for a well of width a and depth V_0 if:

$$\frac{m^* V_0 a^2}{\hbar^2} > 1. \tag{4.37}$$

In this case, the overall energy of the polaron is negative and the electron will be self-trapped. If a is equal to the lattice constant, the polaron is called a *small polaron*. The small polaron is strongly localised. It should be noted that, even in situations where mobile carriers do not form polarons, carriers in traps could polarise the surrounding lattice and form a polaronic state of lower energy. Polarons can also be formed with properties intermediate between those of the strongly localised small polaron and the unperturbed electron. These intermediate or *large polarons* have wave functions that extend over a few lattice repeat units.

As noted above, the band model fails as a model for charge transport if the bandwidth becomes too small and the mean free path of the charge carriers becomes comparable to the lattice spacing. This will also be true if the carriers become self-trapped as small polarons, if the concentration of impurities and defects leads to very high densities of deep traps, if the crystalline order is destroyed, or if a combination of these factors occurs. Charge transport is then no longer possible via extended band states but can still occur by the hopping of charge carriers between localised states.

4.2.3 Hopping conduction

Hopping conduction makes charge transport possible in the situations described above where band conduction no longer occurs. Just as carriers can be generated thermally by excitation of electrons across the band gap, from the valence band to acceptor states or from donors to the conduction band – so it is also possible to move charge between localised states by thermal excitation. Such conduction requires the electrons to execute discrete jumps across an energy barrier and through space from one site to the next (Fig. 4.9). An electron may either hop over or tunnel through the barrier, the relative importance of these two mechanisms depending on the shape of the barrier, the separation of the sites and the availability of thermal energy. To hop between sites the electron must acquire sufficient thermal energy to surmount the barrier, while for tunnelling the site separation must be small enough for the tail of the electron wavefunction to

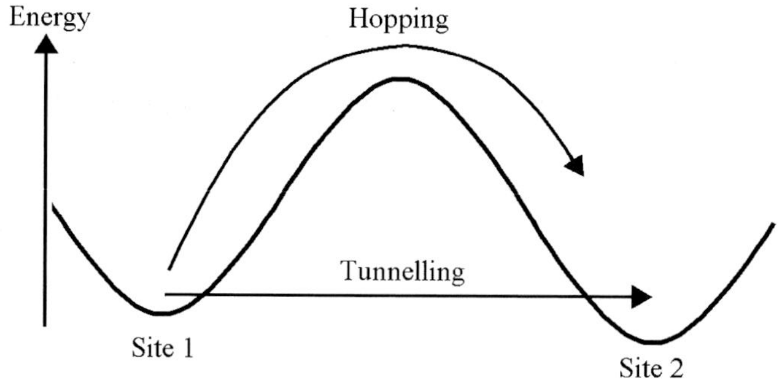

Fig. 4.9 Diagram of electron-transfer mechanisms between adjacent sites separated by a potential-energy barrier.

extend through the barrier. A concerted process, in which the electron is thermally excited sufficiently far up the barrier that tunnelling can occur, is also possible. This process becomes important for sites beyond nearest neighbours, since the probability of long distance hopping will be influenced by the spatial extent of the tails of the wavefunctions. This thermally activated type of mobility will increase with temperature, of course, in contrast with that in band conduction. It is apparent that we learn a lot about the conduction process by determination of carrier mobility, e.g. its temperature dependence provides a good criterion by which we can distinguish band and hopping conduction mechanisms. This is also true for the conductivity, as we shall see below.

In high quality crystals the density of defects will be too small for hopping to occur between defect sites on account of their large physical separation. However, if the defect levels lie close in energy to either the valence or conduction band then transitions to and from the band states are possible. In effect, the defects then act as traps reducing the overall carrier mobility from that which operates *within* the bands. As the density of chemical and structural defects increases, the Fermi level will occur in the distribution of defect states and thermally assisted hopping and tunnelling become possible. As the disorder in the lattice increases, both the energetic and spatial distribution of electronic states are affected. For a random distribution of atoms, i.e. an amorphous material, the electronic energy states spread as a continuous tail into what is normally the forbidden energy zone, and the electrons in these states are localised (Fig. 4.10). Since hopping mobility falls off so rapidly with hopping distance, there is then not so much an energy gap as a mobility gap, see Fig. 4.10(b). In other words, there is an intermediate range of electronic energy states in which mobilities are very low, even though the density of states is not zero. Only when electrons are excited to higher energy states, in which mobilities are higher, can appreciable conduction occur.

Anderson (1958) was the first to consider the effect of disorder on mobility using a tight binding model of a lattice consisting of potential wells with random depths. The random potentials scatter electrons, restricting the mean free path. If the spread of well depths, ΔV, is comparable to the bandwidth of the electron energy states for an array of identical wells, c.f. 2β in Equation (4.22), then the electron mean free path is of the order of the lattice spacing. In this situation the phase coherence inherent in the Bloch functions describing the electrons in the perfect lattice is lost and the total wavefunction is a sum over local wavefunctions with random phase:

$$\Psi = \sum_n c_n e^{i\phi_n} \varphi(r - r_n), \qquad (4.38)$$

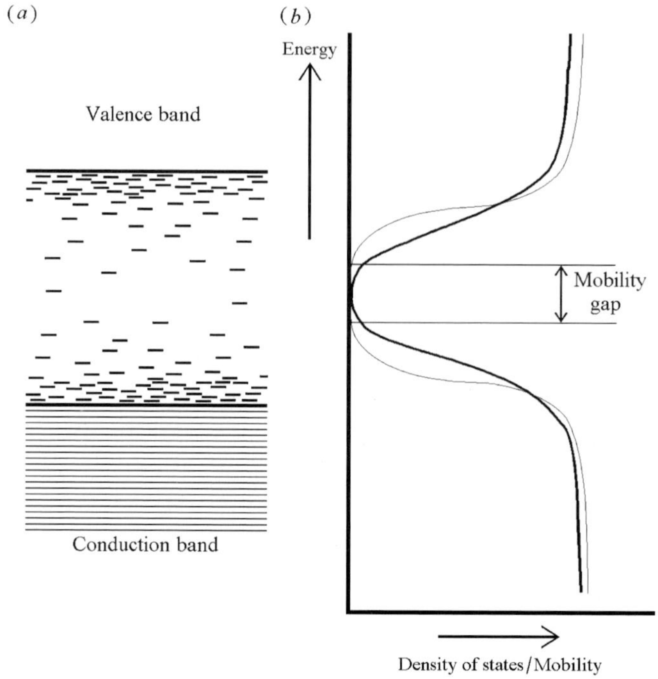

Fig. 4.10 (a) Defect states in a partially disordered semiconductor and (b) plot of the density of states, thick line, and the mobility, thin line, of a disordered solid showing the mobility gap.

where ϕ_n is a random phase and r_n is the co-ordinate of the nth lattice site, see Equation (4.12). Anderson showed that as ΔV became greater than the bandwidth the electron wavefunction took the form:

$$\Psi = \mathrm{e}^{-\{(r-r_0)/\xi\}} \sum_n c_n \mathrm{e}^{\mathrm{i}\phi_n} \varphi(r - r_n), \tag{4.39}$$

where ξ is a localisation length, which is infinite at the onset of localisation but decreases as ΔV increases. The electrons are then localised at the positions given by the values of r_0.

If the Fermi level lies above the mobility edge the electron wavefunctions are sufficiently delocalised that transport is possible and the material is metallic. If the Fermi energy lies below the mobility edge the electrons are localised and the material is insulating. In this case conduction can occur by either:

a. thermal excitation from the Fermi level to the mobility edge when the conductivity will be thermally activated with a well-defined activation energy, i.e.

$$\sigma = \sigma_0 \mathrm{e}^{-(E_\mathrm{m} - E_\mathrm{F})/kT}, \tag{4.40}$$

where E_m is the energy of the mobility edge, or

b. by thermally activated hopping if the density of states within the localised levels at the Fermi energy is non-zero.

(a) will dominate at high temperature, while (b) will become important when the thermal energy is insufficient to excite electrons above the mobility edge. At first sight it might appear that in case (b) the most likely hops would be between localised sites that are nearest neighbours. However, Mott (1968) was the first to realise that, as the range of the hop increases, the greater is the probability of the electron finding a site with a small energy barrier, and that such long-range hops depend on the extent of the tails of the electron wavefunctions. This leads to a characteristic temperature dependence for *variable range hopping*, which differs from the Arrhenius law normally anticipated for a thermally activated process. For hopping over a range R_h, the number of available sites per unit energy is just the density of states per unit energy at the Fermi energy, $N(E_F)$, multiplied by the volume of the sphere of radius R_h. The average separation in the energy of the available site is, therefore:

$$\Delta E = \frac{1}{\frac{4}{3}\pi R_h^3 N(E_F)}. \tag{4.41}$$

This is approximately the minimum energy barrier that the electron will encounter for a range R_h. The probability of hopping will then be a Boltzmann factor determined by ΔE and a tunnelling factor given by the amplitude of the electron wavefunction at a distance R_h from the original site. The optimum hopping distance is then given by the maximum of the probability function:

$$\Phi = C e^{-2\alpha R_h} e^{-\Delta E/kT}, \tag{4.42}$$

where $1/\alpha$ is the decay length of the wavefunction. This will occur when the sum of the two exponents has a minimum value, i.e. when:

$$2\alpha = \frac{1}{4\pi R_h^4 N(E_F)kT}, \quad \text{or } R_h = \left(\frac{1}{8\pi N(E_F)\alpha kT}\right)^{\frac{1}{4}}. \tag{4.43}$$

Thus the hopping probability and hence the conductivity has the form:

$$\Phi = A e^{-B/T^{\frac{1}{4}}}. \tag{4.44}$$

Experimentally, the temperature dependence of the exponent has been found to take the form T^{-n} rather than $T^{-1/4}$, where n lies between one half and one quarter. This distinct temperature dependence is a signature of hopping conduction and distinguishes it from conduction due to the thermal excitation of carriers across an energy gap.

In a semiconductor, as discussed in the previous section, localisation can also occur as the width of the allowed energy band is reduced, and this was defined in terms of a limiting mobility. The Anderson model shows that disorder can lead to localisation in metals as well as semiconductors. In metals, since conduction is due only to electrons within a partially filled band, the energy in the band tail that separates localised from delocalised electron states is termed the mobility edge. The onset of localisation in a metal occurs at a minimum conductivity. This can be seen as follows. For an electron at the Fermi energy its mean free path, l, is just the scattering time, τ, multiplied by the electron velocity at the Fermi energy, v_F. Then, from Equations (4.1) and (4.2) it follows that:

$$\sigma = \frac{ne^2 l}{mv_F} = \frac{ne^2 l}{\hbar k_F}, \tag{4.45}$$

where k_F is the wavevector at the Fermi energy. The electron concentration can be written in terms of k_F as the total volume in k-space contained within the spherical Fermi surface divided by the space occupied by each electron state, which is $8\pi^3$. Allowing for spin degeneracy, this gives:

$$\frac{n}{2} = \frac{(4\pi/3)k_F^3}{8\pi^3} \text{ and } \quad \sigma = \frac{k_F^2 e^2 l}{3\pi^2 \hbar}. \tag{4.46}$$

Setting l equal to the lattice spacing a and $k_F a \cong \pi$, which follows since $n = 1/a^3$, the limiting conductivity at the onset of localisation is:

$$\sigma_{min} = \frac{e^2}{3\hbar a}. \tag{4.47}$$

A more complex relation is found when multiple scattering and the reduction of the density of states in the energy band by disorder are taken into account (Mott, 1987). This is usually reduced to the simpler form of Equation (4.47) with the factor 1/3 replaced by a constant usually taken to be ~0.03.

4.2.4 *The metal–insulator transition*

We have seen in the previous section that disorder results in the localisation of charge carriers and that the conductivity will fall as a consequence of this. There is a minimum metallic conductivity, which corresponds to the electron mean free path being equal to the lattice repeat distance. The occurrence of the mobility edge means that in an amorphous metal the conductivity can switch from that of an insulator, when $E_F < E_M$, to that of a metallic conductor, when $E_F > E_M$. This is an example of a metal–insulator transition. In this instance the transition occurs as the level of disorder is increased. There are other circumstances where a similar switch in conductivity can occur.

The Peierls distortion in a 1-dimensional metal leads to a lower energy semiconducting state at low temperature. The energy gained by opening up an energy gap at the Fermi level arises as a consequence of the difference in interaction energy between the closely spaced pairs of atoms, or dimers, and between the dimers. As discussed in Section 4.2.1, this leads to an energy gap $2\Delta(= 2E'_D - 4E''_D)$. As the temperature of the system is raised, the mean separation of the atoms in the dimer will increase as stretching vibrations are thermally excited. This will lead to a gradual reduction in the separation of the valence and conduction bands as E'_D is reduced. There will also be an increase in the bandwidths as E''_D increases. The band gap will therefore be reduced as the temperature is increased. When the thermal energy is of the order of Δ, the band gap closes and the system reverts to the metallic state. Hence, in systems where a Peierls distortion occurs a semiconductor–metal transition occurs at a characteristic transition temperature. Since the conductivity at low temperatures is usually very small, this transition is also referred to as an insulator–metal transition. This transition is accompanied by a crystallographic phase transition as the lattice reorders from a dimerised structure to one with equal separations. The transition occurs as a consequence of the electron–phonon interaction in the 1-dimensional system. It has been observed in a number of organic charge transfer salts. In these compounds planar, aromatic, molecular ions form stacks in which the π-electron orbitals orthogonal to the plane of the ions overlap, allowing extended electron states to form along the stacking axis. The strength of molecular interaction within the stacks is such that the insulator–metal transitions typically occur in the range 30–300 K. These materials are discussed briefly in Section 8.3.2.

Metal–insulator transitions can also occur as a result of the electron–electron interaction. In the discussion of band theory in Section 4.2.1, the usual one-electron model has been presented in which the repulsive interaction of two electrons, usually referred to as *electron correlation*, has been ignored. This means that a given lattice site is equally likely to be occupied by either one or two electrons. For a 1-dimensional lattice this means that there is no difference in energy between the configurations where there is one electron per site and that where one site is empty and the neighbouring site is doubly occupied, as shown in Fig. 4.11(a) and (b). If the Coulomb repulsion of the two electrons on the same site is taken into account, there will be a finite energy difference between these configurations. This means that the ground state of the 1-dimensional

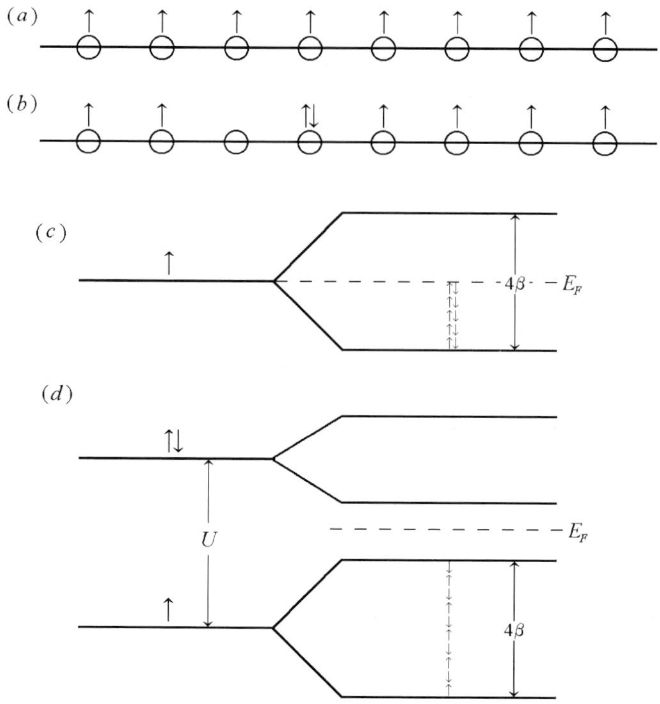

Fig. 4.11 Configurations of a 1-dimensional metal with (a) one electron per site and (b) one doubly occupied site. (c) and (d) show the band schemes when $U \ll 4\beta$ and $U \gg 4\beta$, respectively.

metal will be insulating at low temperature, as finite activation energy is required to move electrons between sites. Hence inclusion of electron correlation has the same effect as the static lattice distortion produced by the Peierls transition. In real materials both electron–phonon and electron–electron interactions will occur and either or both may play a role in determining the actual behaviour of the material. The effects of electron–electron interaction are described in terms of the *Hubbard U* (Hubbard, 1964). This parameter is a measure of the electrostatic repulsion of two electrons on the same site. The electron correlation has no effect on the 1-dimensional metal if U is much less than the bandwidth 4β, see Equation 4.22 and Fig. 4.11(c). However, if U is much larger than 4β, the interaction leads to a band scheme, shown in Fig. 4.11(d), similar to that shown in Fig. 4.3, and the right-hand side of Fig. 4.7. The Hubbard U gives rise to the initial splitting of the configurations with one electron per site and one site doubly occupied. As the lattice repeat distance is unchanged the band-width of the lowest band remains the same, but the number of available states is halved so the band is full rather than half-filled. The Fermi level now falls in the band gap rather than at mid-band. Variation in the value of U relative to the bandwidth will result in a transition from a metallic to an insulating state when U is equal to the bandwidth. This transition is known as the *Mott transition*. In the insulating state, assuming that the filled and empty bands in Fig. 4.11(d) have equal width, the activation energy is given very approximately by $(U-4\beta)$. Charge transfer between adjacent lattice sites will occur at a slightly lower energy than $(U-4\beta)$, as the repulsion energy will only fall to zero when the electrons are separated by many lattice repeat units. For electrons on adjacent lattice sites the repulsion energy is V_{nn}, and the energy required to transfer an electron to an adjacent site is approximately $[(U-4\beta-V)_{nn}]$.

The concept of the Mott transition was first introduced in the context of impurity bands in semiconductors (Mott, 1949). In lightly doped semiconductors the Fermi level lies in the gap between either the valence band and the acceptor levels or the donor levels and the conduction band. As the concentration of dopants is increased, interaction between the dopants causes the impurity levels to spread out into a band and the Fermi level occurs in this impurity band. The activation energy for conduction in such a heavily doped semiconductor should be zero, and the conductivity finite at absolute zero. However, inclusion of electron correlation will split the impurity band into lower and upper Hubbard bands, as shown in Fig. 4.11. In general these bands will be of unequal width. The system will undergo a Mott transition from an insulating to a metallic state when the Hubbard bands start to overlap, that is when U is of the order $(W_u + W_l)/2$, where W_u and W_l are the bandwidths of the upper and lower Hubbard bands. With increased doping the semiconductor should switch from an insulating (semiconducting) to a metallic state as the level of doping increases, rather than have a gradually increasing conductivity as the carrier concentration increases. No experimental evidence for such a transition in a doped semiconductor has ever been observed. This is attributed to the disorder present in the distribution of dopants in the semiconductor, which leads to localisation of the type described by Anderson, and the transition will be controlled by the relative positions of the Fermi level and the mobility edge, as described in the first paragraph of this section. It can be shown (Mott, 1987) that the transition is still determined approximately by U but it will be a continuous rather than a discontinuous transition.

Although direct evidence of the effects of electron correlation has not been forthcoming for inorganic semiconductors, it has proved to be an important factor in organic materials. In organic charge-transfer salts that have complete charge transfer between the anions and cations, electron correlation results in the appearance of a band gap and the materials are insulators. If the charge transfer is not complete, the materials are metallic since electrons can transfer to adjacent neutral molecules without incurring an energy penalty. These materials are metallic but subject to the Peierls instability, as discussed above. Electron correlation is an important factor in determining the electronic properties of polymers with either saturated or conjugated backbones, as discussed below and in Chapter 9.

4.2.5 Band theory applied to polymers

The essential new feature here is that groups of atoms are chemically bonded together in discrete molecules, which in turn are held together by relatively weak van der Waals forces. We must therefore distinguish intra- and intermolecular types of electronic motion. We can immediately foresee that transfer of electrons from one molecule to another may be the main stumbling block for good conduction on the macroscopic scale. In a system of long polymeric molecules one could assert with some justification that the intermolecular conduction problem is much less important, because relatively few intermolecular transfers are required. In applying the band model to polymers we are concentrating our attention on the intramolecular motion of charge carriers.

If we regard each molecule for the moment as a 1-dimensional lattice we can make the following initial assumptions:

a. A definitely spaced series of atoms with fixed distances between nearest neighbours. (Unless the molecules are in a crystalline phase, considerable twisting about bonds may be present.)
b. Small separation of atoms giving good overlap of atomic orbitals.
c. A full valence shell, analogous to a full valence band.
d. A relatively large excitation energy to the lowest excited electronic state, corresponding to strong chemical binding.

In the case of a polymer with a saturated chemical structure, such as polyethylene, the strength of σ-bonding is such that the band gap will be comparable to that in diamond. However, for a polymer with a conjugated structure, such as polyacetylene, the chemical binding of the π-electrons is much weaker and a gap of a few eV, comparable to those in inorganic semiconductors, is anticipated.

There are a number of different theoretical approaches to the calculation of the band structures of polymers. These are extensions of the methods employed to calculate the electronic states of molecules, which obtain molecular orbitals from a linear combination of atomic orbitals. In this case the states in question are those of an infinitely long molecule, which is approximated by a finite length system with cyclic boundary conditions, i.e. the right-hand end of the chain is, in effect, joined to the left-hand end of the chain. This is the method used for band structure calculations of crystals. Alternatively, the electronic states of chains with n repeat units, oligomers, can be calculated and extrapolated to the infinite repeat unit limit. This approach is limited by the maximum value of n that can be utilised in the calculation, but has the advantage that the results can be compared with experimental data for a number of oligomers as well as for the polymer. As noted in Section 4.2.1, the basic approach in either case is that of *Hartree–Fock self-consistent field theory* (HF–SCF). The use of this method for molecules and polymers is not a simple problem, and so some approximations must be introduced. The calculation methods then fall into two general classes, the semi-empirical and *ab initio* methods.

In semi-empirical methods experimental data are used as a basis for the approximation of the integrals that occur in the quantum mechanical problem, and that are treated as parameters. A wide variety of approximations are used, affording different levels of simplification. The *Hückel* method is applicable to conjugated structures and considers one π-electron for each carbon atom and the interactions between these electrons; the σ-electrons are simply considered to contribute to the background potential. *Extended Hückel theory* (EHT) incorporates all valence electrons, taking into account the $2s$ and $2p$ electronic states of carbon atoms and the $1s$ state of hydrogen atoms. The *valence effective Hamiltonian* (VEH) method incorporates all valence electrons but uses a model Hamiltonian to simulate a particular *exact* Hamiltonian. The approximations made in these three approaches mean that they are not self-consistent field models, since the underlying potential is determined from the experimentally observed molecular structure. The *Pariser–Parr–Pople* (PPP) method uses the SCF approach with a combination of approximation and parameterisation to simplify the problem, but is again limited to consideration of π-electrons. The principal approximation in the PPP method is that the overlap between atomic wavefunctions on adjacent atoms is zero. *Complete neglect of differential overlap* (CNDO) makes the same approximation for all integrals involving two electrons but includes all valence electrons. The approximation is relaxed in the *intermediate neglect of differential overlap* (INDO) and *modified neglect of diatomic overlap* (MNDO) methods. The choice of initial wavefunctions, i.e. the basis set of atomic orbital functions, leads to variants of these methods. Alternatives to hydrogenic orbitals are Slater orbitals with an exponential radial dependence and Gaussian orbitals with a Gaussian radial dependence. Slater orbitals represented as a combination of Gaussian orbitals (STO–3G) are commonly used. Extensions of this approach use more that one basis function for each atomic orbital. The CNDO, MNDO and INDO methods are widely used for calculations of the ground state geometry of molecules, i.e. the structure with minimum energy, as well as the energies of the electronic states and other properties.

The *ab initio* approach includes all σ- and π-electrons and seeks to use either analytical or numerical solutions to the integrals that occur in the quantum mechanical problem. This procedure was initially carried out within the framework of the one electron HF–SCF method using the basis sets described above. Subsequently it has been implemented using *density functional theory* (DFT). If the electron density in the ground state of the system is known, then in principle this knowledge can be used to determine the physical properties of the system. For instance, the locations of the nuclei are revealed by discontinuities in the electron density gradient, while the integral of the density is directly related to the number of electrons present. *Ab initio* methods are obviously computationally intensive.

In general terms the semi-empirical methods offer fast computation. Many of the established routines can be run in real time on a PC, but are limited in applicability. This results from the fact that the parameterisation employed approximates different quantum mechanical integrals for different physical properties, i.e. the parameters are not transferable. *Ab initio* methods are more accurate and have greater generality, but are much more computationally intensive. While this was a problem in the past, the availability of more powerful, modern computers has made possible their more widespread use for molecular systems. A difficulty encountered with single electron theories is that, although the overall form of the bands is accurately predicted, the calculated values of bandwidths and band gaps differ significantly from the experimental data. In particular, the HF theory using either semi-empirical or *ab initio* approaches overestimates energies, while DFT underestimates them. Some success in improving theoretical estimates has been achieved using the *tight-binding* approach. This has been principally implemented for the modelling of crystalline materials where the computational cost of *ab initio* methods is particularly high. The method uses an atomic-like basis set, which retains the quantum mechanical aspects of bonding, but the exact many-body Hamiltonian is replaced by a parameterised Hamiltonian matrix. It is usually employed without attempting to achieve self-consistency. The quality of the parameterisation is directly reflected in the quality of the calculation.

An important defect of all the single electron calculations is the neglect of electron correlation. Various methods have been employed to rectify this problem. Its inclusion using second order perturbation theory has been found to provide much better agreement between theory and experiment in some instances. The inclusion of electron correlation has a profound effect since the electrons and holes in the first excited state are bound by their Coulomb interaction to form a localised state, an exciton. This state is separated from the conduction band by the exciton binding energy. That this model is appropriate for the description of conjugated polymers is strongly supported by experimental evidence, as discussed in Chapter 9.

Saturated backbone polymers

Polyethylene has been the focus of most studies of saturated backbone polymers because of its simple chemical structure and its technological importance. On the basis of the assumptions set out above, McCubbin and Gurney (1965) applied band theory to a single, long polyethylene molecule in the planar zigzag conformation found in the crystalline phase (Fig. 4.12). Although these calculations predicted the anticipated very wide band gap (greater than 5 eV), they also gave an estimated carrier mobility of about $5 \times 10^{-3} \, m^2 V^{-1} s^{-1}$ (for holes), which is comparable to that in metals and substantially larger than that observed experimentally (Fig. 4.2). So far it has proved impossible to prepare a high quality single crystal of polyethylene with the degree of chain regularity required to preserve the band structure and avoid trap formation.

Since these first calculations, a variety of theoretical methods have been studied and applied to ideal chains of polyethylene and other polymers. The books by Ladik (1988) and André *et al.* (1991) describe theoretical methods employed, and the literature published prior to 1990. While these calculations provided some insights into the microscopic origins of the electronic properties of the polymers modelled, the correlation of predicted properties, e.g. the band gap and bandwidths, with experimental data was poor. The quality of the experimental data was also variable.

Experimental data can be obtained by ultra-violet absorption spectroscopy, electron energy loss spectroscopy and photoelectron spectroscopy. UV absorption and EELs have been described briefly in Chapter 3. The former provides information only about the band-gap, while EELs gives more general information about the conduction bands. Both X-rays and UV photons can be used to generate photoelectrons: these two methods are given the acronyms XPS and UPS. The energy spectrum of the emitted electrons provides information about the density of electron states in the valence bands. In principle the size of the band gap can be obtained, but care must be taken as the absolute energy

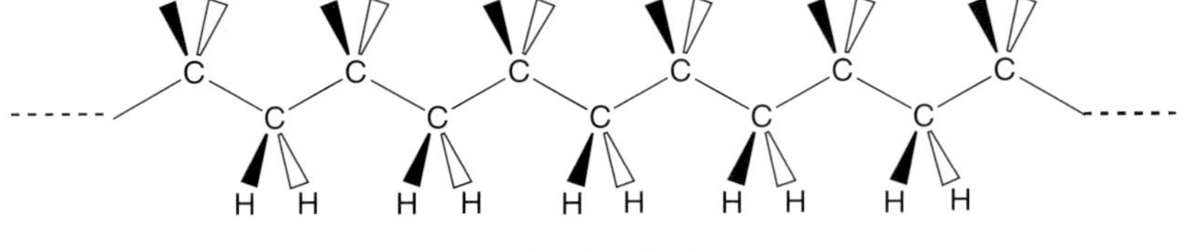

Fig. 4.12 Chemical structures of polyethylene.

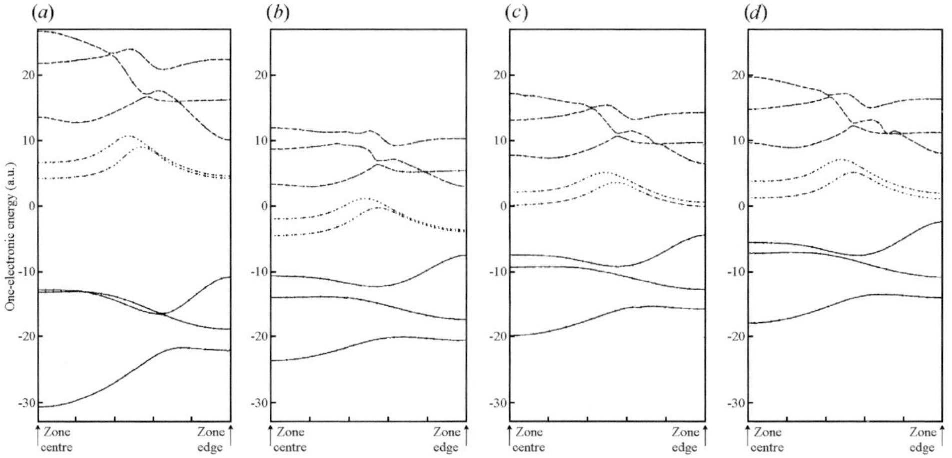

Fig. 4.13 Theoretical band structures for polyethylene, solid lines: filled states (valence bands), dashed lines: empty states (conduction bands); for details see text. Reproduced with permission of the American Institute of Physics from Süle *et al.* (2000).

of the photoelectrons is affected by charge accumulation on the surface of insulating samples. XPS and UPS have been more extensively employed than UV spectroscopy and EELs. In consequence there are more data available on the electron densities of states for the valence bands than for the conduction bands. In the case of polyethylene there is also little precise data on the value of the band gap. The situation is complicated by the complex morphology of polyethylene samples which gives rise to the presence of chain defects. Because of the non-ideal nature of real polymer samples, experimental studies have also employed oligomers, e.g. the *n*-alkanes have been used as models of polyethylene. The *n*-alkanes are crystalline solids in which the molecules are fully extended and defect free. Thus, they provide model systems for comparison with calculations made for ideal defect free chains. Seki *et al.* (1986) have provided a summary of XPS and UPS data for polyethylene and *n*-alkanes.

Recent calculations of the band structure of polyethylene have employed variations of the *ab initio* method incorporating electron correlation. Sun and Bartlett (1996) utilised many-body perturbation theory to encompass electron correlation in the *ab initio* framework. Süle *et al.* (2000) and Serra *et al.* (2000) have employed variants of DFT. These calculations involved the optimisation of local effective potentials and a local-density approximation respectively. Figure 4.13 shows a comparison of the band structure obtained by Süle *et al.*, Fig. 4.13(d), with those obtained by other *ab initio* DFT calculations using the Hartree–Fock (HF), Fig 4.13(a), and Slater approaches, Figs. 4.13(b) and (c).

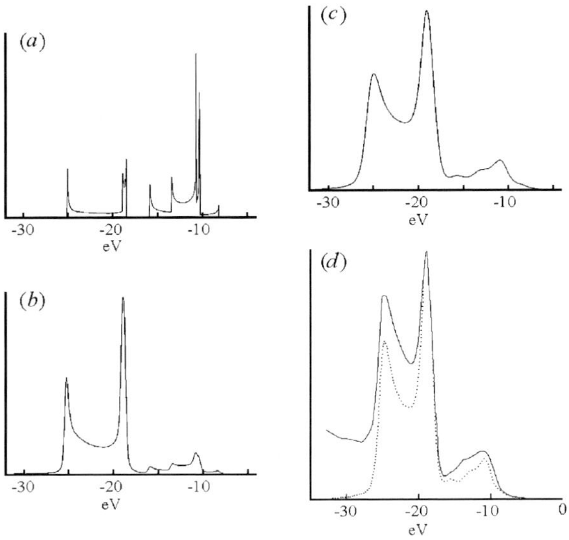

Fig. 4.14 (a) Calculated density of states for polyethylene, (b) and (c) calculated photoelectron spectrum with instrumental half-widths of 0.2 and 0.75 eV respectively, and (d) comparison of the calculated spectrum (c), dotted line, and the observed XPS spectrum, solid line. Reprinted with permission from Sun and Bartlett (1996). Copyright 1996 by the American Physical Society.

While the general shapes of the bands as a function of wavevector are similar, the HF approximation gives a much larger band gap and greater widths for the valence and conduction bands. The band gap and bandwidths are very similar for the other calculations, but the Slater approach produces an overall shift in the energy of the bands. The valence band structures obtained by Sun and Bartlett are also very similar to those shown in Fig. 4.13, and the densities of states they obtain are in good agreement with XPS photoelectron spectra. The density of states must be convoluted with a broadening function simulating the experimental resolution. The effect of changes of the half-width of this function is illustrated in Figs. 4.14(b) and (c). The width and shape of the theoretical spectrum generated with the larger half-width is in excellent agreement with the experimental data, as shown in Fig. 4.14(d). This calculation also gives good alignment of the energy of the calculated and observed features. As can be seen from Fig. 4.13, this is often not the case, and although a match between theory and experiment similar to that shown in Fig. 4.14 (d) can be obtained, this requires an arbitrary shift of the calculated spectrum. Thus, agreement of the theoretical and experimental ionisation potentials and band gaps in not so good.

Unfortunately, there is some uncertainty as to the value of the band gap in polyethylene. Measurements in the UV show an absorption peak at 7.6 eV and an onset of photoelectron emission at either 8.8 or 9.1 eV (Bloor, 1976). Ueno *et al.* (1986) have obtained poorly defined EELs spectra for polyethylene and find broad peaks at *c.* 7 and 11 eV for n-alkanes. They conclude that the conduction band lies at 0.5 ± 0.1 eV above the vacuum level in crystalline polyethylene but falls below the vacuum level in molten polyethylene. They assert that this will also be true for amorphous polyethylene. Serra *et al.* (2000) and Righi *et al.* (2001) have addressed this problem theoretically. They perform calculations for crystalline polyethylene and show that, while the valence bands are insensitive to the inter-chain spacing, the conduction bands are strongly affected. Thus, the valence bands have a predominantly intra-chain character and the conduction bands an inter-chain character. Calculation by Righi *et al.* for two crystals separated by a narrow gap shows the vacuum level falling below the conduction band edge and that there are surface states $\sim$1.2 eV below the conduction band edge. These results show that electrons in crystalline polyethylene have a negative electron affinity. A further consequence is that electrons in real polyethylene samples will be expected to migrate to regions of low density, i.e. amorphous regions, as the conduction band states are lower in energy in such regions of the material. The exact nature of the absorption in the UV remains an open question. The surface states have an appropriate energy, but it is not clear whether there are enough surface states to produce the strong UV absorption and the peak in the EELs data. The calculated inter-band transition appears to lie too high in energy, leaving the possibility that there is an exciton level in the band gap. However, Hirata *et al.* (1999) calculated the exciton energy to be greater than the calculated band gaps, at odds with the experimental data.

While there have been extensive experimental studies of other saturated polymers, they have been subjected to less intensive theoretical study. The experimental data are predominantly valence band spectra and the features observed can be interpreted using the theoretical techniques set out in the books by Ladik and André *et al.*, e.g. the analysis of UPS data for perfluorotetracosane performed by Miyamae *et al.* (2000).

Conjugated backbone polymers

The earliest band theory calculations for conjugated polymers were focussed on *trans*-polyacetylene (*t*-PAc). It was the first system in which metallic levels of electrical conductivity were observed (Shirakawa *et al.*, 1977) and has by far the simplest molecular structure of this class of polymers. There are two possible structures, either:

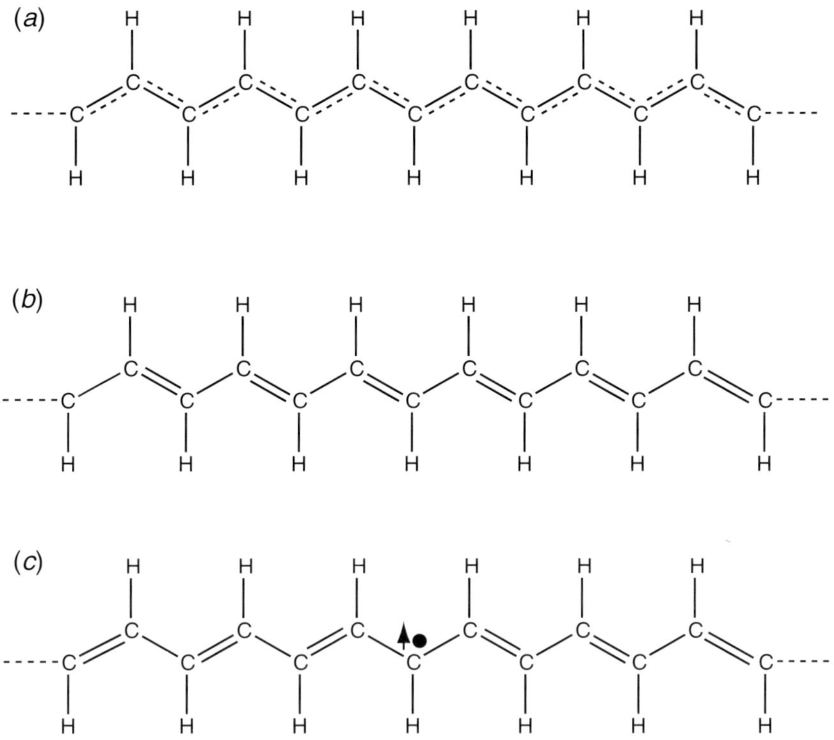

Fig. 4.15 Chemical structures of *trans*-polyacetylene (a) with equal bond lengths, (b) bond alternation and (c) a bond alternation defect.

a. with backbone C–C bonds of equal length and a two-fold screw axis in the direction of the polymer backbone, as shown in Fig. 4.15(a), or
b. with alternating C–C bond lengths and a mirror plane in the plane of the polymer backbone, Fig. 4.15(b).

The question of whether the bond lengths were equal or alternated in an infinite *t*-PAc chain was first answered satisfactorily using a *Hückel* model nearly twenty years before the discovery of metallic behaviour in PAc (Longuet-Higgins and Salem, 1959). This work was carried out at a time when the Hückel model was being used to model the evolution of the optical spectra of the polyenes, i.e. polyacetylene oligomers, as the length of the molecules increased.

As noted above, the Hückel model assumes that the underlying σ-electrons in the polymer backbone provide a framework that supports the π-electrons and it is the latter which give rise to the molecular orbitals that determine the optical and electrical properties of the polymer. Structure (i) has one π-electron per carbon atom, and if the extended chain is modelled by an equivalent 1-dimensional lattice this gives one electron per lattice site. Thus, structure (i) corresponds to the model discussed in Section 4.2.1 of a 1-dimensional metal. An equal bond length structure results in π-electron wavefunctions that extend throughout the polymer backbone and the electrons are able to move freely along the polymer chain. However, structure (i) will be unstable against a Peierls distortion, which introduces shorter and longer bonds and doubles the unit cell length. This structure localises the π-electrons in the double bonds in structure (ii). An energy gap is introduced into the electronic states which lowers the energy of the π-electrons. This is achieved at the expense of the elastic energy required to distort the equal length σ-bonds. The energy gained by lowering the π-electron energy is greater than that needed to distort the σ-bonds. On this basis the more stable structure of *t*-PAc is expected to be the bond-alternated structure (ii). *t*-PAc is, therefore, predicted to be a semiconductor with a finite energy gap between valence and conduction bands. In this model the band states in *t*-PAc can be visualised as evolving from the molecular states of polyene molecules as the length of the molecules increased. The highest occupied and lowest unoccupied π-electron molecular orbitals (HOMO and LUMO levels) eventually broaden into the valence and conduction bands following the Pauli exclusion principle, as shown in Fig. 4.3.

Another concept introduced in the 1960s was that of the *bond-alternation defect* (Pople and Walmsley, 1962). In a finite *t*-PAc chain where the terminal repeat units are forced by the terminating groups to be single bonds, two situations can occur. Either (a) the molecule is of the correct length to accommodate the bond-alternating structure, Fig. 4.15(b), or (b) there is a mismatch between structures propagated from either end of the molecule. The only structure that can accommodate the discontinuity resulting in the backbone structure, while keeping the molecule electrically neutral, is that shown in Fig. 4.15(c), where there is a carbon atom with two single C–C bonds. The carbon at this *bond-alternation defect* has one unpaired π-electron and the defect is paramagnetic. Electron spin resonance studies of PAc and other conjugated polymers carried out at this time showed the presence of unpaired electron spins and it was thought that these could arise from bond alternation defects. This model was revived in a modified form after the discovery of the metallic behaviour of PAc. It was realised that the dramatic change on bond length in the original bond-alternation defect, Fig. 4.15(c), required too much energy for such defects to be thermodynamically stable. But if the changes in bond lengths between the two conflicting bond-alternation structures occurs gradually, over several repeat units, the energy required to distort the σ-bonds is greatly reduced. The extended defects were termed *solitons*, since the equation describing the extended defect is similar to that for a solitary wave. These extended defects were predicted to exist in *t*-PAc and to be able to move along the polymer backbone.

More sophisticated molecular orbital models, described earlier in this section, have been employed to model the properties of PAc and other conjugated polymers. Although most of these studies either excluded electron correlation or treated it as a negligible perturbation, its potentially significant role was recognised by some (Pugh, 1973), since it could contribute to the band gap and could also result in the lowest excited states being excitons. Importantly, the energy required to form an exciton, which may be regarded as a bound electron–hole pair, is less than that to generate a free electron–hole pair by excitation across the band gap. Although the band picture without electron correlation was initially taken to be the best model of the electronic properties of conjugated polymers, experimental evidence gradually mounted against this. It is now generally acknowledged that electron correlation cannot be neglected when modelling the behaviour of conjugated polymers, and that excitons play an important part in determining their electronic behaviour.

A more detailed discussion of the various theoretical models that have been used to describe conjugated polymers will be given in Chapter 9, Section 9.3, alongside the discussion of the physical properties.

4.2.6 Superconduction

The subject of electrical conduction in organic compounds cannot be complete without mention of superconduction. Superconduction was first observed in metals by Kamerlingh Onnes in Leiden in 1911, soon after his success in liquefying helium. The phenomenon occurs when certain metals are cooled below a characteristic transition temperature where they spontaneously enter what is known as the superconducting state. In this state electrons can flow without any resistance at all. The technological potential of this is immediately apparent – lossless power transmission and powerful electromagnets would be obvious applications. The main snag is that the critical temperature for superconduction in metals and metallic alloys is always below *c.* 20 K. Little, 1964 presented a model for superconduction in polymers that offered the possibility of superconduction up to very high temperatures, well above room temperature. Although Little's theoretical prediction has never been realised in practice, much higher transition temperatures have been obtained in both organic and inorganic materials. The first organic superconductor was discovered in 1979; this was a selenium organic salt and had a transition temperature, T_c, of 2 K under a pressure of 12 kbar (Bechgaard *et al.*, 1980). Since then, many organic salts have been discovered that superconduct at ambient pressure, culminating in the fullerene (C_{60}) salts with transition temperatures as high as 40 K (Williams *et al.*, 1992). As noted above, partial charge transfer between stacks of aromatic anions and their accompanying cations leads to metallic conduction along the stack that is mediated by the overlap of the π-electron orbitals along the axis of the stack. Superconductivity in these materials is, therefore, not unexpected. Much higher transition temperatures, over 100 K, are possible in complex copper oxides, although a comprehensive theoretical model of superconduction in these materials has yet to be formulated.

The first polymer to show superconductivity was the inorganic polymer poly(sulphur nitride), $(SN)_x$. This is normally metallic and becomes superconducting at the rather low temperature of 0.26 K (Greene *et al.*, 1975). This polymer is obtained by the ring opening polymerisation of S_2N_2 in the solid state and can be obtained as high quality crystals containing extended polymer chains. The sulphur atoms on adjacent chains are separated by 0.355 nm. This distance is short enough to ensure strong inter-chain interaction. This is sufficient to suppress the Peierls transition, i.e. the polymer crystal is a 3-dimensional solid, not an assembly of non-interacting 1-dimensional chains. It is apparently a unique material as no analogous compounds with similar properties have ever been discovered.

The central feature of the generally accepted model of superconduction, which has been shown to hold for all but the high temperature oxide superconductors, is the occurrence of Cooper electron pairs (Bardeen, Cooper and Schrieffer, 1957). Normally, electrons repel each other because like charges repel, but an effective attraction can develop between conduction electrons in certain metals at low temperature through distortion or polarisation of the atomic lattice (Fig. 4.16(a)). An electron moving from left to right polarises the lattice around itself by its attraction for the metal atom cores, which carry net positive charges. As the electron moves on, the lattice polarisation dies away but slowly; this trailing polarisation will interact with another electron moving from right to left, and at low temperatures this is sufficient to provide an effective coupling between electrons. The interaction will be strongest with an electron with the trajectory shown by the dotted arrow in Fig. 4.16(a). The duration of interaction with the decaying lattice polarisation will be much smaller for electrons moving in other directions, e.g. from top to bottom. The coupling is, therefore, somewhat unusual in that it occurs between electrons with equal and opposite momenta. Thus the Cooper pairs appear to have zero momentum.

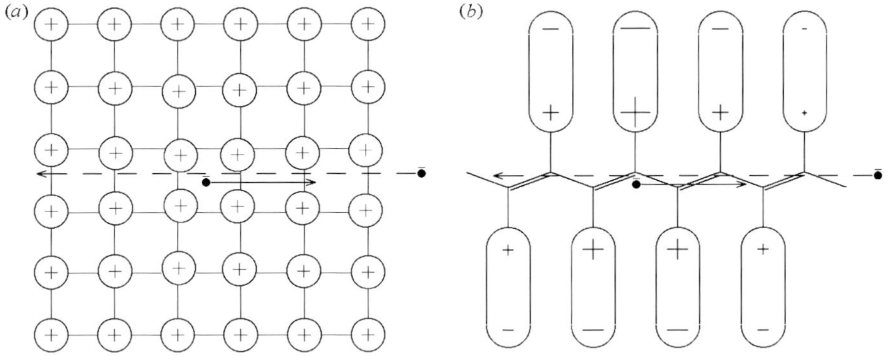

Fig. 4.16 Diagrams of (a) Cooper pair formation via lattice polarisation and (b) Little's model of pair formation by side-group polarisation.

The range of interaction is also large, being of the order of 10^{-6} to 10^{-7} m. The coupling is mediated by the lattice distortion, and in the language of quantum mechanics is via the exchange of virtual phonons between the electrons. Since phonon (lattice vibration) frequencies are inversely proportional to the square root of the mass of the positive ions of the lattice, so is the strength of the electron coupling in superconductors. Thus, if the ions are heavy, the displacement caused by a passing electron is small and vice versa. Experimentally, this relation is observed as an inverse square root dependence of the transition temperatures of metallic superconductors on the isotopic mass. The binding of electrons into Cooper pairs opens up an energy gap at the Fermi level with a width of the order of $3.5\,kT_c$. Below the gap the electrons form a sea of Cooper pairs able to move without scattering, i.e. with zero resistance. Above the gap the pairs are destroyed and the electrons behave normally. Thus, external influences that destroy the gap in the density of states, e.g. magnetic fields and high temperature, destroy the superconductivity. The fact that most superconducting transition temperatures are low reflects the strength of the interaction forming the Cooper pairs and the small energy gap that it produces, which is typically a few meV. This explains the ease with which superconductivity is destroyed in most materials.

Little envisaged a conjugated polymer with highly polarisable side groups, similar to the polar non-linear optical molecules described in Chapter 3, attached regularly along the chain. He suggested that an electron moving along the polymer chain would polarise the electron cloud of the side groups, Fig. 4.16(b). He argued that the residual polarisation would interact with an electron counter-propagating along the chain, giving rise to an electron–electron interaction mediated by the polarisation, leading to the formation of Cooper pairs. The conceptual similarity to the model of Cooper pair formation via the electron–phonon interaction is apparent by comparison of Figs. 4.16(a) and (b). Taking into account that the electronic mass is much smaller than any atomic mass, the coupling might be expected to be very large indeed in comparison with the lattice vibration mediated coupling in conventional superconductors. A stronger interaction would lead to a much higher transition temperature. However, another consequence of the lighter electron mass is that the electronic polarisation decays much more rapidly than the lattice polarisation, which would restrict the range of the interaction. Indeed, in a later paper (Davis *et al.*, 1976), Little and co-workers subjected this model to further scrutiny and concluded that the polarisable side group must be either very close to the polymer backbone or completely surround it. Hirsch and Scalapino (1985) tested the validity of Little's model by a detailed simulation of a wide range of potentially superconductive systems. They concluded that while it might be possible in very special circumstances for an isolated polymer chain to be superconductive, this outcome was highly unlikely for an array of polymer chains even when the coupling between chains was weak. On top of these difficulties there are relatively few polymers with the highly perfect extended chains on which the model is based. Even when well-ordered samples are available the conjugated backbones are intrinsically semiconductors rather than metals. Thus there are very few conduction band electrons available and their mutual Coulomb repulsion is only very weakly screened. The introduction of dopant ions to render the polymers conductive tends to disrupt the ordered structure and introduces charged species adjacent to the polymer chain in addition to the polarisable side groups. Thus on both theoretical and practical grounds it appears that, while this is an appealing theoretical concept, the realisation of the mechanism in a real material is extremely unlikely.

Hopes that superconduction was possible in polymers were raised briefly in 2001, when Schön *et al.* (2001) reported superconductivity at 2.35 K in a thin film of polythiophene with an excess of electrons induced by an adjacent gate electrode. These and a number of other remarkable results obtained with thin films of organic materials could not, however, be duplicated by other workers, and were declared suspect by an independent committee set up by Bell Laboratories to investigate inconsistencies in the papers reporting these findings (Brumfiel, 2002).

4.3 Further reading

More extensive discussions of the band theory of semiconductors are given in standard texts such as Kittel (1996). Taylor and Heinonen (2002) is a more advanced text on condensed matter theory. Conduction in non-crystalline materials is dealt with at length by Mott and Davis (1971) and in a more succinct form by Mott (1987). The application of band theory to polymers in general is presented in Ladik (1988) and André, Delhalle and Brédas (1991).

5

Measurement of electric properties

5.1 Introduction

One inherent advantage of studying materials through their electrical properties is that electrical measurements can be made over a very wide range of frequencies with a high degree of precision. Thus, the relative permittivity and dielectric loss of solid polymers are often known from 10^{-4} Hz up to optical frequencies. Taken together with the possibility of varying temperature and pressure, the breadth of the experimental technique becomes very impressive, and researchers have made full use of this tool in probing underlying molecular behaviour. The detailed way in which dielectric properties may be mapped is well exemplified by the data for poly(ethyleneterephthalate). Figure 5.1 shows relief models of the variation of relative permittivity and dielectric loss with temperature and frequency for this polymer (Reddish, 1950).

The main improvements in measurement techniques in recent years are largely linked to the application of digital computers. With their aid, measurements may be made much faster, enabling real-time, dynamic monitoring of dielectric response in materials that are in the process of change. As a result, dielectric measurements over a wide frequency range may be scanned as, say, temperature or pressure are raised or lowered. It also becomes trivial to programme the experiment to collect extensive sets of data covering a wide range of conditions automatically. Measurements of this kind are often referred to as *dielectric spectroscopy*. Very importantly, the increase in speed of measurement also makes it possible to follow the course of chemical reactions, especially polymerisation, crosslinking and degradation, which affect dielectric properties. This affords a new method of industrial process control.

In considering the methods of measurement we must distinguish two basic types, namely *lumped-circuit* and *distributed-circuit* methods. The aim of

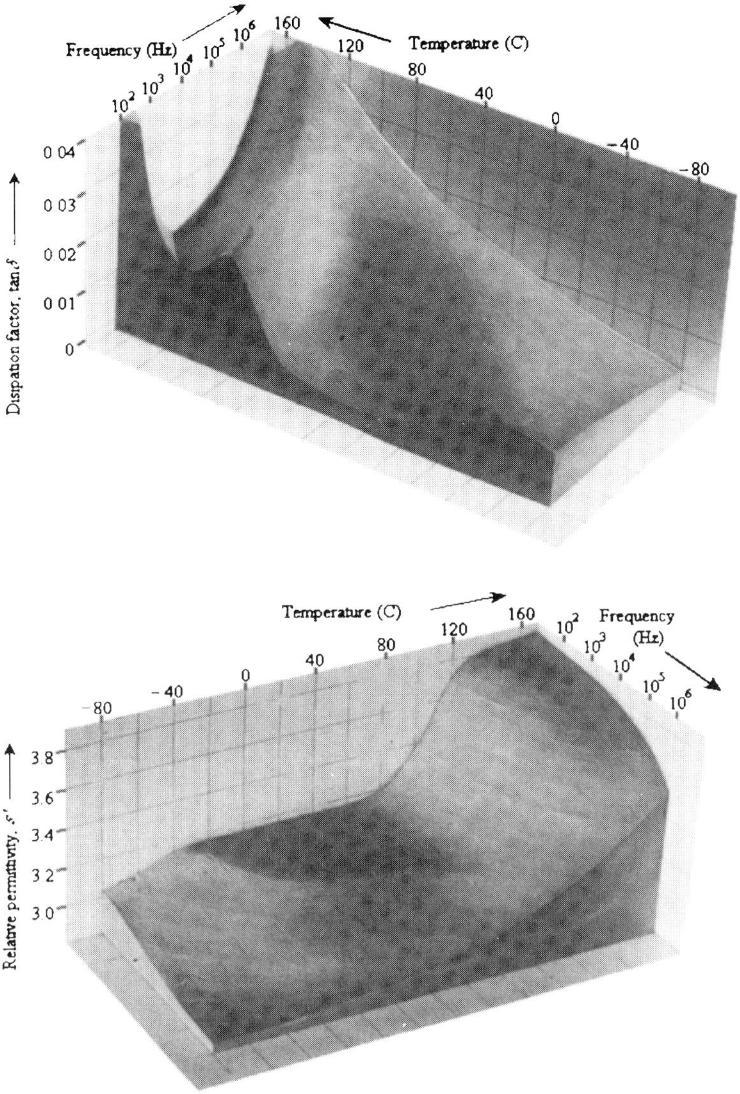

Fig. 5.1 Photographs of relief models showing the variation of relative permittivity and dielectric loss of poly(ethyleneterephthalate) with temperature and frequency of measurement. From Reddish (1950) with permission of the Royal Society of Chemistry.

lumped-circuit methods, which are always used at lower frequencies, is to determine the equivalent electrical circuit of a specimen at a given frequency. From the previous discussion in Section 3.1 of the current–voltage relationship for a dielectric material between the plates of a capacitor, we can see directly how the components of the complex permittivity of the material may be

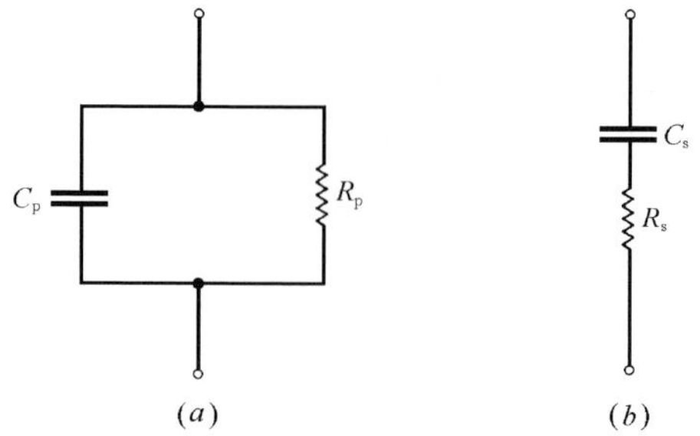

Fig. 5.2 Equivalent electrical circuits of dielectric specimens: (a) parallel, (b) series.

expressed in terms of an equivalent parallel circuit. Thus, let the specimen be represented by a capacitance C_P in parallel with a resistance R_P, as shown in Fig. 5.2(a). The total impedance Z will then be given by

$$\frac{1}{Z} = \frac{1}{R_P} + i\omega C_P. \tag{5.1}$$

Application of the alternating voltage represented by the real part of $V = V_e^{i\omega t}$ will produce an out-of-phase or capacitive current I_C:

$$I_C = \text{Imaginary part}\left[\frac{V}{Z}\right] = i\omega\, C_P V, \tag{5.2}$$

and an in-phase or resistive current I_R:

$$I_R = \text{Real part}\left[\frac{V}{Z}\right] = \frac{V}{R_P}. \tag{5.3}$$

Comparing Equations (5.2) and (5.3) with Equations (3.8) and (3.9), respectively,

$$\varepsilon' = \frac{C_P}{C_0}, \tag{5.4}$$

$$\varepsilon'' = \frac{1}{R_P C_0 \omega}, \tag{5.5}$$

and

$$\tan\delta = \frac{\varepsilon''}{\varepsilon'} = \frac{1}{R_P C_P \omega}. \tag{5.6}$$

Alternatively, the specimen may be regarded in terms of a series circuit. If the equivalent series components of capacitance and resistance are C_S and R_S, respectively (see Fig. 5.2(b)), the total impedance will be given by

$$Z = R_S + \frac{1}{i\omega C_S}. \tag{5.7}$$

If we again compare out-of-phase and in-phase currents after application of the alternating voltage, we obtain

$$\varepsilon' = \frac{C_S}{C_0(1 + \tan^2 \delta)}, \tag{5.8}$$

$$\varepsilon'' = \frac{R_S C_S}{C_0(1 + \tan^2 \delta)}, \tag{5.9}$$

and

$$\tan \delta = \frac{\varepsilon''}{\varepsilon'} = R_S C_S \omega. \tag{5.10}$$

We can therefore calculate the relative permittivity and dielectric loss of a material from measured values of either equivalent series or parallel circuit components of a specimen.

The reciprocal of the specimen resistance in the equivalent parallel circuit for a given frequency is sometimes called the specimen conductance G_P. It is a combination of DC conductance, by which we mean any real flow of charge through the sample under the influence of the applied field, and the *anomalous* conductance due to any time-dependent polarisation processes. The contribution that a true DC conductivity σ_{DC} (conductivity is defined as the reciprocal of resistivity, which is the resistance between opposite faces of a unit cube of material) will make to the dielectric loss at an angular frequency ω can be readily calculated as follows for the material in a parallel-plate capacitor. If the capacitor plates have area A and separation s:

$$G_P = \frac{1}{R_P} = \frac{\sigma_{DC} A}{s} \text{ and } C_0 = \frac{\varepsilon_0 A}{s}, \tag{5.11}$$

and substitution for resistance in Equation (5.5) gives:

$$\varepsilon'' = \frac{\sigma_{DC}}{\varepsilon_0 \omega}. \tag{5.12}$$

This shows how DC conductivity causes ε'' to rise rapidly at low frequencies.

The term AC *conductivity* is also commonly used. It is defined by analogy with DC conductivity, and it expresses conduction-like processes that only produce in-phase currents at high frequencies, e.g. localised hopping of carriers that does not allow through-going DC conduction. By analogy with DC conductivity:

$$\varepsilon'' = \frac{\sigma_{AC}}{\varepsilon_0 \omega}, \quad \text{or} \quad \sigma_{AC} = \omega \varepsilon_0 \varepsilon'', \tag{5.13}$$

where the frequency ω is sufficiently high that σ_{DC} has very little effect.

At high frequencies the electromagnetic wavelength inevitably becomes comparable with sample dimensions, and lumped-circuit methods must be abandoned in favour of distributed-circuit methods in which the sample becomes the medium for propagation of electromagnetic waves. Relative permittivity and loss must then be obtained from the observed wavelength and attenuation characteristics.

As an alternative to methods that use an applied sinusoidal voltage, called *frequency-domain* measurements, a step change in voltage across a specimen may be made and the ensuing current transient observed. Such time-domain measurements may be translated into frequency-domain terms by Fourier integral transformations.

The methods which are appropriate for the various frequency regimes are charted in Fig. 5.3.

There are some general principles concerning the design of specimen holders for lumped-circuit methods, which should be discussed before going on to

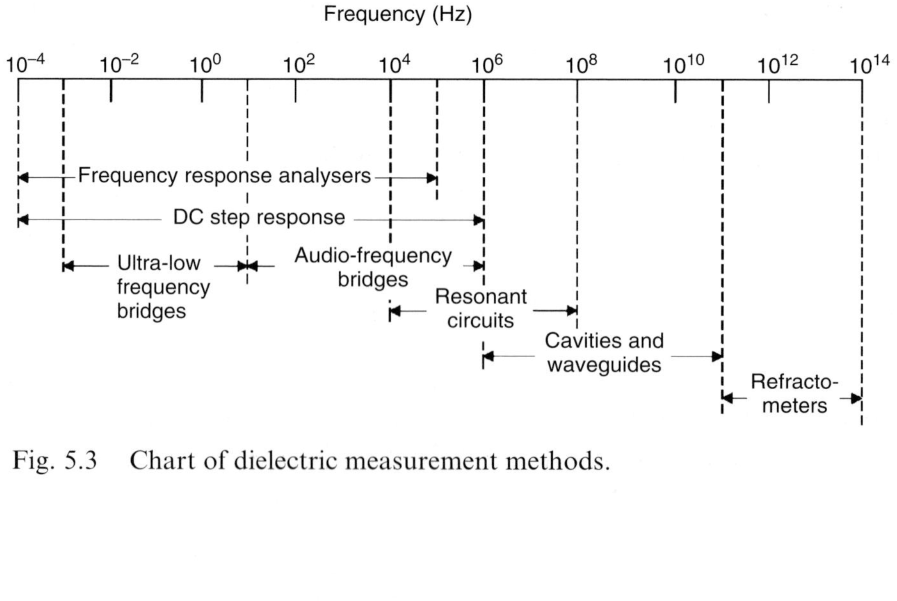

Fig. 5.3 Chart of dielectric measurement methods.

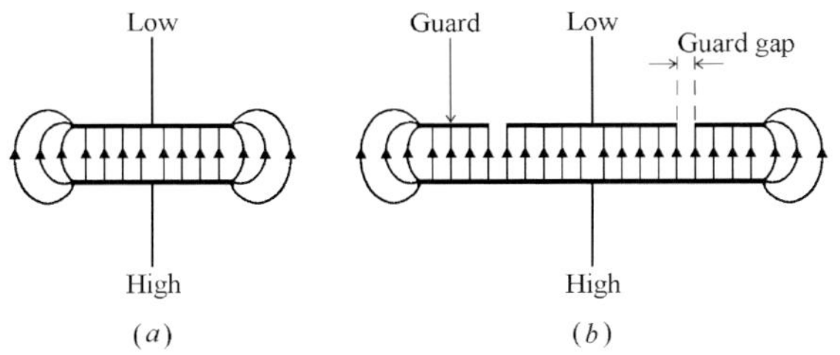

Fig. 5.4 Electric field patterns in parallel-plate electrode systems: (a) un-guarded, (b) guarded.

describe individual measurement methods. Most specimen holders are based on a simple parallel-plate capacitor, taking a thin disc of solid material between their electrodes or being filled with liquid.

The primary drawback of a parallel-plate capacitor is the fringing of the field at the edges (Fig. 5.4(a)). Whatever the arrangement of a 2-electrode system, it is difficult to make an accurate edge correction, because it changes in the presence of the test material. Semi-empirical corrections have been derived for various arrangements, but it is better, if possible, to avoid the difficulty altogether by using a guard electrode, as shown in Fig. 5.4(b). The guard is held at the same potential as the guarded electrode but is not connected to it. Provided the specimen and guard electrode extend beyond the guarded electrode by at least twice the specimen thickness, and provided the guard gap is small in comparison with the specimen thickness, the field distribution in the guarded area is identical with the specimen in place and with nothing between the electrodes; the relative permittivity is the direct ratio of the capacitances between the guarded electrode and the counter electrode in the two cases. Furthermore, the geometry is defined by the guard electrode, so that the vacuum capacitance may be accurately calculated from the electrode dimensions. The guard electrode may also serve to eliminate the effect of surface conduction round the edges of the specimen, which can be a nuisance with hygroscopic materials.

Stray capacitances, e.g. between a lead and the opposite electrode, must be avoided by proper shielding. For this reason it is preferable to run the guarded electrode at zero potential. Then the guard electrode can be earthed and a completely shielded 3-terminal cell constructed (Fig. 5.5), where the only stray capacitances are between the electrodes and earth, giving an equivalent circuit as shown.

Unfortunately, with most high-frequency measurement methods it is not possible to use a guarded electrode system and, in addition, lead inductances become appreciable at high frequencies. To meet the high-frequency requirement

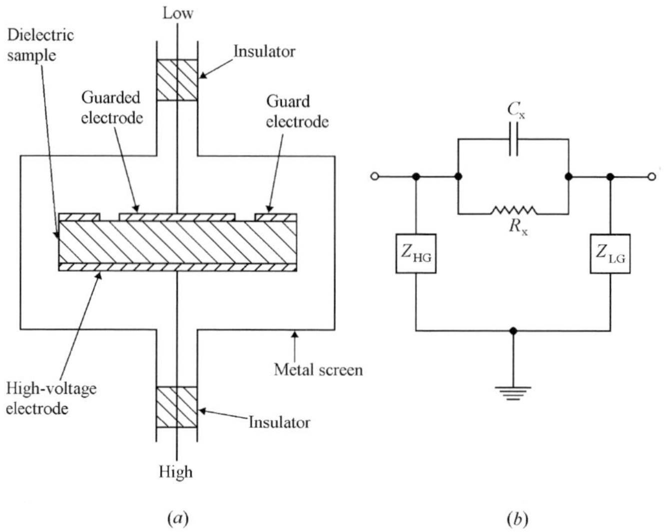

Fig. 5.5 A 3-terminal specimen holder: (a) diagram of shielded electrode system with guard ring, (b) equivalent electrical circuit of the cell.

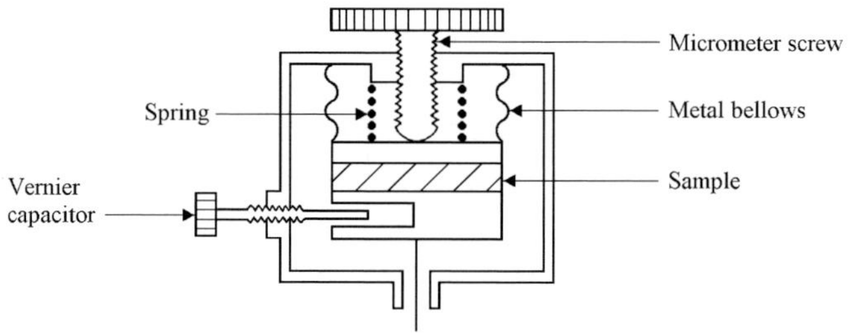

Fig. 5.6 Diagram of a 2-terminal micrometer electrode cell.

a special 2-electrode system, as shown in Fig. 5.6, has been developed to eliminate errors due to lead impedances and to reduce edge effects as much as possible. (A built-in side capacitor is also provided for use in resonance methods – see Section 5.3.) The cell capacitance is first measured with the specimen clamped between the electrodes. The specimen is then removed and the electrodes adjusted by means of the micrometer to give the same capacitance. The edge correction is reduced to a minimum by using a specimen of the same size as the electrodes. For accurate capacitance measurements the edge effect may be reduced even further by immersing the specimen in a fluid whose relative permittivity is approximately matched to that of the samples.

Any air gaps between the cell electrodes and specimen surfaces will introduce a series capacitance. Accidental gaps are usually avoided by applying thin foil electrodes, e.g. tin foil applied with a minimum quantity of pure vaseline or

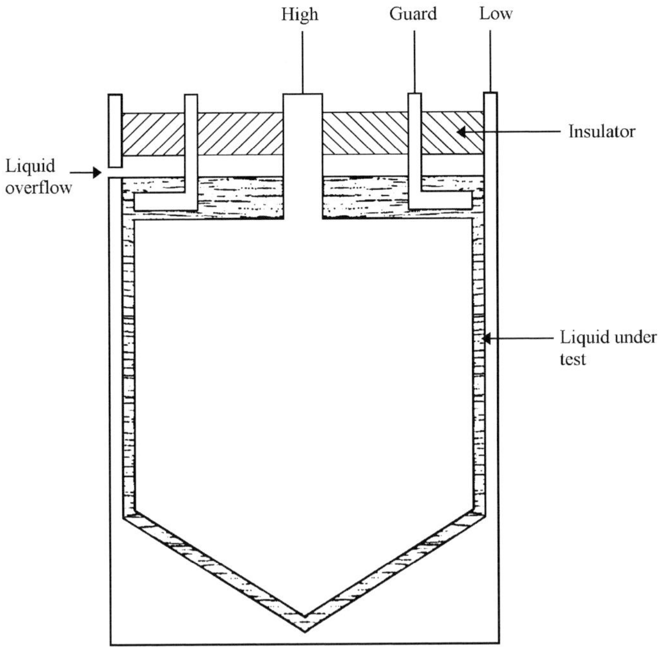

Fig. 5.7 Diagram of a 3-terminal electrode cell for liquids.

silicone grease, or by evaporating metal on to the specimen. Alternatively, when very low losses are to be measured and it is desirable not to introduce any foreign matter into the sample, a known air gap may be deliberately left above the specimen and allowed for in the calculation of specimen capacitance and loss.

Liquid cells are designed along the same lines as those for solids. A typical 3-terminal version is shown in Fig. 5.7.

5.2 Bridge methods

The most widely used way of determining the equivalent capacitance and resistance of a specimen is to use a type of Wheatstone bridge network to compare the unknown with standard components. The most versatile form for measurements of dielectric materials is the Schering bridge, with which it is possible to make very accurate measurements over the frequency range 10 to 10^5 Hz (the audio-frequency range). The basic circuit for a practical form of the bridge is shown in Fig. 5.8. At balance, when the detector indicates a null, we have the usual relation between the impedances of the arms:

$$\frac{Z_1}{Z_2} = \frac{Z_3}{Z_4}. \tag{5.14}$$

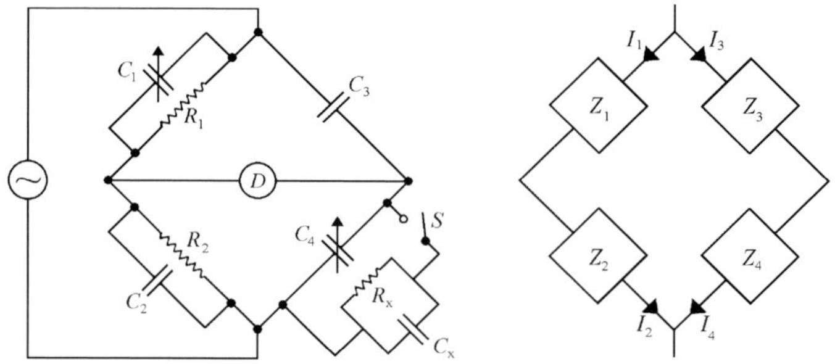

Fig. 5.8 Circuit diagrams of a conjugate Schering bridge.

Equal resistive ratio arms are used and a small variable capacitor is applied across resistance 1 to balance the conductance of the unknown in arm 4. A substitution procedure is generally used. First, the bridge is balanced with the sample cell out (switch S open). Then the sample cell is connected in (switch S closed) and balance regained by compensating for the cell capacitance C_X by decreasing C_4 and for the cell conductance G_X by increasing C_1. From the two sets of balance conditions we obtain, for tan $\delta_x < 10^{-2}$,

$$C_x = (C_4)_{\text{out}} - (C)_{\text{in}} = \Delta C_4, \tag{5.15}$$

$$\tan \delta_X = \frac{G_X}{\omega C_X} = \frac{\omega R_1 (C_4)_{\text{out}} \Delta C_1}{C_X}. \tag{5.16}$$

The success of an audio-frequency bridge depends on being able to deal with stray impedances which couple various parts of the bridge by unknown amounts, depending on component settings and the position of the operator. Thus the foregoing balance conditions, (5.15) and (5.16), really only hold true when $I_1 = I_2$ *and* $I_3 = I_4$, which means that there must be no stray capacitances from the detector terminals to earth. From this point of view the Schering bridge is particularly good, because almost complete shielding of the bridge components from one another may be achieved. First, the absence of series connections in any bridge arm between capacitance and resistance components simplifies the shielding. Second, a voltage divider may be connected across the generator terminals to ensure that the detector terminals are at earth potential at balance, as shown in Fig. 5.9. With this arrangement, called a Wagner earth, the modified measurement procedure is as follows. The bridge is first balanced as normal, and then the detector connected from one bridge corner to earth and the Wagner circuit adjusted to give another null. This disturbs the bridge slightly, which is then rebalanced and so on until, simultaneously, the bridge is balanced and the detector terminals are at earth potential.

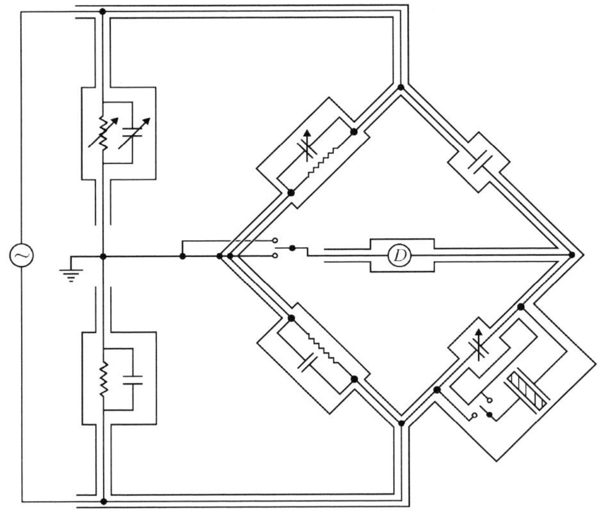

Fig. 5.9 Circuit diagram of a shielded conjugate Schering bridge with a Wagner earth and a guarded specimen holder.

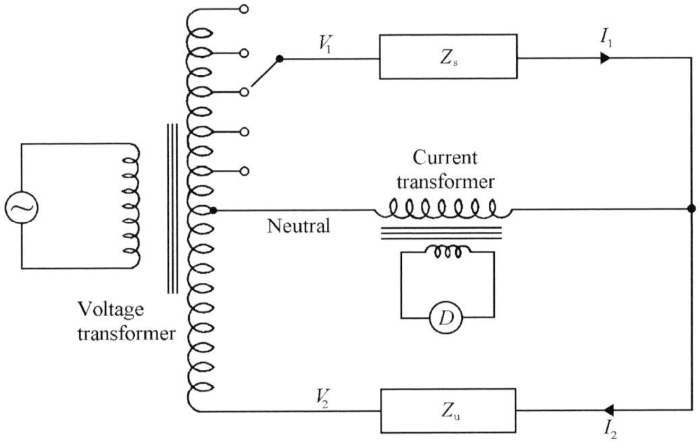

Fig. 5.10 A simplified circuit diagram of a transformer-ratio-arm bridge.

In this condition, earthed screens can be used around all bridge components, and a guard electrode can be used in the sample cell.

Using these techniques great precision can be achieved with a Schering bridge over all of its working range. Differences in δ as small as 300 nrad may be resolved (Astin, 1936).

Another kind of audio-frequency bridge, which has come to the fore recently and which is very convenient to use, is the transformer-ratio-arm bridge, shown schematically in Fig. 5.10. This bridge uses inductive ratio arms to compare the unknown directly with standard components. The voltage transformer is energised by an oscillator connected to its primary winding and develops one voltage V_1 between the unknown impedance Z_u and the neutral (earthed) line, and a second voltage V_2, 180° out of phase with the first, across the standard Z_s. The currents $I_1$and I_2 flowing through Z_u and Z_s are then V_1/Z_u and V_2/Z_s, respectively. When these currents are equal they combine, on account of their phase difference, to produce zero core flux in the current transformer, and the detector indicates a null. The circuit has two big advantages: (a) impedances between the unknown and earth merely shunt the low resistances of the voltage and current transformers and do not affect the bridge balance, so that long screened leads and guarded electrodes can be used; (b) the voltage transformer can be tapped very accurately to obtain decade ratios, so that only a few standards are required. The principal drawback is that full sensitivity is restricted to the small frequency range for which the transformers are designed.

Below about 20 Hz transformers become very inefficient and more specialised bridges are necessary. Below about 10^{-2} Hz bridge balancing also becomes a very slow business, because the period of each cycle is so long that one has to wait a long time after each readjustment of the bridge to see the resulting change in the amplitude of the output signal at the detector. The frequency range can, however, be extended down to about 10^{-3} Hz by balancing the bridge during just part of a cycle using a phase-sensitive detector.

Bridge methods can be speeded up to some extent by the introduction of automatic balancing systems, where a phase-sensitive detector is used in a feedback loop to make adjustments to the reference arm. In spite of this, bridge techniques have been superseded in many applications by computer-based methods (see below), although they still offer better performance in terms of their ability to detect small dielectric losses.

At high frequencies, bridges can still be used, provided that special precautions are taken to eliminate the effects of stray inductances which become very large, but it is usually better to adopt a resonance method above 10^6 Hz.

5.3 Resonance methods

At radio frequencies a medium- to low-loss material can be most sensitively examined by making it part of a resonant circuit. The Hartshorn and Ward (1936) method, for which the basic circuit is shown in Fig. 5.11, has been highly developed (Reddish *et al.*, 1971) for very accurate measurements in the 10^5 to 10^8 Hz range. Here a disc specimen is held in a micrometer electrode system (see Fig. 5.6, where the specimen is represented by a capacitance C_x in parallel with a resistance R_x) and connected directly to a coil of fixed

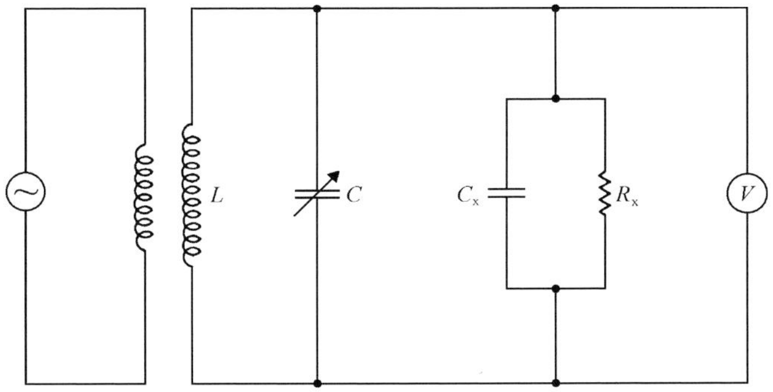

Fig. 5.11 The basic electrical circuit of the Hartshorn and Ward (1936) resonance method of dielectric measurement.

inductance L. The circuit is energised by a loosely coupled oscillator, whose frequency is tuned to resonance as indicated by the voltmeter, and the maximum voltage V_i noted. The specimen is now removed, the circuit retuned to resonance by decreasing the inter-electrode gap by an amount Δx, and the new maximum voltage V_0 noted. Finally, with the aid of the side capacitor C, the capacitance width ΔC_0 of the resonance curve at half-power height is determined with the specimen absent. Since the circuit capacitance must be the same for the two resonance conditions, we obtain immediately for the relative permittivity of the material

$$\varepsilon' = \frac{t}{(t - \Delta x)}, \tag{5.17}$$

where t is the specimen thickness. This ignores any change in edge capacitance due to the presence of the specimen. Although some correction may be made for this, a more reliable alternative is to use a liquid immersion cell if the ultimate in accuracy is wanted.

Circuit analysis shows that the loss tangent of the material is given by

$$\tan\delta = \left(\frac{V_0}{V_i} - 1\right)\frac{\Delta C_0}{2C_0}, \tag{5.18}$$

where C_0 is the calibrated air capacitance of the main electrode set at resonance with the sample out. In its most refined form the method is capable of measuring a δ of 50 μrad to ±1 μrad.

Measurements may be made at different frequencies by changing the inductance coil. In going to high frequencies the required inductance eventually becomes impractically small, however, and stray inductances take over. It is then necessary to use a resonant cavity of some kind, and a re-entrant cavity of

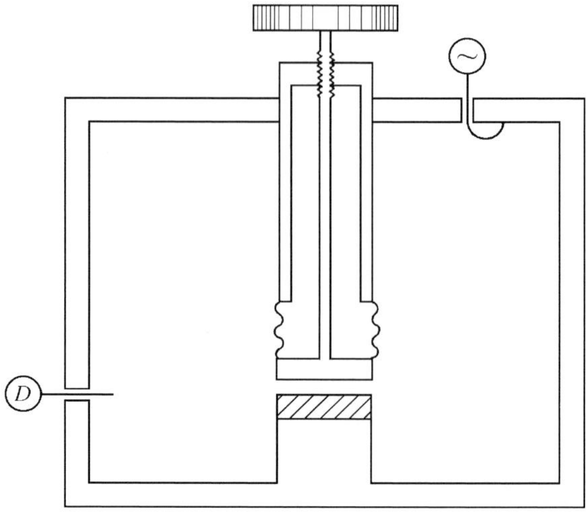

Fig. 5.12 Diagram of a re-entrant cavity for dielectric measurements.

the type shown in Fig. 5.12 is often used in the 10^8 to 10^9 Hz range. The circuit is a hybrid one in the sense that its inductance and fixed capacitances are distributed along a short-circuited, coaxial transmission line, whereas the electrode gap in the centre conductor forms a lumped, variable capacitor, which may be calibrated in the normal way in air at low frequencies.

Measurement with the re-entrant cavity is similar to the above Hartshorn and Ward method, except that a decrease in the electrode gap necessarily incurs an increase in inductance, and this must be taken into account (Works, 1947). The latter effect may be expressed as an equivalent capacitance change ΔC_L and, assuming that the inductance increases like that of a loss-free transmission line, we may calculate that

$$\frac{dC_L}{dl} = \frac{1}{cZ_0} \operatorname{cosec}^2 \frac{2\pi l}{\lambda}, \tag{5.19}$$

where l is the length of the line, Z_0 its characteristic impedance, λ the wavelength, and c the velocity of light. The capacitance width of the resonance curve with the sample absent may be determined by fine adjustment of the electrode gap, again taking into account the accompanying change in inductance. The tan δ value is then calculated from Equation (5.18).

5.4 Frequency response analyser methods

The entry into the measurement scene of efficient, commercial frequency response analysers, combined with digital signal processing capability,

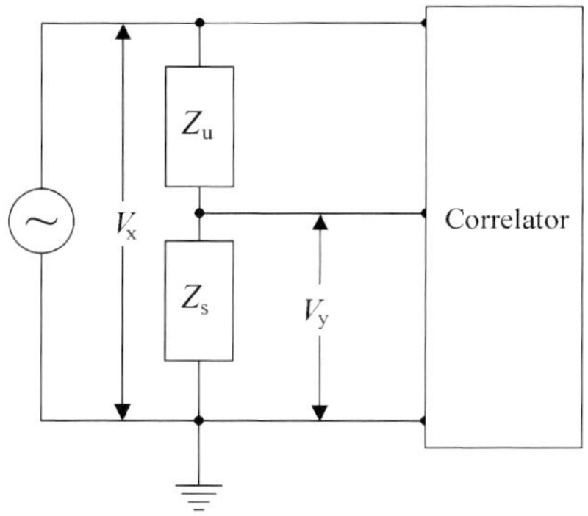

Fig 5.13 Block diagram showing frequency response analyser method of measurement.

makes possible a radically different approach to measurements in the frequency domain. Morse (1974) described an early example of instrumentation of this kind. In principle, the method is rather simple, as illustrated in Fig. 5.13. The frequency response analyser has two main parts: a generator which provides a digitally-synthesized alternating voltage $V_0 \cos \omega t$ over a range of frequencies, and a correlator which measures the in-phase and quadrature components of any input signal with respect to the output of its generator. The correlator works by multiplying the input signal by cosine and sine reference signals derived directly from the generator, and then integrating these products over a specified number of cycles of oscillation. Integrating over a large number of cycles reduces the noise level. For an input signal of $V_\mathrm{i} \cos(\omega t + \varphi)$ the correlator outputs are:

$$\frac{1}{NT}\int_0^{NT} V_\mathrm{i} \cos(\omega t + \varphi) \cos \omega t \,\mathrm{d}t = \frac{1}{2} V_\mathrm{i} \cos \varphi = \frac{1}{2} a, \tag{5.20}$$

and

$$\frac{1}{NT}\int_0^{NT} V_\mathrm{i} \cos(\omega t + \varphi) \sin \omega t \,\mathrm{d}t = -\frac{1}{2} V_\mathrm{i} \sin \varphi = -\frac{1}{2} b, \tag{5.21}$$

where $V_\mathrm{i} = (a^2 + b^2)^{1/2}$, $\varphi = \tan^{-1}(a/b)$ and NT is the integration time of N cycles of periodic time $T(= 2\pi/\omega)$.

The integration over many cycles of the generator is highly effective in averaging out random noise from the input signal. Also, only the response at the fundamental frequency is detected, since terms like $\cos(n\omega t + n\varphi) \sin \omega t$ with $n = 2, 3, 4 \ldots$ average to zero over a complete cycle; harmonics are totally

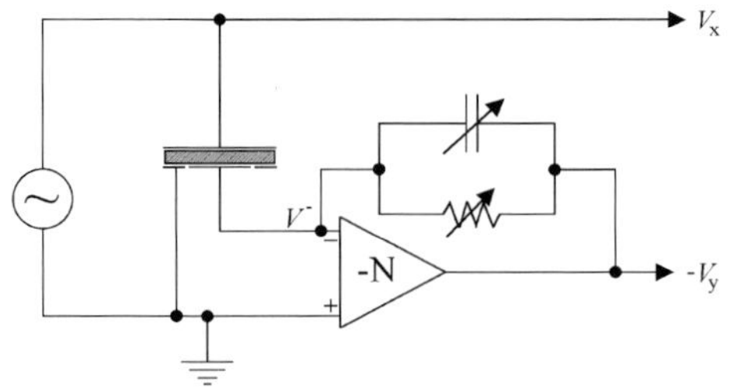

Fig 5.14 Circuit for connecting sample to a frequency response analyser.

rejected. Harmonics in the input signal can be measured by synthesizing the appropriate harmonic of the generator signal, the n_ith, say, and multiplying the input signal by $\cos(n_i\,\omega t)$ and $\sin(n_i\,\omega t)$. This can be used in studying non-linear effects and, possibly, as a way to access additional frequencies.

In making the measurement of the unknown impedance Z_u of a sample, the signal from the generator is applied across the sample in series with a standard impedance Z_s. The correlator is then used to determine the amplitude and phase of the voltage established across the sample $(V_x - V_y)$ and the current flowing through the sample, $Z_u V_y$. From these the analyser is able to compute the capacitance and conductance of the unknown sample. In practice, some correction is required to compensate for stray capacitances across a standard resistor or for leakage of a standard capacitor, so the standard is never a pure resistor or a pure capacitor.

In the simple arrangement of Fig. 5.13 the input of the correlator also appears across the standard impedance, requiring another correction. This loading of the standard can, however, be avoided by using a circuit like that shown in Fig. 5.14, which utilises a high-gain, high-input impedance, operational amplifier in negative feedback mode to maintain a *virtual earth* at its negative input. If the gain of the amplifier is A, then then $V_y = AV^-$ (where V^- is the potential of the negative input). The potential difference across the standard impedance is actually $V_y - V^-$, but for an amplifier of sufficiently high gain, V^- can be made negligible in magnitude in comparison with V_y. Thus measurement of V_y by the correlator effectively measures the voltage across the standard without the correlator being placed physically in parallel with it. There are further benefits of using this kind of circuit. Firstly, the virtual earth at the lower terminal of the sample allows the use of a 3-terminal specimen holder. Secondly, the low impedance output of the amplifier allows long leads to be used for connection to the analyser. In order to optimise the accuracy of the measurement it is important to make V_x and V_y of comparable size, and this is done by introducing a standard impedance made up of standard resistors and capacitors in parallel, whose range may be switched to match that of the sample's impedance. Typically, with modern instruments, the method can be used over the frequency range $10^{-4}$to 10^5 Hz to measure loss angles as small as 100 μrad.

5.5 Wave-transmission methods

As the frequency is taken above about 10^9 Hz, electromagnetic wavelengths become comparable with typical specimen dimensions. Applied fields then vary from place to place within a specimen and we must analyse the dielectric response in terms of Maxwell's wave equations. Experimentally, it is convenient to confine the waves by using a coaxial transmission line, consisting of a central conductor enclosed in a hollow conducting tube, or, at very high frequencies, a simple waveguide of rectangular or circular cross-section. The wavelength range over which this technique can be used, approximately 1 to 300 mm, is normally called the microwave region.

A preliminary step to dielectric measurement by wave-transmission techniques is to relate the basic wave parameter, called the propagation factor, γ_s^* of the material, to permittivity. In terms of the propagation factor the equations for the electric and magnetic fields of a plane wave travelling in the x-direction in a uniform, infinite material are:

$$\boldsymbol{E} = \boldsymbol{E}_0 \exp(\mathrm{i}\omega t - \gamma_s^* x), \tag{5.22}$$

$$\boldsymbol{H} = \boldsymbol{H}_0 \exp(\mathrm{i}\omega t - \gamma_s^* x), \tag{5.23}$$

where the respective field vectors $\boldsymbol{E}$ and $\boldsymbol{H}$ orthogonal (a transverse electromagnetic, or TEM, mode). In accordance with Maxwell's electromagnetic field equations, the complex propagation factor of the material is given by:

$$\gamma_s^* = \mathrm{i}\omega(\varepsilon_s^* \mu_s^*)^{\frac{1}{2}} = \alpha_s + \mathrm{i}\beta_s, \tag{5.24}$$

where ε_s^* and μ_s^* are the absolute (complex) permittivity and permeability of the material. The real part α_s defines the attenuation (zero in a vacuum) of the wave, and the imaginary part β_s defines the wavelength λ_s in the material:

$$\lambda_s = \left(\frac{2\pi}{\beta_s}\right). \tag{5.25}$$

In a vacuum the propagation factor simplifies to

$$\gamma_0 = \mathrm{i}\omega(\varepsilon_0 \mu_0)^{1/2} = \mathrm{i}\beta_0, \tag{5.26}$$

where ε_0, μ_0 are the permittivity and permeability of free space, respectively. The complex relative permittivity of a non-magnetic ($\mu^* = \mu_0$) dielectric material at angular frequency ω is then related to the propagation factor of a plane wave of that frequency travelling through the material by

$$\varepsilon^* = -\gamma_s^{*2}\left(\frac{\lambda_0}{2\pi}\right)^2, \tag{5.27}$$

where λ_0 is the free-space wavelength at that frequency. Equating real and imaginary parts:

$$\varepsilon' = \left(\frac{\lambda_0^2}{\lambda_s^2}\right)\left(1 - \frac{\alpha_s^2}{\beta_s^2}\right) = n^2\left(1 - \frac{\alpha_s^2}{\beta_s^2}\right), \tag{5.28}$$

$$\varepsilon'' = \left(\frac{\lambda_0^2}{\lambda_s^2}\right)\left(\frac{2\alpha_s}{\beta_s}\right) = n^2\left(\frac{2\alpha_s}{\beta_s}\right), \tag{5.29}$$

n being the refractive index of the material at the particular frequency concerned. Taking into account boundary conditions, it may be shown that Equations (5.28) and (5.29) apply not only to plane waves propagating in the body of an isotropic material, but also to the principal (or TEM) mode of propagation in a coaxial transmission line. For all modes in other waveguides or for higher order modes along a coaxial line the basic relation (5.27) must, however, be modified as follows (von Hippel, 1954):

$$\varepsilon^* = \frac{(1/\lambda_c)^2 + (\gamma_{sg}^*/2\pi)^2}{(1/\lambda_c)^2 + (1/\lambda_{0g})^2}, \tag{5.30}$$

where γ_{sg}^* is the propagation factor of the particular mode in the guide filled with the material in question, λ_{0g} is the mode wavelength in vacuum, and λ_c is the cut-off wavelength of the guide. For a wave to pass along a waveguide, its wavelength must be less than a certain critical value, the cut-off wavelength, which is related to the cross-sectional dimensions of the guide. If the dimensions are chosen correctly, transmission can be limited to a single transverse mode TE_{01}. In a coaxial-line only the principal TEM mode (for which $\lambda_c = \infty$) will propagate, provided the wavelength is above a critical value, which is of the order of the separation between the inner and outer conductors. Clearly, for measurement purposes, multiple modes must be excluded, and consequently practical frequency ranges for the application of coaxial-line and waveguide methods are 10^8 to 5×10^9 Hz and 3×10^9 to 6×10^{10} Hz, respectively.

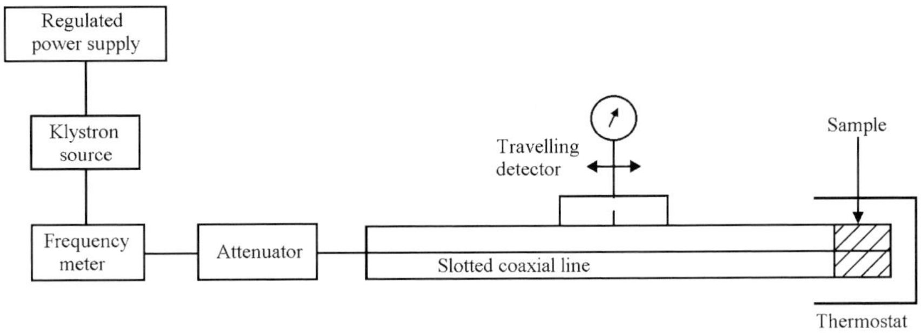

Fig. 5.15 Diagram of a slotted-line apparatus.

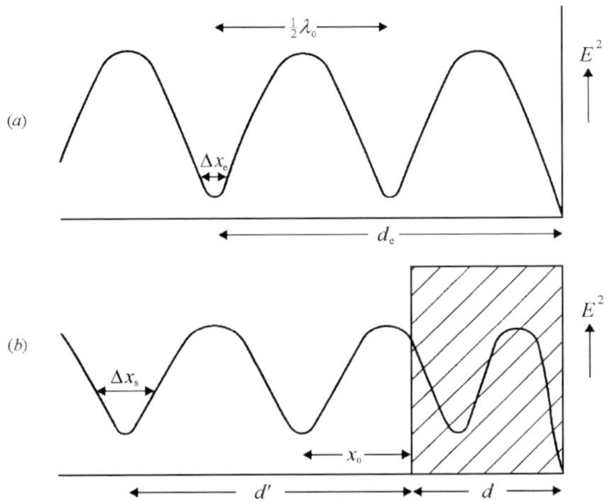

Fig. 5.16 Diagram of standing waves (a) in an empty waveguide, and (b) with a dielectric specimen in place at the closed end.

In the basic method of Roberts and von Hippel (1946), standing waves are established in a waveguide or coaxial line with a closed (shorted) end and with the dielectric sample filling the space inside to a distance d from the closed end. A diagram of a suitable apparatus is shown in Fig. 5.15. If the sample is a liquid it is necessary, of course, to mount the tube vertically. The standing wave takes the form shown in Fig. 5.16. The wave in the space in front of the samples is characterised by measuring (a) the inverse standing wave ratio E_{min}/E_{max}, i.e. the ratio of the standing wave amplitudes at node and antinode positions, (b) the distance x_0 of the first node from the surface of the sample, and (c) the wavelength λ_{0g} given by twice the separation of adjacent nodes. This is done by means of a small travelling probe inserted into the guide through a longitudinal slot. Roberts and von Hippel were able to show that the propagation factor γ^*_{sg} of the wave in the sample is related to the experimental observables in the following way:

$$\frac{\tanh(y_{sg}^* d)}{y_{sg}^* d} = -i\frac{\lambda_{0g}}{2\pi d} \cdot \frac{E_{min}/E_{max} - i\tan(2\pi x_0/\lambda_{0g})}{1 - i(E_{min}/E_{max})\tan(2\pi x_0/\lambda_{0g})}. \tag{5.31}$$

Numerical procedures have been devised for solving this complex transcendental equation.

When the dielectric loss of the sample is low ($\tan\delta < 0.1$), the real part of Equation (5.31) reduces to

$$\frac{\tan\beta_s}{\beta_s} = -\frac{\lambda_{0g}}{2\pi d}\tan\frac{2\pi x_0}{\lambda_{0g}}, \tag{5.32}$$

which may be solved by reference to tables of $(\tan\theta)/\theta$. For a coaxial line ($\lambda_{0g} = \lambda_0$) we may then obtain the following expressions for the relative permittivity and loss of the sample:

$$\varepsilon' = \left(\frac{\beta_s \lambda_0}{2\pi}\right)^2, \tag{5.33}$$

$$\tan\delta = \frac{E_{min}}{E_{max}} \cdot \frac{\lambda_0}{\pi d} \cdot \frac{\beta_s \delta[1 + \tan^2(2\pi x_0/\lambda_0)}{\beta_s d(1 + \tan^2 \beta_s d) - \tan\beta_s d}. \tag{5.34}$$

The experimental procedure may also be modified for a low-loss sample. It may be shown that the inverse standing wave ratio may be determined very accurately by measuring the width of a node:

$$\frac{E_{min}}{E_{max}} = \frac{\pi \Delta x}{\lambda_0}, \tag{5.35}$$

where Δx is the distance between double-power points on either side of a minimum. When the loss due to the sample is very low, it also becomes necessary to take into account wall losses in the transmission line. The width Δx_e of a node distance d_e from the end of the empty line is a measure of these, and may be used to correct the results. First the observed width Δx_s of a node distance d' in front of the sample is corrected for the wall losses in the empty section in front of the sample:

$$(\Delta x_s)_{corrected} = (\Delta x_s)_{observed} - \frac{d'}{d_e}\Delta x_e. \tag{5.36}$$

The total $\tan\delta$ value for the sample-filled section is then calculated by Equation (5.34) and the wall-loss component $\Delta x_e/d_e$ (assuming $\mu = \mu_0$) subtracted to give the net value for the sample alone.

5.6 Time-domain methods

Sudden application of a steady voltage across a capacitor containing a dielectric will produce a transient charging current whose form depends on the rate at which equilibrium polarisation is attained in the dielectric. In principle, the equivalent point-by-point *frequency-domain* behaviour of the dielectric can then be obtained from this time-dependent response by a Fourier integral transformation which extracts all the separate harmonic components from the one transient signal in the time domain. For a linear dielectric in a capacitor (vacuum capacitance C_0) subjected to a step voltage V_0 at time $t=0$, the complex relative permittivity is given in general by

$$\varepsilon^*(\omega) = \frac{1}{C_0 V_0}\int_0^\infty I(t)\mathrm{e}^{-\mathrm{i}\omega t}\mathrm{d}t, \tag{5.37}$$

where $I(t)$ is the transient charging current at time t. In practice, the current transient will consist of an initial pulse corresponding to the atomic and electronic polarisations, which are effectively instantaneous, followed by a decaying current from any relaxing, dipolar orientation processes. If the dielectric is appreciably conductive (DC conductance G), there will also be a constant current component. We may therefore rewrite Equation (5.35) as

$$\varepsilon^*(\omega) = \varepsilon_\infty + \frac{1}{C_0 V_0}\int_0^\infty I(t)\mathrm{e}^{-\mathrm{i}\omega t}\mathrm{d}t - \mathrm{i}\frac{G}{\omega C_0}, \tag{5.38}$$

where the integration now excludes the initial current pulse and the constant conduction current. The form of current and charge transients for simple step voltage changes are shown in Fig. 5.17, where

$$q = \int_0^t I\mathrm{d}t.$$

The complete Fourier transformation of current-transient data involves very extensive numerical computation, and two approximate methods have often been used instead. The first method, known as the Hamon (1952) approximation, is based on the assumption that the current transient conforms to the equation

$$I(t) = At^{-n} \quad (0.5<n<1), \tag{5.39}$$

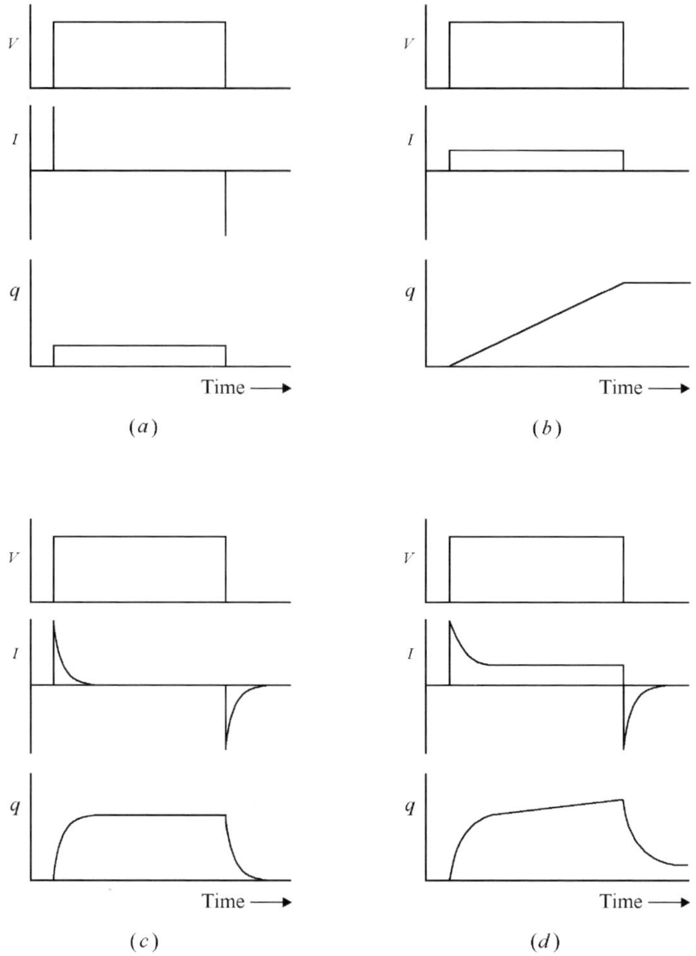

Fig. 5.17 Transient currents from step voltages across (a) a loss-free dielectric, (b) a conductor, (c) a dielectric with dipolar relaxation, and (d) a dielectric with dipolar relaxation and conduction.

which is indeed found to be true experimentally over limited portions of a transient curve. Hamon was able to show that the dielectric loss was in that case given approximately by the simple formula

$$\varepsilon^*(\omega) = \frac{I(t)t}{0.63C_0V_0}, \tag{5.40}$$

with $\omega t = \pi/5$. This result, which reflects the fact that the main contribution to the Fourier transform for frequency ω is related to the behaviour of the current transient at times near $t = 1/\omega$, is quite accurate for typical Debye-like relaxation processes in polymers.

A better alternative has recently been given by Hyde (1970) who bases his method on the experimental fact that the rise in the time integral $q(t)$ of the transient current is always a gradual one with respect to log t. Accordingly, an economic record of the transient over a wide frequency range can be obtained by measuring q at equal intervals on a logarithmic timescale. Considering the intrinsic width of a Debye-like relaxation process (frequency half-width = 1.14 decades) a log 2 interval (i.e. $t_{n+1}/t_n = 2$) between data points gives adequate time resolution, and is convenient for data processing on a computer. Defining a general time interval, $2^{n-1/2}t_1$, to $2^{n+1/2}t_1$, over which the increment in charge is $\Delta q(n)$, and choosing $t_1 = 1/\omega$, the fast Fourier transforms then take the forms:

$$\varepsilon'(\omega) = \sum_{p=-\infty}^{+\infty} \Delta q(n)x(n), \tag{5.41}$$

$$\varepsilon''(\omega) = \sum_{p=-\infty}^{+\infty} \Delta q(n)y(n), \tag{5.42}$$

where $n = 2^p$ and $x(n)$ and $y(n)$ are constant coefficients for the given frequency ω. In practice only the first few terms of the transforms are needed: just five terms are sufficient for 1% accuracy in $\varepsilon^*(\omega)$.

In the past the DC step-response technique has been used primarily to measure relative permittivity and loss at very low frequencies (down to 10^{-4} Hz), which were not accessible in any other way. The basic experimental technique is very simple. A step voltage in the range 1 to 500 V is applied across a sample held in a normal guarded-electrode system. The resulting charging current is detected by the voltage that it produces across a standard resistor in the circuit. For the very small transient polarisation currents normally encountered, typically in the 10^{-12} to 10^{-9} A range, a very high-value standard resistor 10^9 to $10^{12}\,\Omega$, say) has to be used. Since the input capacitance C_i of the electrometer used to measure the voltage is usually about 10^{-11} F, and since it shunts the standard resistor, the time constant (R_sC_i) for the response of the measuring circuit is long – about 10 s for a resistance of $10^{12}\,\Omega$. This means that transients cannot be directly measured at short times after application of the step voltage by the simple technique. Consequently, a circuit like that shown in Fig. 5.18(a) is often used. Here the electrode connected to the electrometer amplifier is virtually kept at earth potential by negative feedback, which counteracts the voltage developed across the standard resistor R_s by the current I flowing through it. The input capacitance of the amplifier no longer

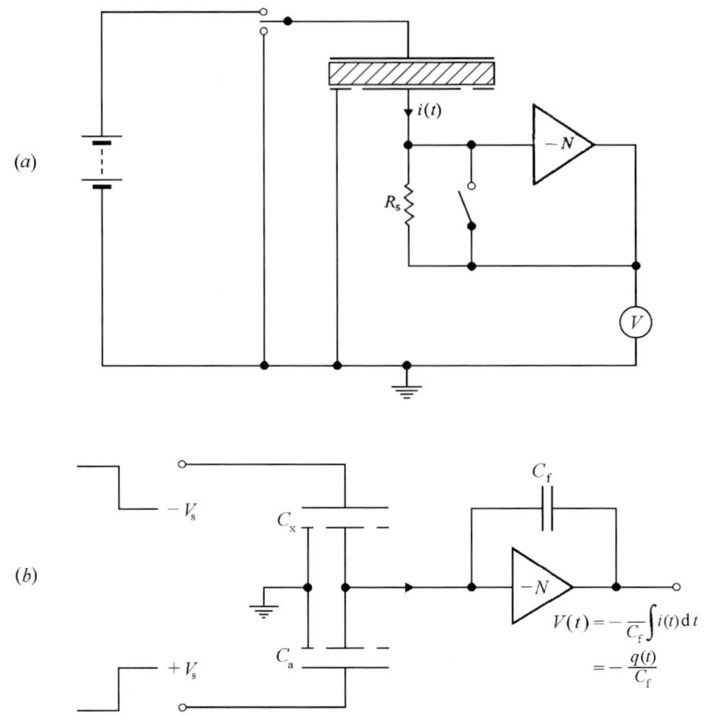

Fig. 5.18 Electrical circuits for DC step-response methods: (a) measurement of a current transient, (b) measurement of the time integral of current, with compensation for the large instantaneous component of the response.

shunts R_s and the time constant is reduced by a factor equal to the amplifier gain N, which may be 1000 or more. Then at times long in comparison with the effective time constant (i.e. $t > R_x C_i / N$) the polarisation current is given by

$$I = -\frac{V}{R_s}, \tag{5.43}$$

where V is the output voltage of the amplifier. The discharging current is also recorded as the voltage is reduced to zero by switching the upper electrode to earth.

In a more recent version of the method, the time integral of the current is measured by an electrometer amplifier with negative capacitive feedback, and the required dynamic range of the measuring circuit is much reduced by backing off the instantaneous response of the sample with a signal from an equivalent air capacitor fed with a voltage step of opposite sign (see Fig. 5.18(b)). This technique has been extended, using on-line computation, to provide a very rapid means of making measurements of relative permittivity and loss over the frequency range 10^{-4} to 10^6 Hz (Hyde, 1970).

The Fourier integral transform technique has also been used recently to obtain very high frequency (10^8–10^{10} Hz) data. The experimental method, usually referred to as time-domain spectroscopy, involves subjecting a sample in a waveguide to a train of very short radiation pulses, with fast rise times, and then observing the reflected or transmitted pulses with a sampling oscilloscope (Suggett, 1972).

5.7 Measurement of resistivity

In this section we shall consistently refer to resistivity, rather than to its reciprocal, conductivity. The choice is arbitrary, and commercial instruments are sometimes calibrated in units of resistance and sometimes in units of conductance.[1] Unless otherwise stated, when we refer to resistivity we shall mean volume resistivity ρ_V (Ωm), the resistance between opposite faces of a unit cube. A surface resistivity ρ_S is often used to characterise current flow over a surface, as with an antistatic coating, and is defined as the resistance between opposite edges of a unit square. We note first that the resistance across a square is independent of the size of the square and that the unit of surface resistivity is simply the ohm (Ω), occasionally written rather superfluously as ohm per square. Secondly, a conducting surface must in reality be a layer with a finite thickness t, and we have only an *effective* surface resistivity, which is related to the true volume resistivity of the layer by

$$\rho_S = \frac{\rho_V}{t}. \tag{5.44}$$

In a similar way resistivities of homogeneous thin sheet or film materials are often quoted as effective surface resistivities.

Experimental determinations of resistivity have used several different arrangements of the electrical connections to the sample, variously termed electrodes, terminals or probes, as illustrated in Fig. 5.19. The simplest volume resistivity measurements use two electrodes, a 2-terminal method, in two basic arrangements. One is a rectangular or cylindrical block with electrodes on the ends, Fig. 5.19(a). The other is like that used for measurements of relative permittivity where electrodes are applied to either side of a thin disc, Fig. 5.20. The latter is more appropriate to high-resistivity materials. In both cases the volume resistivity is related to the measured resistance R between the electrodes by

[1] The unit of conductance, the reciprocal ohm (Ω^{-1}), is often called the Siemen (S).

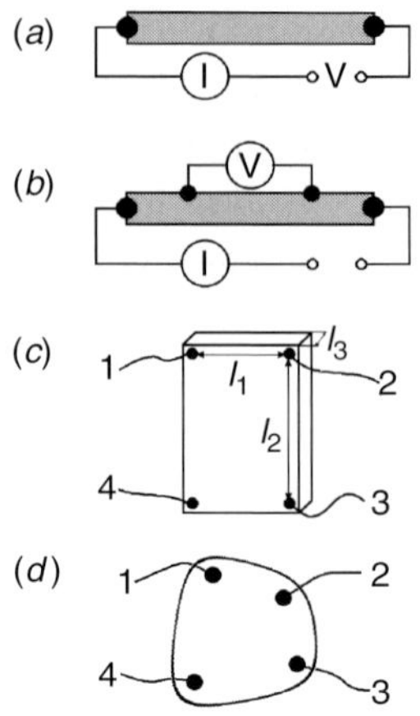

Fig. 5.19 Electrode configurations for resistivity measurement (a) 2-terminal, (b) 4-terminal in line, (c) 4-terminal Montgomery and (d) 4-terminal van der Pauw.

$$\rho_{\mathrm{v}} = \frac{RA}{l}, \tag{5.45}$$

where A and l are the cross-sectional area and the length of the specimen between the electrodes, respectively.

Concentric ring electrodes are the easiest to use for measurement of surface resistivity. The resistance R between them is the sum of the resistances of the elemental annuli in series:

$$R = \int_{r_1}^{r_2} \left(\frac{\rho_s}{2\pi r}\right) \mathrm{d}r, \tag{5.46}$$

where r_1, r_2 are the radii of the inner and outer electrodes, respectively. Hence

$$\rho_{\mathrm{s}} = 2\pi r \ln\left(\frac{r_1}{r_2}\right). \tag{5.47}$$

The main problem in accurate measurement of resistivity is one of contact resistance between the measurement electrodes and the specimen. This is clearly the case for samples with low resistivity, but can also be a problem for more resistive samples if either the contact resistance is high or the contact is non-ohmic. Contact resistance may be reduced by painting electrodes directly on to the surface of the specimen instead of relying on pressure contact with metal plates or foils. Suitable paints are silver dispersions or Aquadag (an aqueous dispersion of colloidal graphite). A much better way to deal with contact resistance is to use a 4-terminal, potentiometric method. The simplest configuration for use with bulk samples is that of two current electrodes and two potential electrodes, as shown in Fig. 5.19(b). A known current density j is established in the central region by a battery connected to the outer electrodes. If the electric field E is then determined by measuring the potential drop ΔV across two inner electrodes, separation Δx, the resistivity is given by

$$\rho_V = \frac{j}{E} = \frac{I/A}{\Delta V/\Delta x}. \tag{5.48}$$

Any effect of contact resistance is then avoided, providing the contact resistance is much smaller than the input resistance of the voltmeter.

When making measurements of a high resistivity, the main problem becomes one of leakage of current between the input terminals of the ohmmeter via routes other than the intended one through the specimen. Surfaces often provide a low-resistance path through the accumulation of dirt and moisture on them. One can usually overcome the problem to a large extent by the use of an extra guard electrode on the specimen. Any leakage currents over the surface of the specimen are then collected by the guard electrode and are not included in the measured current. In its most convenient form (see Fig. 5.20)

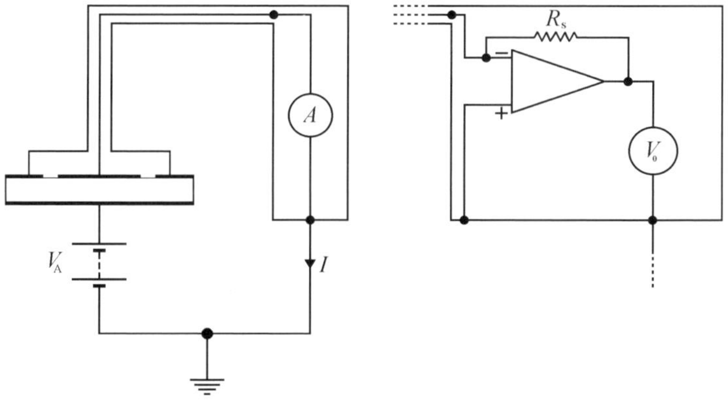

Fig. 5.20 Circuit diagram for resistivity measurements by the 2-terminal method with guarding.

the guard is connected to earth and surrounds that electrode which is connected to the highly insulated side of the ohmmeter, whilst the other electrode is maintained at a fixed, high voltage V_A. The guard may be extended to shield the connecting lead too. So long as the guarded electrode itself stays near earth potential, there will be no significant current flow between it and the guard; this will appertain if the internal resistance of the ohmmeter is much less than the sample resistance. In extreme cases, where the sample resistance is very high, the current I that flows is very small and it can only be determined by measuring the voltage drop across a standard high resistor R, with an electrometer (see Fig. 5.20 again). In that case the effective input resistance of the electrometer may be kept relatively low, so as to maintain the virtual earth condition at the input, by connecting the standard resistor in a negative-feedback loop of the high-gain ($-N$), high-input-resistance amplifier. If the output of the amplifier, measured by a relatively low-resistance voltmeter, is V_0,

$$I = V_0/R_s, \tag{5.49}$$

and

$$R = R_s V_A/V_0. \tag{5.50}$$

In the foregoing methods of resistivity measurement, specially shaped specimens and electrodes are required, the arrangements being chosen to give uniform, or uniformly divergent, fields in the material. An alternative approach which is useful for rapid measurements on small specimens or for *in situ*, non-destructive tests on a moulded item is to use some kind of electrode probe applied to the outside of the material in question (Blythe, 1984). When pointed electrodes are used in this way, however, the electric fields and current densities are no longer uniform. Fortunately, there is a strict analogy between the field equations of current flow and those of electrostatic charges, since they both comply with Laplace's equation:

$$\frac{\partial^2 V}{\partial x^2} + \frac{\partial^2 V}{\partial y^2} + \frac{\partial^2 V}{\partial z^2} = 0, \tag{5.51}$$

where V is the potential at any point in the field. Thus electrodes, which inject or drain current, are equivalent to positive and negative charges. For example, the potential at a distance r from a point electrode supplying a current I inside a material is given by

$$V = \frac{I\rho_v}{4\pi r}, \tag{5.52}$$

which may be compared with the potential due to a point charge q in a dielectric medium:

$$V = \frac{q}{4\pi\varepsilon_0\varepsilon' r}. \tag{5.53}$$

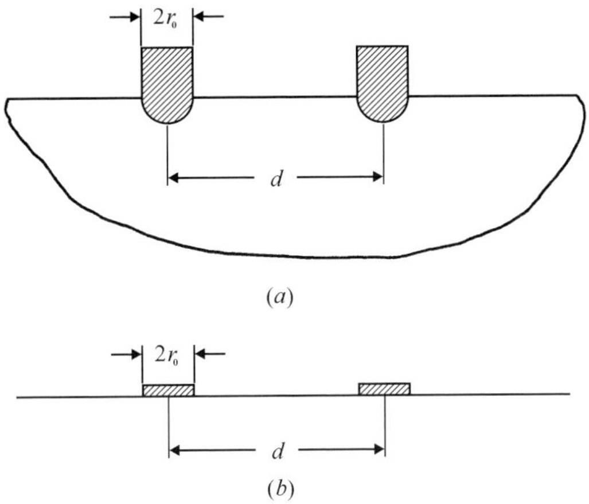

Fig. 5.21 Types of 2-terminal resistance probes: (a) hemispheres in a conductive solid, (b) discs on a conductive surface.

We may therefore make use of well-known solutions of electrostatic equations in order to relate resistance measured by a point probe to volume resistivities.

Suppose that a pair of hemispherical electrodes makes contact with the surface of a semi-infinite sample in the way shown in Fig. 5.21(a). If the separation of the electrodes d is much greater than the hemispherical radius r_0, analogy with the field equations for charged spheres (Moon and Spencer, 1961) shows that the measured resistance R between the electrodes (neglecting contact resistances) will be given by

$$R = \frac{\rho_V}{\pi r_0} \quad (r_0 \ll d). \tag{5.54}$$

Here we see that the resistance is independent of the electrode separation. The physical reason is that the major part of the voltage drop occurs in the immediate vicinities of the tips of the electrodes, and we may think of the resistance as the sum of a spreading resistance from the source electrode and a convergence resistance to the drain electrode. From this we conclude that 2-point probe measurements will not be very reliable or reproducible, because they will depend so sensitively on the small sample of material at the electrode tips. On a pliable material the form of indentation and contact area will be especially difficult to control with precision, and, in addition, we recognise that the material will be most distorted by the electrode pressure in just that region to which the resistance measurement is most sensitive. Therefore 2-point volume-resistivity measurements are not recommended.

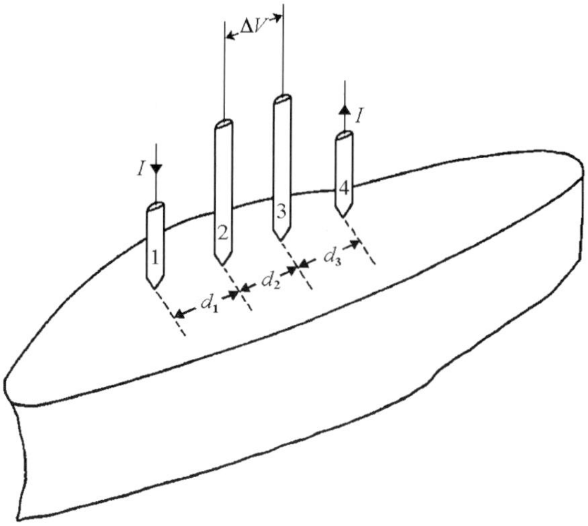

Fig. 5.22 Diagram of volume resistivity measurement with a 4-point probe.

The resistance R measured on a conductive surface by a 2-point probe will depend both on the radius r_0 of the circular contact at each electrode and on their separation d (see Fig. 5.21(b)):

$$R = \frac{\rho_s}{\pi} \cosh^{-1} \frac{d}{2r_0}. \tag{5.55}$$

Unlike the case of volume resistivity, the resistance never becomes independent of electrode separation. Again, measurement with finely pointed electrodes will be over-sensitive to the precision of contact.

The situation is quite different with 4-point probe measurements, shown in Fig. 5.22. In this type of measurement, a known current I is injected by electrode 1 and collected at electrode 4, whilst the potential difference ΔV between electrodes 2 and 3 is measured. Using Equation (5.52) for the potential of a current source, one can obtain the net potentials at electrodes 2 and 3 due to the current electrodes on the semi-infinite block, whence (Valdes, 1954):

$$\Delta V = \frac{I\rho_v}{2\pi} \left(\frac{1}{d_1} - \frac{1}{d_2 + d_3} - \frac{1}{d_1 + d_2} + \frac{1}{d_3} \right). \tag{5.56}$$

If the electrodes are equispaced distance d apart, Equation (5.56) reduces to

$$\Delta V = \frac{I\rho_v}{2\pi d}. \tag{5.57}$$

The result is independent of the electrode contact area if the size of the contacts is much smaller than the inter-electrode spacing.

When the electrodes are applied to a specimen with finite boundaries, the apparent resistivity ρ'_v that is measured is higher than the true value. This is usually expressed in terms of a correction divisor (CD):

$$\rho_v = \frac{\rho'_v}{CD}. \tag{5.58}$$

Uhlir (1955) has derived expressions for CD for various geometries by the method of images. For a conductive sheet of thickness t where $t \ll d$ the expression for CD reduces to

$$CD = 2\frac{d}{t}\ln 2. \tag{5.59}$$

Equation (5.57) for the 4-point probe now becomes

$$\Delta V = \frac{I\rho_v \ln 2}{\pi t} \text{ or } \frac{I\rho_s \ln 2}{\pi}. \tag{5.60}$$

In general, for single crystal samples the resistivity is anisotropic and will be represented by a second rank tensor of the form

$$\tilde{\rho} = \begin{bmatrix} \rho_{11} & \rho_{12} & \rho_{13} \\ \rho_{21} & \rho_{22} & \rho_{23} \\ \rho_{31} & \rho_{32} & \rho_{33} \end{bmatrix}. \tag{5.61}$$

This tensor is symmetric and requires a maximum of six independent parameters for low symmetry samples. The number of parameters is reduced for more symmetric samples. In general, polymeric samples will either be isotropic, with one resistivity parameter, or have axial symmetry generated by mechanical processes, see Section 1.3.6, with two resistivity parameters. Montgomery (1971) introduced a more general 4-point probe method suitable for the measurement of anisotropic solids. This method utilises a sample in the form of a rectangular prism with electrodes attached at the corners, illustrated in Fig. 5.19(c). Measurement of the voltage/current ratios for opposite pairs of electrodes,

$$R_1 = V_{12}/I_{34} \text{ and } R_2 = V_{23}/I_{41}, \tag{5.62}$$

enables the components of the resistivity tensor to be calculated. This method works well for materials described by two resistivity parameters. It is based on the mapping of an anisotropic sample of arbitrary shape on to an equivalent isotropic sample, as described by van der Pauw (1961). For a rectangular sample of anisotropic material described by three resistivity parameters, ρ_{11}, ρ_{22} and ρ_{33} with x-, y- and z-axes aligned with the edges of the sample and of dimensions l_1, l_2 and l_3, the dimensions of the equivalent isotropic solid are given by

$$l_i^* = l_i\sqrt{\rho_{ii}/\bar{\rho}}, \tag{5.63}$$

where the subscript i runs over x, y and z and the resistivity $\bar{\rho}$ is given by $\bar{\rho}^3 = \rho_{11}\rho_{22}\rho_{33}$. The equivalent resistivity is found experimentally from

$$\bar{\rho} = HER_1, \tag{5.64}$$

where H is a function of the ratio l_2^*/l_1^*, that can be found in the paper by Montgomery (1971), and E is the effective thickness of the equivalent isotropic sample, which is the same as l_3^* if $l_3^* \ll \sqrt{l_1^* l_2^*}$. For a thin sample, Equation (5.64) leads to the result

$$\sqrt{\rho_{11}\rho_{22}} = Hl_3R_1. \tag{5.65}$$

It follows from Equation (5.63) that

$$\sqrt{\rho_{22}/\rho_{11}} = (l_1/l_2)\left(l_2^*/l_1^*\right). \tag{5.66}$$

The ratio $\left(l_2^*/l_1^*\right)$is obtained from the measured ratio R_2/R_1 using the data of Montgomery (1971). This allows Equations (5.65) and (5.66) to be solved to obtain ρ_{11} and ρ_{22}. If ρ_{22} is equal to ρ_{33} the determination is complete.

The analysis of van der Pauw can be used for arbitrary electrode arrangements on thin samples of anisotropic materials of any shape and has been widely used in studies of conjugated polymers, where only small and irregular samples are available. Four electrodes are attached, as shown in Fig. 5.19(d). The resistance values R_1 and R_2 are measured as for the Montgomery method. The volume resistivity is then calculated from these values as

$$\rho_v = \frac{\pi t}{\ln 2}\frac{R_1 + R_2}{2}f\left[\frac{R_1}{R_2}\right], \tag{5.67}$$

where t is the thickness of the sample and f is a function, which is tabulated in the paper of van der Pauw (1961). This function has values in the range 1 to 0.2 for values of the resistance ratio between 1 and 10^3 respectively. When the sample is isotropic and is in the form of a square with the electrodes attached in the corners, the two resistances are equal and Equation (5.62) simplifies to

$$\rho_v = \frac{\pi t}{\ln 2}\frac{R_1}{2}. \tag{5.68}$$

Four-point probe measurements are very reliable, but the technique has limitations, because, as more highly resistive materials are examined, the point contacts are incapable of supplying currents that are high enough to make ΔV readily measurable. Furthermore, the necessary input resistance of the voltmeter, which must be greater than the resistance between points 2 and 3, becomes very high. Modern high input impedance electrometers set a practical upper limit of resistivities that can be measured by the 4-point probe technique of about $10^{12}\,\Omega\,\mathrm{m}$.

Observation of the rate of decay of charges by leakage to earth has formed the basis for other methods of determining high resistivities, especially surface resistivities of film or sheet material. The specimen is first charged to a high voltage by one of several methods: (a) exposing the surface to a source of air ions, (b) rubbing with another material to produce charge transfer, or (c) connecting the edges to a high voltage source. The charge is then monitored as a function of time by means of an electric-field meter (see Section 7.2). The decay rate is difficult to interpret exactly, for it depends on the product of the system's effective resistance and capacitance to earth; these are related in a rather complex way to the geometrical arrangement employed (Henry, Livesey and Wood, 1967). As a rough guide, a time constant (assuming an exponential decay) of one second corresponds to a surface resistivity of $10^{11}\,\Omega$ in most practical situations.

5.8 Further reading

Measurements of dielectric properties by bridge, resonance and waveguide methods are described in the basic text edited by von Hippel (1954). Experimental methods have been more recently reviewed by Hedvig (1984), and in the book edited by Kremer and Schönhals (2002). Much practical information, including detail on the design of sample holders and the preparation of specimens, is also given in national and international publications of standard test methods, e.g. those published by the *American Society for Testing and Materials* (ASTM), the *International Electrotechnical Commission* (IEC) and *Comité Européen de Normalisation* (SEN). The National Physical Laboratory (2003) has also published a very helpful practical guide to dielectric measurements at radio and microwave frequencies.

6

Dielectric breakdown

6.1 Introduction

If the voltage across a piece of dielectric material is steadily increased, there must come a point at which the insulating barrier properties will be exceeded, and total breakdown will ensue. Characteristically, the final event is localised, sudden and catastrophic. At the high voltage involved the quick release of so much electrical energy usually means that the material burns out in the breakdown region between the electrodes. Although the course of the breakdown is usually determined by some localised imperfection or non-uniformity in the material or its surroundings, we still try to define a relevant material property. Thus the existence of a maximum voltage which an insulator will support for a long time without failing leads to the concept of a dielectric strength, defined as the breakdown voltage divided by the thickness of the insulator, i.e. a maximum electric field which the material can sustain indefinitely. The intrinsic dielectric strength of a homogeneous solid is evidently very high, usually in excess of $100\,\mathrm{MVm^{-1}}$, and proves to be a very elusive fundamental property. One reason for this is the rather irreproducible nature of breakdown behaviour: the results on a set of samples which have been carefully prepared to be as similar as possible will always show a large scatter. In other words, the breakdown process, and its initiation in particular, are inherently stochastic in nature, depending critically on the statistical variation of the composition and structure of each individual sample. A second reason is that a sample may often more easily fail in many different ways which have more to do with its environment, its physical state and purity and the type of electrode used than with its basic constitution. These alternative breakdown mechanisms are the ones which generally limit the effective strength of an insulator in a practical situation and they are difficult to avoid altogether. The quest for an intrinsic strength has consequently become somewhat academic. In this context we must be wary of standard tests of dielectric strength of solids – they usually do not in fact measure an intrinsic value, because they allow premature discharges to occur in the surrounding gaseous or liquid medium. The main approach that has been adopted in industry for the necessary assessment of breakdown behaviour of materials in exercises of product improvement and replacement has been to design special tests. These simulate the conditions of the applications concerned. In this context the ageing or deterioration of the material during service, e.g. chemical degradation in strong sunlight and cracking under prolonged mechanical stress, is vitally important too, since changes brought about in this way almost always introduce electrical weaknesses. Furthermore, under the application of high electric fields, there are additional modes of progressive and cumulative ageing that lead on to final breakdown. The spatial distribution of these failure modes reflects the stochastic nature of breakdown, and tree-like patterns of destruction are consequently a characteristic feature of them.

In the following sections we discuss briefly the principal mechanisms of electrical breakdown in solid polymers and some of the consequences for the use of these as insulating materials. A brief outline of the statistical treatment of breakdown data is given in the appendix (Section 6.9).

6.2 Electronic breakdown

Insofar as there is such a thing as an intrinsic breakdown strength of an insulating material, the most likely source of the instability will be the small numbers of electrons which are available for acceleration by the applied field. By analogy with the mechanism of sparking in gases, one can imagine that a Townsend-like avalanche may occur whenever the field is high enough for a conduction electron to gain sufficient energy to excite more electrons by collision. Actually, the breakdown strengths calculated for insulating solids on the basis of the simple gas model for the cumulative process are much too high. More successful approaches take into account the high electron mobilities in regular lattice structures, and are rooted in band theory. In one of the first theories of electronic breakdown von Hippel considered the situation in a pure, crystalline material at low temperatures. In such a system electron–electron collisions are rare, and electrons are principally scattered by interaction with the vibrations of the lattice (electron–phonon collisions). Continual thermalisation of the electrons by this process sets the limit to the kinetic energy that an electron can acquire through acceleration in the electric field. One immediate prediction is that a rise in temperature, which enhances the scattering of electrons by the lattice (the mean free path of electrons between collisions with phonons is shortened), will increase the rate at which electrons lose the energy that they have gained from the electric field. The dielectric strength will therefore increase with temperature. Careful measurements at low temperatures on simple ionic compounds, e.g. sodium chloride, have largely substantiated this. At high temperatures and in crystals containing many impurities and imperfections, as well as in molecular and amorphous solids, including polymers, Fröhlich recognised that the combined number of electrons in localised excited states and in a conduction band, if it exists, will be relatively high, so that electron–electron scattering must predominate. Since the transfer of energy from these electrons to the lattice is rather slow, the electron temperature T may rise above the lattice temperature T_0 when an electric field feeds energy directly to the conduction electrons. Since we only have rather limited knowledge of conduction processes in polymers, a good model for detailed calculations is not available and we can only make a reasonable guess at a generally appropriate one.

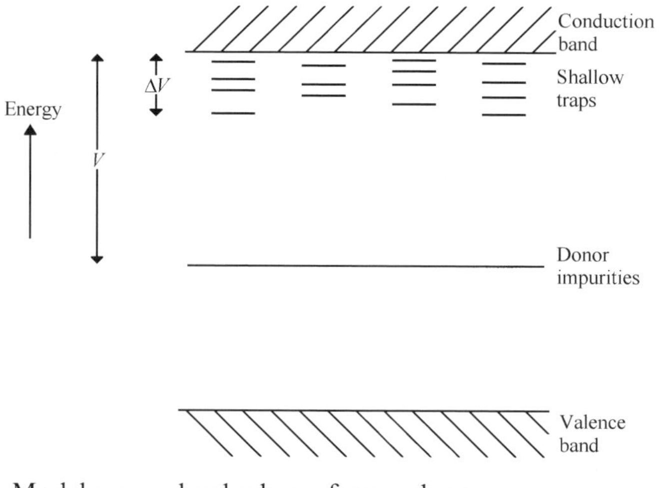

Fig. 6.1 Model energy-level scheme for a polymer.

Fröhlich (1947) based his calculations on the hypothesis of the energy-level scheme shown in Fig. 6.1, where conduction electrons are derived from impurity levels lying deep (V = 1 eV or more) in the forbidden zone. There is also a set of shallow traps spread below the conduction-band edge ($V \gg \Delta V \gg kT$). In outline, the theory of breakdown is then as follows. In an applied electric field E, energy is transferred directly to the conduction electrons (charge e, mass m) at a rate $A = jE$, where j is the current density. If we suppose that each electron is accelerated in the field direction for an average time 2τ between collisions at which its energy is completely randomised, then the mean drift velocity of the conduction electrons in the field direction is $-(eE/m)\tau$. The current density due to drift of all the conduction electrons (concentration n) is $ne^2E\tau/m$ and we consequently have

$$A = \frac{e^2\tau n E^2}{m}. \tag{6.1}$$

From semiconductor theory the concentration of conduction electrons is given by

$$n = C(T)\mathrm{e}^{-V/2kT}, \tag{6.2}$$

where $C(T)$ is a factor which depends on the effective density of states in the conduction band and which varies only slowly with temperature. Substituting for n in Equation (6.1),

$$A = \frac{e^2\tau E^2}{m} C(T)\mathrm{e}^{-V/2kT}. \tag{6.3}$$

The energy thus gained from the field will be rapidly shared amongst the conduction and trapped electrons by electron–electron collisions.

Let us now consider the loss of energy from the electrons to the molecular lattice. The concentration of trapped electrons will be given approximately by

$$n_{\mathrm{T}} = C_{\mathrm{T}}(T)\mathrm{e}^{-(V-\Delta V)/2kT}, \tag{6.4}$$

where the factor $C_{\mathrm{T}}(T)$ now relates to the density of trapping states and again varies slowly with temperature. If we assume a high density of traps, the concentration of trapped electrons will greatly exceed the concentration of conduction electrons, and it will be mainly trapped electrons that are responsible for transferring energy to the lattice.

For simplicity, suppose that energy is transferred to a single vibrational mode of frequency ν in single phonon jumps of energy $h\nu$. The number of transitions per unit volume per second W_{e} of an electron from a level of energy U to one of $U + h\nu$ involving phonon emission by the lattice is the product of the electron transition probability $P(U)$, the concentration $f(U)$ of electrons in levels of energy U, and the average number density $N_{\mathrm{v}}(T_0)$ of vibrational quanta (phonons) present in the lattice at temperature T_0, i.e.

$$W_{\mathrm{e}} = P(U)f(U)N_{\mathrm{v}}(T_0). \tag{6.5}$$

Similarly, for transitions in the reverse direction with phonon absorption:

$$W_{\mathrm{a}} = P(U)f(U + h\nu)[N_{\mathrm{v}}(T_0) + 1]. \tag{6.6}$$

Hence, the rate of energy loss B per unit volume to the lattice from trapped electrons is

$$\begin{aligned} B &= h\nu \sum (W_{\mathrm{a}} - W_{\mathrm{e}}) \\ &= h\nu \sum P(U)\{f(U + h\nu)[N_{\mathrm{v}}(T_0) + 1] - f(U)N_{\mathrm{v}}(T_0)\}, \end{aligned} \tag{6.7}$$

where the summation extends over all trap levels. Assuming that an equilibrium exists in the trap population,

$$\frac{f(U + h\nu)}{f(U)} = \mathrm{e}^{-h\nu/kT}. \tag{6.8}$$

Then, replacing $P(U)$ for all levels by a mean value P, noting that $\sum f(U) = n_{\mathrm{T}}$, and using the Planck expression $(\mathrm{e}^{h\nu/kT_0} - 1)^{-1}$ for the energy distribution of photons, we obtain from Equation (6.7)

$$B = h\nu P\left(\mathrm{e}^{h\nu/kT} - 1\right)^{-1} C_{\mathrm{T}}(T)\mathrm{e}^{-(V-\Delta V)/2kT}\left(\mathrm{e}^{h\nu/kT_0 - h\nu/kT} - 1\right). \tag{6.9}$$

Taking into account that τ and P will depend only weakly on temperature, Equations (6.3) and (6.9) show that the rates of energy gain and loss per trapped electron, given by A/n_{T} and B/n_{T} respectively, vary with temperature in the manner indicated in Fig. 6.2, where a family of energy-transfer curves for different applied fields is shown. At low field strengths, an equilibrium electron temperature will be attained, energy lost to the lattice just balancing energy gained from the field; but at fields above a critical value E_{c}, no balance is

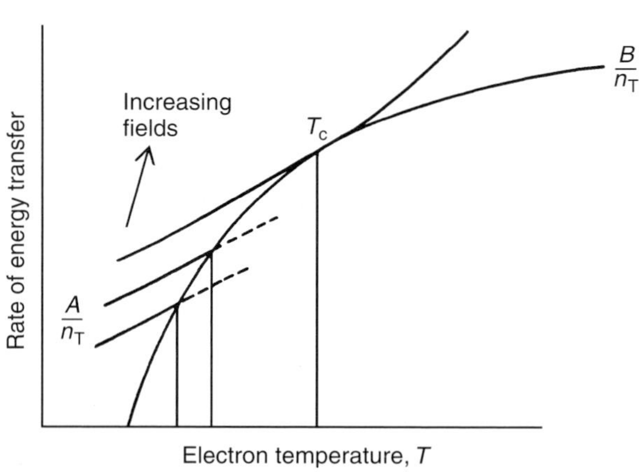

Fig. 6.2 Energy-transfer curves for increasing applied fields, showing the onset of electron temperature runaway. After Fröhlich (1947), by courtesy of the Royal Society.

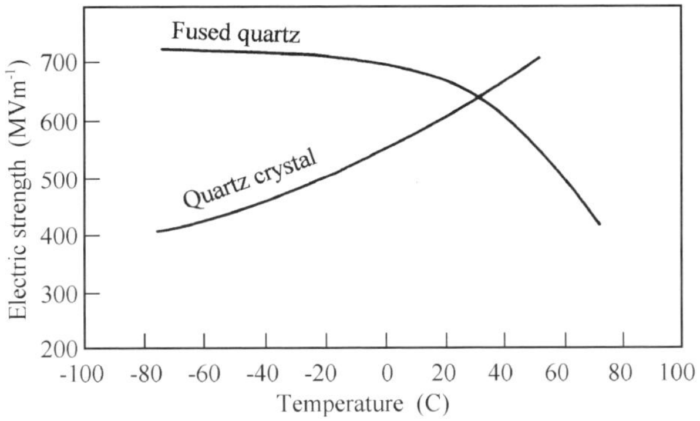

Fig. 6.3 Comparison of breakdown behaviour of crystalline and amorphous (fused) quartz. Reprinted with permission from von Hippel and Maurer (1941). Copyright 1941 by the American Physical Society.

possible and this defines the breakdown field strength. Substituting for the critical condition in the energy balance equation ($A = B$), and taking the first approximation in exponential expansions, we obtain the relation

$$E_c \propto e^{\Delta V/4kT_0}, \tag{6.10}$$

which predicts a decrease in breakdown field strength as temperature rises. The physical reason for such behaviour is the increase in the relative number of conduction, as opposed to trapped, electrons which can gain kinetic energy from the field. Von Hippel and Maurer (1941) found confirmatory evidence of this effect in fused (amorphous) quartz, in sharp contrast to the results for pure crystalline quartz (Fig. 6.3).

As we intimated in the introduction, it is arguable whether truly intrinsic breakdown has ever been observed in polymers. In this sense an intrinsic breakdown strength must represent an upper limit to any value that can ever be realised experimentally. The most reliable measurements, judged by the high values that were obtained in comparison with those from other measurements, have been made with recessed specimens, illustrated in Fig. 6.4, with evaporated aluminium electrodes. The recess design neatly places the high stress just where it is required across a thin layer, whilst at the same time avoiding excessive stresses in the air or other medium surrounding the edges. Results for polyethylene (Lawson, 1966), shown in Fig. 6.4, definitely indicate a fall in dielectric strength with temperature in accord with Fröhlich's theory. The extra, steep fall at higher temperatures is probably not electronic in origin, as we shall see in the next section. Somewhat higher values still for the breakdown strength of low-loss polymers have been obtained with a more complicated arrangement, where the specimen is embedded in epoxy resin

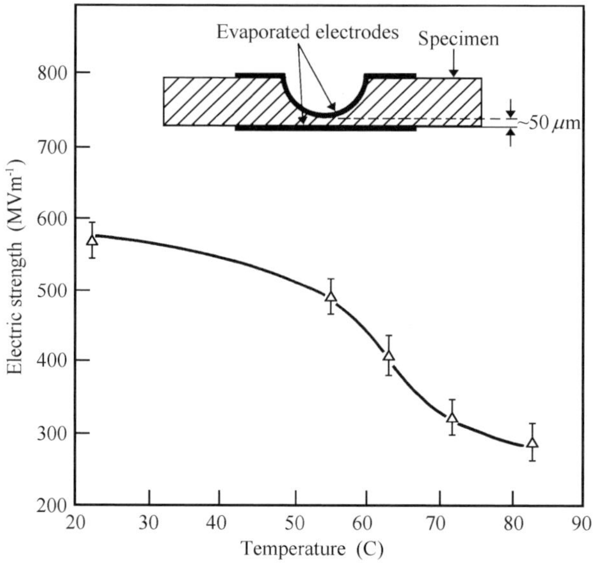

Fig. 6.4 The variation of the electric strength of polyethylene with temperature. The inset diagram shows the type of recessed specimen used. Reproduced from Lawson (1966) with permission of the Institution of Electrical Engineers.

(McKeown, 1965), but there is some uncertainty as to whether this result is an experimental artefact or not.

In order to predict absolute dielectric strengths we need to have more detailed information than is yet available about electronic states and mobilities in polymers. For the present we can only conclude that there is satisfactory agreement between the form of the theoretical results, based on a rather general electronic model, and the best experimental results. To the extent that the model is a very reasonable one, we can say that we can understand intrinsic breakdown behaviour. Measurement of pre-breakdown currents, especially with pointed electrodes which impose regions of very high field strength at their tips when embedded in the material, suggests that electronic carrier production either by injection from the electrodes (Schottky emission) or from impurities (Poole–Frenkel effect) may play a part in the breakdown process. More work is required, however, before this can be fully understood.

6.3 Electromechanical breakdown

Electrodes attached to the surface of a specimen during a dielectric breakdown test will exert a compressive force on the specimen by mutual Coulombic attraction of the electrodes as the voltage V is imposed. If this is sufficient to cause appreciable deformation at fields below the intrinsic breakdown value, the dielectric strength will be reduced. The attractive force F is given by the differential of the energy U stored in the system with respect to the thickness d of the material at constant applied voltage:

$$F = \left(\frac{\partial U}{\partial d}\right)_V = \frac{\partial}{\partial d}\left(\frac{1}{2}CV^2\right)_V. \quad (6.11)$$

Substituting for the parallel-plate capacitance C of the specimen (cross-sectional area A, relative permittivity ε'), we obtain an expression for the compressive force per unit area:

$$\frac{F}{A} = -\frac{1}{2}\varepsilon_0\varepsilon'\left(\frac{V}{d}\right)^2. \quad (6.12)$$

At equilibrium the electrical force causing compression is balanced by the elastic restoring force, as expressed by the equation:

$$\frac{1}{2}\varepsilon_0\varepsilon'\left(\frac{V}{d}\right)^2 = Y \ln\frac{d_0}{d}, \quad (6.13)$$

where Y is Young's modulus for the material,[1] and d_0 is the initial thickness of the specimen. Solving this equation for d then gives the equilibrium thickness of the slab under the applied voltage. Since $d^2 \ln (d_0/d)$ has a maximum value when $d/d_0 = \exp(-1/2) \approx 0.6$, no real value of V can produce a stable situation for values of d/d_0 less than 0.6. Further increase of V produces mechanical collapse, the critical electric stress being

$$E_c = \left(\frac{Y}{\varepsilon_0\varepsilon'}\right)^{\frac{1}{2}}. \quad (6.14)$$

The highest apparent dielectric strength E_a which can be observed is therefore

[1] The logarithmic form of the mechanical stress–strain relation is required, because the strains may be large in thermoplastic polymers. Young's tensile modulus is defined as:

$$Y = \frac{\text{tensile stress increment, } \Delta\sigma}{(\text{increase in length, } \Delta d)/(\text{length, } d)} = \frac{d\Delta\sigma}{\Delta d}.$$

Hence, for a linear, elastic material the stress σ at extended length d is given by:

$$\sigma = \int_{d_0}^{d} \frac{1}{d}\,\mathrm{d}d = \log_e \frac{d}{d_0}.$$

$$E_a = \frac{V_c}{d_0} = \frac{d}{d_0} E_c \approx 0.6 \left(\frac{Y}{\varepsilon_0 \varepsilon'} \right)^{\frac{1}{2}}. \tag{6.15}$$

(We should note that the above simple treatment totally ignores any departure from linear elastic behaviour at large strains.)

Stark and Garton (1955) recognised this behaviour in polyethylene when they compared ordinary polyethylene with an irradiated sample whose Young's modulus (and also dielectric strength) did not decrease so much with temperature.

Apparently low values of breakdown strength of many rubbery materials are described quantitatively by Equation (6.15), and most plastics fail by the electromechanical mechanism at high temperatures.

6.4 Thermal breakdown

Whenever there is sufficient conductivity present in a dielectric to produce appreciable Joule heating in an applied field, the possibility of thermal runaway exists, for the accompanying rise in temperature will increase the conductivity still further. In alternating fields there may be additional heat generated through one or more relaxation processes, as described in Chapter 3, and this will hasten the onset of any thermal runaway condition. Whether thermal breakdown will eventually develop in this way or not will also depend on the rate at which heat is conducted away to the surroundings. The heat balance equation is expressed by the following continuity equation:

electrical power dissipated in material per unit volume	=	rate of increase in heat content	+	rate at which heat is conducted away,

$$\text{i.e.}\quad \sigma E^2 = C_v \frac{dT}{dt} - \mathrm{div}(\kappa\, \mathrm{grad}\, T), \tag{6.16}$$

where σ is the electrical conductivity of the material, C_v is its heat capacity per unit volume and κ is its thermal conductivity. It is not possible to obtain a general solution of this equation, because σ, C_v and κ are all functions of temperature. Approximate numerical solutions have been obtained, however, for a semi-infinite dielectric slab between electrodes maintained at a constant temperature, and these predict that the temperature of the hottest part of the dielectric at the centre of the slab varies with time for different constant applied voltages in the way shown by the curves in Fig. 6.5 (Whitehead, 1951). The main feature is a critical voltage V_c above which the temperature increases indefinitely. This

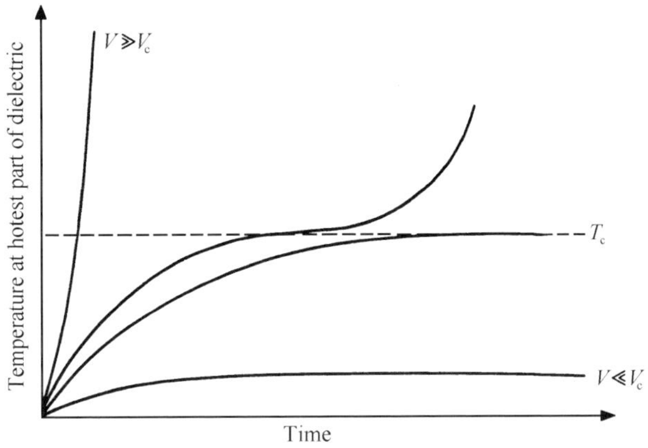

Fig. 6.5 Calculated heating curves for increasing voltages applied across an infinite slab, showing the onset of thermal runaway. Reproduced from Copple *et al.* (1939) with permission of the Institution of Electrical Engineers.

will reduce the intrinsic dielectric strength until breakdown ensues, unless melting or chemical decomposition intervenes. We can appreciate that the critical voltage V_c will be independent of thickness in this particular case, because an increase in thickness will decrease the electric field for a given applied voltage, but also reduce the rate at which heat can escape from the central region by conduction. For this reason thermal breakdown *voltages* are sometimes quoted. Thermal breakdown in polymers is most liable to occur (a) at high ambient temperatures where electrical conductivity may become appreciable, e.g. Nylon-6,6 above 90 C, or (b) at high frequencies where the rate of heat dissipation by dipolar relaxation processes may be high.

6.5 Breakdown caused by gas discharges

The dielectric strength of a gas is very much less than that of a solid insulator, being of the order of 3 $\mathrm{MVm^{-1}}$. Consequently, during the application of a high voltage to a solid specimen, discharges are likely to occur at an early stage in any gas that is at the edges of the electrodes or may be occluded as bubbles in the solid. Such external or internal discharges tend to damage the solid and repeated discharges lead to dielectric failure. Organic polymers are especially prone to this type of failure, because bombardment of the polymer molecules by ions in the discharge is able to break chemical bonds. The reactive free radicals thereby produced, together with any available oxygen (often dissolved in the polymer), enter into a degradative chain reaction, causing damage to the insulator. Each discharge in a cavity is extinguished as charges are deposited on the end walls, thereby reducing the field within the cavity. With an applied

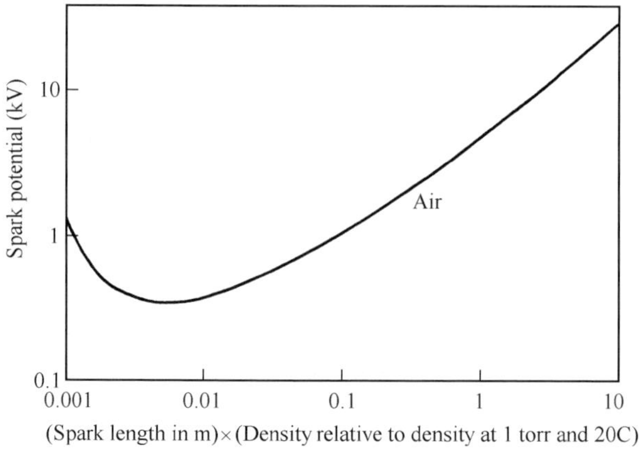

Fig. 6.6 Paschen's law curve for air. Reproduced with permission from Whitehead (1951).

direct voltage, discharges will recur only after surface charges deposited by the previous discharge have had time to leak away, but under an alternating voltage, discharge will be repeated with every half-cycle and the problem becomes much more severe.

Before we can recount the effects of gas discharges in more detail we need to say something about the nature of gas discharges themselves. In a uniform field, breakdown is a cumulative Townsend electron avalanche process. Breakdown voltages follow Paschen's law, which says that the minimum voltage V_g necessary to produce a spark across a gap depends on the product $l\rho$, where l is the length of the gap and ρ is the density of the gas. Paschen's law for air is shown in Fig. 6.6. From the latter we may deduce that (a) at constant gas density the breakdown field strength increases towards shorter gaps, and very steeply so for gaps smaller than 1 mm at atmospheric pressure, and (b) for a given gas there is a minimum voltage required for breakdown, regardless of gap or density. (For air it is about 350 V at room temperature.)

6.5.1 Internal discharges and electrical treeing

Internal discharges are a common feature of polymeric insulators, because voids are very easily left in the material at the moulding stage, or are associated with irregular, particulate contaminants. Since the relative permittivity of the gas filling a void ε_g is generally less than that of the polymer ε', the electric field E_v in the void will be greater than the field E in the surrounding medium, depending to some extent on the shape of the void. The effect is greatest for a disc-like cavity lying perpendicular to the field, when

$$E_{\mathrm{v}} = \frac{\varepsilon'}{E\varepsilon_{\mathrm{g}}}. \tag{6.17}$$

The maximum voltage which can be applied to the specimen without a discharge in the cavity is then, taking $\varepsilon_{\mathrm{g}} = 1$, given by

$$V_{\mathrm{i}} = E_{\mathrm{g}}d'[1 + (d/d' - 1)/\varepsilon'], \tag{6.18}$$

where E_{g} is the dielectric strength of the enclosed gas and d, d' are the thicknesses of the specimen and the cavity, respectively. This kind of relationship, in combination with Paschen curves for gaseous breakdown, suggests that the voltage at which discharges begin decreases with increase in cavity size, and this has been verified for air in artificially-made cavities. To achieve long service life in an insulator, it is clear that gas discharges must be completely absent at the working voltage. For this reason the voltage at which discharges start, commonly referred to as the discharge inception voltage (DIV), is an important characteristic of a product for high-voltage applications. Very sensitive DIV detectors incorporating visual displays are commercially available for testing purposes.

When the applied field is sufficient to initiate a discharge in a pre-existing cavity, the resulting damage produces a tubular channel in the polymer with a diameter of 1–5 μm, progressing generally in the field direction. The process develops more readily under AC conditions, because discharges continue to be triggered at each field reversal. The extending tubules of destruction progress in a way that seeks out pathways of predisposed weakness, e.g. interfacial boundaries in non-homogeneous materials. There is also a strong tendency for branching to occur, and this results in a pattern of growth called an *electrical tree*. There are two generic forms of electrical tree: vented and bow-tie, as shown in Fig. 6.7. Vented trees start at the boundary of the insulation next to an electrode and grow in tree- or bush-like fashion towards the opposite electrode. Bow-tie trees grow simultaneously towards the two electrodes from an interior discharge initiation point. The inception and propagation of the two types of trees vary in their detail and there is a large literature on the subject.

Vented trees are most easily induced artificially by means of needle electrodes inserted into a sample. On application of a high voltage between the needle and an electrode on the other side of the material, discharges start at the needle tip where the divergent field is very high and where a small void normally exists. Figure 6.8 shows a photograph of an electrical tree grown in this way. Experiments using this technique have been used extensively to investigate the many factors that affect tree growth and to search for tree-resistant materials.

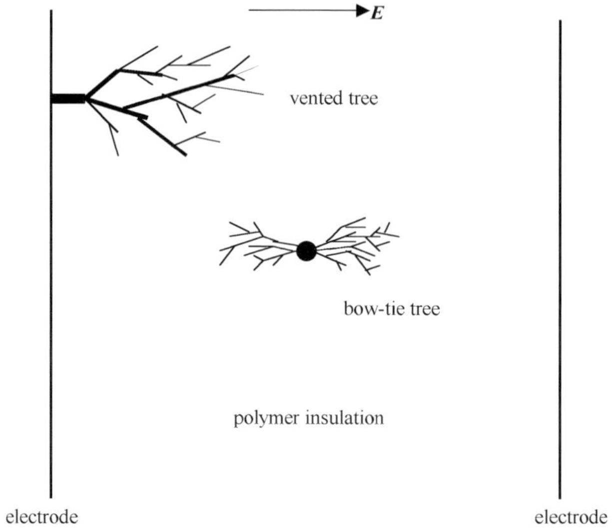

Fig. 6.7 Types of tree pattern: vented and bow-tie.

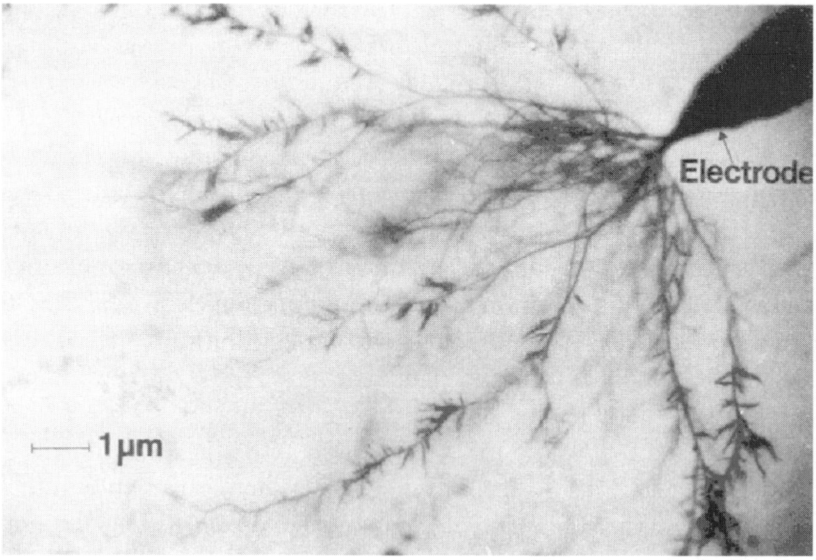

Fig. 6.8 Photograph of electrical discharge channels in low-density polyethylene after application of 20 kV (rms) at 50 Hz for 200 min. Reproduced from Billing and Groves (1974) with permission of the Institution of Electrical Engineers.

Sometimes there will be a significant delay, or tree initiation time, before growth propagates. This depends on the size and shape of the initial cavities. Needle-shaped cavities give rise to rapid tree initiation, but spherical cavities may require a longer time, during which the cavity wall is eroded before tree growth proper begins.

The gas discharge process will continue to drive the growth of the tree until the tree structure joins the electrodes. Large destructive currents can then flow, and complete breakdown of the insulation will ensue.

There are various ways of inhibiting tree growth so as to improve the ability of the material to withstand higher voltages, notably reducing the incidence of voids within the material and/or filling interstices with an insulating liquid.

When voids are substantially absent in an insulating material, electrical trees may still occur, though generally at higher fields. In these circumstances there are two distinct stages in the treeing process: a discharge-free, induction period during which an initial cavity is formed, by a different kind of mechanism (Shimizu and Laurent, 1998), followed by the usual period of tree propagation driven by gas discharge activity. A crucial element of the tree inception mechanism is injection of charge carriers in regions of enhanced field, e.g. near an asperity on an electrode or near a contaminant particle. Any trapping of the charges further increases the local field when the voltage is reversed under AC. Under these conditions some injected electrons are accelerated by the field, so that they acquire sufficient kinetic energy to break chemical bonds on impact with the polymer. Further degradation ensues, eventually leading to the formation of a microcavity where gas discharge activity may begin. The same hot electrons are also able to cause collisional excitation of molecules, which in turn produces luminescence – often seen as a precursor to treeing. Tree propagation then ensues. At higher fields, tree initiation may proceed via localized electron-avalanche breakdown. This is a much faster process than the cumulative degradation that occurs at lower fields. It is a process that is responsible for electrical trees under high voltage impulses.

6.5.2 External discharges and surface tracking

It is of importance to be able to design high-voltage insulation systems that do not fail by the premature onset of air discharge at electrode boundaries. With this in mind consider a uniform slab of insulation material (relative permittivity ε') of thickness d between a smooth-edged electrode and an earth electrode (see inset in Fig. 6.9). At the edge, the applied voltage V will divide capacitively across a number of parallel air gaps of gradually increasing thickness with

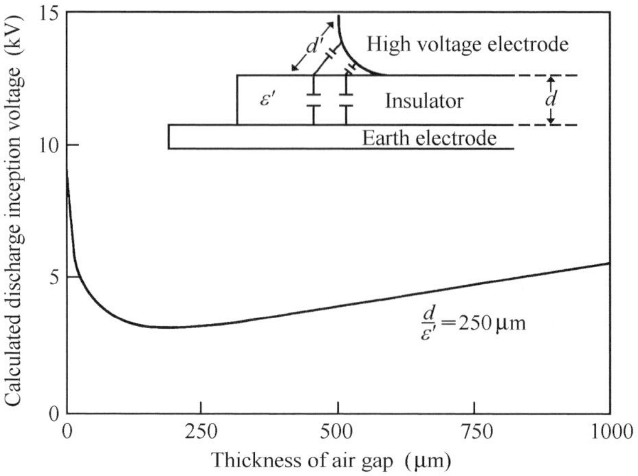

Fig. 6.9 Calculated inception voltages for edge discharges. The inset diagram shows the electrode arrangement. Reproduced from Halleck (1956) with permission of the Institute of Electrical and Electronics Engineers, Inc.

elements of the insulator in series with them. For each pair of series elements, involving an air gap d',

$$V = E_a(d' + d/\varepsilon'), \tag{6.19}$$

where E_a is the field in the air gap. Substituting the breakdown strength of air for different air gaps in Equation (6.19), a discharge-inception curve may be calculated, as shown in Fig. 6.9, and the minimum inception voltage determined by inspection. In this way the possibility of breakdown by edge discharges at a given voltage may be checked theoretically.

In many practical situations an insulator is most likely to fail through deterioration in its surface insulation properties. Deposition of dirt and moisture must inevitably allow some conduction over the surface, although good insulators will recover quickly as the heating effect of the leakage current tends to clean up the surface. Under these conditions some polymers tend to suffer permanent damage, called *tracking*, leading to complete electrical breakdown. Tracking starts as the surface dries out and a narrow, dry band forms. Most of the voltage is then dropped across the higher resistance of the dry band and this can cause sparks to pass through the surface layer. If these sparks char the polymer, a conductive track may develop across the surface of the insulator and this will finally result in flash-over, the polymer sometimes bursting into flames at this stage. Standard tracking tests, designed to meet various sets of operating conditions, are used to rank polymers in an order of susceptibility to tracking. In one such test, the dust-fog test (ASTM Test Method D2132–6T), the polymer is coated with a solid contaminant

Table 6.1 *Results of a dust-fog test*

Material	Time to track (hr)
Polyethylene	33
Poly(methylmethacrylate)	162
Polypropylene	191
Polytetrafluoroethylene	600
Poly(vinylchloride)	0.3
Polystyrene	0.9

containing 3% sodium chloride, and is exposed to an artificial fog. A potential difference of 1500 V is applied across the surface via copper electrodes one inch apart. The criterion for failure by tracking is the time taken for the electrodes to be bridged by a track across the surface. (Non-tracking materials fail by erosion through the thickness of the sample to an earthed plate underneath.) Table 6.1 lists typical lives of polymers in the test (Billings, Smith and Wilkins, 1967).

One generalisation that can be made by examining the results from different tests on many organic polymers is that polymers based on aromatic compounds or with weakly bonded or easily oxidised pendant groups are very liable to pyrolise in a way that deposits carbon, and are especially prone to tracking.

6.6 Long term endurance under electrical stress

The previous sections have described a variety of breakdown mechanisms that may operate in different kinds of materials and situations. All of them take a finite time to produce the final breakdown event, although a relatively short timescale is expected in each case, with the exception of electrical treeing. What happens in the long term under levels of electrical stress that are to be used in electrical applications, and which are generally set well below the short term breakdown levels, is a major concern for manufacture of reliable electrical equipment. In this regime, below the extremes of electrical strength, the breakdown process is, however, complex and multi-staged, with elements of several of the previous mechanisms involved, and with additional important features that depend on the very distinctive ageing processes in polymeric materials. For practical purposes, engineers have used various empirical formulae to relate the characteristic time to breakdown t_c to the applied voltage V. The one that is most commonly used is a 3-parameter inverse power law relationship

$$\begin{aligned} t_c &= \infty & V &< V_{th} \\ &= \text{const.}(V - V_{th})^{-n} & V &> V_{th}, \end{aligned} \tag{6.20}$$

where V_{th} is a threshold voltage below which breakdown does not occur. Only now are convincing molecular interpretations of this kind of ageing behaviour of polymeric materials under electrical stress emerging.

6.6.1 Endurance experiments

Characterising the fundamental endurance of a material under the application of a high electric field must take into account the intrinsically stochastic nature of the breakdown process, so that large sets of samples are required, with rigorous control of experimental conditions to avoid unwanted side-effects, e.g. external discharges, which can lead to premature failure. The statistical method of dealing with times to failure developed by Weibull (1951) is most commonly used to characterise the results – see Section 6.9. Recent experiments (Griffiths *et al.*, 1998) used multi-arrays of McKeown-type test cells connected in parallel to an AC (50 Hz) high-voltage supply via a special high-voltage circuit-breaker. When a sample failed, the circuit-breaker automatically disconnected that cell from the high-voltage supply, whilst maintaining the connection of the supply to the rest of the cells. This meant that sufficient data for sound statistical analysis on the times to breakdown as a function of applied field were collected. An example of results for a high-grade polyethylene with an applied field of 100 MVm^{-1} is presented in Fig. 6.10, using a Weibull function to fit the data. We can see that the data conform well to this

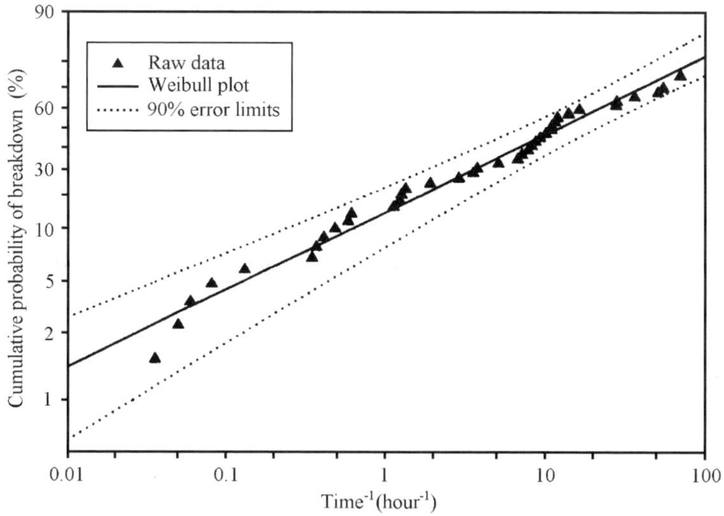

Fig. 6.10 Weibull plot of the cumulative probability of breakdown for a set of polyethylene film samples as a function of time under an applied field of 100 MVm^{-1} at room temperature.

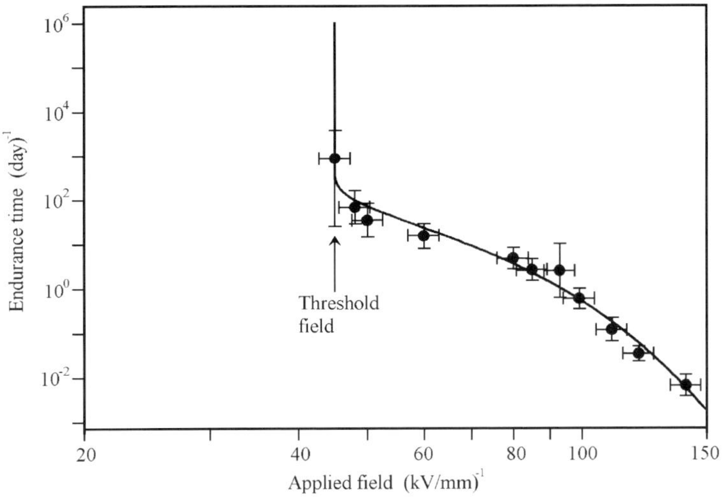

Fig. 6.11 Electrical endurance curve for polyethylene film.

function. For the same material Fig. 6.11 shows the endurance or life curve obtained by plotting representative or characteristic failure times (corresponding to 63.2% cumulative probability values from Weibull plots) as a function of applied stress at room temperature. These refined endurance data can be used as an empirical basis for predicting the life of the material at any particular applied field for engineering design purposes. Scientifically, the challenge is to explain the behaviour in terms of the underlying molecular mechanisms. The data clearly show that the electric field strongly affects the ageing process, either by accelerating processes that are normally present or by introducing new ones. It is also apparent from the data that there is a threshold field: only above this field is there any appreciable effect of the field on ageing or failure.

6.6.2 *Electrokinetic modelling of dielectric endurance*

An electrokinetic model of breakdown in polymers aims to explain the kind of breakdown process that occurs over a relatively long timescale in high-quality (uncontaminated, defect-free) materials, in contrast to the quick and violent modes of breakdown that are often experienced, when materials are over-stressed. Undoubtedly, we require a multi-stage model, which will vary somewhat, depending on the exact composition of the material and its environment.

First, consider the mechanical stresses developed in a material as an electric field E is applied to it. The material will deform until the Maxwell electric stresses are balanced by changes in the mechanical stresses. For an

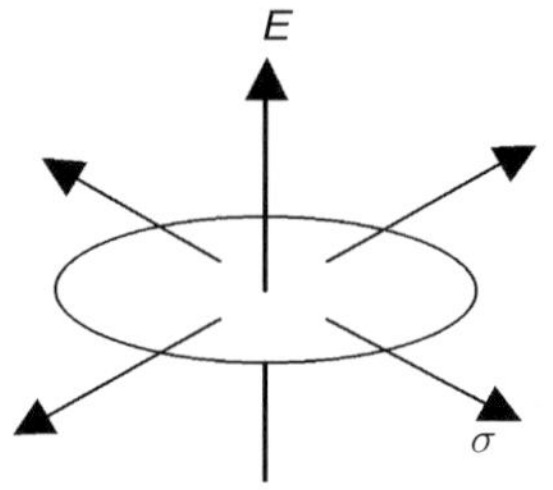

Fig. 6.12 Planar tensile stress orthogonal to applied electrical field.

unconstrained, isotropic elastic material (neglecting changes in relative permittivity with deformation) the resulting mechanical stresses consist of a compressional stress in the direction of the applied field and a planar tensile stress in the orthogonal direction. The latter stress, tending to pull the material apart, is illustrated in Fig. 6.12. The magnitude of the tensile planar stress σ caused by an electric field E in the z-direction is given by[2]

$$\sigma = \sigma_{xx} = \sigma_{yy} = \frac{1}{2}\varepsilon_0 \varepsilon E^2. \tag{6.21}$$

The next step is to enquire how the tensile stress imposed by an electric field will operate at the molecular level in a way that might lead to dielectric failure. Probably the simplest approach is to assume that the mechanical tensile stress begins to rupture the material, opening up cracks in the field direction by breaking chemical bonds in the polymer chains. Chemical bonds are, however, very strong, and the energy barrier to break a covalent C–C bond is of the order of 300–500 $\mathrm{kJmol^{-1}}$. Therefore direct bond breaking is unlikely to be responsible for the early stages of the failure process. A clue to what may lie behind the distinctive time dependence of the breakdown process in polymers comes from recent work on *slow brittle fracture* observed in polyethylene pipes under mechanical stress. The interpretation of this phenomenon calls on our knowledge of polymer morphology, and explains how a fracture can develop from a gradual untangling of polymer chains. The energy barriers for this kind of mechanism depend on intermolecular van der Waals forces, and are much lower that those for chemical processes. On this basis we may develop at least a

[2] The mechanical effects of an electric field on a material are usually expressed by means of the *electrostriction* tensor defined by:

$$u_{ij} = \gamma_{ijkl} E_k E_l ,$$

where u_{ij} is a component of the strain tensor, E_k and E_l are components of the electric field vector, and γ_{ijkl} is the fourth-rank electrostriction coefficient tensor. Davies *et al.* (2002) have recently analysed the electrostrictive response of a polar rubber in detail. They derived expressions for the mechanical stresses developed under an applied electric field, and have compared theoretical strain responses with those measured experimentally.

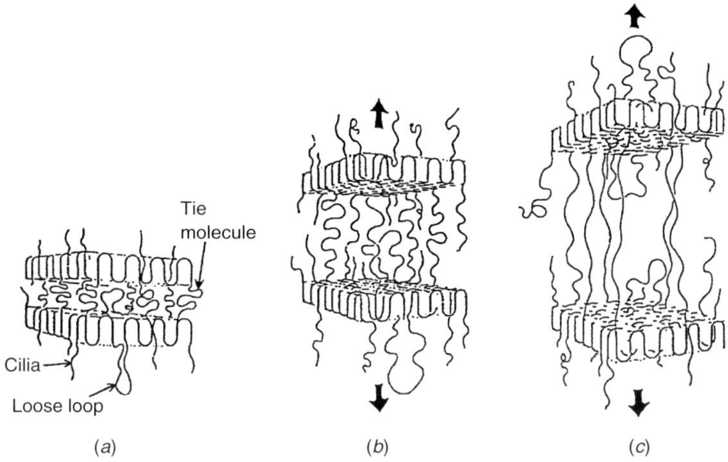

Fig. 6.13 Initial steps in the deformation of polyethylene. Reprinted from Lustiger and Markham (1983). Copyright 1983, with permission from Elsevier.

qualitative account of the sequence of molecular events that explain the typical course of degradation under an applied electrical stress.

In a semi-crystalline polymer like polyethylene we expect that the initial effect of the tensile force is to deform the softer, amorphous phase, enabled by the reptational motion of the loosely entangled chains. In effect, the amorphous region is extended, moving the crystalline lamellae further apart, and, at this early stage, the more tightly held chains within the lamellae remain largely undisturbed. As the stretching progresses, more and more of the chains which pass from one lamella to another, the *tie molecules*, become taut, and further separation is only allowed if the tie molecules can be pulled out of the lamellae. The situation is illustrated in Fig. 6.13. At this stage, the tie molecules play a key role, as they also do in the case of slow brittle mechanical fracture. It is worth noting that the process differs from the usual macroscopic, ductile yield behaviour, which requires larger tensile forces to shear and fibrillate the crystalline lamellae to give much larger overall strains ($\gg 10\%$).

In polyethylene the α_c-relaxation process (see Section 3.4) enables the movement of chains into and out of the crystalline lamellae. Theoretical treatments have demonstrated that it most probably proceeds by propagation of a localized twist (180° rotation) about the chain axis extending over 12 CH_2 units (Fig. 6.14). As the twist defect travels along the chain, it rotates and translates the chain by half a unit cell (i.e. by one CH_2 unit) – this is termed the c-shear process (Mansfield and Boyd, 1978). The activation energy for this process is about 110 kJmol^{-1}, corresponding to the extra energy required to introduce the twist defect into the crystal. Once formed, the twist can freely

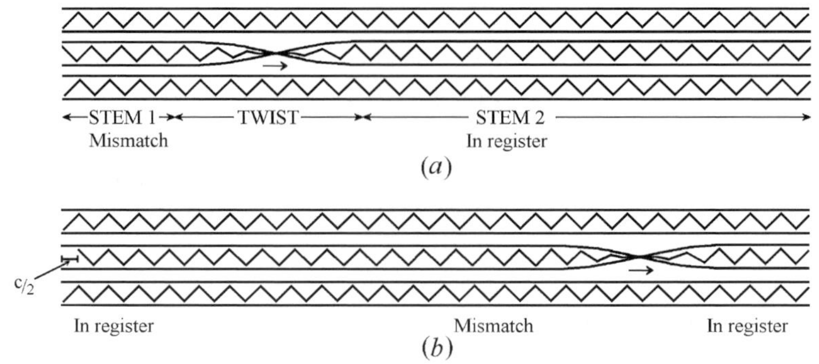

Fig. 6.14 Illustration of the α_c (c-shear) relaxation process in polyethylene.

travel through the crystal without incurring further significant expenditure of energy. As the tie molecules are disengaged one by one, all the tension will be sustained by fewer and fewer of them, with bond breakage almost certain to occur in the last remaining tie molecules. Indeed, spectroscopy has provided evidence of this. Bond scission inevitably produces reactive free radicals, the notorious precursors of oxidative degradation of organic polymers.

Oxidation generally plays a major role in the way that degradation proceeds in polymers. It is the initial generation of free radicals which begins the process, and which is usually the rate-determining step. The electrokinetic mechanism for their generation, as just described above, is therefore a significant means of hastening oxidative degradation. There are, however, several other ways by which initiating free radicals are produced, most notably by photochemical reaction. Photochemical oxidation, it should be noted, also tends to be autocatalytic in the sense that oxygen-containing functional groups that are produced are chromophores with enhanced absorbance for UV light, so that the oxidation itself increases the rate at which photochemical reactions proceed. We must therefore take into account that some degree of degradation of polymer insulation will ensue as soon as it is made, regardless of whether it is subjected to a high electric field, which merely accelerates the process.

The ground state of an oxygen molecule is unusual in that it is a triplet state, i.e. in effect, it is a diradical:

$$^{\bullet}O{-}O^{\bullet}.$$

Since any radical-pairing process has a low activation energy, any available oxygen (usually present in a polymer as a dissolved species) reacts swiftly with any free radicals that are produced in its vicinity. The result is a chain reaction:

$$R^{\bullet} + O_2 \rightarrow ROO^{\bullet}. \tag{6.22}$$

$$ROO^{\bullet} + RH \rightarrow ROOH + R^{\bullet}. \tag{6.23}$$

The second step (6.23) involves the breaking of a carbon–hydrogen bond and has a higher activation energy. The rate of this reaction is therefore much slower, and the rate of this step in the chain reaction generally determines the rate of oxidation, once it has started. Oxidative chain scission also occurs subsequently, according as the following reactions for polyethylene:

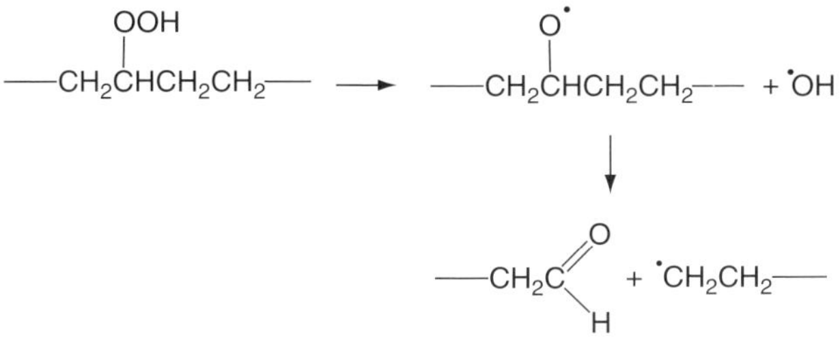

(6.24)

The overall result of the oxidative degradation is a fragmentation and weakening of the structure, allowing voids to open up and cracks to propagate, with the direction of the propagation being in the general direction of the field. Eventually the size of the voids initiated in this way will become large enough to allow discharge activity, finally leading to overall electrical breakdown. It is easy to appreciate that a crack, once started, will propagate erratically, following the pathways offering the least resistance, depending on the local morphology, e.g. staying within the most disordered regions and circumnavigating spherulites. The consequence is a growth that is tree-like in form. The rate-determining step in the breakdown will be the disengagement of the tie molecules; the succeeding processes will occur precipitately. Confirmation of this model comes from the thermal activation energy that is observed in thermal endurance studies. The value of around 120 $kJmol^{-1}$ matches that of the α_c-relaxation, which is the enabling mechanism for the disengagement.

We can understand how the observed threshold effect may arise as follows. Movement of the tie molecules in and out of the lamellae is a thermally activated rate process, and some chains will always be entering at the same time as others are leaving. The effect of the tensile stress is to bias the movement towards a greater rate of leaving. Only when a sufficient stress is applied to shift the equilibrium, such that the reduced number of tie molecules is unable to sustain the stress without breaking, will breakdown ensue.

In principle, it is possible to use the model to estimate the time to failure, but more detail of the morphology of tie molecules is required than is available at present.

6.6.3 Alternating fields

The above electrokinetic model has implicitly only considered endurance under a steady electric field. Obviously many applications of polymeric insulation concern alternating fields, and it is well known that AC conditions are much more conducive to ageing and treeing than are DC ones. Hence it is important to extend our considerations to alternating fields.

In an alternating field $E = E_0 \sin\omega t$ there will be a cyclic mechanical tensile stress

$$\sigma(t) = \varepsilon_0 \varepsilon E_0^2 \sin^2 \omega t = (1/2)\varepsilon_0 \varepsilon E^2 (1 - \cos 2\omega t). \tag{6.25}$$

This is equivalent to a steady stress $(1/2)\varepsilon_0 \varepsilon E_0^2$ and a cyclic one $(1/2)\varepsilon_0 \varepsilon E_0^2 \cos 2\omega t$; note that the direction of the tensile stress does not reverse. The deformation and morphological reorganisation induced by the steady stress will be the same as that which might be induced by a steady field alone and will lead to elastic energy storage. The movement of the polymer chains, including the tie molecules, under the cyclic stress will cause an oscillating (frequency 2ω) elastic energy storage and release in each half-cycle accompanied by some energy dissipation. This, in effect, provides extra heat energy in just the right region to facilitate the molecular motion. The system will attain a steady-state under cyclic stress in which a fraction of the polymer chains will be in conformations other than their equilibrium one, and this will have the effect of also increasing free volume in the system. We can appreciate that an alternating field will therefore accelerate the formation of voids in the amorphous regions. Other fatigue processes, like buckling of the chains under compression, may enhance this further.

Taking the usual equation for the rate of energy dissipation per unit volume, W, under a cyclic stress $\sigma_0 \sin \omega t$,

$$W = (1/2)\sigma_0^2 \omega J''(\omega), \tag{6.26}$$

where $J''(\omega)$ is the loss compliance at frequency ω, the dissipation caused by the electric field will be given by

$$W = (1/8)(\varepsilon_0 \varepsilon)^2 E_0^4 \omega J''(2\omega). \tag{6.27}$$

We note firstly that the energy dissipation depends on E_0^4 and so will be particularly strong where the field is locally enhanced, e.g. at electrodes and where space charges are present. Secondly, the appropriate value of the loss compliance will be that for the craze region in the amorphous phase, and this is likely to be greater than the average macroscopic value, which includes the crystalline phase.

6.6.4 Water treeing

A most significant case of tree-like breakdown failure arose in the 1970s when a number of underground medium-voltage power cables, with extruded polyethylene insulation, failed in service. The polyethylene insulation had been introduced in the previous decade as a replacement for traditional, and more expensive, oil-impregnated, wrapped-paper insulation. Upon examination, the failing polyethylene insulation showed many tree-like patterns, superficially like an electrical tree described in Section 6.5.1 generally *growing* from one electrode towards to the other. This clearly indicated a serious degradation phenomenon during service under electrical stress. The problem stimulated a vast programme of investigation, because the trees were clearly implicated as precursors of the cable failures. Experiments using needle electrodes inserted into polyethylene samples were devised to enhance local fields and thereby accelerate the inception of tree growth: the timescale can be reduced from years to a matter of a few minutes. In many instances water was found to play a major role, and, indeed, the way in which water could access the insulation in the original problem cables (they had no seal against water) certainly contributed to their early demise.

Detailed microscopic examination of the trees developed in insulation exposed to water whilst also under a high electric field shows that they are, in fact, rather different from the usual electrical trees. This type of tree is called a *water tree*. They appear to consist of strings of microcavities (1–5 μm in diameter) filled with electrolyte, possibly with some chemically modified inner surfaces, and in semi-crystalline polymers, with some disordering of lamellae. There is no direct connectivity between the cavities, although electrolyte appears to be able to percolate among them through amorphous material. Like electrical trees, they grow both in vented and bow-tie forms. Water trees can sometimes bridge the entire insulation between the electrodes without producing the final breakdown. It is usually an electrical tree that is responsible for the final breakdown, often starting from within a water tree.

Initiation and propagation of water trees depend on a multitude of parameters: electric field strength, mechanical stress, type and concentration of ions in the water, constitution of the polymer and its additives, etc. At this stage there is no satisfactory model that explains the process of water treeing. In a comprehensive review, Crine (1998) carefully considers the various types of possible mechanism in relation to a vast amount of experimental investigation, but is unable to come to any definite conclusion about primary mechanisms. Rather surprisingly, oxidation appears to play no major role (Bulinski *et al.*, 1998). It seems only likely that ingress of water, aided by the presence of (solvated) ions under the influence of the electric field, may induce some kind of mechanical cracking, which is partially stabilised by filling with water, and which continues to propagate relatively slowly.

The profound effect of water on tree growth that is so widely reported may be expected with the electrokinetic model on the basis of three principal effects. Water filling the crazes as they develop will help prevent their collapse. Water, as a good solvent for ionic species, will be an excellent medium to facilitate entry of surface-active agents, which, by a process similar to that of environmental stress cracking, will advance the void and craze formation caused by the electric field. Water, with its high relative permittivity, will distort and locally enhance electric fields in the neighbourhood of the voids and crazes where it accumulates. Whether one or other of these effects dominate in a particular situation depends on the exact nature of the specimen and its environment.

6.6.5 Effect of space charges

Especially under the influence of high electric fields, charges may enter a dielectric by injection from electrodes, and become trapped there as *space charges*. Space charges may also arise by charge separation from ionic impurities within the material. The fields from these space charges add to, or subtract from, the externally applied electric field, and can generate high-field distortions in their immediate vicinity. Consequently, one must expect that space charges will have a bearing on breakdown behaviour. It was not until techniques became available for measuring space charges that their role could be properly assessed; these are described in Section 7.2.2.

Under DC conditions space charges build up gradually over time, and Fig. 6.15. shows the developing profile of charge density across a sample of polyethylene subjected to a high voltage between the electrodes. When the field is turned off, these charges decay over time scales that vary from seconds to hours. In the case illustrated, the space charges injected from the electrodes have the same sign as that of the injecting electrode, and in effect they transfer field to the central region of the dielectric. Thereby the space charges may be the cause of premature breakdown. Charge separation originating in the body of the dielectric would have the effect of increasing the field near to the electrodes, and this may be even more damaging to the life of the material in service. In developing materials for high-voltage insulation it then becomes important to check the susceptibility of a material to build up of space charges, and, if necessary, to take steps to reduce it. Attention to cleanliness in order to avoid the presence of ionisable impurities in the material and at its interface with the electrodes is very important.

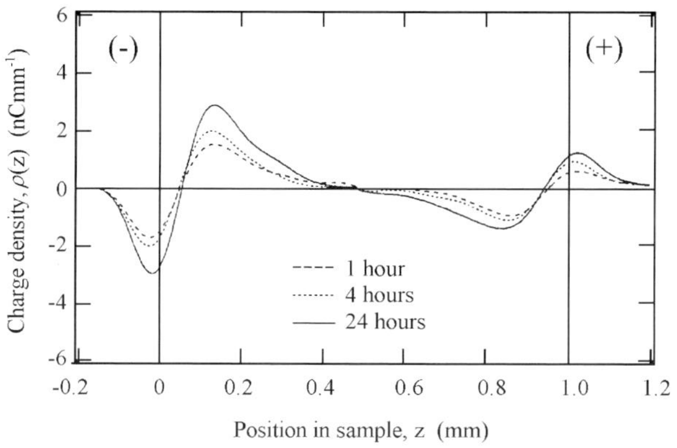

Fig. 6.15 Profiles of space charge in polyethylene under an applied field of 40 MVm^{-1} using conductive polyethylene electrodes. Reproduced from Sanden (1998) with permission of the Institute of Physics.

In the case of AC fields, one might expect that the build up of space charges would be much less than that under steady DC conditions, because the field reversals should continually cancel the effects of the previous half-cycles. Measurements do, however, show that there can be some space charge development even under AC conditions. This is possibly explained by variations in composition across the layer and over the interface layer at the electrodes.

6.7 Examples of high-voltage design

6.7.1 Power cables

Thermoplastics are natural candidates for use as insulants in the construction of electric power cables, because (a) the cable can be made by a simple, continuous extrusion process, (b) plastics meet the mechanical requirement for toughness combined with flexibility in the final product, and (c) they have excellent insulation qualities. When a cable with a thermoplastic insulant is to be used for high voltages, however, problems arise from the presence of minute voids and imperfections, often generated by differential thermal expansion effects, at the interfaces between the insulation and both the inner metal (usually aluminium or copper) core and any outer (earthed) metal sheath (Fig. 6.16(a)). Gas discharges would occur in the voids, especially in the higher field region next to the inner core, and the cable would have a low DIV (see Section 6.5.1). It is very difficult to prevent voiding altogether, but the associated electrical breakdown problem may be cured by interposing a layer of conductive plastic between the metal members and the insulation. The conductive plastic, usually a carbon-black composite of the same polymer as the

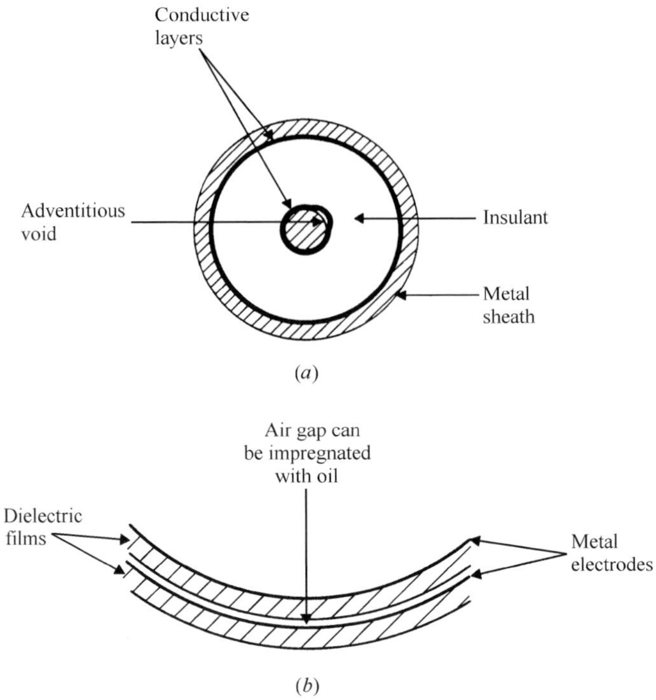

Fig. 6.16 Illustrations of high-voltage designs: (a) cross-section of a power cable, (b) cross-section of a thin-film capacitor, showing a small segment of one capacitive layer in the roll.

insulation itself, bonds firmly to the insulation. The conductive layer takes up substantially the same voltage as the adjacent metal, and so no electrical stress is imposed on the voids, which are therefore rendered innocuous. The cable industry uses this technique in all medium- or high-voltage cables using plastics and rubbers as insulating materials. A further refinement can also be employed. During its life a cable is liable to suffer inadvertent current overloads from time to time, when quite high, though temporary, temperatures may occur through Joule heating. In order to reduce the risk of thermal or electromechanical breakdown on these occasions, the strength of the polymer at high temperatures may be improved by crosslinking. In the usual method a crosslinking agent is incorporated in the polymer before extrusion and extra heating is applied afterwards to complete the cure.

6.7.2 Thin-film capacitors

One of the chief requirements for capacitors is that they shall be small, which implies that a large capacitance-to-volume ratio (C/v) is desirable. For a parallel-plate configuration, neglecting the volume of the electrodes, it is easy to show that:

$$\frac{C}{v} = \frac{\varepsilon_0 \varepsilon'}{d^2}, \tag{6.28}$$

where ε' and d are the relative permittivity and thickness of the dielectric layer, respectively. High ratios are therefore principally dependent on having a thin film. (The available range of relative permittivity is rather limited and is of secondary importance in this context.) For operation at a given voltage, the field across the dielectric will be inversely proportional to thickness, however, so that breakdown characteristics become very important if advantage is to be taken of thin films.

Many medium-voltage polymeric film capacitors are essentially rolled-up parallel-plate arrangements, made from two tapes, each of which carries one metal electrode (Fig. 6.16(b)). The metal electrodes are usually evaporated on to the film and are very thin (about 0.1 μm). The polymeric films that are used, e.g. polypropylene, polycarbonate, polystyrene, can be made as thin as 10 μm or even less, across which mains voltage (240 V) would produce a field of about 24 $\mathrm{MVm^{-1}}$. In order to withstand this high field the film must be of very high quality, i.e. free of voids and impurities. Fortunately, the situation is alleviated to some extent by a *self-healing* mechanism inherent in this type of capacitor. Breakdown across a small imperfection usually does nothing more than evaporate the electrode metal away from the affected area, and the breakdown is arrested. Each self-healing event incurs a small decrease in electrode area though, so that too many faults would soon lead to an intolerably large degradation in the value of the capacitance. Discharges are liable to start in any air trapped between the layers, and this becomes a serious problem when higher working voltages are needed. A common remedy adopted in high-voltage capacitors made by interleaving metal foil electrodes with polymeric film is to displace the air by impregnation of the capacitor with a liquid having a high break-down strength. A high relative permittivity for the impregnant is also an advantage because this reduces the field in any cavity it fills. In the past the most commonly used liquids for this purpose have been polychlorinated biphenyls, which have advantages of low viscosity, high relative permittivity (in the range 5–6), high dielectric strength, and good fire resistance. Unfortunately, these substances have been found to be very dangerous biochemically. More acceptable substitutes, based on phthalate esters, for example, are presently being sought. Overheating of a capacitor under AC conditions can occur if the dielectric has a high dielectric loss factor ε'' at the working frequency, and this too must be taken into account if breakdown is to be avoided.

6.8 Further reading

Classic texts on dielectric breakdown phenomena are by Whitehead (1951) and O'Dwyer (1973). A more recent book by Dissado and Fothergill (1992) provides a comprehensive review of advances since then. Sillars (1973) gives a good review of test methods and materials. Discharge testing techniques are described by Kreuger (1964).

6.9 Appendix – breakdown statistics

The random variation of dielectric breakdown data presents a problem for engineers who want to design electrical products where the insulation must have guaranteed performance and life. The situation is particularly severe in the case of high-voltage cables, which require reliable operation over long distances underground for long time periods (it is not uncommon for 40 years to be specified). When items of insulation that would normally be regarded as identical are tested for electrical breakdown under identical conditions, the resulting breakdowns that are observed typically occur neither at the same voltage (as voltage is ramped up) nor after the same time (for constant applied voltage). In order to make practical assessments of the performance of insulants and to predict their life in service, it then becomes necessary to apply statistical analysis to the test data.

The investigation of failures of manufactured components and systems, especially in the electronics and aerospace industries, has generated a variety of statistical models on which data analysis may be based. Each model uses a specific distibution of failure probabilities, and it is important to select a model that matches the actual distribution inherent in the product concerned. In the case of dielectric breakdown, where a large number of quite different modes of failure are known to occur, sometimes even together, the application of a particular statistical failure model must be approached with great caution. Nevertheless, one treatment, based on a Weibull distribution of failure probability, has taken root, and is most generally used in practice. For a dielectric, the Weibull failure probability function has the form

$$P_{\mathrm{f}} = 1 - \mathrm{e}^{-(E/\alpha)^{\beta}}. \tag{6.29}$$

Here P_{f} is the probability of failure and E is the applied electrical stress (field strength) that progressively increases and eventually causes the item to fail. (The *stress* could also be taken as the time to failure at constant applied voltage.) The parameters α and β define the scale and shape of the distribution; in this they are analogous to the mean and standard deviation of the normal

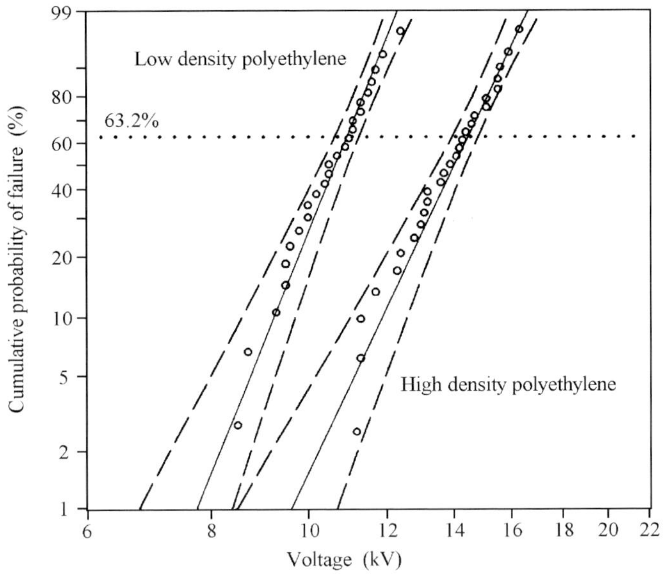

Fig. 6.17 Weibull plot of breakdown data for low-density and high-density polyethylenes.

distribution. The Weibull distribution is an *extreme-value distribution*, corresponding to failure by the weakest link in a system, and is generally suited to electrical breakdown situations.

The Weibull Equation (6.27) may be transformed to

$$\log_{10}[-\ln(1-P_{\mathrm{f}})] = \beta \log_{10} E - \beta \log_{10} \alpha. \tag{6.30}$$

Here a plot of $\log_{10}[-\ln(1-P_{\mathrm{f}})]$ against $\log_{10} E$ is a straight line. The shape parameter β is equal to the slope of the line, and the scale parameter α is equal to the field at the intercept with the abscissa (i.e. where $\log_{10}[-\ln(1-P_{\mathrm{f}})] = 0$). The straight-line plot is the most convenient way of plotting experimental data, and the goodness of the fit to a straight line indicates how well the data conform to Weibull statistics. If data points lie outside 90 % confidence limits, further investigation would be required. Special graph paper is used to make this plot easy, with the ordinate accordingly labelled directly as P_{f}, otherwise known as the cumulative probability of failure, which corresponds to the proportion of samples that have failed up to a particular field (stress) level. Figure 6.17 shows an example for a series of tests made on a set of plaques of low-density polyethylene tested under standard conditions. The applied field (stress) E when $\log_{10}[-\ln(1-P_{\mathrm{f}})] = 0$, i.e. at a cumulative failure probability of $P_{\mathrm{f}} = 1 - 1/\mathrm{e} = 0.632$ or 63%, gives the scale parameter α. The slope of the straight line gives the shape parameter β.

Although the scale parameter α is the primary characteristic value of failure in a set of samples, representing the applied field by which 63% of the samples have broken down, the shape parameter β is also extremely important. β is required in order to calculate when the first 1%, say, of samples will have failed – a more important parameter for a design engineer than the 63% figure.

In practice, the two-parameter Weibull distribution function fits a wide range of data sets, and it can be used to approximate a wide range of other functions. (The formula is rather forgiving in this sense, and this is one reason why it is used so widely.) It may even give a good fit to the data where more than one breakdown mechanism (each with its own characteristic failure distribution) may be operating. There is then the danger that the Weibull plot will be used for extrapolation and prediction of time to, or stress at, breakdown, where it is not really justified. Nevertheless, the equation has been used effectively in breakdown analysis, especially in product development procedures, where the failure mode of a class of materials or designs tends to remain the same. If the experiment aims, however, to compare different materials or different configurations with respect to values of breakdown time or stress, it may be better to consider using distribution-free statistical methods.

7
Static charges

7.1 Introduction

A locally high concentration of electric charge q inside any material will decay exponentially with time t, current flowing away under the influence of the self-field of the charge (initial concentration q_0):

$$q = q_0 \exp(-t/\tau), \tag{7.1}$$

where the time constant τ, called the relaxation time for decay of free charge, depends on the product of the material's inherent charge-storage capacity, as expressed by its permittivity $\varepsilon_0\varepsilon'$, and its bulk resistivity ρ:

$$\tau = \varepsilon\varepsilon'\rho. \tag{7.2}$$

For metals τ is so short that it is hardly measurable (e.g. for copper τ is of the order of 10^{-18} s), but for common, insulating polymers, e.g. polyethylene, which have particularly high resistivities, it can be very long. In other words, polymers may retain charges for very long periods, sometimes for several years, and this presents an important aspect of their insulating character.

Although charges may in certain circumstances become trapped within the bulk of a polymer specimen, it is the surface of a polymer which is the most vulnerable towards charging effects. As with conduction in polymers, the charging process can be either electronic or ionic in origin. Out of the many different ways in which charges can be imparted, the commonest, and most basic, is simple contact charging: whenever any two different materials merely come into contact with each other there is always some redistribution of electrons or ions across the interface. Our understanding of even these elementary processes has been slow to develop, partly, no doubt, as a consequence of contact charging being essentially a surface effect, suffering from the usual complications of high concentrations of

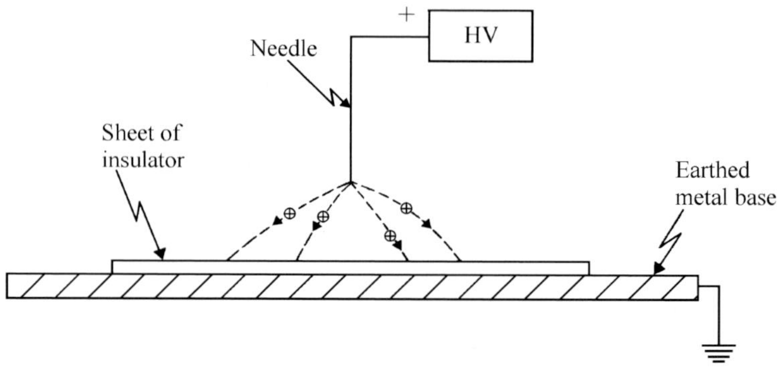

Fig. 7.1 Diagram of a corona charging system.

impurities and structural defects which tend to be more prevalent at surfaces. Recently progress has, however, been made in this area, and the work is described in this chapter. As is well illustrated by the classical experiments in which insulators are charged by rubbing with cat's fur etc., transfer of charge between two surfaces is usually enhanced by friction. The complete explanation of this effect is still obscure, although we know that several factors are often involved, including an increase in the true area of contact, a localised temperature rise, and surface abrasion.

Controlled charging is most easily accomplished by spraying a surface with ions, and this lies at the heart of many industrial electrostatic processes, e.g. xerography. Figure 7.1 shows the way in which the surface of an insulating film may be charged by a corona. A high voltage (10 kV, say) applied to the metal needle produces a very high field just in the vicinity of the needle tip, causing the air to ionise smoothly in that region. The air ions of the same sign as the applied voltage are repelled and move along the divergent field lines towards earth. They are intercepted by the insulator surface, and the surface charge density builds up until the field at the needle tip is so reduced that the corona is extinguished, the final charge density depending on the magnitude of the voltage applied to the needle. (The ions of the opposite sign are attracted to the needle and are discharged there.) In order to charge large areas of film the needle may be replaced by a fine wire stretching across the full width of the film, which may also be moving.

The density of charge that can be obtained on a surface is often limited by the breakdown strength of air, rather than by the basic charging mechanism itself and, as mentioned in Section 1.1, this has often confused the interpretation of electrostatic experiments in the past. Thus, if we take the typical value of 3 MV m^{-1} for the breakdown strength of air, the maximum uniform charge density that can be maintained on a large flat sheet with large air gaps on both sides will be 54 μCm^{-2}. In practice, it is difficult to achieve much more than 10 μCm^{-2}, because nearby earthed objects tend to focus the field and initiate sparks. When high fields in air are avoided, by evacuating the apparatus for instance, charge densities as high as 10 μCm^{-2} can be achieved by simple contacts. In these cases the field approaches the breakdown strength of the polymer.

Charge patterns on insulating surfaces may be easily revealed by dusting with charged powders. A suitable mixture is obtained by shaking talc together with jewellers' rouge, when the talc particles become negatively charged, and the rouge particles positively charged. When the mixture is lightly blown towards an insulating surface, the particles are attracted to oppositely charged areas, to give a detailed picture of the charge distribution. If the surface has been previously charged with one sign of charge and then suffered air discharges, very intricate charge patterns, called Lichtenberg figures, indicate where the localised breakdown events have occurred and have deposited compensating charges. Most noteworthy is the difference in charge pattern according to whether the surface was originally positively or negatively charged. As can be seen in the photographs of Fig. 7.2, discharges to a positive surface give predominantly circular patches (of negative charge), whereas discharges to a negative surface give filamentary, star-shaped deposits (of positive charge). These differences may be explained (Bertein, 1973) by the fact that the electrons produced in the air breakdown are very much more mobile than the associated positive ions. This affects the way in which the individual breakdowns proceed and are eventually extinguished in the two cases.

The principle of charge limitation by air breakdown also applies to the behaviour of insulating powder particles. The particles become readily charged by contact with the walls of their containers and the pipes through which they pass, but their surface charge density cannot exceed the air limit. Of course, the smaller the particle size, the larger is the surface area per unit mass of powder and the larger will be the charge which can be carried by the powder.

Static charge phenomena on polymers prove to be a distinct nuisance in many situations. Unless special precautions are taken, plastics are always liable to accumulate charges in use, and these spoil the appearance of moulded items by attracting dirt and dust, make handling of fibre and film products difficult by causing them to cling to metal-work, and cause discomfort and ignition hazards through sparking. On the other hand, the ability of polymeric materials to hold charges has also proved to be of great benefit and is extensively exploited in such industrial applications as electrostatic powder coating and electret transducers.

(a)

(b)

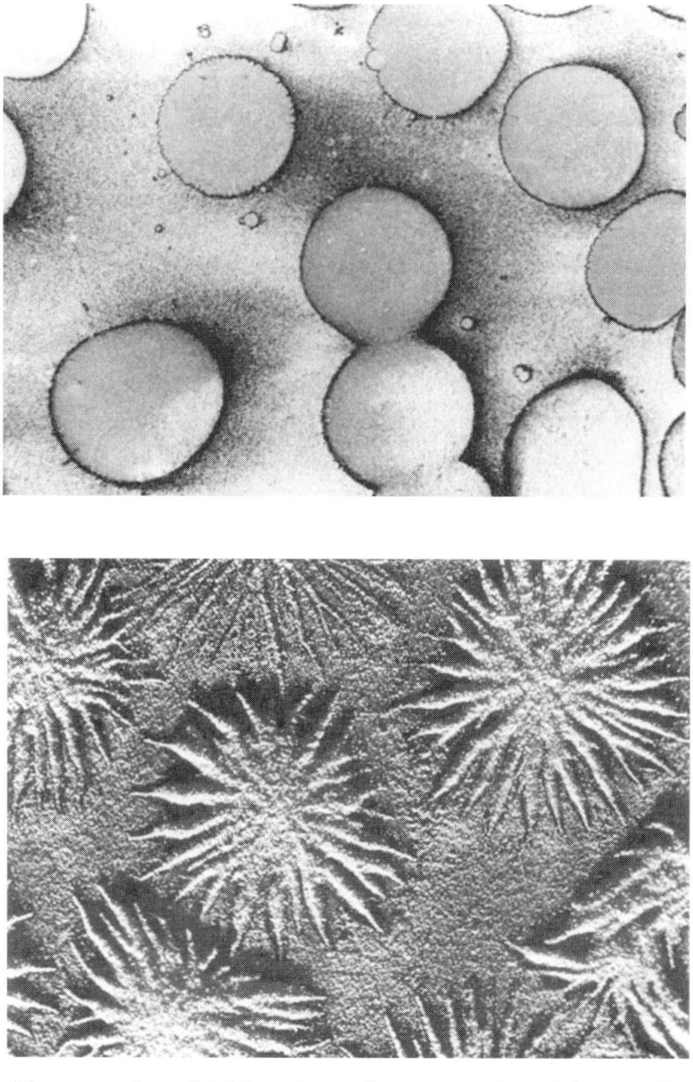

Fig. 7.2 Photographs of Lichtenberg figures produced by sparks to an insulator charged (a) positive and (b) negative. (Courtesy of Mlle H. Bertein.)

7.2 Measurement of static charges

Measurement of static charges requires rather special techniques since the actual amounts of charge involved are usually very small, precluding the use of instruments which depend primarily on current for their operation. Historically, the gold-leaf electroscope was most commonly used for measurement of static charges. To measure the total charge on an object, the object was introduced by means of an insulating thread into a metal can (usually referred to as a Faraday ice pail) standing on the insulated terminal of the electroscope, the charge thereby induced on the electroscope causing a deflection of the gold leaf. Alternatively, the electroscope could be used as a potential probe by holding an extension of the terminal near to a charged surface. Although these techniques are satisfactory in the sense that they involve no extra degradation of the charge under investigation – the only 'connection' to the electroscope is via the electrostatic field – an electroscope is very delicate, and quantitative measurements are very difficult to make with it.

The modern phase of electrostatic measurements began with the use of the electrometer valve (or tube), which is essentially a very well-insulated triode. The induced potential on a probe can be measured directly by connecting it to the control grid of an electrometer valve and observing the resulting change in anode current. The leakage resistance from the grid to earth can be made sufficiently high that the voltage does not decay appreciably during the course of the measurement. Latterly, the solid state equivalent of the electrometer valve, the insulated-gate field-effect transistor, has become available with insulation resistances also in excess of $10^{14}\,\Omega$. On account of the usual advantages of solid-state devices the transistor has largely taken over from the electrometer valve, and the metal oxide-semiconductor variety (MOSFET), which typically has a very low input bias current, is now used in many commercial instruments for electrostatic measurements.

The measurement of surface charges requires special capacitive probes that define the actual area being measured. In the measurement of bulk charges, often referred to as space charges, the extra dimension that defines the depth of charge in the material is usually *probed* by means of some kind of pulse that travels through the material.

7.2.1 Measurement of surface charges

The use of a probe for charge measurement may be exemplified by the case shown diagrammatically in Fig. 7.3(a). A probe in the form of a plate (area A) is brought to a position opposite the charged surface of an insulating sheet (relative permittivity ε') lying on an earthed metal base, and the potential of the probe measured. The equivalent electrical circuit for this arrangement is shown in Fig. 7.3(b). Assuming that the probe remains perfectly insulated from earth (electrometer with an infinitely high input resistance), the potential V_m induced on it is related to the potential V_s of the charged surface by

$$V_m = \frac{C_g}{C_g + C_m} V_s, \tag{7.3}$$

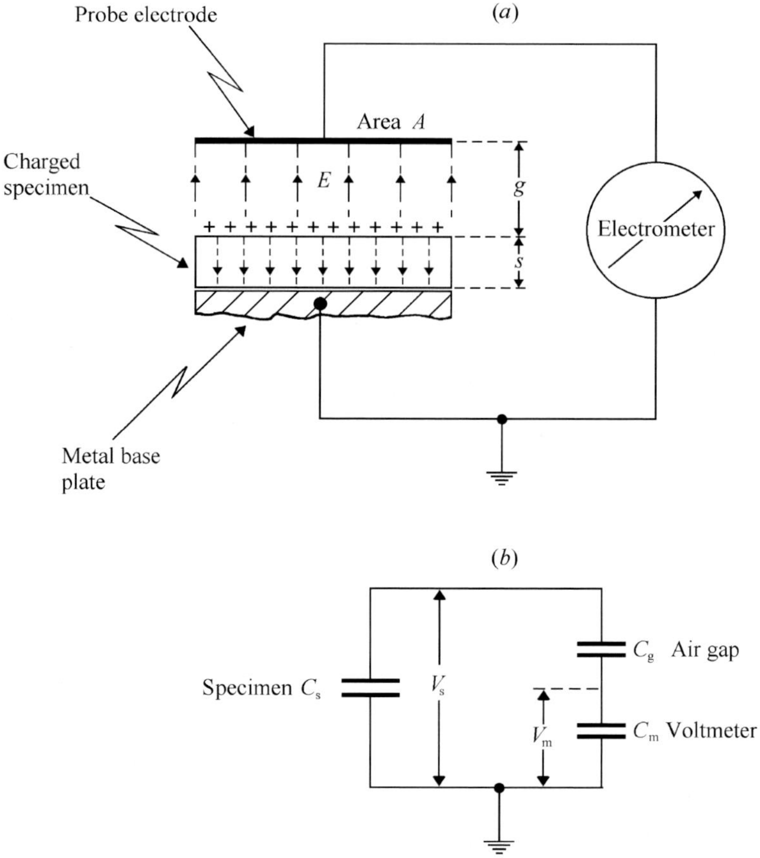

Fig. 7.3 Measurement of surface charge density with a field probe: (a) diagram of experimental arrangement, (b) equivalent electrical circuit.

where C_g is the effective air capacitance between the charged surface and the probe (not a true capacitance in that the stored charge does not actually reside on a conducting electrode), and C_m is the input capacitance of the electrometer. Now the total effective capacitance C between the charged surface and earth is given by

$$C = C_s + \frac{C_g C_m}{C_g + C_m}, \tag{7.4}$$

where C_s is the effective specimen capacitance between the charged surface and the underlying earthed base. Hence, we may relate the surface charge density σ to the induced potential on the probe as follows:

$$\sigma = \frac{CV_s}{A} = \left(C_s + C_m + \frac{C_s C_m}{C_g}\right)\frac{V_m}{A}. \tag{7.5}$$

For parallel-plate geometry the capacitances of the specimen (thickness s) and air gap (thickness g) are given by the following formulae:

$$C_{\mathrm{s}} = \frac{\varepsilon_0 \varepsilon' A}{s}, \tag{7.6}$$

and $$C_{\mathrm{g}} = \frac{\varepsilon_0 A}{g}. \tag{7.7}$$

If the input capacitance of the electrometer is not already known, it may be obtained from a calibration in which the charged surface is replaced by a metal plate held at a fixed potential V_{s}' (e.g. by connection to a battery), when the probe potential V_{m}' will be given by

$$V_{\mathrm{m}}' = \frac{C_{\mathrm{g}}}{C_{\mathrm{g}} + C_{\mathrm{m}}} V_{\mathrm{s}}'. \tag{7.8}$$

From the example described above we can appreciate that we must have a well-defined geometry in order to be able to deduce charge densities from measurements with probes. The situation is necessarily more complicated when charge is distributed through the thickness of a specimen rather than just residing on a single surface (Blythe, 1975b).

The probe arrangement really behaves as a field sensor, since the electrometer reading is directly proportional to the average field E at the probe surface. To see this, we may consider again the probe opposite the metal plate held at a fixed potential. The field at the probe is then

$$E = \frac{V_{\mathrm{s}}' - V_{\mathrm{m}}'}{g}. \tag{7.9}$$

Substituting for V_{s}' and C_{g} from Equations (7.9) and (7.7), respectively, in Equation (7.8), we obtain

$$V_{\mathrm{m}}' = \frac{\varepsilon_0 A}{C_{\mathrm{m}}} E = \text{constant} \times E. \tag{7.10}$$

An important case to consider is that where the earthed base of Fig. 7.3 is removed completely, so that the probe is looking at an otherwise isolated charged surface, i.e. $C_{\mathrm{s}} = 0$, when Equation (7.5) shows that

$$\sigma = \frac{C_{\mathrm{m}}}{A} V_{\mathrm{m}}. \tag{7.11}$$

Here the electrometer reading is independent of C_{g} and therefore not sensitive to the distance between the charged surface and the probe. Physically, the reason is that in this situation all the flux from the charged surface is directed towards the probe whatever the separation, so that the field at the probe is invariant. Measurements of net charge density on plastic films may be conveniently made in this way. (Net charge in this context means the algebraic sum of the charges on both sides and within the film per unit area of film.)

If we take into account the finite input resistance R_m of any real electrometer, we recognise that any reading of a probe potential will decay exponentially with a time constant $R_m C_m$. If C_m is of the order of 10 pF (it cannot be made high otherwise V_m will be too small to measure) the input resistance must be kept above about 10^{13} Ω to keep the time constant long compared with the time required to take a reading.

The basic probe system has been refined to avoid the effects of finite input resistance as much as possible and to increase accuracy, sensitivity and spatial resolution in field and charge investigations. Three versions are briefly described below.

Negative-feedback electrometer

The effective input resistance of a DC amplifier may be greatly increased by the use of negative feedback. This principle is illustrated for charge measurement in Fig. 7.4, where a feedback capacitor is connected across an amplifier with a high input resistance R_m and gain $-N$. To make a measurement, the amplifier is first unshorted whilst the input probe faces an earthed shield. The shield or cap is then removed and the probe taken near to the charged surface in

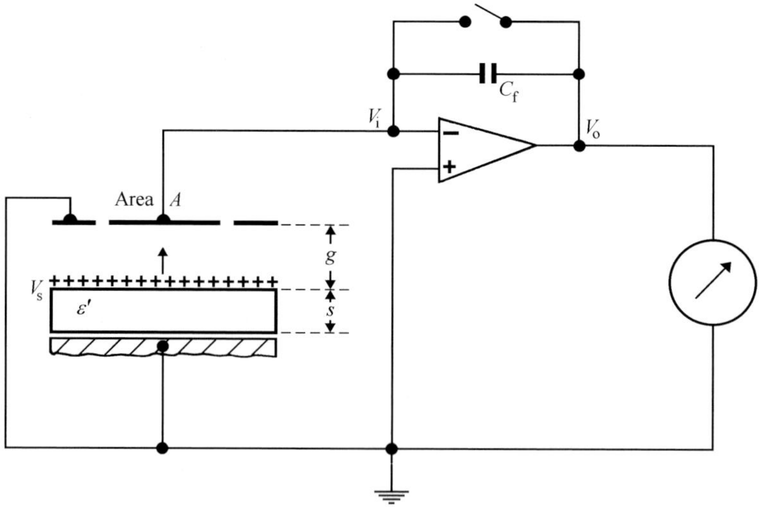

Fig. 7.4 Block diagram of a method of measuring surface charge density with a guarded probe connected to a negative-feedback electrometer.

question, now at potential V_s. The action of the negative-feedback loop is to keep the input probe virtually at earth potential (N is typically of the order of 10^3). Considering the probe-to-surface capacitance C_g in cascade with the feedback capacitance C_f, we have

$$C_g V_s = C_f (V_i - V_o), \tag{7.12}$$

where V_i and V_o are the input and output voltages of the amplifier. But $V_i \approx 0$, so that

$$V_s = -\frac{C_f}{C_g} V_o. \tag{7.13}$$

If the probe and surface form a parallel-plate capacitor, spacing g, area A, it follows that

$$V_s = -\frac{gC_f}{\varepsilon_0 A} V_o. \tag{7.14}$$

The surface charge density σ is related to V_s by

$$\sigma = \frac{CV_s}{A}, \tag{7.15}$$

where C is the sum of the capacitance C_g to the probe and the capacitance C_s to the underlying earthed plate. Using Equations (7.6) and (7.7) the final expression for the surface charge density then becomes

$$\sigma = -\frac{C_f}{A}\frac{g + \frac{s}{\varepsilon'}}{\frac{s}{\varepsilon'}} V_o. \tag{7.16}$$

When the base plate is removed altogether,

$$\sigma = -\frac{C_f}{A} V_o. \tag{7.17}$$

The principal advantages of the negative-feedback electrometer are:

(a) the input resistance of the amplifier is effectively increased by a factor equal to the open-loop gain N;
(b) it is a very simple and compact system, with no moving parts;
(c) the action of the feedback loop, which renders the probe a virtual earth, means that an earthed guard ring may be conveniently set around the probe so as to define very accurately the area which the probe *sees*.

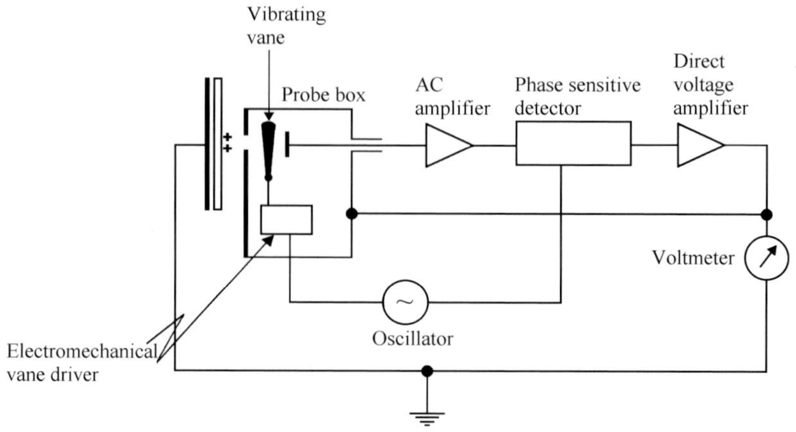

Fig. 7.5 Block diagram of a method of measuring surface voltage with a positive-feedback electrometer

The principal disadvantage is that the probe may acquire charge, for example by accidentally touching a charged surface, when a permanent false reading is obtained until the amplifier is re-zeroed.

Positive-feedback electrometer

Another type of feedback electrometer works in a quite different manner, and Fig. 7.5 illustrates the method of using it to measure charge density on the surface of an insulating sheet or film lying on an earthed base. The probe *peeps* at the surface in question through a hole in a metal box, and, in addition, the field to the probe is interrupted by a vibrating shutter. The periodic signal on the probe is amplified and phase-sensitively detected, and the output used to drive an amplifier whose own output in turn raises the potential of the metal box and the shutter.

When this potential reaches the same potential as the portion of surface which the probe sees, the AC signal falls to zero, because the probe is now exposed to the same potential whether the shutter is open or closed. If the voltage V applied to the probe box is measured in this condition, the surface charge density is simply given by

$$\sigma = \frac{\varepsilon_0 \varepsilon' V}{s}, \tag{7.18}$$

where ε' and s are the relative permittivity and thickness, respectively, of the sheet. This instrument may be made very sensitive with a very fine spatial resolution. One drawback, however, is that it cannot be used in a simple fashion to make measurements on isolated sheets, because the voltage of the charged surface will increase indefinitely as the voltage of the probe increases.

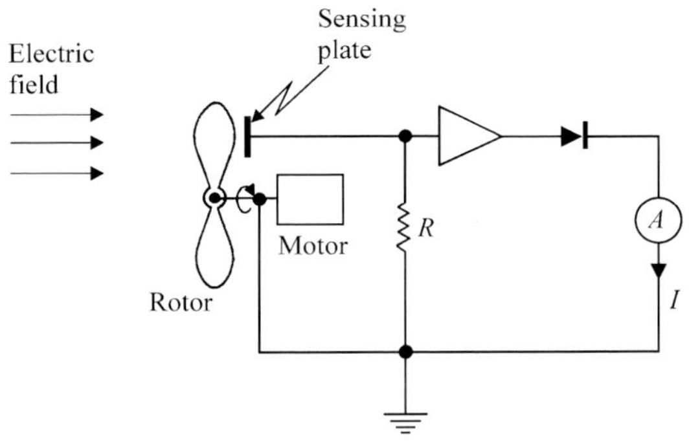

Fig. 7.6 Basic design of a field mill.

Field mill

Many commercial instruments are based on the field mill principle, shown in Fig. 7.6. Here the electric field falling on the probe is interrupted by the rotation of the shaped sector. The periodic voltage generated on the input in this way may be readily amplified by an AC amplifier, and the rectified output displayed on a meter. The principal advantages of this system are:

(a) The input resistance R_m does not have to be so high as in the simple electrometer, since the decay rate only has to be slow with respect to the periodic time of the alternating input. For a 2-sector blade rotating at 5000 rpm and an input capacitance of 100 pF, R_m need only be of the order of 10 MΩ.
(b) The sensitivity can be very high due to the efficiency of AC amplification techniques. If phase-sensitive detection is used, the signal-to-noise ratio may be greatly enhanced and the sign of the charge also determined.

The main disadvantage is that the probe, necessarily incorporating a mechanical field-chopping mechanism, is somewhat complicated and fine spatial resolution is difficult to achieve.

7.2.2 Measurement of space charges

As we have already commented in Section 7.1, static charges are most commonly manifested as surface charges on everyday polymers or plastics, simply because surface charging occurs readily by merely making contact with other materials, and also because the insulating nature of these materials prevents charges from migrating into their interior. Nevertheless, charges in the bulk of a polymer, usually called space charges, do occur. They may get there during a fabrication process, when temperatures are raised and conductivity becomes appreciable, or they may be injected under the influence of very high electric fields. For a long time our knowledge about the distribution of space charges in insulating materials remained rather speculative; there was no direct method of measurement. Then, over the last fifteen years, major advances were made in experimental techniques, revolutionising our capabilities in this area. Consequently, it is now possible to make detailed measurements of space charges, their distribution and even their time evolution under various conditions, with a high degree of precision and resolution. There are two different types of measurement, and these are described separately below. Both of them are non-destructive – they leave the space charges undisturbed – and are very fast.

The pulsed electroacoustic method

This technique, usually referred to as the PEA method, was originally devised by Maeno, Takada and co-workers (Takada & Sakai, 1983 and Maeno *et al.*, 1989) in order to study the way in which space charge develops near electrodes in a solid polymer under the application of a high voltage and deposits during electron beam irradiation.

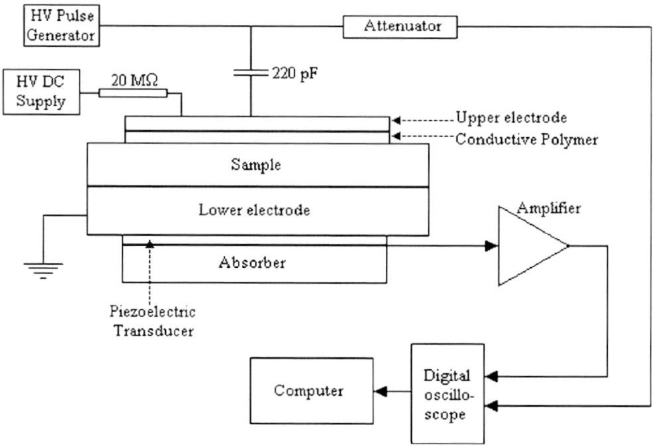

Fig. 7.7 Apparatus for space charge measurement by the PEA method.

Figure 7.7 shows a block diagram of a typical experimental arrangement. In order to make a measurement, a short, high-voltage pulse, typically 0.1 to 4.0 kV for 5–30 ns, is applied across the planar sample, thickness 0.1–6.0 mm. The resulting electric field E_p acts simultaneously on the whole distribution of space charge $\rho(z)$ within the sample, and on the induced charges, $\sigma(0)$, $\sigma(d)$ at the electrode surfaces, to produce mechanical stresses proportional to the local charge densities, as indicated in Fig. 7.8. This mechanical stimulation generates two acoustic pulses, which propagate in opposite directions, with amplitude profiles proportional to the net charge distribution. The one travelling downwards traverses the lower electrode to reach the piezoelectric

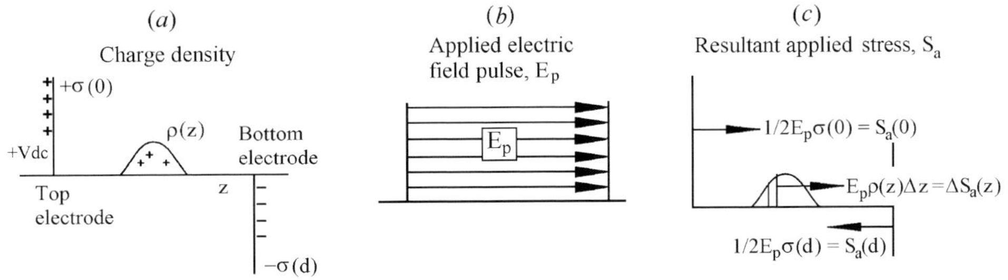

Fig. 7.8 Acoustic excitation by an electric field pulse acting on a charge distribution.

detector, which in turn generates an electric signal proportional to the amplitude of the pressure wave passing through it. The voltage trace, recorded on the output as a function of time, then gives an image of the spatial, one-dimensional charge density distribution across the sample with a resolution of 5–10% of the sample thickness. In practice, the output is amplified by a wide-band amplifier, and repeated scans are averaged by a high-speed digital oscilloscope. Connections to the upper electrode allow for simultaneous application of a steady (DC) voltage and a high-voltage pulse. The thick metal of the lower electrode acts as a delay line to shift the required signal from the noise generated by the initiating electrical pulse. An absorbing block of material, matched to the acoustic impedance of the piezoelectric element and placed beneath it, prevents any acoustic reflection of the pulse that might interfere with the measurement.

The response of the detector/amplifier system distorts the signal from the acoustic pulse. This can be overcome by deconvolution of the received signal, using as reference the signal generated from an induced surface charge obtained by application of a known steady voltage across the sample.

The trace depicted in Fig. 7.9 shows an example of space charge measured in a film of PVC (with no applied DC voltage). The internal, negative space charge is clearly defined, together with the induced compensating positive charges on the sample electrodes. The repeated scans show the effect of removing successive layers beneath the right-hand electrode. The internal space charge had been trapped in the specimen during the film manufacturing process.

The pressure wave propagation method

The second technique, usually referred to as the PWP method, is really complementary to the first one in the sense that an acoustic wave is now used as the primary stimulus to displace the space charge and thereby produce an electrical signal. In the previous PEA method the primary stimulus was the electrical pulse whose direct effect on the space charge produced an acoustic

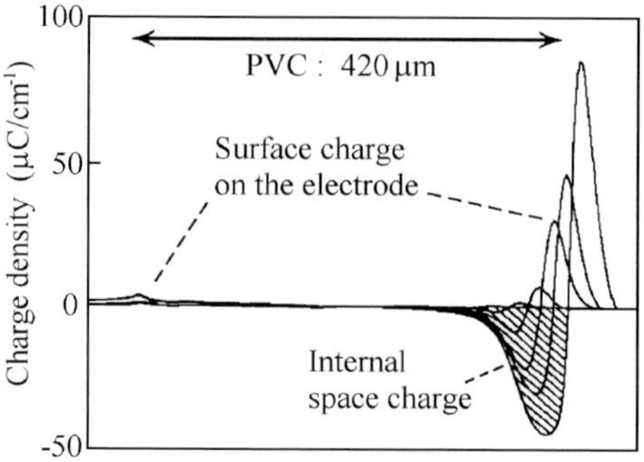

Fig. 7.9 PEA output trace showing the distribution of space charge in a sample of PVC film (420 μm thick) without any applied voltage. Reproduced from Maeno and Fukunaga (1996) with permission of the Institute of Electrical and Electronics Engineers, Inc.

wave that was detected. In one embodiment of the PWP method a laser pulse impacting on the surface of the sample electrode is used to generate the pressure pulse that propagates through the sample at the velocity of sound. This variation is called the laser induced pressure pulse or LIPP method, and was pioneered by Lewiner and his co-workers (Alquié *et al.*, 1981). Alternatively, the pressure pulse may be generated by means of a piezoelectric transducer resting on the sample. As the pressure pulse travels through the material of the sample, the compression distorts the local molecular structure and causes a slight displacement of any trapped charge that is present at that point. The movement of the charge induces a signal on the adjacent electrodes. Figure 7.10 shows a block diagram of a typical experimental arrangement.

Provided that the incident pulse is short compared with the time taken for the pressure pulse to traverse the sample, and that attenuation and dispersion of the pressure pulse are negligible, the transient current which flows in the external circuit (under virtually short-circuit conditions) will give a representation of the distribution of space charge in the sample, as illustrated in Fig. 7.11. The current peaks detected as the pressure pulse enters and leaves the sample at the electrodes give a measure of the induced surface charges that are induced at the electrode surfaces by the space charges within the sample; their relative magnitudes reflect the position of the centroid of the space charge with respect to the electrodes.

In the absence of space charges, but where an electric field is applied across the insulating sample, the current transient reflects the induced charges at the electrode/insulator interface. Since these charges occupy a negligible depth, the profile of the current peaks reflects the profile of the pressure pulse itself. This enables a calibration of the measurement system to be made.

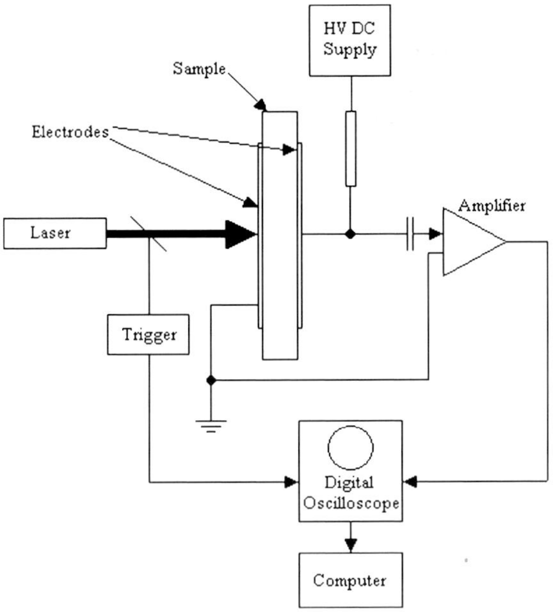

Fig. 7.10 Apparatus for space charge measurement by the PWP or LIPP method.

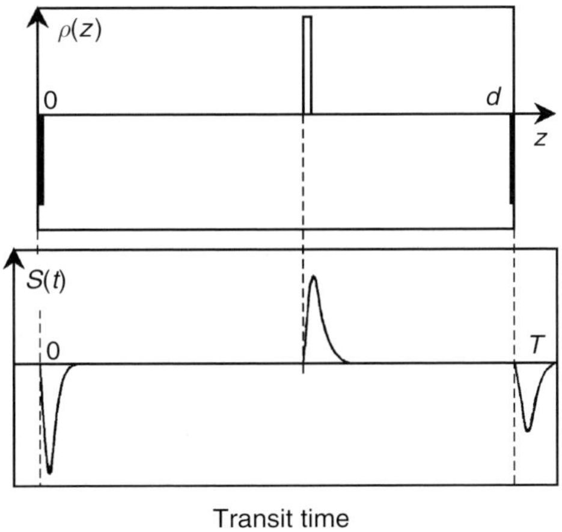

Fig. 7.11 Example of a PWP trace for a simple charge distribution.

In a typical experimental measurement a Nd:YAG laser fires a 5 ns pulse on to the target electrode. It is convenient to use a polymeric material for the electrode, made conductive by the addition of conductive carbon black, to achieve intimate contact with the specimen. Adding a thin layer of a volatile liquid to the electrode surface just prior to firing the laser greatly enhances the magnitude of the signal, by helping to increase the size of the pressure pulse that is generated. The transient signals are recorded in terms of the voltage across a small (50 Ω) sampling resistance in the external circuit.

More rigorously, the current flowing in the external circuit during propagation of the pressure pulse across the sample (z-direction) is expressed by:

$$I_{cc}(t) = \chi G(\varepsilon_r) C_0 \int_0^{z_f(t)} E(z) \frac{\partial p(z,t)}{\partial t} \, dz, \tag{7.19}$$

where I_{cc} (t) is the measured short-circuit current, χ is the compressibility of the material, $G(\varepsilon_r)$ is a coefficient which depends on the relationship between the relative permittivity and pressure on the material, C_0 is the capacitance of the pressure-free sample, $E(z)$ is the electric field distribution before the action of the pressure wave, $p(z, t)$ is the pressure wave profile as a function of time and position, and $z_f(t)$ is the position of the wave front at time t. Deconvolution of a current transient, using Equation (7.19), then completely determines the electric field distribution $E(z)$ across the sample, and the space charge distribution $\rho(z)$ may be obtained by differentiation of $E(z)$ with respect to z.

Application of space charge measurement

One of the most useful applications of the PEA and PWP measurement techniques has been to follow the development of space charge in an insulator with time under the influence of an applied electric stress as happens in high-voltage cables. The importance of these studies lies in the role of space charges in triggering premature dielectric breakdown and failure. As described in Section 6.6.6, this effect of space charges stems from the way in which they may increase the electric field in certain regions of the insulating material beyond that which would normally occur. Being able to monitor space charging enables remedial action to be taken by appropriate control of the composition of high-grade insulation materials and the conductors in contact with them.

7.3 Contact charging of polymers

7.3.1 Charge transfer by electrons

The spontaneous redistribution of electrons across an interface between two materials is most completely documented and understood in the case of metal–metal contacts, and this provides a good starting point for discussion of contact charging in general. Figure 7.12(a) shows the energy-level diagrams for electrons in two different metals which are initially uncharged and

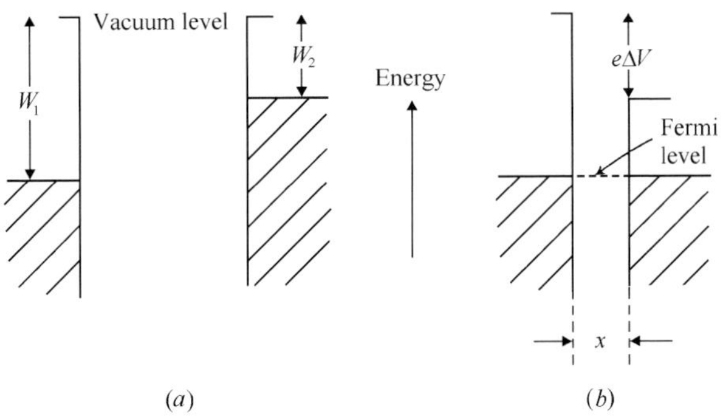

Fig. 7.12 Energy-level diagrams for two different metals: (a) before contact, and (b) after contact.

separated. The energy zero in these diagrams is taken to be the vacuum level corresponding to a stationary, isolated electron. The work function W of a metal is defined as the minimum energy that has to be expended to remove an electron from the metal and corresponds to the depth of the Fermi energy level (for a metal this corresponds to the topmost filled electronic energy state) below the vacuum level. As indicated in Fig. 7.12(b), when the two metals are allowed to make contact with each other, electrons will flow from metal 2, which has the lower work function, to metal 1, leaving metal 2 positively charged and vice versa. The flow will continue until a new equilibrium is reached, where the electrostatic contact potential difference ΔV established between the two metals just balances the difference in work function, and the Fermi energy levels are equalised. Hence,

$$e\Delta V = W_1 - W_2. \tag{7.20}$$

Metal work functions typically fall in the range 4–5 eV.

Imagine now that the two metals are slowly separated again, supposing that they are kept well insulated so that charge cannot leak away to earth. As the gap x widens between them, the capacitance C between the two metals decreases and less charge transfer is required to maintain the same potential difference ΔV. Consequently, as long as electrical contact is maintained through electron tunnelling, there is a back-flow of charge. At the critical gap x^* (typically $\sim$5 nm), where electrical contact is finally lost, the charges are *frozen*. On this basis the final charge transferred per unit area of contact is given by

$$\sigma = \frac{C}{A}\Delta V = \frac{\varepsilon_0}{x^*}(W_1 - W_2), \tag{7.21}$$

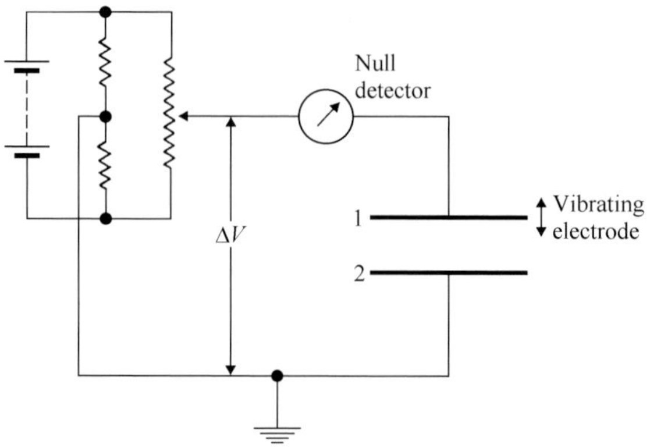

Fig. 7.13 Diagram of a modified Kelvin method for measurement of the contact potential difference between two different metals, 1 and 2.

where A is the area of contact.

Although metallic work functions may be obtained in various ways, including studies of thermionic emission, measurement of contact potential difference is the most convenient way of measuring relative values. In the Zisman (1932) modification of Kelvin's original method, shown diagrammatically in Fig. 7.13, parallel plates of the two metals, electrically connected together, are vibrated with respect to each other. The changing capacitance across the gap causes an alternating current to flow in the connecting wire (to keep ΔV constant). The magnitude of the bias potential which must be applied to one metal to produce a null on the current monitor then gives the contact potential difference. As may be expected, metal work functions are very sensitive to the condition of the surface, and in experiments where work function values are required it is usually best to measure them *in situ*.

Coming closer to a case which may be more relevant to the situation with insulators, let us now consider contact between a metal and a semiconductor. The energy-level scheme for the simple model usually adopted to explain this is shown in Fig. 7.14. In the particular case shown, the Fermi level of the n-type semiconductor is higher than that of the metal, so that when contact is made, electrons flow to the metal until the equilibrium contact potential difference is established:

$$e\Delta V = W_m - W_s, \tag{7.22}$$

where W_{m}, W_{s} are the respective work functions of the metal and semiconductor. In this process all the semiconductor's donor levels, concentration n_{d}, are emptied up to a depth x, and the result is a layer which carries a positive

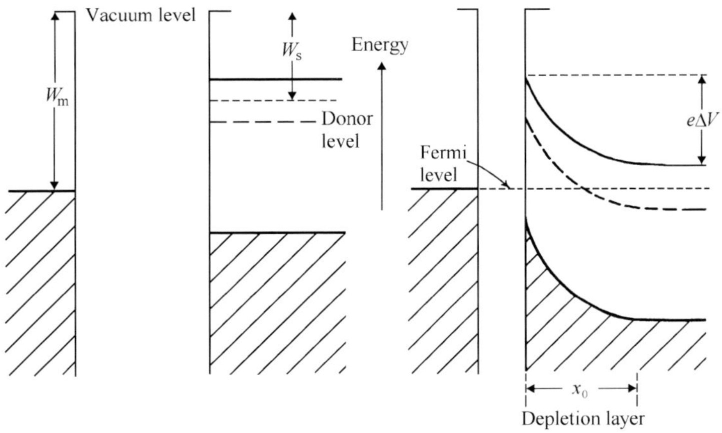

Fig. 7.14 Energy-level diagrams for a metal and an *n*-type semiconductor: (a) before contact, and (b) after contact.

volume density of charge $n_d e$ (ionised donor centres) and which is depleted of carriers (electrons). The potential V which develops progressively on traversing the space charge of the depletion layer must satisfy Poisson's equation:

$$\frac{d^2 V}{dx^2} = \frac{n_d e}{\varepsilon_0 \varepsilon'}, \qquad (7.23)$$

where ε' is the relative permittivity of the semiconductor. As a consequence, a degree of *band bending* occurs near the interface. Integrating Equation (7.23) between the surface and the inner boundary of the charge layer, we obtain the following expression for the contact potential difference:

$$\Delta V = \frac{n_d e x_0^2}{2\varepsilon_0 \varepsilon'}. \qquad (7.24)$$

For a typical inorganic semiconductor $\Delta V = 1$ V, $\varepsilon' = 10$ and $n_d = 10^{25}$ m^{-3}, so that $x_0 \approx 10$ nm.

During contact the charge transferred per unit area is given by

$$\begin{aligned} \sigma &= n_d x_0 e \\ &= [2\varepsilon_0 \varepsilon' n_d (W_m - W_s)]^{\frac{1}{2}}. \end{aligned} \qquad (7.25)$$

The actual charge left on the semiconductor after breaking the contact will again depend on the back-flow of charge.

What may we then expect to happen in the case of a metal–insulator contact, which is a case of prime importance in the context of contact charging of polymers? The essential feature of a perfect insulator is a very wide band gap that is free of any impurity levels. In that case there can be no appreciable charging by transfer of electrons into or from the bulk of the insulator to equalise the Fermi levels. The reason is simply that thermal energies would be insufficient for the necessary promotion of electrons from the valence band to the conduction band. We expect, however, that localised states, due to impurities or defects, will be present to some extent in all real polymers and that a hopping-type process may allow some electronic mobility between such trap sites. Some charging analogous to the semiconductor case should then occur, although it would be slow to develop. If contact times were sufficient for equilibrium to be established, we should predict that the charge transferred per unit area of contact would again conform to Equation (7.25).

Based on the practical observation that some polymers behave as near-perfect insulators, with any charge failing to dissipate over periods of many years, it is reasonable to take the view that a mechanism for contact charging which involves electron transfer to or from the bulk will not always apply. We are left with only one other possibility to explain charging on an electronic basis, and that is to suppose a predominant dependence on surface states. Bauser, Klöpffer and Rabenhorst (1971) considered the energy-level scheme for a simple model of metal–insulator contact charging, involving only surface states on the insulator, and assuming that the energy density of surface states per unit area, D_s, is uniform. In the particular case shown in Fig. 7.15, the Fermi level in the metal W_m is lower in energy than the limit W_i (surface Fermi

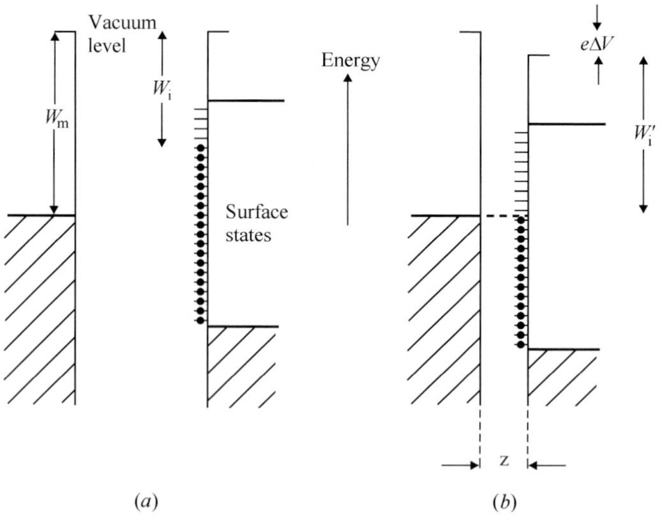

Fig. 7.15 Energy-level diagrams for a metal and an insulator: (a) before contact, and (b) after contact.

level) to which the surface states on the insulator are filled. Upon contact there will be an immediate flow of electrons from the insulator surface states to the metal, the charge transferred per unit area of contact being given by

$$\sigma_s = eD_s \Delta W_i, \tag{7.26}$$

where ΔW_i is the shift $(W_i' - W_i)$ to the new surface Fermi level W_i' caused by the surface charge. Equilibrium requires that the Fermi levels for electrons in the metal and on the insulator are equalised. Hence, denoting the consequent contact potential difference developed between the metal and insulator surfaces by ΔV, we also have

$$W_m - e\Delta V = W_i + \Delta W_i. \tag{7.27}$$

But the geometrical capacitance per unit area between them for a separation x is given by

$$C_A = \frac{\varepsilon_0}{x}, \tag{7.28}$$

so that

$$\Delta V = \frac{\sigma_s}{C_A} = \frac{\sigma_s x}{\varepsilon_0}. \tag{7.29}$$

Combining Equations (7.26), (7.27) and (7.29),

$$\sigma_s = \frac{eD_s(W_m - W_i)}{1 + e^2 D_s x/\varepsilon_0}. \tag{7.30}$$

We may assume that the density of surface states on a molecular solid such as a polymer is relatively low (i.e. of the order 10^{16} eV^{-1} m^{-2} or less), because there is no high concentration of unsatisfied chemical bonds at the surface, unlike the case of a covalent solid such as silicon. Therefore we may assume that

$$\frac{e^2 D_s x}{\varepsilon_0} \ll 1,$$

and Equation (7.30) simplifies to

$$\sigma_s = eD_s(W_m - W_i). \tag{7.31}$$

This formula predicts that the surface charge density imparted by metal contact to the insulator surface is directly proportional to the work function of the metal. Also, the charge is not sensitive to any critical separation at which electrical contact is lost; this is a consequence of the effective capacity of a surface with a

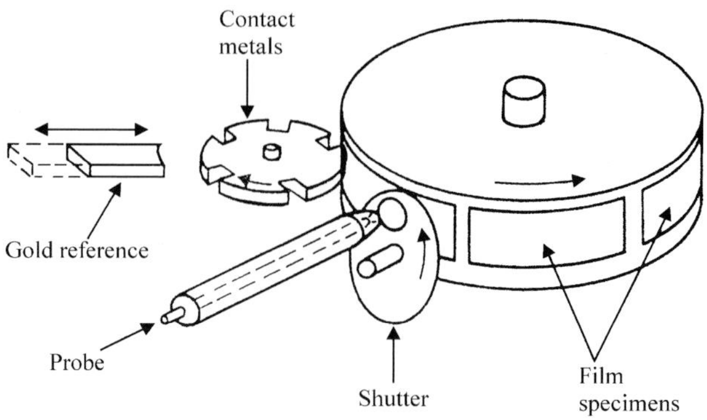

Fig. 7.16 Diagram of apparatus for metal–polymer contact charging experiment Reproduced from Davies (1969) with permission of the Institute of Physics.

low concentration of electronic states available to receive charge being much less than the capacitance of a parallel-plate capacitor of equivalent geometry.

We must now take a look at the results of experimental investigations of contact charging of polymers. Probably the first definitive work in the modern era on this subject was that of Davies. In his first series of experiments Davies (1969) studied the charging of polymers by contact with metals of varying work function. A diagram of the essential parts of the apparatus which he used is shown in Fig. 7.16. Rotation of the drum brings the earthed contact wheel, consisting of segments of five different metals, into rolling contact with each of six different polymeric films, synchronism being maintained during continuous operation by a gear system. As the drum rotates, the charge density is measured by a shuttered probe system and the contact potential difference of each segment of the contact wheel is determined with respect to an oscillating gold reference electrode, using the Kelvin method. The apparatus is contained in a vacuum chamber to avoid any discharge through air.

The results for Nylon-6,6 are shown in Fig. 7.17, where the surface charge density produced on the polymer is plotted as a function of the work function of the contacting metal. (The work function of the gold reference was taken to be 4.6 eV.) We can see that the charge density depends in an essentially linear fashion on the work function, as was also established by Arridge (1967) using Nylon-6,6 threads. The strong dependence on the electronic work function provides convincing evidence that contact charging of this type is an electronic process. The linear relationship strongly suggests that surface states govern the charging behaviour of the polymer in accord with Equation (7.31). Contact charging via surface states is also consistent with the observation that contact

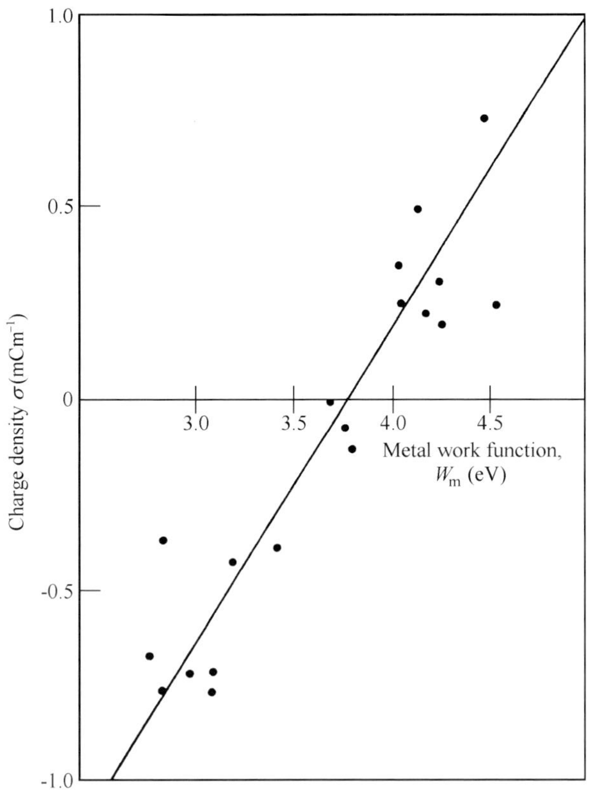

Fig. 7.17 Plot of charge density against work function of the contacting metal for Nylon-6,6. From data of Davies (1969).

charging is a fairly rapid process in most cases. If Equation (7.31) then holds, we can determine the effective work function of each polymer surface by the intercept on the work function axis (zero charge transfer), and the results are given in Table 7.1, which also includes the calculated density of surface states. From this table of polymer work functions we can then predict the direction of charge transfer which will occur when contact is made between any two of these polymers.

In principle, the surface state model may be readily extended to insulator–insulator contacts. There are two limiting cases, shown in Fig. 7.18(a) & (b), depending on the density of surface states. In both of them electrons will tend to flow from the surface with the lower work function to the surface with the higher one, so as to equalise the Fermi energies of the electron populations in the adjacent surface states, thereby instigating the charge transfer on contact. In the case of low surface state densities, D_{s1} and D_{s2}, the electron transfer will not significantly alter the energy levels themselves on the two surfaces. For each electron transferred per unit area the work function gap will be

Table 7.1 *Effective work functions and surface-state densities of polymers.*

Polymer	Work function, W_i(eV)	Surface-state density, D_s ($eV^{-1}m^{-2} \times 10^{-16}$)
Poly(vinylchloride) (PVC)	4.85	1.34
Polyimide (PI)	4.36	1.89
Polycarbonate (PC)	4.26	1.80
Polytetrafluoroethylene (PTFE)	4.26	0.85
Poly(ethyleneterephthalate) (PET)	4.25	1.29
Polystyrene (PS)	4.22	0.69
Nylon-6,6	4.08	0.76

Work functions taken from Davies (1969).

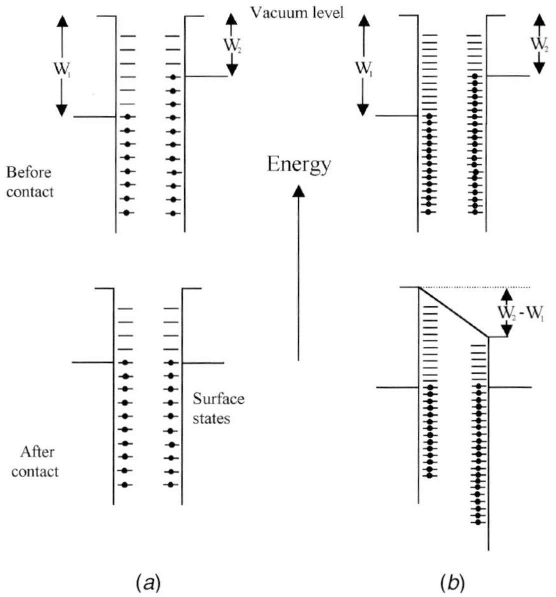

Fig. 7.18 Energy level diagrams for two different insulators, before and after contact in the two limiting cases: (a) low surface state densities and (b) high surface state densities.

decreased by an amount $(1/D_{s1} + 1/D_{s2})$. Hence, the charge transferred for total equalisation will be given by:

$$\sigma_{sL} = \frac{e(W_1 - W_2)}{1/D_{s1} + 1/D_{s2}}, \tag{7.32}$$

where W_1, W_2 are the respective surface work functions of insulators 1 and 2. When $W_1 < W_2$, insulator 1 will acquire a positive charge and vice versa. The equation predicts that the contact charging will depend linearly on the difference of the effective work functions of the two surfaces, these values being determinable from observations of the contact charging behaviour of each of the individual surfaces against metals.

In the limit of high density of surface states, the electronic charge flow creates a high enough electric field to alter the energy levels themselves, resulting in an interfacial potential difference of $e(W_1 + W_2)$. Assuming that the surfaces are separated by a distance x during effective electrical contact, the surface charge density can then be expressed in terms of the electric field between the surfaces as:

$$\sigma_{\mathrm{sH}} = -\varepsilon_0 \frac{W_1 - W_2}{ex}. \tag{7.33}$$

This equation also predicts a linear dependence of charge exchange on difference in work functions. Using typical expected values for the variables, e.g. $(W_1 + W_2) \approx 1$ eV and $x = 1$ nm, Equation (7.33) predicts surface charge densities that are higher than are ever observed experimentally. This does not necessarily mean that the basic treatment of the contact charging is wrong, but rather that there are additional effects taking place, like back-tunnelling and/or dielectric breakdown upon separation of the surfaces, which limit the final net charge transferred.

In a second series of experiments, Davies (1970) confirmed this deduction, and the results are given in Table 7.2, which includes values calculated using the data from Table 7.1 in conjunction with Equation (7.32). In all cases the

Table 7.2 *Charge densities (mCm^{-2}) transferred in polymer–polymer contacts (theoretical values are given in parentheses).*

	PI	PTFE	PS	Nylon-6,6
PVC	+0.20 (0.61) -0.19	+0.80 (0.73) -0.49	+0.13 (0.46) -0.12	+0.05 (0.60) -0.19
PI		+0.10 (0.15) -0.21	+0.11 (0.07) -0.02	+0.27 (0.25) -0.93
PTFE			+0.14 (0.03) -0.16	+0.17 (0.15) -0.48
PS				? (0.08) ?

Experimental values taken from Davies (1970).

correct sign of charging is predicted. Quantitatively, the agreement between theory and experiment is reasonable considering the amount of experimental scatter.

The question of whether the high or low surface state density limit applies remains an issue, although detailed analysis of the extensive data set relating to charging of toners used in xerographic (or electrostatic) copiers and printers suggests that perhaps an intermediate density is operative. In the toner situation contact charging occurs between the polymeric toner particles themselves (approximately 10 μm in diameter) and the larger carrier particles (approximately 100 μm in diameter) as the two are mixed together, and expressions for the equilibrium charge-to-mass ratios have been derived for the low and high surface state density cases. Castle (1997) has reviewed recent results.

The possibility of ranking polymers in a self-consistent way for the purposes of predicting the relative direction of charge transfer when any two of them are allowed to make contact with each other has been recognised for a long time, and many so-called triboelectric series, obtained in an empirical way, have been published. The physical basis for such series is now evident.

Thus surface-state theory provides a satisfactory basis for discussing contact charging of polymers by electron transfer. The model is, however, probably oversimplified, and it would be unwise to exclude the possibility of penetration of charge into the bulk polymer altogether in view of conduction effects that can usually be observed at high fields in polymers. We must also remember that the results we have been talking about refer to very clean polymer surfaces and that the contact charging in most practical cases can be very different due to the presence of dirt and moisture on the surface.

It is perhaps important to note here that PE exhibits a negative electron affinity (Bloor, 1976), a feature shared with only a very restricted set of materials. This means that excess electrons prefer energetically to reside outside PE rather than be in any way bound to the PE molecular structure. Nevertheless, recent calculations show that there are electronic surface states lying below the vacuum level in the forbidden band gap of PE (Righi *et al.*, 2001). These surface states will certainly be expected to act as traps for transferred or injected electrons, and they will therefore be involved in contact charging. Their resonance interaction with negative ion states of typical PE dopants (O_2 and H_2O) may be very important too.

7.3.2 Charge transfer by ions

Although most modern research on contact charging, as described in the previous section, generally assumes that the underlying process is essentially one of electron transfer, the possibility of contact charging by movement of ions across the interface cannot be ignored, especially in polymers which have ionic structures or which are liable to contain ionic ingredients. Common polymer additives, especially antistatic agents, tend to *bloom out* on the surface, and in that case ionic-charge transfer may be expected to dominate. Whenever ions are present at surfaces, they do inevitably play a role in contact charging, although there are many factors to take into account in predicting their effect on the resulting net transfer of charge. Factors that affect the charging are:

(a) The degree of dissociation of the ionic component. This in turn will depend on the environment that the ions experience. The presence of water with its high relative permittivity will promote ionisation, making more ions available for transfer.
(b) The mobility of the ions. Ion size will affect this, and solvation (attachment of solvent, e.g. water, molecules to the ions) will reduce mobility.
(c) The relative stability of the positive and negative ions on the two surfaces. This would be expected to determine the final equilibrium situation.

This complexity explains why definitive, quantitative verification of ionic contact charging has proved to be so elusive. Nevertheless, a body of convincing evidence for the role of ions in the charging process now exists for special categories of polymers with ionic features in their compositions.

The clearest instances of contact charging by ion transfer are exhibited by ionomers, e.g. poly(styrene-co-N-methyl-4-vinyl-pyridinium toluenesulphonate),

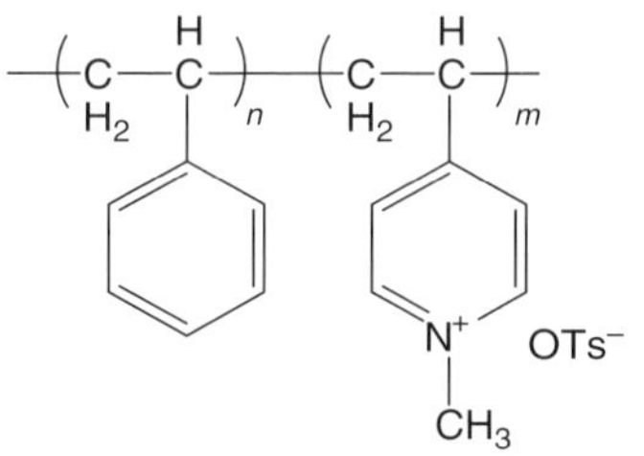

or partially sulphonated polystyrene

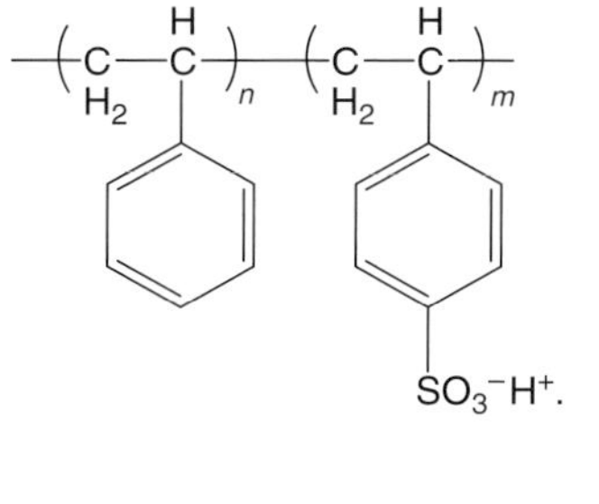

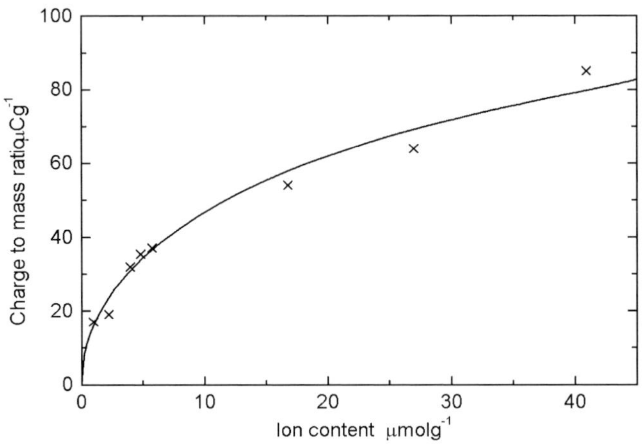

Fig. 7.19 Graph of charge-to-mass ratio versus ionic content for particles of a blend of styrene-co-N-methyl-4-vinyl-pyridinium toluenesulphonate ionomer with a styrene-co-butyl methacrylate copolymer, charged by rolling with coated beads. Reprinted with permission from Diaz *et al.* (1992). Copyright of the American Chemical Society.

In these cases, the sign of the ion which is covalently bonded to the polymer, and which is consequently locked in position with very little mobility, dictates the sign of the charge on the polymer after contact with another surface. Also, the magnitude of the charge transferred increases monotonically with the concentration of ionic ingredient in the polymer, though not linearly, and there is a close correspondence between the amount of material transferred as ions and the resulting surface charge density. Experimental results for a blend of styrene-co-N-methyl-4-vinyl-pyridinium toluenesulphonate ionomer with styrene-co-butyl methacrylate copolymer, charged by rolling with coated ferrite beads, are shown in Fig. 7.19. The charge transfer process will depend on the availability of dissociated ions on the surface. The concentration of dissociated ions will be affected by a number of factors, including any ion pairing and any higher degrees of ionic aggregation, and this explains the departure from non-linearity of the charging with respect to the stoichiometric concentration of ionic ingredient.

When a polymer simply contains ionic salts that are not bound covalently to the polymer chains themselves, e.g. tetraalkylammonium toluenesulphonate, the situation is more complicated, because ions of both sign are mobile and are available for transfer. The magnitude of the net charge transferred will then depend on the relative transfer aptitude of the two ions in the salt. Both of the ions tend to transfer in excessive amounts, which contaminates the second surface and produces a relatively low net surface charge density.

The chief industrial importance of ionic contact charging has been in electrophotography. Ionic ingredients are used there to ensure effective charging of the developer systems, consisting of charged toner powder particles carried by beads charged with the opposite sign. Such electrostatic developers are activated simply by rolling the two components together, relying on the contact charging to produce the appropriate charge-to-mass ratios. Detailed models for the charging process, enabling enhanced materials design, have been very successful.

7.4 Electrets

Heaviside was apparently the first to speculate about using permanent polarisation to make the electrical counterpart of a magnet. He coined the name *electret*, and foresaw that it would have practical value in providing a convenient, portable source of electric field, needing no batteries. Eguchi was the first to experiment with such systems. He took a slab of Carnauba wax, heated it up until it melted, applied a strong electric field to polarise it, and cooled it down before switching the field off again. In this way he expected to freeze into position displaced charge carriers or oriented dipoles, thereby producing positive and negative surface charges opposite the negative and positive electrodes, respectively. He examined the surface charges periodically by an induction method, rather like the old electrophorus experiment, in what is often called a *dissectible capacitor* (see Fig. 7.20).

To make the measurement the electret is placed between capacitor plates which are shorted together, when the electret's surface charges induce equal and opposite charges in the adjacent electrodes. The top electrode is then electrically isolated and lifted away, and the charge on it measured with an electrometer. By this method Eguchi discovered that if the newly formed electret was stored with its surface shielded by metal plates, its surface charge decreased in magnitude, eventually reversing in sign and increasing to a large permanent value of opposite sign to the original surface charge. If the electret was subsequently left unshielded for some time, the charge decayed, but if the surfaces were shielded again, the charges partly recovered. This typical but rather surprising electret behaviour, illustrated in Fig. 7.21, was explained by Gross (1949) as an interplay between surface heterocharges, due to volume polarisation, of opposite sign to the potential of the adjacent electrode during polarisation, and homocharges, due to the presence of free charges, of the same sign as the potential of the adjacent electrode.

The origin of the homocharges was explained in the following way. As the volume polarisation increases at the beginning of poling (stage A) the field in

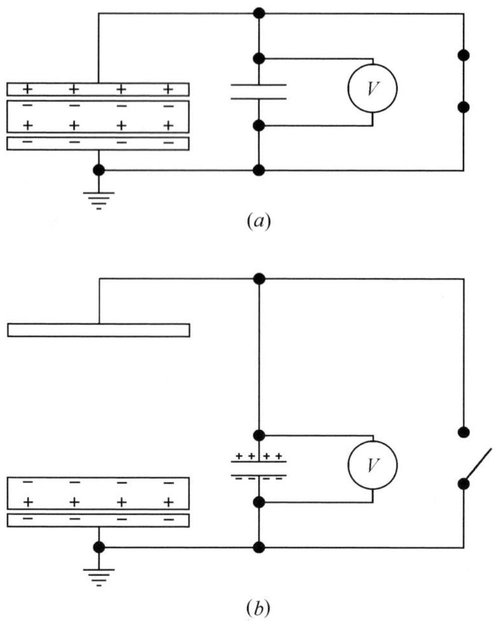

Fig. 7.20 The dissectible capacitor: (a) both electrodes on the electret and shorted to earth, (b) top electrode disconnected from earth and lifted away from the electret.

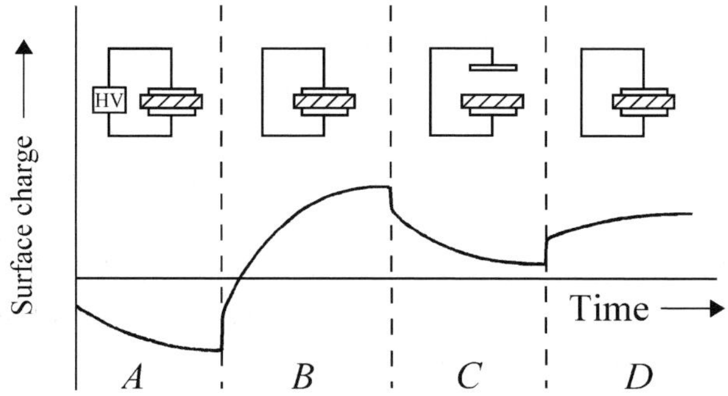

Fig. 7.21 Typical behaviour of an electret at various stages: *A*, poling by an applied field; *B*, electrodes shorted; *C*, upper electrode removed; *D*, electret re-shielded. Reproduced with permission of the American Institute of Physics from Perlman and Meunier (1965).

the gap between the dielectric and an electrode increases until discharges occur across the air gap, depositing ions of the same sign as the electrode potential. As further polarisation develops, repeated discharges occur, further increasing the homocharges. At the end of the poling stage a mixture of hetero- and homocharges will then be present at the dielectric surfaces. Any experimental

Table 7.3 *Time constants τ for decay of electret charges.*

Polymer	Decay constant, τ(yr)	
	Homocharge	Heterocharge
PET	1.5	0.5
FEP	50	0.5

detector of the surface charge density will give the algebraic sum (or net) value. When the poling stage is over, and shorted, shielding electrodes are applied (stage *B*), the internal field in the dielectric is much reduced, so the volume polarisation relaxes and the induced heterocharge decays, leaving only homocharge. If an electrode is now removed (stage *C*), the homocharge is allowed to regenerate an internal field in the same direction as the original poling field and volume polarisation begins to develop, again producing some heterosurface charge, and at the same time the homocharge decays under the influence of its own field. Upon reshielding (stage *D*) the polarisation relaxes once again. Confirmation of the origin of the homocharge has been obtained by experiments at low air pressures, when smaller homocharges were found, in accord with Paschen's law for discharges in air.

It took a long time for this somewhat complicated behaviour of electrets to be sorted out, and in the meantime high-grade insulating polymers had been developed. The question then to be settled was what system would give the most stable electret, and this prompted a study of the relevant decay processes in polymers.

From the above interpretation of the behaviour of electrets we can see that decay of hetero- and homocharges can be separately evaluated by making measurements at both stages *B* and *C*. In addition, since dipolar relaxation and conduction, which govern the decay of hetero- and homocharges respectively in electrets, are known to be thermally activated processes, it is possible to avoid inordinately long room-temperature tests. The temperature dependence of the time constants can be measured at high temperatures, where the rate is fast, to establish an activation energy, and then the value at room temperature obtained by extrapolation. Typical values for the time constant τ in the surface charge decay equation $q_s = q_0 \exp(-t/\tau)$ are given in Table 7.3 for poly(ethylene terephthalate) (PET) and fluorinated ethylene-propylene copolymer (FEP).

The conclusion from these results is that homocharges are much longer lasting than heterocharges, and this has led to the recent production of electrets

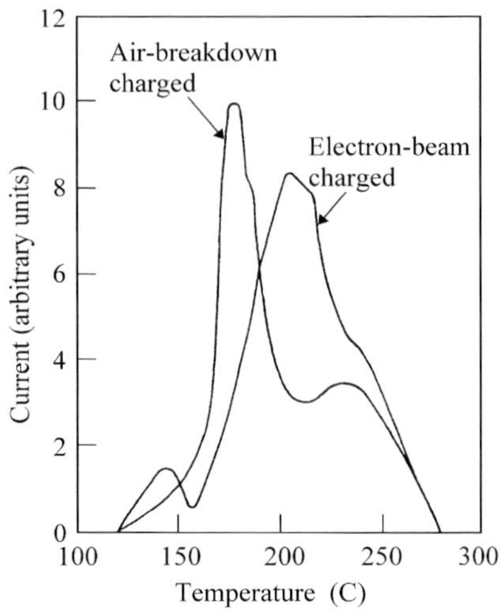

Fig. 7.22 Thermally stimulated current spectra of FEP electret foils (aluminium, coated on one side) immediately after polarisation. The heating rate was 4 C min^{-1}. Reproduced with permission of the American Institute of Physics from Sessler and West (1971).

with excellent long-term stability. Thin (12 μm) FEP films, metallised on one side, are charged by depositing electrons directly in them by exposure to a low-energy (10 to 40 keV) beam (Sessler and West, 1970). By adjusting the beam energy, the average penetration depth can be selected to be equal to half the film thickness. In this context it is worth mentioning that such an electret cannot be fully characterised by the normal measurement of apparent surface charge density – the measurement cannot distinguish between a plane of high charge density lying deep in the film near to the metallised backing from a plane of lower charge density near to the top side. In order to check the mean depth of the charge it is necessary to make supplementary measurements of charging currents flowing to electrodes on both sides of the specimen during electret formation, or to use pulse-scanning methods, as described in Section 7.2.2.

Another way of investigating the depth in energy to which charges are trapped (a measure of the stability towards thermal decay) may be obtained by thermally stimulated currents (TSC). In this technique, electrodes on the two sides of the electret are connected via a sensitive current meter and the specimen is then heated at a constant, slow rate (1 Cmin^{-1}, say). Discrete current peaks are observed as a function of temperature as successively more deeply trapped charges are released (Fig. 7.22). Dipolar relaxation may also give peaks in the TSC spectrum (van Turnhout, 1975).

Charge densities as high as 1 mCm^{-2} can be obtained with a time constant for decay in excess of 20 years. The principal application for such electrets is

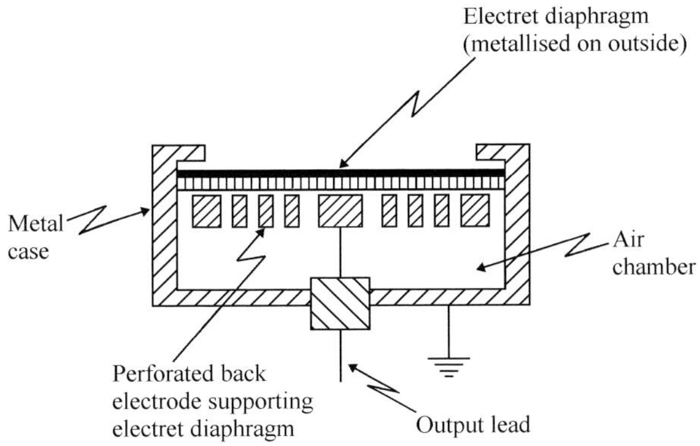

Fig. 7.23 Basic design of an electret microphone

the electret microphone, whose construction is shown in Fig. 7.23. The electret film forms the diaphragm on which the sound waves impinge. As the film moves, the electric field from the electret to the counter electrode varies and induces a signal in the circuit connected to it; a pre-amplifier (usually comprising an integrated circuit with a field-effect-transistor input) is incorporated in the microphone unit. The frequency response of the electret microphone is comparable with that of a condenser microphone without the requirement of any high bias potential. They are well suited to miniaturisation and they are now made in large numbers for telephones and tape recorders.

7.5 Further reading

Textbooks devoted to the subject of electrostatics and its applications have been published by Cross (1987) and by Taylor & Secker (1994). Harper's (1967) book covers older work. One of the best sources of information about recent work on static electrification of polymers continues to be the Institute of Physics conference series: Electrostatics, see Taylor (1999).

8

Ionic conduction and particulate and molecular composites

8.1 Introduction

Great advances have been made in recent years in developing organic molecular systems which have high intrinsic electronic conductivity. In spite of this, there are still only a few inherently conductive polymers where the necessary combination of electrical and mechanical properties and stability offer the potential to provide a truly conductive plastic material to replace metals for general electrical applications. There are, however, other aspects to conduction which are of scientific and technical interest where polymer based materials are employed. Thus ionic conduction is particularly important in electrochemical applications. Here materials that allow ion transfer, whilst disallowing electronic conduction, are required for cell separators and membranes. Where only a modest level of conductivity is required, a major practical solution is often to make a conductive composite material by dispersing a conductive material, like a metal or carbon black, in an insulating polymer in some way. Similarly, semiconducting materials can be obtained by the incorporation of a sufficient concentration of conjugated molecular species into insulating polymers. In molecules and oligomers with conjugated structures the energy gap between the highest occupied molecular orbital (HOMO) and the lowest unoccupied molecular orbital (LUMO) is much smaller than that of the chemically saturated analogues. Electrons which are injected from external electrodes will naturally gravitate to the lowest-lying, unoccupied electronic states in the material, and electrons that are removed at an electrode to form holes will come from the highest-energy occupied electronic states. These states will be located in the largest conjugated structures in the material. The conjugated regions need not be continuous throughout the material, since conduction can occur between physically separated conjugated structures by hopping conduction, as discussed in Section 4.2.3. Thus conduction can occur in polymers with saturated backbones, provided there are conjugated groups either pendent to, or incorporated in, the backbone. Alternatively, conjugated molecules can be dispersed as a solid solution in a saturated polymer at a concentration sufficiently high that hopping conduction can occur. Such materials have been widely used in the xerographic process. In this chapter we describe these alternatives.

8.2 Ionic conduction

Ionic conduction processes come into play in a variety of ways in polymeric materials. In conventional plastics for electrical applications, which exploit insulating properties, control of ionic impurities becomes especially important. On the other hand, the extremely low conductivity that is a natural characteristic of many polymers, e.g. polyethylene, proves to be troublesome in certain products. The problem is the accumulation of static charges which are unable to leak away. The presence of charges leads to the attraction of dust, cling of films, and in some circumstances even to generation of electric shocks and sparks. Antistatic agents that are used to alleviate these problems are normally based on ionic conduction, providing just the minimum of conductivity to allow charge to flow to earth effectively. At the other end of the scale, polymers that are tailored to have high levels of ionic conductivity are now finding application as solid electrolytes for batteries and fuel cells. The various categories of ionic conduction are discussed separately below.

8.2.1 Ionic impurities

The most definitive evidence for ionic conduction is the detection of electrolysis products formed on discharge of the ions as they arrive at the electrodes. Unfortunately, the very low level of conductivity in ordinary polymers generally precludes such detection. Even at a conductivity of $10^{-9}\,\Omega^{-1}\text{m}^{-1}$, and we must bear in mind that many polymers exhibit conductivities several orders of magnitude lower than this, 100 V applied across a specimen 100 mm^2 in area and 1 mm thick would only produce about $10^{-11}\,\text{m}^3$ of gas at NTP per hour. We therefore have to rely on rather more indirect means of elucidating the mechanism of conduction.

When the conductivity in question is very low, there are other experimental difficulties too. On application of a step voltage across a specimen, the initial current may be dominated by a displacement current due to polarisation of the material. Since some dipole orientation may be very slow to reach equilibrium, the displacement current can swamp a small conduction current for a long time. Taking into account the drift characteristics of any practical measurement system, we can appreciate that in extreme cases a steady current reading will never be obtained, and a direct measurement of conductivity will not be possible.

Some physical insight into what a conductivity of around $10^{-9}\,\Omega^{-1}\mathrm{m}^{-1}$ or less entails may be gauged through the basic equation for ionic conductivity σ:

$$\sigma = e\left(\sum_i m_i^+ n_i^+ \mu_i^+ + \sum_j m_j^- n_j^- \mu_j^-\right), \tag{8.1}$$

where m_i^+ is the number of units of electronic charge on the ith positive ion, n_i^+ is its concentration and μ_i^+ is its mobility, where m_j^-, n_j^-, μ_j^- refer to the corresponding quantities for the jth negative ion, and where the summation covers all the ionic species that are present. We can tentatively use a mobility value for all ions of the order of $10^{-9}\,\mathrm{m^2V^{-1}s^{-1}}$, as found for ionic carriers in hydrocarbon liquids at room temperature. Although we would expect the true value for solid polymers to be lower, it should not differ by more than a factor of 100 or so for small ions. Then, assuming that the ionic ingredient is just dissociated into a pair of positive ions (cations) and negative ions (anions), each just carrying one unit of electronic charge:

$$AB \rightarrow A^+ + B^-, \tag{8.2}$$

substitution in Equation (8.1) gives a carrier concentration of around $3.10^{18}\,\mathrm{m}^{-3}$. Comparing this with the typical number density of molecular repeat units (nominal molar mass $= 100\,\mathrm{gmol}^{-1}$) of about $10^{28}\,\mathrm{m}^{-3}$, we can see that an exceedingly low carrier concentration is implied. This means that degrees of ionic impurities which may be totally ignored in the context of other properties may have a significant effect on conductivity.

In some instances evidence favouring ionic conduction is provided by a strong correlation between relative permittivity and conductivity, which is readily explained by the reduction of the Coulombic forces between ions in a high relative permittivity medium. This renders the dissociation energy of an ionic compound inversely proportional to the static relative permittivity ε_s. Consider the dissociation reaction:

$$\begin{array}{ccc} AB & \longleftrightarrow A^+ & + B^-, \\ (1-f)n_0 & fn_0 & fn_0 \end{array} \tag{8.3}$$

where the original concentration of the ionic compound is n_0 and its fractional degree of dissociation is f at equilibrium. Applying the Law of Mass Action, we may then define an equilibrium constant K in terms of the concentrations of the reactants and resultants, as follows:

$$K = [A^+][B^-]/[AB] = f^2 n_0/(1-f). \tag{8.4}$$

The equilibrium will be governed by the change in free energy ΔG for the reaction, so that

$$K \propto \exp\left(-\frac{\Delta G}{kT}\right) = K_0 \exp\left(-\frac{\Delta W}{\varepsilon_s kT}\right), \tag{8.5}$$

where ΔW is the energy required to separate the ions in a medium of unit relative permittivity, and entropy terms are taken into the constant K_0. If AB is the only ionisable species present, the conductivity will be given by

$$\sigma = f n_0 e(\mu^+ + \mu^-), \tag{8.6}$$

where μ^+, μ^- are the mobilities of the positive and negative ions, respectively, and e is the magnitude of the charge on an electron. For a small degree of dissociation, Equation (8.4) becomes

$$f \approx \left(\frac{K}{n_0}\right)^{1/2}, \tag{8.7}$$

and substitution from Equations (8.5) and (8.7) in (8.6) gives:

$$\sigma = (K_0 n_0)^{1/2} e(\mu^+ + \mu^-) \exp\left(-\frac{\Delta W}{2\varepsilon_s kT}\right). \tag{8.8}$$

The presence of the relative permittivity in the exponent of Equation (8.8) means that it will exert a strong influence on conductivity. In this way the absorption of water, which has a relatively high relative permittivity, generally enhances the conductivity of a polymer greatly, and polymer–water systems frequently conform to the equation

$$\log \sigma = -A/\varepsilon_s + B, \tag{8.9}$$

where A and B are constants (Hearle, 1957).

Equation (8.8) also shows that (a) $\sigma \propto n_0^{1/2}$, and (b) the apparent activation energy obtained from the slope of the Arrhenius plot of $\log\sigma$ against $1/T$ is $\Delta W/2\varepsilon_s$. The square-root dependence of σ on n_0 and the occurrence of the factor ½ in the activation energy both stem from the control of the ionic dissociation equilibrium by the Law of Mass Action.

Further evidence of ionic conduction may be obtained from studies of the dependence of current on applied voltage, for which a theoretical expression

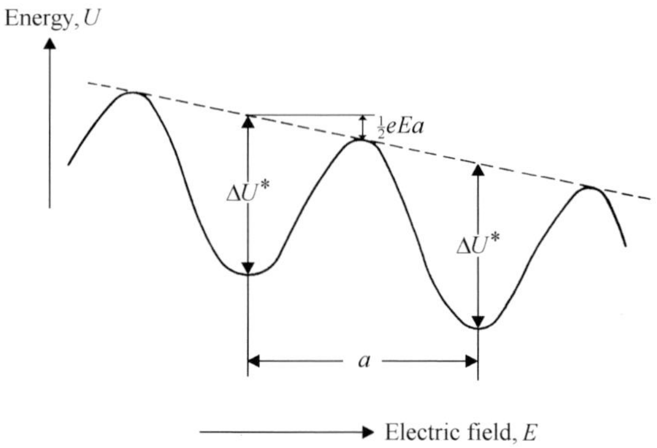

Fig. 8.1 Diagram showing the deformation of ionic potential-energy wells by an applied electric field.

may be derived on the basis of a simple model (Mott and Gurney, 1948). Suppose that the unit motion of an ion in the absence of a field is a jump within the matrix of polymer molecules to a neighbouring position of exactly equal energy, passing over a potential-energy barrier of height ΔU^*. The ion will be in a constant state of vibration (frequency ν) when lodged in a potential well, and we can assume that the probability that it will pass over the barrier is $\exp(-\Delta U^*/kT)$ in each vibration, or $\nu \exp(-\Delta U^*/kT)$ per second. Now consider the effect of applying a uniform electric field E. In the direction perpendicular to the field the potential-energy barrier will be unaffected, but in the direction of the field, and against the field, the barrier heights will be changed by $\pm 1/2eEa$, respectively, where a is the distance between neighbouring potential wells (see Fig. 8.1). The probability that the ion will move in the direction of the field will now be

$$\nu \exp\left[-\frac{(\Delta U^* + eEa/2)}{kT}\right],$$

and in the opposite direction

$$\nu \exp\left[-\frac{(\Delta U^* - eEa/2)}{kT}\right].$$

The mean drift velocity u in the field will therefore be

$$u = a\,\nu\,\exp\frac{-\Delta U^*}{kT}\,2\,\sinh\frac{eaE}{2kT}. \tag{8.10}$$

Taking into account that the concentration of ions is proportional to $\exp(-\Delta W/2\varepsilon_s kT)$, the current density j flowing through a specimen across which an electric field E is applied will be given by

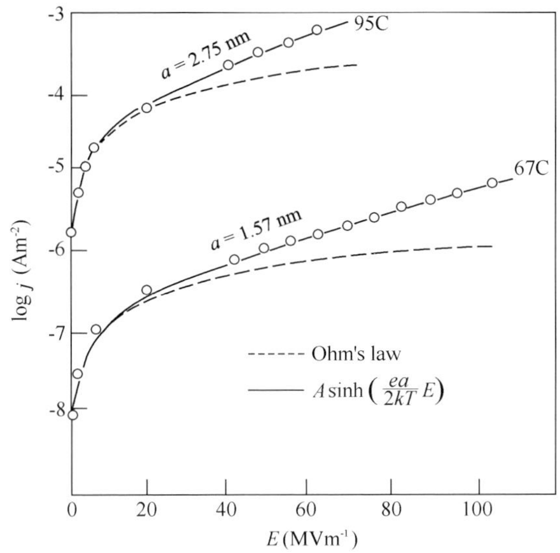

Fig. 8.2 Graphs showing the effect of electric field strength on conduction in poly(vinylchloride). Reproduced with permission of the American Institute of Physics from Kosaki, Sugiyama and Ieda (1971).

$$j \propto \exp \frac{-(\Delta W/2\varepsilon_s + \Delta U^*)}{kT} \sinh \frac{eaE}{2kT}, \tag{8.11}$$

or, more simply, at constant temperature,

$$j \propto \sinh \frac{eaE}{2kT}. \tag{8.12}$$

The graph in Fig. 8.2 shows how well the results (Kosaki *et al.*, 1971) for a sample of PVC fit this formula over a wide range of fields. From this curve one can also derive a value for the jump distance a. It is found that below T_g (87 C for the PVC specimen) the apparent jump distance is about 1.2 nm, a value which is quite consistent with typical molecular spacing.

Although it is not possible to identify the ions experimentally, we may reasonably assume that they are mainly derived from fragments of polymerisation catalyst, degradation and dissociation products of the polymer itself, and absorbed water. On this basis a polymer like PVC most probably contains H_30^+, Na^+, K^+ cations and OH^-, Cl^-, Br^- anions.

A feature which is common to most polymers is a change in temperature dependence of conduction as the temperature is raised through the glass transition region. Now we may expect that free volume, which affects the rate of molecular motions at the glass transition (see Section 3.3), will be equally important to the mobility of ions. Miyamoto and Shibayama (1973) have taken this into account by assuming that the probability per second of a unit ion jump in the absence of an electric field is proportional to $\nu P_{\rm v}\exp(-\Delta U^*/kT)$, where $P_{\rm v}$ is the probability that a hole of sufficient size is available for the ion to jump into, i.e.

$$P_{\rm v} = \exp\left(-\frac{\gamma\nu_{\rm i}^*}{\nu_{\rm f}}\right), \tag{8.13}$$

where $\nu_{\rm i}^*$ is the critical free volume for an ion jump, $\nu_{\rm f}$ is the average free volume per molecular segment, and γ is a numerical factor to correct for the overlap of free volume. The current density equation, (8.11), may then be modified to become

$$j \propto \exp\left(-\frac{\gamma\nu_{\rm i}^*}{\nu_{\rm f}} - \frac{\Delta W/2\varepsilon_{\rm s} + \Delta U}{kT}\right)\sinh\frac{eaE}{2kT}. \tag{8.14}$$

Taking free volume characteristics from WLF data, see Section 3.3, a straight line is obtained when $(\log\sigma + \gamma\nu_{\rm i}^*/2.303\nu_{\rm f})$ is plotted against $1/T$ for polystyrene through the glass transition region, as shown in Fig. 8.3, in accord with the behaviour expected from Equation (8.14).

The greater ease with which ions may be reasonably expected to move through amorphous regions, as compared with crystalline regions, ties in with the general fall in conductivity with crystallinity observed in most polymers.

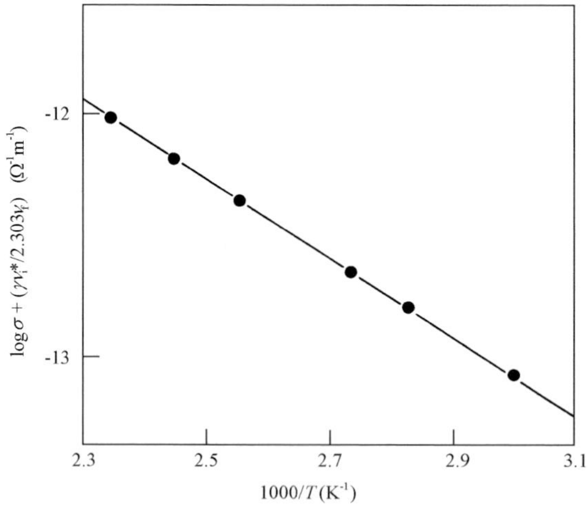

Fig. 8.3 The temperature dependence of the conductivity of polystyrene through the glass transition region. Reproduced with permission of the American Institute of Physics from Miyamoto and Shibayama (1973).

8.2.2 Antistatic agents

Electrostatic problems on insulating plastics (see Chapter 7) are often alleviated by the use of so-called antistatic agents. These are usually surface-active chemicals, which form a thin (10 nm or less) layer on the outer surface of the plastic moulding, film etc. In most cases their operation additionally requires the uptake of water in the surface layer to promote ionic dissociation and to provide a medium for ionic flow. Dissipation of charges by conduction through the surface layer of such systems therefore works best in moist atmospheres. Although performance may be satisfactory at normal ambient relative humidity of 50% or so, conduction may fail in very dry conditions, depending on the exact nature of the surface layer. Figure 8.4 shows a graph of the dependence of the surface conductivity of a polyethylene film containing an antistatic agent. The response of surface conductivity to changes in relative humidity is remarkably rapid, with a time constant of as little as a few seconds. This is indicative of a phenomenon that is truly restricted to the surface. Sometimes antistatic agents are applied directly to the surface by a dip or spray method. Alternatively, they are added (at ~0.1%) to the bulk of the

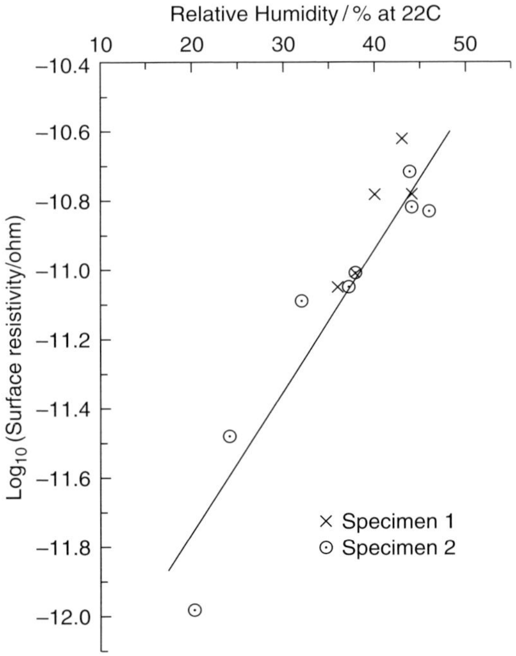

Fig. 8.4 The dependence of surface conductivity of discharge-treated polyethylene film (95μm), containing 425 ppm Ethomeen® T12 as an antistatic agent, on ambient relative humidity.

polymer powder before processing, with the development of surface conductivity then relying on the agent migrating to give a surface *bloom*.

In the case of low-density polyethylene for film products, a typical antistatic agent is a long-chain aliphatic amine, e.g. Ethomeen ®T12, with the formula:

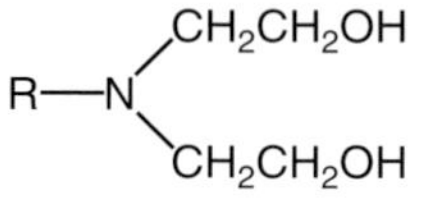

$R = C_{12}$ alkyl group from tallow.

The long alkyl group makes this compound compatible with the olefinic polymer, and it is readily dispersed uniformly into the bulk, to serve as a reservoir of antistatic agent. On the final film product, the development of surface conductivity depends on the surface condition and characteristically is rather slow. Clearly, it must depend on the antistatic agent migrating to the surface, and this is slow, because (a) the diffusion rate of the bulky Ethomeen® molecules is low, and (b) the preference of the free amine molecule for a surface location, with its olefinic tail in the polymer and its more polar hydroxyethyl group on the outside, is only slightly favoured energetically. Furthermore, production of ions requires a quaternisation reaction with an acid:

$$R(CH_2CH_2OH)_2N + HA \rightarrow R(CH_2CH_2OH)_2NH^+ + A^-. \qquad (8.15)$$

Acidic molecules from the ambient air are sufficient for this reaction to proceed over a long period of time, but it is not very effective. Adsorbed water will also promote the reaction by providing a high relative permittivity medium for the ions. In practice, corona discharge treatment of the film surface is usually employed to accelerate and enhance the process. The effect of this is two-fold. Firstly, it generates a large amount of nitric acid in the vicinity of the surface to be available for reaction with free amine. Secondly, it oxidises the surface (Blythe *et al.*, 1978), thereby providing a polar environment to welcome the polar part of the Ethomeen® molecules; in other words the free energy of the Ethomeen® molecules will be reduced at the treated surface, so that they will tend to accumulate there. In this way, if the antistatic agent is introduced at the 500 ppm level in the bulk, surface conductivities of $\sim 10^{-12}$ to $10^{-11}\Omega^{-1}$ can be guaranteed within a few hours.

Improved, more permanent and durable protection against static charges can sometimes be obtained by the incorporation of an ionic copolymer.

8.2.3 Polyelectrolytes and protonic conductors

Our considerations above have laid emphasis on conduction by ions present by default or added as a separate ingredient to the polymer. Some polymers are,

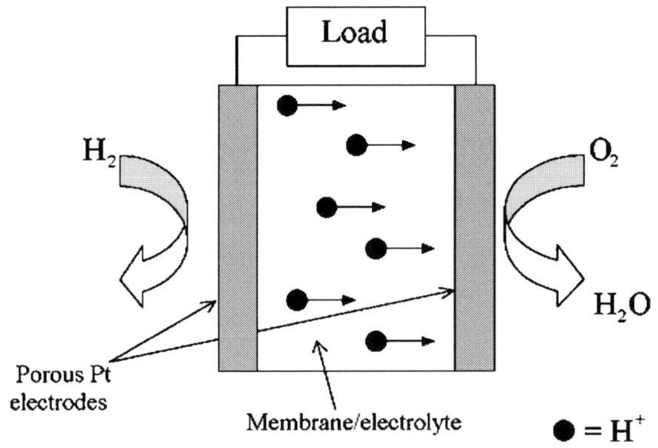

Fig. 8.5 Diagram of a hydrogen fuel cell.

however, themselves inherently conductive in having at least one ionisable group per monomer unit, one of the most common being cellulose and its derivatives. Such a polymer is called a polyelectrolyte. In the ionised state of these structures one set of ions (either cations or anions) is held fixed on the main polymer structure, whilst the counter ions are free to move under the influence of an applied field. One important characteristic of the conduction in these materials is its strong dependence on water content. The reasons for this behaviour are the same as those discussed above in Section 8.2.1 concerning the role of water with respect to ionic impurities. In the dry state there is no appreciable ionic conductivity at all, but as water is taken up in a humid atmosphere, the conductivity may rise to levels in the region of $10^{-8}\,\Omega^{-1}\mathrm{m}^{-1}$. Polyamides also show pronounced ionic conduction effects at elevated temperatures. (The conductivity of Nylon-6,6 intrinsically exceeds $10^{-8}\,\Omega^{-1}\mathrm{m}^{-1}$ above 100 C, evidently as a result of the dissociation of amide groups to give protons.) There has been much discussion about the possibility of a particularly fast mechanism for the transfer of these protons through a lattice of hydrogen-bonded polyamide molecules, analogous to the conduction process in ice, but the case is difficult to prove unambiguously.

A prime need for a solid ionic conductor arises in the design of electrochemical fuel cells (Alberti and Casciola, 2001). Perhaps the most important type is the hydrogen fuel cell, shown diagrammatically in Fig. 8.5. Here the membrane which separates the two electrodes must be able to transfer protons efficiently. A material which combines the flexibility and toughness of a plastic with high protonic conductivity would be an ideal candidate. Prototype cells were successfully operated with membranes made from poly(styrenesulphonate),

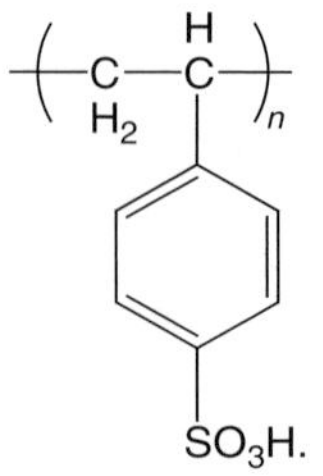

In the presence of water this polyelectrolyte is an excellent protonic conductor, although it suffers some degradation in a working fuel cell. Probably the most successful polyelectrolytes developed so far are based on perfluorinated polymers, the first of which was Nafion®, with the structure:

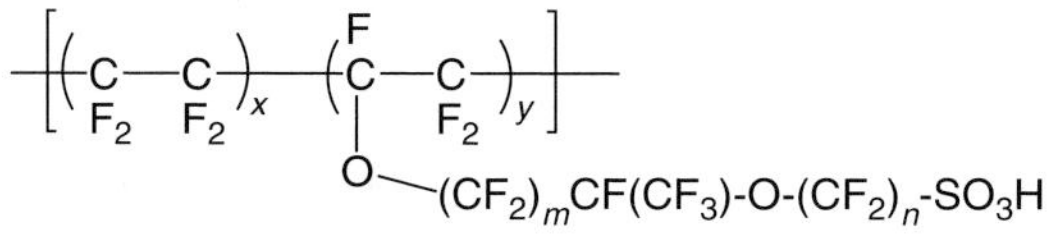

$$x = 5 - 13.5,\ y = 1,\ m \geq 1,\ n = 2.$$

There is a range of similar polyelectrolytes with the same generic structure, differing only in the values of x, y, m and n and either with or without $-(CF(CF_3))-$ moieties in the pendent chain. All of them are thermoplastic and possess good thermal, chemical and mechanical properties due to their perfluorinated polymer backbones. They only become, however, appreciably conductive on uptake of water, which they do surprisingly easily, in spite of the mainly hydrophobic nature of their composition. They absorb water to a degree in excess of 15 H_2O molecules per $-SO_3H$ group, when they attain conductivities as high as 1 $\Omega^{-1}m^{-1}$. The conductive, hydrated form is limited in high temperature performance, because above the boiling point of water the absorbed water is rapidly lost. It is widely accepted that the high ionic conductivity of the hydrated systems depends on a network of ionic clusters about 4 nm in diameter, centred on the hydrophilic $-SO_3H$ groups, scattered in the hydrophobic fluorocarbon matrix. Interconnection of the clusters via narrow channels in the main polymer matrix then provides a three-dimensional conductive array. A more realistic model for Nafion® may contain three structural regions, with ionic hydrophilic clusters co-existing with crystalline and amorphous domains. The crystalline domains inhibit dissolution of the polymer, making the material suitably resilient for fuel cell applications.

One shortcoming of the perfluoro-based materials is their high cost, and this has led to the development of alternative proton-conducting electrolytes using aromatic polymers, which exhibit high thermal and chemical stability. Thus, for example, poly(etheretherketone) (PEEK) may be sulphonated using chlorsulphonic acid or fuming sulphuric acid to produce a polyelectrolyte,

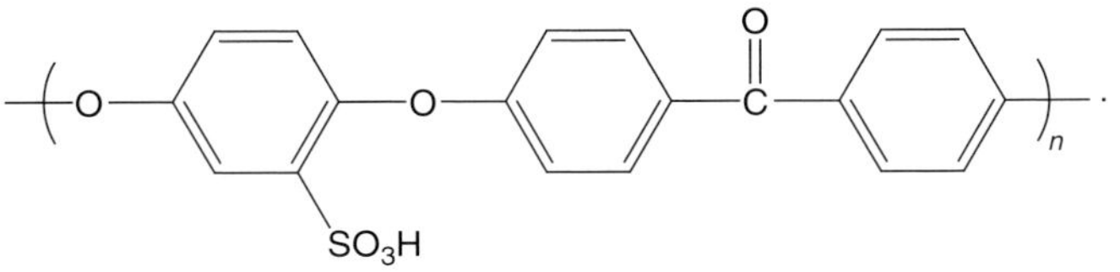

Sulphonated PEEK, in its hydrated form, displays a conductivity as high as $5\,\Omega^{-1}m^{-1}$ even at 100 C, and the sulphonation is stable up to 250 C.

The upper temperature at which polyelectrolytes can be utilised can be extended by substitution of a polar organic solvent for water in the electrolyte. Conductivities of up to $1\,\Omega^{-1}m^{-1}$ have been obtained for sulphonated PEEK using anhydrous pyrazole and imidazole at 200 C (Kreuer *et al.*, 1998). Solvents used to form electrolytes in Nafion® membranes have included alcohols, amines, propylene carbonate, dimethoxyethane, dimethyl formamide, dimethyl sulphoxide and *N*-methyl pyrrolidone. Typical conductivities for the latter three solvents are 10 to $1\,\Omega^{-1}m^{-1}$ at room temperature, and the conductivity is thermally activated with activation energies in the range 9 to 24 kJ/mol (Doyle *et al.*, 2001).

Further development of polyelectrolytes is proceeding rapidly, and membranes made from them are expected to fulfil an important role in advanced fuel cell technology.

8.2.4 Solid polymer electrolytes

The development of solid polymer electrolytes (SPEs) really stems from the discovery by Fenton, Parker and Wright in 1973 of a remarkably high degree of conductivity in alkali metal salts in polyethylene oxide (PEO). Alkali metal salts can be dissolved in PEO up to a maximum molar ratio of one of the salt to every four ether monomer units $\text{-(}CH_2\text{-}CH_2\text{-}O\text{)-}$, which for simplicity we write as MX-PEO_4. Typical salts are $LiClO_4$ and NaCNS. The presence of the lone pair of electrons on the oxygen atoms of the polyether imparts a good solvating power to the polymer chain, and their interaction with the alkali metal cations resembles the well-known formation of *crown ether* complexes by certain cyclic oligomers of ethylene oxide. Armand *et al.* (1979) originally speculated that the high conductivity of the MX-PEO_n systems derived from crystalline regions allowing fast transfer of the alkali metal ion down PEO helices. The idea was reinforced by (a) the possibility of arranging the oxygen atoms so as

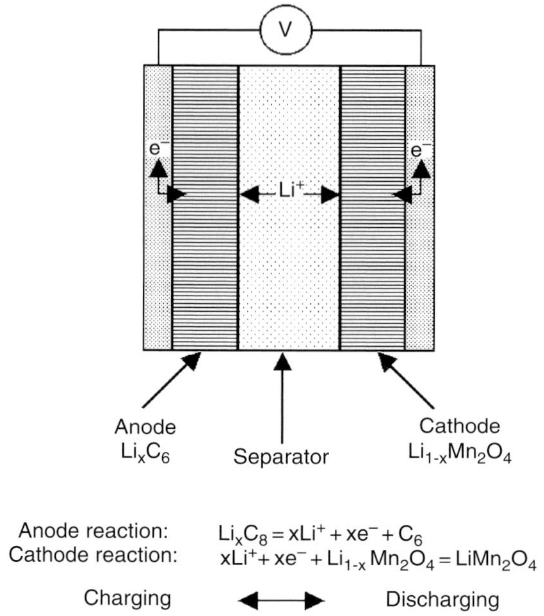

Fig. 8.6 Diagram of a lithium ion battery.

to provide a sequence of complexing sites along the axis, and (b) the cations Li^+, Na^+ or K^+ being small enough to fit inside the polymer helix. Armand also recognised that a *superionic* conductor of this type would make an excellent SPE for a lithium ion battery.

Lithium, with the highest oxidation potential of any element, is in principle ideally suited for high-energy density batteries. But highly reactive lithium metal in contact with a liquid electrolyte is hazardous, and therefore lithium metal has to be replaced with a safer lithium ion source. In the so-called *rocking-chair* cell, lithium ions move between lithium–carbon and lithium–metal oxide intercalation compounds, which act as the electrode couple, as shown schematically in Fig. 8.6. In a cell of this type the two electrodes must be electronically separated by a medium which allows easy transfer of metal ions during the charging and discharging processes. In the usual arrangement a separator frame of insulating ceramic or plastic is filled with a liquid electrolyte. The chief advantage of a solid polymer electrolyte is that it would combine the functions of separator and electrolyte in one, easily fabricated, flexible element. The MX–PEO systems with their supposed fast ion (M^+) transport seemed ideal for the purpose, and such materials became the focus of intense research (Meyer, 1998).

It eventually became clear that the particular fast ion mechanism which had M^+ travelling down the inside of PEO helices did not apply. Indeed, the ionic

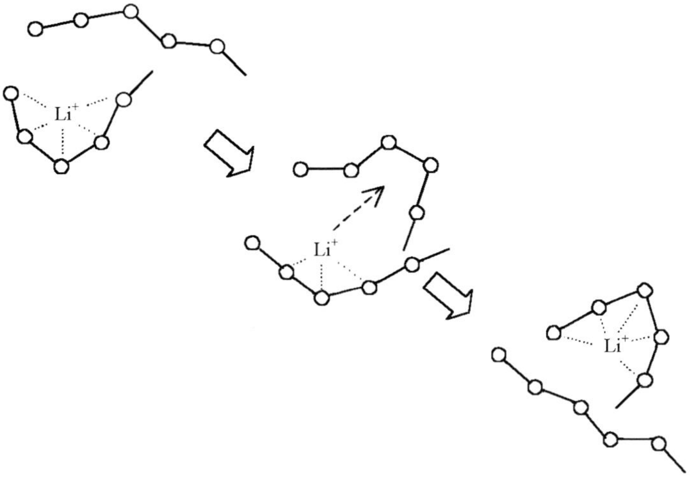

Fig. 8.7 Illustration of segmental motion assisted movement of Li^+ in a PEO matrix, showing the interaction of a lithium ion with ether oxygen atoms.

conduction takes place only in the amorphous phase, aided by segmental motion of the polymer chain, as depicted in Fig. 8.7. Furthermore, measurement of the relative movement of ions, e.g. by means of radio-tracer methods, shows that the transport numbers for the cations and anions are roughly equal. Nevertheless, with conductivities reaching $1\times10^{-2}\,\Omega^{-1}\mathrm{m}^{-1}$ at room temperature for optimum compositions (MX:EO ratios of $\sim$0.04), the materials are still real contenders for battery use. However, in addition to high Li^+ conductivity and low interfacial resistance between the electrolyte and electrode, the potential at which decomposition of the electrolyte occurs must be above the maximum operating potential of the battery. For many electrode combinations the PEO electrolyte decomposes below the maximum operating voltage, and the reversible lithium intercalation capacity is much lower than that of cells using liquid electrolytes. The addition of sub-micron size particles of Al_2O_3, TiO_2 and $BaTiO_3$ have been found to improve the properties of the PEO electrolyte. The addition of $BaTiO_3$ increases the decomposition voltage of the electrolyte to over 4 V, enabling high voltage cells to be cycled without decomposition and loss of capacity (Li *et al.*, 2002).

Learning that efficient ion transport depended on segmental motion of the host polymer chains, it became clear that one approach to increasing conductivity would be to search for *low glass transition (low-T_g)* materials. T_g may be reduced by incorporation of plasticising components in various ways, e.g. by using copolymers of ethylene oxide and dimethylsiloxane for the host polymer:

$$\text{-}\!\!(\mathrm{Si(CH_3)_2{-}O{-}(CH_2{-}CH_2{-}O{-})_4})\!\!\text{-}_n.$$

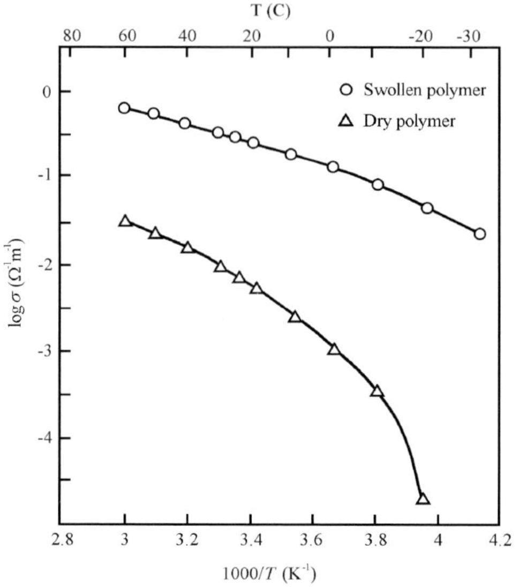

Fig. 8.8 Comparison of the temperature dependence of the ionic conductivity of a dry and plasticiser-swollen crosslinked SPE. Reproduced with permission of Wiley-VCH from Meyer (1998).

A number of more complex structures were also devised, based on comb-branched architectures. Although conductivities were indeed improved, there was a severe drawback: the mechanical properties were compromised with the materials becoming very soft and weak. One solution is to combine plasticisation with crosslinking, so that the polymer matrix retains some strength above T_g. Chemical crosslinking of the polymer inhibits crystallisation, which is beneficial to the conductivity, but crosslinking of the network also restricts segmental mobility (T_g rises as the crosslink density is increased), which works against conductivity. Effective segmental mobility can, however, be regained by *swelling* the crosslinked network with a plasticiser, e.g. propylene carbonate. The resulting *gel* combines the mechanical properties of a swollen crosslinked polymer network with the high ionic conductivity of a liquid electrolyte. Figure 8.8 compares the temperature dependence of a dry crosslinked SPE with that of the same polymer network swollen with a plasticiser (Tada *et al.*, 1994). The crosslinked polymer was obtained by UV exposure acting on the double bonds in the side-chain, end-groups of a phosphazene polymer with the structure:

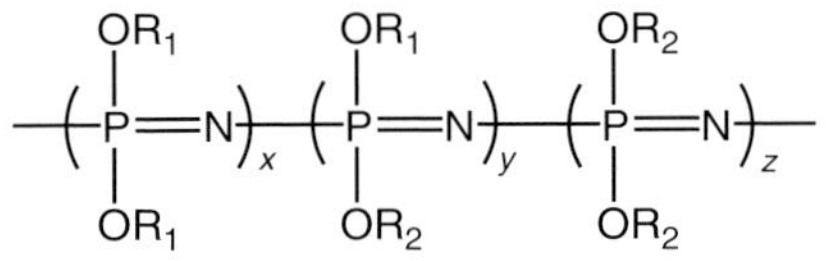

$R_1 = -(CH_2CH_2O)_7CH_3$

$R_2 = -(CH_2CH_2O)_7-CH_2CH{=}CH_2$

$x + y + z = n$ (random)

$R_1/R_2 = 1.5.$

The alkali metal salt was $LiClO_4$, with Li/O ratios of 0.028 and 0.043 for the dry and swollen cases, respectively; the swollen network is able to accommodate a greater concentration, because the plasticiser (propylene carbonate, concentration 60% by mass) is itself a good solvent for the salt.

An alternative to crosslinking, as a means of achieving good mechanical strength in combination with high conductivity, is *molecular reinforcement*, using rigid main chains with flexible side chains to provide the medium for ionic mobility. These are examples of *hairy-rod* molecules, which display an ability to self-organise into semi-ordered layers, when cast from solution, and which behave as molecular composites. Lauter *et al.* (1997) used poly-(*p*-phenylene) (PPP) as the rigid polymer with oligo(ethylene oxide) side groups:

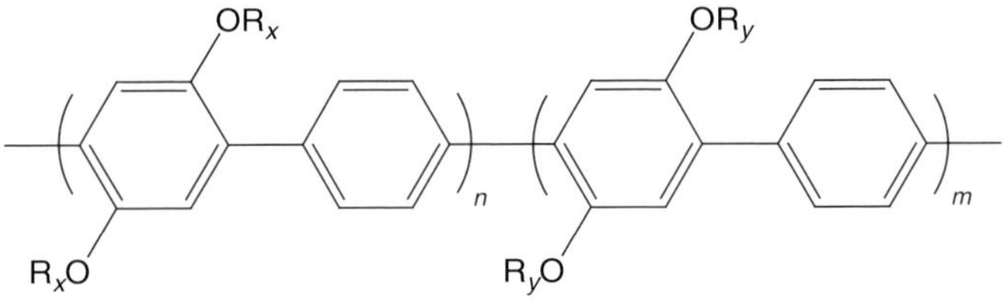

$R_x = (CH_2CH_2O)_5CH_3$

$R_y = (CH_2CH_2O)_6CH_3$

$n \sim m$ (random).

This polymer assumes a smectic B-type liquid-crystalline phase when cast from solution. Then, using $LiCF_3SO_3$ (LiOTf) as the alkali metal salt (Li/O = 0.4), a conductivity in excess of $1 \times 10^{-2}\,\Omega^{-1}m^{-1}$ is achieved at room temperature, the amorphous EO matrix effectively remaining above its T_g. The molecular architecture of the whole structure is depicted in Fig. 8.9.

The SPE gels and molecularly reinforced materials described above have gone a long way to meet the requirements of battery components. Rechargeable lithium batteries with composite electrodes and a solid electrolyte developed at Bellcore utilise a copolymer of vinylidene fluoride and hexafluoropropylene as a matrix for the composite electrodes and the electrolyte. The electrodes are formed from a solution of polymer and particles of Li_xC_6 and $LiCoO_2$ for the anode and cathode respectively. The battery structure is formed by lamination of a solvent containing electrodes and spacer. The solvent is removed, leaving a porous structure which is impregnated with liquid electrolyte (Tarascon *et al.*, 1996). However, although lithium batteries are entering the market place,

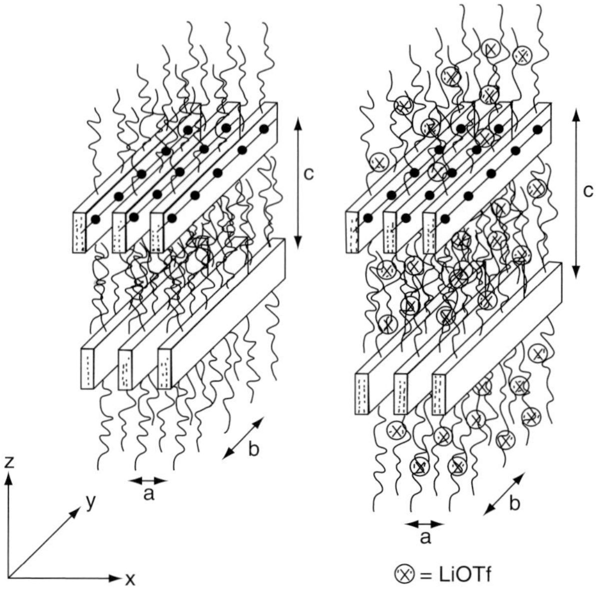

Fig. 8.9 Model of the molecular architecture of a PPP-reinforced SPE. Reproduced with permission of Wiley-VCH from Meyer (1998).

further development of SPE is required to obtain the long-term chemical stability that is necessary in many commercial applications.

8.3 Particulate composites

Probably the simplest conducting composite consists of a fine metallic powder dispersed uniformly throughout an insulating plastic matrix. The main drawback of this system, apart from the obvious *dilution* by the plastic of the conductivity of the metallic phase, is that the conductive particles remain isolated from each other, and therefore contribute no through-going conductivity to the composite at all, unless they are present in very high concentration. But high concentrations of particulate material tend to destroy the desirable mechanical properties of the plastic, the material typically becoming stiff and brittle. It is, therefore, necessary to tailor the composite carefully to achieve an acceptable combination of electrical and mechanical properties for targeted applications.

The *all or nothing* feature of metal powder composites is very much a feature of conductive composite systems. In order to understand this behaviour, most theories borrow from *percolation theory* (Broadbent and Hamersley, 1957), which was originally developed as a model for predicting fluid permeation through porous media. The percolation model is based on having a medium with a set of randomly distributed sites into which fluid may flow. The density of sites then determines the probability that interconnecting paths form by sites joining together, thereby allowing transfer of fluid through the sample. By analogy, the theory also describes the passage of current through a polymer matrix containing conductive particles. In this situation, the volume fraction of the conductive filler determines the probability of particles touching and thus forming conductive pathways through the material. The statistical analysis of percolation theory then predicts that DC conductivity σ_{DC} in a composite will depend on the volume fraction of the conductive filler ν_{f} according to the equation:

$$\sigma_{\mathrm{DC}} \propto (\nu_{\mathrm{f}} - \nu_{\mathrm{fc}})^{t}, \tag{8.16}$$

where ν_{fc} is a critical volume fraction, the percolation threshold, at which a 3-dimensional conductive network first appears, corresponding to a sudden increase in conductivity with respect to the volume fraction of conductive filler. Monte Carlo simulations indicate that the index t has a value between 1.6 and 2 and that for a system of spherical particles ν_{fc} is 0.16. Although percolation theory describes the form of the conductivity *versus* filler level curve very well above the percolation threshold, it fails below ν_{fc}. The behaviour below ν_{fc} appears to be described by an equation similar to Equation (8.16), i.e.

$$\sigma_{\mathrm{DC}} \propto (\nu_{\mathrm{fc}} - \nu_{\mathrm{f}})^{-s}. \tag{8.17}$$

Unfortunately, the percolation approach is also not really able to predict accurately the critical volume fraction in real composites, because so many different factors, like filler shape, size, distribution and particle agglomeration, come into play. Lux (1993) has reviewed the various percolation models that have been proposed in theoretical treatments of the problem.

Two other approaches have been taken to modelling the conductivity of composites, effective medium theories (Landauer, 1978) and computer simulation. In the effective medium approach the properties of the composite are determined by a combination of the properties of the two components. Treating a composite containing spherical inclusions as a series combination of slabs of the component materials leads to the Maxwell–Wagner relations, see Section 3.6.1. Treating the composite as a mixture of spherical particles with a broad size distribution in order to minimise voids leads to the equation:

$$\nu_{\mathrm{f}} \frac{\sigma_{\mathrm{h}} - \sigma_{\mathrm{m}}}{\sigma_{\mathrm{h}} + A\sigma_{\mathrm{m}}} + (1 - \nu_{\mathrm{f}}) \frac{\sigma_{\mathrm{l}} - \sigma_{\mathrm{m}}}{\sigma_{\mathrm{l}} + A\sigma_{\mathrm{m}}} = 0, \tag{8.18}$$

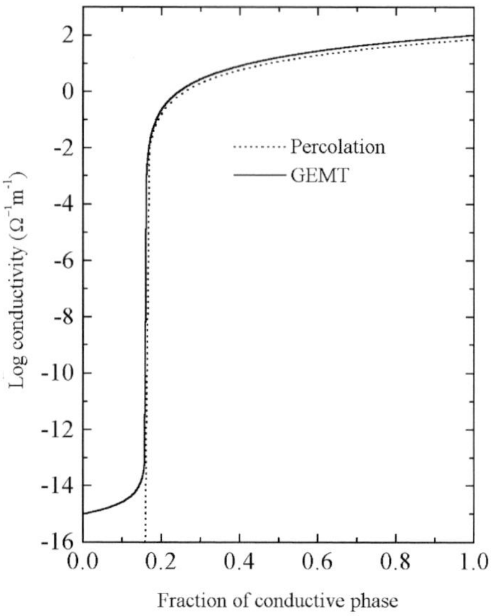

Fig. 8.10 Comparison of the calculated conductivity of a composite given by the percolation and general effective medium theories. Parameters are as described in the text.

where σ_h, σ_l and σ_m are the conductivities of the filler, matrix and composite and A is a factor defining the local field concentration at the conductive particles ($A = 2$ for spherical particles). ν_{fc} is related to A since $A = (1 - \nu_{fc})/\nu_{fc}$. Hence this model gives ν_{fc} as 0.33, which is higher than both the percolation model and experimental values. McLachlan (2000) has developed an extension of this model, the general effective medium theory (GEMT). The conductivity of the composite is then obtained by solving the equation:

$$\nu_f \frac{\sigma_h^{1/t} - \sigma_m^{1/t}}{\sigma_h^{1/t} + A\sigma_m^{1/t}} + (1 - \nu_f)\frac{\sigma_l^{1/s} - \sigma_m^{1/s}}{\sigma_l^{1/s} + A\sigma_m^{1/s}} = 0, \quad (8.19)$$

where s and t are the exponents from Equations (8.16) and (8.17). This phenomenological equation has three fitting parameters s, t and A and gives good fits to experimental data for a wide range of systems. Figure 8.10 illustrates the results obtained from percolation theory, Equation (8.16), and Equation (8.19) with the parameters $\nu_{fc} = 0.16$, $s = 1$, $t = 2$, $\sigma_h = 100\,\Omega^{-1}m^{-1}$ and $\sigma_l = 10^{-15}\,\Omega^{-1}m^{-1}$. The model can also be utilised to calculate the AC properties of the composite although analytical solutions require that the ratio s/t is an integer (McLachlan, 2000).

Computer simulations have also been used to model the properties of composites for specific particle distributions and geometry, e.g. spherical particles with either a bimodal size distribution (Fu *et al.*, 1999) or an anisotropic spatial distribution (Han and Choi, 1998 and Lu and Lin, 2002). These simulations can be helpful in understanding the properties of specific types of composite.

The characteristic threshold behaviour of conductive composites is an unsatisfactory feature for many applications, both because the poor mechanical properties incurred at concentrations where conduction is obtained cannot be tolerated, and because the high level of conductivity above the threshold is not required or even not desirable. Thus a common requirement is for an antistatic material which has good plastics properties, sufficient conductivity to allow charges to leak away, and sufficient resistivity to prevent dangerous shocks to personnel who may become accidentally connected to mains electrical supplies through it. Unfortunately, the very steep slope of the conductivity versus filler concentration curve in the region of the threshold makes it very difficult to manufacture materials with reliable intermediate conductivities.

Undoubtedly, the art of making a good conducting composite is to be able to use the minimum quantity of conductive component to achieve the required degree of electrical performance, and in this context it is important to know more detail about the factors which control the formation of conductive networks with a given conductive component in the chosen matrix. There are two main factors:

a. *Quality of interparticle contacts.* Conductive networks in a 2-phase system depend on the particles of the conductive phase being able to make good electrical contact when they touch or come close to each other. Since a colossal number of interparticle contacts are involved, any changes in contact properties will be highly significant to the conductivity that may be realised. Oxidation of metal surfaces is detrimental to electrical contact and for this reason many metal powders (including copper) will not conduct even on their own unless highly compressed. The noble metals have the advantage in this respect, and some highly conducting composites employ silver-coated copper powder which combines the excellent contact properties of silver with the cheapness of copper.
b. *Shape and size of conductive particles.* The resistivity of a compaction of carbon powder has been successfully explained on a simple model where the constrictions at interparticle contacts are assumed to have a dominating influence (Mrozowski, 1959). The theory uses Holm's (1946) expression $\rho\pi/a$ for the contact resistance between two spheres when the radius a of the circular area of contact between them is small with respect to the radius of the spheres, ρ being the resistivity of the material of the spheres. Thus, taking into account the way in which compressive forces cause elastic and yielding deformation, thereby altering a, the dependence of the resistivity of the compaction on pressure may be predicted quantitatively. The same theory may be expected to apply also to composites containing high loadings

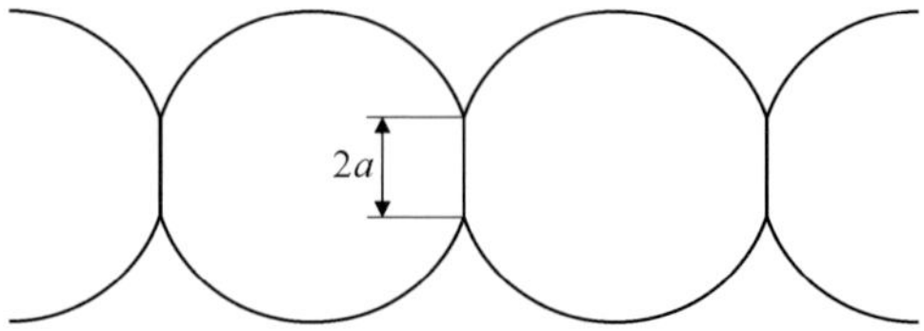

Fig. 8.11 Model form of intersphere contacts.

of conductive fillers where contacts amongst a certain proportion of particles form the conducting paths through the material. On this basis, where contact resistances dominate, spherical particles will be rather inefficient at imparting conductivity, since most of the material in the spheres is wasted as far as electrical conduction is concerned (see Fig. 8.11). Relatively more is wasted when the sphere diameter is increased. Elongated shapes, especially either fibres or needles and even flakes, are much more efficient in this respect. The greater surface-to-volume ratio of these shapes also makes interparticle contact somewhat more likely at lower concentrations, resulting in lower percolation thresholds for needles and flakes than for spherical particles. Furthermore, fibrous fillers, e.g. carbon fibres, can enhance the mechanical properties of the composite. In practice, conductive composites fall into two general categories; first, those containing approximately spherical particles and, second, those containing elongated particles, typically fibres.

8.3.1 Conductive powder composites

Carbon-black composites

The additive used more than any other to make conducting plastics for both commercial applications and for scientific study is carbon black produced by incomplete combustion of hydrocarbon vapours. It should be mentioned that the addition of particulate fillers has been used for a long time in the rubber industry primarily to give mechanical reinforcement, and carbon black has generally been preferred for this, because (a) it is a compatible material, mixing in, and adhering to, the matrix rather well, (b) it does not change the overall density very much, and (c) it is cheap. Rather higher concentrations of special grades are necessary, however, for electrical purposes. The dependence of conductivity on concentration of a polyethylene furnace black composite is shown in Fig. 8.12; the circles and triangles are two sets of measurements made by the four-probe method and the diamonds are the DC limit extrapolated from AC measurements on metallized sheet samples. The upper limit of conductivity that can be achieved in a useful composite is about $100\,\Omega^{-1}\text{m}^{-1}$.

Particles of carbon black are very small, with diameters in the range 10–300 nm, and densities of about $1.8\,\text{Mgm}^{-3}$. They consist of roughly spherical clusters of small fragments, each of which is similar to a small bit of graphite

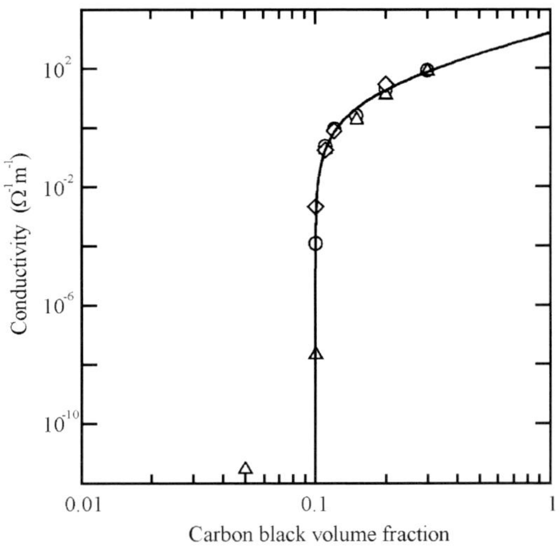

Fig. 8.12 The dependence of the conductivity of a polyethylene furnace black composite on carbon-black content. Reprinted from Hindermann-Bischoff and Ehrburger-Dolle (2001). Copyright 2001, with permission from Elsevier.

(a few parallel layers), but with the hexagons in successive layers oriented in different directions. Although the shape of the graph of log(conductivity) against carbon-black concentration is similar for all grades of carbon black, the conductivity for a given concentration depends very much on the way the black has been made. The so-called *furnace* blacks give the highest values. Blacks which give high conductivity are characterised by small particle size, but, more importantly, by a propensity to agglomerate into chains to give a pseudo-fibre arrangement, usually referred to as *structure*, which may persist even when the black is mixed into another material. The reason for this structure is not fully understood. Some of it may be explained by the amalgamation of droplets of high-molecular-mass hydrocarbons into *necklaces* at an intermediate stage in the manufacture, but this does not explain the fact that some structure which is disrupted by shearing action will reform on resting. The condition of the surface of a carbon black affects conductivity, probably through its influence on structure. Thus treatment of particles to remove surface hydrogen and oxygen greatly increases the conductivities that may be obtained in composites. An indication of the degree of structure inherent in a particular black can be obtained by various empirical methods.

The most common index of structure is oil absorption, defined as 'the minimum volume of oil (usually dibutyl phthalate) which will give, under conditions of controlled mixing, a mix having no voids'. Spherically shaped powders pack closely, whereas branched-chain aggregates in a high-structure black give rise to more voids and thereby to a high absorption of oil. One type of highly conductive carbon black consists of hollow particles. These are made by a controlled oxidation process by which the more reactive, amorphous carbon is burnt out of the inside of the forming particles. The hollowness effectively increases the structure of the black, as measured by its oil absorption index, and in the most conductive cases, can reduce the threshold concentration for conductivity by as much as a factor of 2. Probst and Grivei (2002) have presented mechanical and electrical data for representative types of carbon black.

In the face of the complicated nature of carbon black it is not surprising that it proves to be difficult to predict theoretically the accurate conductivity of a composite with a given filler concentration. For one thing, the all-important *structure* is affected by the process of mixing the black into the polymer. Nevertheless, there are two theories which go some way in explaining the behaviour of the composites in certain regimes.

The first theory (Voet *et al.*, 1965) sets out to treat those systems which show non-Ohmic conduction. This kind of behaviour is indeed often found at the low-conductivity end of the scale, and it implies that some kind of electron emission process controls the conduction, probably a tunnelling of electrons from one particle to the next across gaps of up to 5 nm, say. It is relevant to note here that the very small particle size of carbon blacks means that Brownian motion might be responsible for momentarily bringing particles close enough together for electron tunnelling to occur. Emission current density j is related to applied voltage V by an equation of the form

$$j = AV^n \exp\frac{-B}{V}, \tag{8.20}$$

where A, B and n are constants; n usually lies between 1 and 3. A is a function of the tunnelling frequency, i.e. the *attempt-to-escape* frequency of the electron, and the factor $\exp(-B/V)$ defines the transition probability through the energy barrier of the gap. B will be proportional to the gap between particles, so that, neglecting the pre-exponential factor, the effective conductivity at a particular voltage will be given by

$$\ln \sigma \propto 1/l_{av}, \tag{8.21}$$

where l_{av} is some average interparticle separation. Now it may be shown that the interparticle separation for an array of spherical particles (radius a) on a cubic lattice (repeat distance l) where $l \gg a$ is given by

$$l = a\left(\frac{4\pi}{3f_v}\right)^{1/3}, \tag{8.22}$$

where f_v is the volume-fraction of particles. For a random, low-density dispersion of particles we may therefore expect that

$$l_{av} \propto f_v^{-1/3}. \tag{8.23}$$

Combining this with Equation (8.21) we then obtain

$$\ln \sigma \propto f_v^{-1/3}. \tag{8.24}$$

To assess the validity of this equation we can compare it with the purely empirical equation (Bulgin, 1945) which has a similar form:

$$\ln \sigma = \left(\frac{A}{f_w}\right)^{-p}, \tag{8.25}$$

where f_w is the mass-fraction of the carbon black in the composite (for low carbon-black concentrations $f_w \infty f_v$), and A and p are constants. Most experimental data may be fitted by the latter equation, although the constants A and p vary for different kinds of blacks. Values of p in fact lie in the range 0.6 to 5.0 (Norman, 1970), so that we can only say that the theoretical Equation (8.25) gives rather poor quantitative agreement with experiment.

The second theory, due to Scarisbrick (1973), treats the high-conductivity cases where Ohmic behaviour is encountered. He tacitly assumes that inter-particle contacts are Ohmic, and goes on to calculate the probability of formation of conductive chains. Supposing the composite to consist of a random mixture of conducting and non-conducting elements, the probability of obtaining a sequence of n conducting elements is f_v^n. If the average conductive element is a sphere, diameter d, associated with a spherical region, diameter D, of the composite:

$$\frac{D}{d} = \frac{1}{f_v^{1/3}}. \tag{8.26}$$

By random-walk theory, if the diameter D is taken as the mean distance between the ends of a chain of n links of length d,

$$D = dn^{1/2}. \tag{8.27}$$

Combining Equations (8.26) and (8.27),

$$n = f_v^{-2/3}. \tag{8.28}$$

Hence the probability that the volume associated with one conductive element may be bridged by chain formation is then given by

$$p = f_v^{f_v^{-2/3}}. \tag{8.29}$$

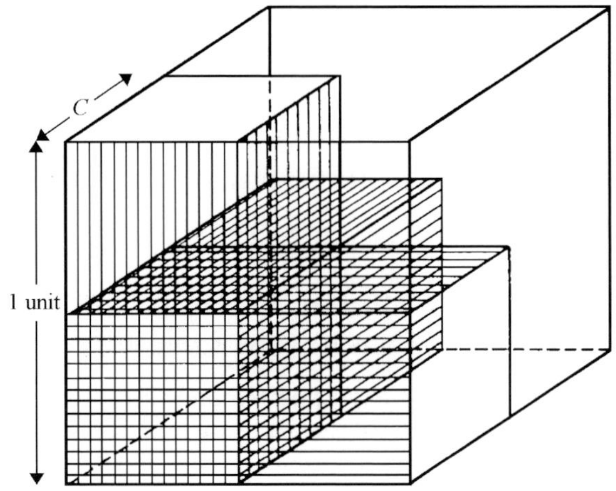

Fig. 8.13 Equivalent distribution of conductive chains along the principal directions of a unit cube of an isotropic conducting mixture, defining the geometrical factor. Reproduced with permission of the Institute of Physics from Scarisbrick (1973).

Scarisbrick also draws attention to the geometrical factor which connects the observable conductivity with the random assembly of conducting chains. Figure 8.13 shows conducting chains *resolved* along the axial directions of a unit cube, and collected together to give a cross-sectional area C^2 in each of the three directions. Taking into account the mutual volume, where the chains along each direction overlap,

$$f_v = 3C^2 - 2C^3. \tag{8.30}$$

The ratio of the conductivity of the composite to that of the conductive component itself is the product of the proportion of conducting to non-conducting elements, the probability of chain formation and the geometrical factor, i.e.

$$\frac{\sigma}{\sigma_c} = f_v \times f_v^{f_v^{-2/3}} \times C^2. \tag{8.31}$$

Comparison of this result with experimental values obtained for a dispersion of carbon black in polyethylene is shown in Fig. 8.14, where the agreement is seen to be fairly good, the discrepancy being at most a factor of about three.

Neither of the above theories takes the macrostructure of carbon black into account in any quantitative way, although in practice this is crucial in selecting a material for a particular application. Sheng (1980) was, however, able to develop the first of these theories and extend its range of applicability. He noted the well-established model for hopping conduction by tunnelling

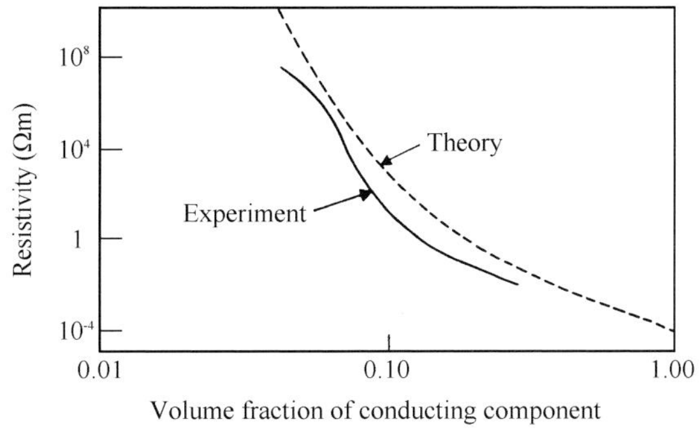

Fig. 8.14 Comparison of the experimental and theoretical dependences of resistivity on carbon-black content in a polyethylene composite. Reproduced with permission of the Institute of Physics from Scarisbrick (1973).

between localised trap sites in disordered semiconductors, see Section 4.2.3, but he recognised that in those systems where the *trap sites* themselves delocalise the charge carriers over large distances, as in metallic particles of a conductive composite, an additional mechanism comes into play for the transfer of carriers between sites. Since tunnelling has exponential distance dependence, the shortest paths between particles will dominate charge transfer. In other than perfectly spherical particles these will occur between small surface features. The effect of thermal fluctuations on these small features, in particular on the voltage across the inter-particle gap, will be large. Thus carrier transport will be dominated by tunnelling induced by these large voltage fluctuations. At very low temperatures the conductivity of a single tunnelling junction will be temperature independent since there is no fluctuation or directly thermally activated tunnelling. At high temperatures the conductivity becomes simply thermally activated and is given by $\sigma = \sigma_0.\exp[-T_1/T]$, where T_1 is a measure of the energy barrier in the junction. At intermediate temperatures fluctuation-induced tunnelling becomes important, leading to a general expression for the temperature dependence of the conductivity of the tunnel junction of the form:

$$\sigma = \sigma_0.\exp\left[-\frac{T_1}{T}(\varepsilon^*)^2 - \frac{T_1}{T_0}\phi(\varepsilon^*)\right], \tag{8.32}$$

where T_0 can be viewed as the temperature above which fluctuation-induced tunnelling becomes important, ε^* is a parameter with a value between 0 and 1 and $\phi(\varepsilon^*)$ is a function with limits $\phi(0) = 1$ and $\phi(1) = 0$. As the field across the tunnel junction increases at a fixed tunnelling distance and temperature the probability of fluctuation-induced tunnelling decreases while that for normal tunnelling increases. $\varepsilon^* = 0$ corresponds to pure fluctuation-induced tunnelling

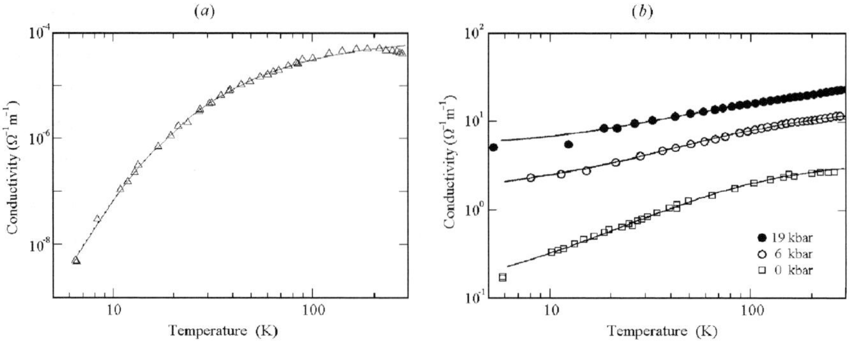

Fig. 8.15 Temperature dependence of the conductivity of composites containing (a) 6% black and (b) 15% black at different pressures. Reproduced with permission of the Institute of Physics from Dawson and Adkins (1996).

and $\varepsilon^* = 1$ to normal thermally activated tunnelling. For intermediate values both mechanisms contribute, and ε^* is such that the sum of the probabilities for fluctuation-induced tunnelling and normal tunnelling is a maximum. At intermediate temperatures the behaviour is determined by ε^* which is dependent on the shape of the potential barrier in the junction. For a parabolic barrier Equation (8.32) takes the simpler form:

$$\sigma = \sigma_0.\exp\left[-\frac{T_1}{T + T_0}\right]. \tag{8.33}$$

The conductivity of a composite is related to the result for a single junction by treating the composite as a random resistor network using an effective medium model (Sheng, 1980). Treating ε^*, T_0 and T_1 as variable parameters, the approach provides excellent fits to the observed temperature dependences of the conductivity of carbon-black composites below 200 K. Examples of the agreement between theory and experiment for carbon black (XC72) in polycarbonate composites with loadings of 6 and 15% black, i.e. below and in the region of the percolation threshold, are shown in Fig. 8.15. The curve in Fig. 8.15(a) is Equation (8.33) with $T_0 = 4$ K and $T_1 = 100$ K. The curves in Fig. 8.15(b) are with $T_0 = 23$ K, $T_1 = 82$ K at 0 kbar, $T_0 = 42$ K, $T_1 = 98$ K at 6 kbar and $T_0 = 46$ K, $T_1 = 78$ K at 19 kbar.

One way which is often used to impart an artificial structure to the dispersion of a carbon black in a composite is to coat the particles of a moulding powder, e.g. 1 mm diameter particles of polyethylene, with a conductive carbon black by mixing them together in a ball-mill. If subsequent moulding does not shear the mixture too much (rotational casting is ideal) the black remains concentrated in a honeycomb-like network, and the overall conductivity is greatly enhanced over that of a uniform dispersion of the same proportion of black.

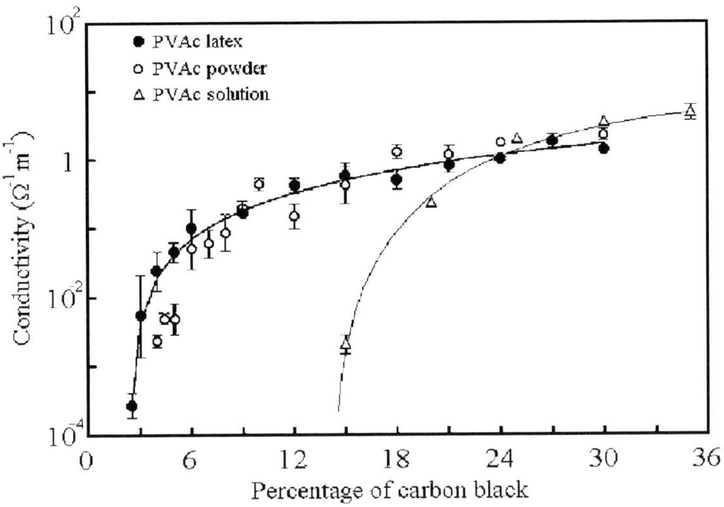

Fig. 8.16 Conductivity of a composite formed from carbon-black coated poly(vinylacetate) pellets (Grunlan *et al.*, 2001). Reprinted by permission of John Wiley & Sons, Inc.

The percolation threshold can be reduced to values as low as 1–2% compared with over 10% for a thoroughly mixed composite, see Fig. 8.16.

The conductivity of composites depends on frequency, temperature, mechanical deformation and solvent swelling of the polymer matrix. These dependencies have been extensively studied and provide the basis for the commercial utilisation of carbon-black–polymer composites.

a. *Frequency dependence.* The frequency dependence of the conductivity of carbon-black composites is typical of conductive composites in general. Data for a variety of composites are shown in Fig. 8.17. The polymer matrices are ethylene butacrylate copolymer (EBA), polyethylene (PE) and an ethylene-propylene-diene rubber (EPDM). At low frequency the AC conductivity is constant and equal to the DC conductivity, while above a critical frequency the conductivity increases with a power law dependence on frequency. Several different models have been used to evaluate the critical frequency, although Jäger *et al.* (2001) have shown that the simple relation:

$$\omega_c = \frac{\sigma_{DC}}{\varepsilon_0 \varepsilon_s}, \tag{8.34}$$

where ε_s is the static relative permittivity, fits the data for a wide range of carbon-black composites. At high frequencies all the composites become lossy and will absorb high frequency radiation. Both carbon-black and metal filled composites have been utilised as absorbers in the construction of anechoic chambers for testing and measurement at high frequency.

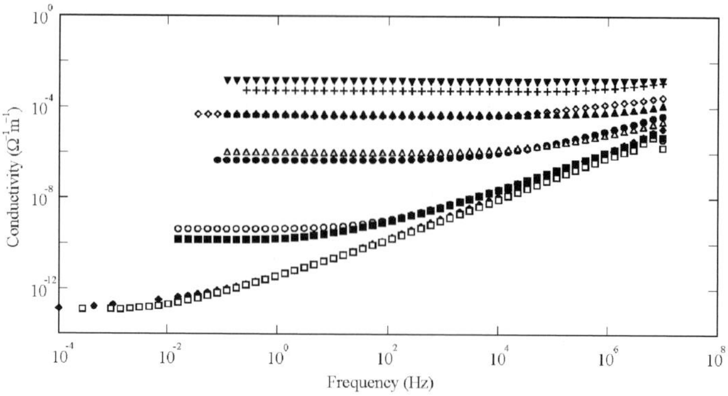

Fig. 8.17 AC conductivity versus frequency for a variety of carbon-black composites: ◇ crosslinked EBA with 8.3% acetylene black (AC), ◆ thermoplastic EBA with 11.3% AC, ✚ thermoplastic EB with 14.5% AC, □ EBA particulate composite with 9.1% AC, ■ EBA particulate composite with 10.7% AC, ○ blended PE with 9.2% AC, ● blended PE with 11.1% AC, Δ EPDM with 15.2% furnace black (FB), ▲ EPDM with 21.1% FB and ▼ EPDM with 26.3% FB (all percentages are by volume). Reproduced with permission of the Institute of Physics from Jäger *et al.* (2001).

b. *Temperature dependence.* As the temperature of a carbon-black composite is raised, its conductivity may either increase or decrease, depending primarily on the degree to which conductive chains are broken up by thermal expansion (Voet, 1980). This in turn depends on the relative magnitudes of the coefficient of thermal expansion of the insulating matrix polymer and that of the conductive carbon-black filler. Thus, although the temperature coefficient of conductivity of graphite is in fact positive, most carbon-black composites exhibit a negative coefficient, reflecting the relatively high thermal expansion that is characteristic of most polymers, see Fig. 8.15(a) above 200 K. This is a very useful property in practice when the composite is to be used as a heating element, because it prevents *thermal runaway*. In this application the operating temperature may be as high as 150 C, so that the temperature dependence of the conductivity of the product is of the utmost importance for stable working. In the case of a silicone-rubber composite, a suitable one for heating panels, the temperature coefficient of conductivity is $-0.0003\ \mathrm{deg}^{-1}$, and this material has been used for under-floor heating. The composite is spread on to a woven fibreglass sheet and then cured *in situ* to the rubbery form. Copper foil electrodes can then be simply sewn along opposite edges for connection to the power supply. Composites which exhibit a decrease of conductivity with temperature are often called PTC materials, referring to their positive temperature coefficient of resistivity, the quantity that electrical design engineers use. In addition to the gradual change of resistivity (or conductivity) with temperature, many composites show a more sudden, large change as the temperature reaches the melting point of any crystalline phase that is present in the matrix, an effect that was first reported by

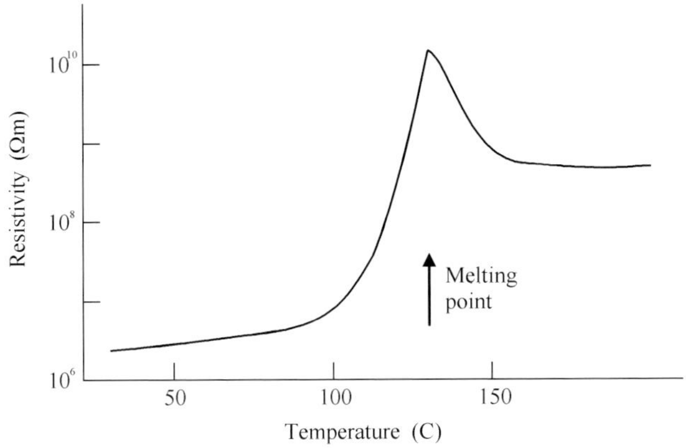

Fig. 8.18 Typical temperature dependence of resistivity in a carbon-black composite exhibiting switching behaviour.

Beuche (1973) for waxes. The step change in resistivity is explained by the disruption of the conductive black network as the crystals melt, the carbon particles within the now more mobile matrix move around, the chains are broken and the resistivity begins to rise. In the fully molten state the filler particles can re-agglomerate, giving a fall in resistivity. This has the effect of giving a peak in the temperature dependence of the resistivity, rather than a rise to a plateau, as illustrated in Fig. 8.18. On returning the composite to a low temperature below the freezing point, the original resistivity is not regained, and resistivity/temperature plots are not reproducible upon repeated temperature cycling. It has been shown experimentally that the increase in resistivity in polyethylene matrix composites results from a loss of connectivity of the conductive black network below the melting point of the polymer (Hindermann-Bischoff and Ehrburger-Dolle, 2001). The carbon particles are located preferentially in the amorphous regions between the crystallites forming percolating chains, and these are disrupted as the amorphous fraction increases. Crosslinking the polymer matrix stabilises the material, so that the network of carbon particle chains remains essentially intact above the melting point (Narkis *et al.*, 1981), with only inter-particle separations being expanded. Resistivity is then reproducible with respect to temperature cycling – the disposition of the particles with respect to each other remains fixed. Typical behaviour for a furnace black in an epoxy matrix is shown in Fig. 8.19. Crosslinked materials of this type have found two important areas of application; self-regulating heaters and circuit protection devices.

The PTC phenomenon, in effect, provides the means of making a thermal *switch*, operated either by an external change in temperature or by an increase in current that generates heat internally, see Fig. 8.20. A common form of self-regulating heater is made by extruding a strip with the cross-section shown in Fig. 8.21, where the conductive component consists of polyethylene/carbon-black composite that is radiation-crosslinked after fabrication. The strip may be extruded in long lengths

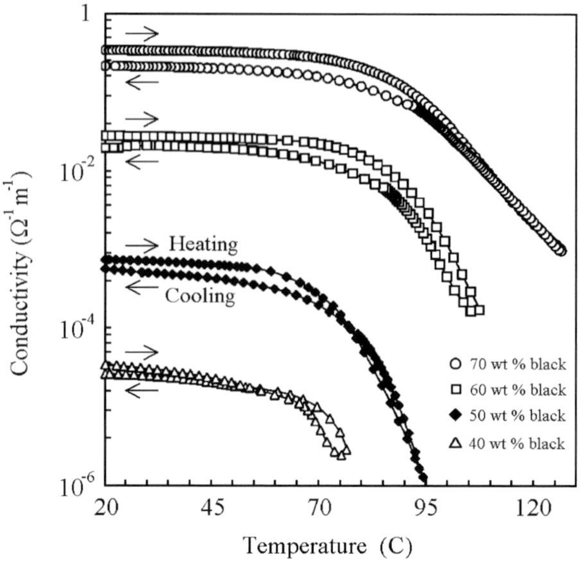

Fig. 8.19 Temperature dependence of the conductivity of carbon-black/epoxy composites at different carbon-black loadings. Reprinted from El-Tantawy *et al.* (2002). Copyright 2002, with permission from Elsevier.

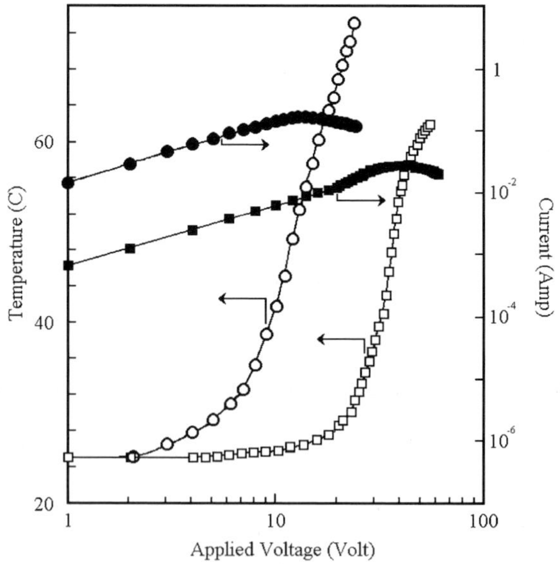

Fig. 8.20 Current–voltage characteristics and temperature of a composite resistor at ● / ○ 70 mass% and ■ / □ 60 mass% carbon black. Reprinted from El-Tantawy *et al.* (2002). Copyright 2002, with permission from Elsevier.

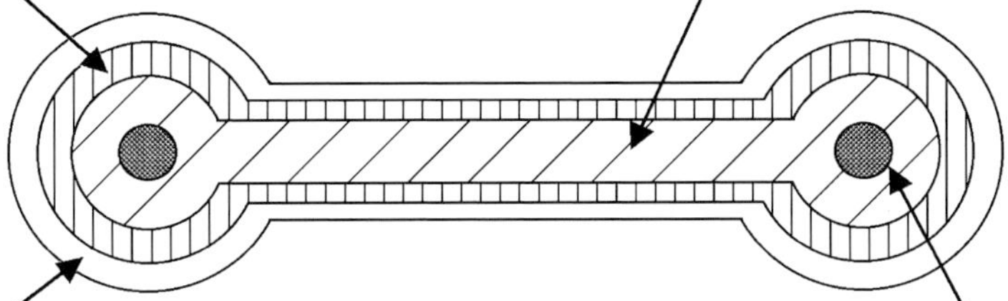

Fig. 8.21 Design of a self-regulating heating element.

for wrapping around pipes, tanks etc. of chemical plants to protect against low temperature conditions. The strip may be cut to whatever length is required, and connected at one end to the power supply. No external control mechanism or thermostat is necessary, making for very easy and reliable installations.

Circuit protection devices consist of small resistors with a high-PTC composite acting as the conductive element. They may be connected in series with other electrical systems, which are liable to be subjected to high, damaging currents caused by accidental application of high voltages or voltage surges. As the current rises in the circuit, the accompanying temperature rise in the PTC element triggers a sudden increase in resistance, thereby switching off the current. The effect is similar to that of a circuit-breaker or fuse, with the added advantage that the circuit is automatically reset to *on* as conditions return to normal and the temperature falls.

c. *Mechanical deformation*. Any mechanical strain can upset the conductive-chain formation, and pronounced effects of this kind are observed in conductive rubbers (Norman, 1970). Generally, compression results in an increase in conductivity as the filler particles are brought into closer contact, and extension results in a decrease in conductivity, as the percolative chains are broken. However, if the deformation is large enough that the sample develops a neck and chain orientation occurs the conductivity becomes approximately constant as a function of elongation (Flandin *et al.*, 2000). Necked samples show considerable hysteresis under cyclic deformation with the conductivity increasing with the number of cycles; this has been attributed to rearrangement of the host and filler to reform percolative chains after repeated cycling.
d. *Solvent swelling*. Swelling of the polymer matrix by solvents, either by immersion in the solvent (Tsubokawa *et al.*, 2001) or exposure to solvent vapour (Lonergan *et al.*, 1996), will also produce changes in conductivity. The change in conductivity for composites with carbon loading just above the percolation threshold can be large but is sensitive to initial loading. More modest but reproducible response is obtained with loadings somewhat above the percolation threshold. Such

composites can form the basis of an electronic nose which utilises an array of sensors with different non-specific responses. When exposed to a single vapour the different responses of the individual detectors produces a *fingerprint* characteristic for the specific vapour. The differences in the swelling behaviour of different polymer matrices mean that carbon-black composites can be utilised in this application (Doleman and Lewis, 2001).

In summary, carbon blacks work well as conductive additives for plastics and rubbers for applications other than power transmission, provided that a high-structure black and a suitable set of processing conditions are chosen, and as long as the black colour is acceptable.

Metal particle composites

Composites containing metal particles provide a clear example of the percolation behaviour described in Section 8.3. By way of illustration we can cite the results shown in Table 8.1 for a dispersion of approximately spherical nickel particles (about 10 μm in diameter) in a low-density polyethylene. Here conductivity is lost entirely at concentrations below about 20% by volume of metal, corresponding to about 70% by mass. Nevertheless, conductive paints, which are frequently used for painting electrodes on to electrical test specimens and devices, work in just this way.

Composites based on silver powder can be made with conductivities as high as $10^{-6}\,\Omega^{-1}\mathrm{m}^{-1}$ at a loading of 85% by mass, when the insulating matrix, which may be an epoxy resin, serves essentially as a glue to hold the metal powder in position without altogether disrupting the metal–metal particle contacts.

Composites with filler concentrations close to the percolation threshold exhibit conductivity which is sensitive to compressive deformation, since this brings the metal particles into contact, thereby forming percolation pathways. This sensitivity has been exploited especially in anisotropic composites. These are made by prealigning the metal particles with either electric or magnetic fields. This alignment is identical with that produced by external fields in electro- and magneto-rheological fluids where at a critical field continuous threads of

Table 8.1 *Conductivities of nickel–polyethylene composites*

Nickel content		Conductivity $\Omega^{-1}\mathrm{m}^{-1}$
% by mass	% by volume	
70	19	10^{-6}
80	29	1.2×10
90	48	5.8×10^3

particles form in the fluid phase. This methodology requires a mobile pre-polymer which allows the alignment to be carried out easily, and which may be subsequently cured or crosslinked to keep the alignment permanently fixed thereafter. Suitable polymer systems are epoxy and silicone resins. The aim of the alignment is to produce a particle distribution which is close to the percolation threshold only in the alignment direction. Uniaxial compression of the composite in this direction will produce a large increase in conductivity, but the composite will remain significantly less conductive in the transverse direction. This effect is utilised to produce electrical contacts between solder bumps on printed circuit boards and the terminals of electrical test rigs. The conductive tracks produced by the local deformation in the vicinity of the bumps eliminates the need for precise alignment while eliminating unwanted electrical leakage between adjacent tracks. The same approach has been used to produce locally conductive adhesives to eliminate the need for high temperature soldering of thermally sensitive circuits. Composites containing aligned elongated particles have also been utilised. The lateral separation of contacts is limited by the size of the metal particles to be greater than 10 μm. Sony developed a novel approach to reduce this, using metal-coated spherical polymer particles that collapse when the composite is locally compressed (Yamada *et al.*, 1999). The disrupted metal coatings form a percolating network in the deformed region, enabling contact to be made between features with lateral separation as small as 3 μm.

The mixing of metal-filled composites usually damages the particles so that, irrespective of initial morphology of the metal powder, the composites contain approximately spherical particles and their behaviour is described by percolation theory. With careful mixing the original morphology can be retained and, in specific cases, this leads to materials with properties somewhat different from those of conventional composites (Lussey, 1998). The combination of spiky nickel particles produced by the decomposition of nickel carbonyl with polymers that adhere strongly to the particles results in a composite which remains insulating even for loadings above the percolation threshold. Electron microscopy shows that each particle is coated and separated by a *skin* of polymer, thus the intimate contact necessary for the formation of percolation chains cannot occur. The composite is, however, sensitive to deformation; the resistance of samples subjected to compression, extension, bending, etc., falls from $\sim 10^{-12} - 10^{-13}\,\Omega$ to $1 - 10\,\Omega$. In the conductive state the composite can carry large currents without degradation. The current–voltage characteristic of stressed samples is extremely non-linear and the resistance shows an exponential dependence on deformation. These facts are explained by a model in which charge transport occurs by field-induced (Fowler–Nordheim) tunnelling from the sharp spikes on the filler particles through the thin polymer layers separating the particles over the whole conductivity range. This class of composites has, therefore, been termed quantum tunnelling composites (QTC) to distinguish them from conventional composites where physical contact of the filler particles is required to obtain high conductivity.

8.3.2 Conductive fibre composites

From the foregoing general considerations, it follows that a conductive additive will be efficiently deployed if it is distributed in such a way that long, interconnected chains exist permanently throughout the matrix. One obvious approach to this ideal is to start with a conductor which is already in fibrillar form, and fine metal wire or carbon fibres and nanotubes can be used in this way. Another variant is metal-coated fibre, silver on glass, for example, where the metal is present as thin-walled tubes. The fibres inherently have a high structure factor which results in a low percolation threshold. Short carbon fibres have ν_{fc} values of 0.01 or smaller while carbon nanotubes have ν_{fc} values of the order 0.001. As noted earlier, the use of mechanically strong fibres will enhance the strength of the composite, thus, while it may be less flexible than the host polymer, the composite does not become mechanically weak at high fibre loading. An alternative approach is possible with metallic organic charge transfer salts which have common solvents with the polymer host. Casting films from a solution of salt and polymer can produce a composite comprising a network of highly elongated crystalline needles in the polymer matrix. Such materials have been termed *reticulate* doped polymers.

The electrical performance of a conducting fibre composite is somewhat analogous to the gelation of a polymer by crosslinking, the conducting network corresponding to the gel fraction. For a segment of fibre to be able to take part in electrical conduction it must be connected to the gel at both ends. At low fibre concentration there is no conducting network at all, and the conductivity of the composite is essentially that of the matrix material. As the fibre concentration is increased, the conductivity remains low until a 3-dimensional network is first established at the gel-point, when it suddenly increases by a large factor. Thereafter the conductivity changes more gradually with fibre concentration. The proportion of fibre which is active in conduction continues to grow, however, since the number of interparticle contacts increases as the square of the fibre concentration.

The upper limit to the conductance between opposite faces of a composite block is the conductance of a uniform wire of the conductor, whose length equals the distance between the faces and whose volume equals the total volume of conductor in the block. In accord with this equivalent-wire limit, the maximum conductivity that can be obtained with a volume fraction f_v of a conductor (conductivity σ_c) in a composite is

$$\sigma_{max} = \sigma_c f_v. \tag{8.35}$$

In a composite containing randomly distributed fibres, orientation of fibre out of the field direction must effectively reduce the conductivity from the maximum. Consider a straight segment of fibre inclined at an angle θ to the applied field direction. Comparing the conductance of this segment with that of the same amount of conductor arranged uniformly, parallel to the field, its contribution to the cross-section of the equivalent wire is reduced by a factor $\cos^2 \theta$. For a random array of fibres the probability of an orientation in the range θ to $(\theta + d\theta)$ is proportional to $\sin \theta$, so that averaging over all orientations gives a mean reduction factor g, thus:

$$g = \frac{2}{\pi}\int_0^{\frac{\pi}{2}} cos^2\theta \sin\theta . d\theta = \frac{2}{3\pi}. \tag{8.36}$$

If we assume that the fibre content is high, inter-fibre contacts will be sufficiently numerous for nearly all of the fibres to be active. Then, neglecting the effects of branch points and dead-ends, the equation for the conductivity of a composite containing randomly oriented fibres becomes

$$\sigma = \frac{2}{3\pi}\sigma_c f_v. \tag{8.37}$$

To cite a practical example, 20% by mass of 12 μm diameter glass fibre coated with silver (70 nm thick) mixed in a thermosetting polyester resin gave a composite conductivity of $2.5 \times 10^4\,\Omega^{-1}\text{m}^{-1}$. From Equation (8.37) the calculated value is $5.3 \times 10^4\,\Omega^{-1}\text{m}^{-1}$, which compares reasonably well with the experiment, if one takes into account that some disintegration of the fibre is inevitable in the mixing process.

Preferential orientation of the fibres in a particular direction will tend, of course, to increase the conductivity in that direction at the expense of the others.

Reticulate doped composites

In all the examples discussed above the conductive fillers have been inorganic materials. As noted in the previous section, it is possible to make conductive composites with *organic metal* fillers. The organic charge transfer (CT) salts are ionic materials which contain linear arrays of conjugated ions with strong overlap of the π-electrons of adjacent ions. If charge transfer is complete and every molecule in the array is charged, the motion of electrons must overcome the Hubbard potential, i.e. the cost of having two electrons on one site. This effect opens up a large energy gap, as discussed in Section 4.2.4, and the material is insulating. If the charge transfer is partial, the arrays will contain both ions and uncharged molecules and charge can move easily from an ion to a neighbouring molecule. In this situation the material will have a partially-filled band and be metallic. These metals are soluble in organic solvents which can also dissolve polymers. This makes it possible to cast the composite from solution by the co-precipitation of the polymeric host and the conductive phase. The anisotropic structure of the CT salts means that they tend to grow as needle or plate morphologies, leading to composites with novel morphology.

Reticulate composites comprising needles of TTT–TCNQ (tetrathiotetracene-tetracyanoquinodimethane) in poly(bisphenol-A-carbonate) were first described in the early 1980s. The composites were prepared by casting thin films from a solution of the components in *o*-dichloromethane. The composite contained a network of fine interconnected needle crystals of the CT salt and had a percolation threshold of 2–3 mass% of the CT salt, as shown in Fig. 8.22.

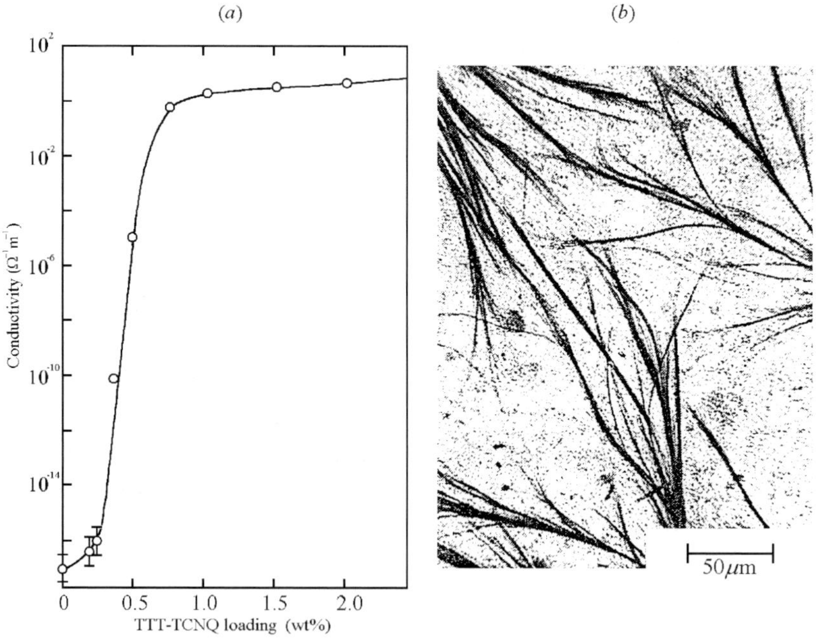

Fig. 8.22 (a) Conductivity of a TTT-TCNQ polycarbonate composite. (Reprinted with permission from *Nature*, Jeszka *et al.*, 1981. Copyright 1981 Macmillan Magazines Limited.) (b) Morphology of a polyethylene film containing 1% TTT-TCNQ. (Reprinted from Ulański, 1990. Copyright 1990, with permission from Elsevier.)

The films obtained at the lower loadings are flexible and the conductivity is stable under reasonable deformation. Films containing a few percent of TTF–TCNQ (tetrathiafulvalene) have been produced commercially as an antistatic film for the wrapping of charge-sensitive integrated circuits.

The same kind of additives can also be used to make polymer films with high surface conductivity, greater than 10^{-3}S, in a two-step process (Jeszka *et al.*, 1999). The precursor organic compound, comprising neutral molecules, is first molecularly dispersed in the polymer film, which is then swelled with solvent vapour at the same time that it is exposed to either iodine or bromine. The CT salt then precipitates close to the surface of the polymer film, producing a localised conductive layer.

8.4 Molecular composites

The processes of band and hopping conduction in semiconductors have been described in Chapter 4. In semiconductors, such as silicon, dopants give rise to distinct energy levels lying in the band gap and carriers in either the conduction or valence bands. The conductivity is thermally activated with a well-defined activation energy. At very high dopant concentrations the overlap of the dopant wavefunctions broadens the dopant levels and gives rise to an impurity band lying in the band gap of the semiconductor. Conduction can then occur within the impurity band. If the semiconductor is disordered, a quasi-continuum of defect states occurs adjacent to the band edges. Conduction then occurs by thermally-activated hopping between states that are close in space and energy. While charge transport in saturated polymers with dispersed, *localised* conjugated moieties also occurs by hopping conduction, and the localised conjugated moieties are often referred to as *dopants*, the underlying physics is different from that in conventional inorganic semiconductors.

The HOMO and LUMO of the conjugated moieties lie within the band gap of the host polymer. The energetic separation of the HOMO and LUMO, typically 2 to 3 eV, is much less than the polymer band gap, typically 6 to 7 eV, so that they are well separated from the polymer band edges and extrinsic conduction is negligible. Even if the concentration of conjugated moieties is too low to produce an impurity band, the HOMO and LUMO states are spread out to form bands due to *solvatochromism*. Solvatochromism, as discussed in Section 3.7.2 for polar molecules, is a general phenomenon and also occurs, though to a lesser extent, for non-polar molecules. The disorder present in the amorphous or partially crystalline polymer host produces variations in the local field experienced by the conjugated moieties and consequently variations of the energies of the HOMO and LUMO. The effect is often referred to as *diagonal disorder* as it is a consequence of differences in the diagonal (energy) terms of the total Hamiltonian describing the system. The contribution of random local environments is called *static diagonal disorder* while that due to thermal fluctuations is called *dynamic diagonal disorder*. The perturbation of the energy levels is enhanced by the interaction of the conjugated moieties, which are randomly distributed, and possess either induced or permanent dipoles. The contribution to the spread in energy levels from this interaction between the conjugated moieties is described as *off-diagonal disorder*.

Electrical conduction will occur by the hopping of either electrons or holes within these distributions of energy levels. Charge transport can be either of holes by transfer between the LUMO states or of electrons between the HOMO states. These correspond to the formation of either a radical cation by the removal of an electron to an adjacent electrode or an anion by the injection of an electron. The nature of the majority carriers will, therefore, be determined by the ionisation potentials and electron affinities of the conjugated moieties. A low ionisation potential will favour hole transport while a high electron affinity will favour electron transport. Most of the conductive polymers reported in the literature have low ionisation potentials and are hole conductors.

Examples of semiconducting polymers are shown in Fig. 8.23. The most widely studied system is poly(vinylcarbazole) (PVK, Fig. 8.23(a)). Poly(1-vinylpyrene), Fig. 8.23(b), is an example of a polymer with a pure hydrocarbon conjugated group. In these examples the conjugated moiety is directly bonded to the polymer backbone. Alternatively, a flexible linkage can be used, as in poly(1-pyrenylmethylvinylether), Fig. 8.23(c), and poly(2-*N*-ethyl-*N*-*m*-tolylaminoethylmethacrylate), Fig. 8.23(d). Larger conjugated moieties such as triphenyldiamine can be incorporated either as pendent groups, e.g. poly([1,1′-biphenyl]-4,4′-diamine-*N*,*N*′-bis(4-methoxyphenyl)-*N*-(3-methyphenyl)-*N*′-(4-phenylvinylene)), Fig. 8.23(e), or integral with the polymer backbone, e.g. the copolymer of tetraphenylbenzidine and diethylene glycol bis(ethyl carbonate), Fig. 8.23(f). These polymers all contain groups with low ionisation potentials and conduct by the transport of holes.

Molecular additives provide a variety of functions. They may act as charge transport agents, as charge transfer agents to promote the generation of the majority carriers and as centres for the recombination of electrons and holes leading to photoemission. While much research has focussed on enhancing charge transport and generation for utilisation in xerography, since 1990 there has been a rapid growth in studies of photo-emissive species for application in organic light emitting diodes (OLEDs). A variety of amines and diamines have

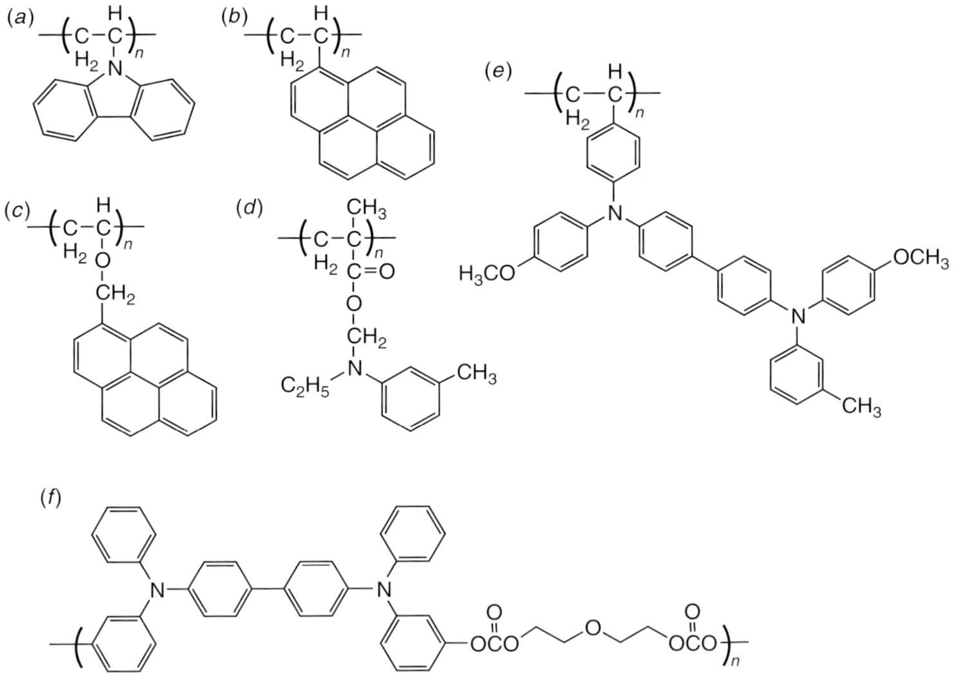

Fig. 8.23 Chemical structures of some semiconducting polymers containing arylamino and aromatic pendent and backbone groups.

been utilised as species providing hole transport, for example 1-*N*-methylcarbazole (*m*Cbz), Fig. 8.24(a), 1,1-bis(di-4-tolylaminophenyl)-*cyclo*hexane (TAPC), Fig. 8.24(d), *N,N*′-diphenyl-*N,N*′-bis(3-methylphenyl)-[1,1′-biphenyl]-4,4′-diamine (TPD), Fig. 8.24(e) and *p*-diethylaminobenzaldehyde-1,1-diphenylhydrazone (DEH), Fig. 8.24(f). The relationship of *m*Cbz to PVK and of TPD to the polymers shown in Figs. 8.23(e) and 8.23(f) is immediately apparent. Examples of molecules that provide electron transport are (4-*t*-butylphenyl)-5-(4-biphenyl)-1,3,4-oxadiazole (PBD), Fig. 8.24(c), 8-hydroxyquinoline aluminium (Alq_3), Fig. 8.24(g) and *fac*tris(2-phenylpyridine) iridium ($Ir(ppy)_3$), Fig. 8.24(h). Alq_3 and $Ir(ppy)_3$ are also light emitting species which act as recombination centres for electrons and holes and have been used in the fabrication of OLEDs, see Section 10.2.2. 2,4,7-trinitro-9-fluorenone (TNF), Fig. 8.24(b), is an example of a species which acts as a charge transfer agent when combined with a hole transporting polymer such as PVK. TNF is an electron acceptor and is used as a *sensitizer*, which enhances the photoconductivity of the host polymer by facilitating the photo-generation of holes on the carbazole moieties.

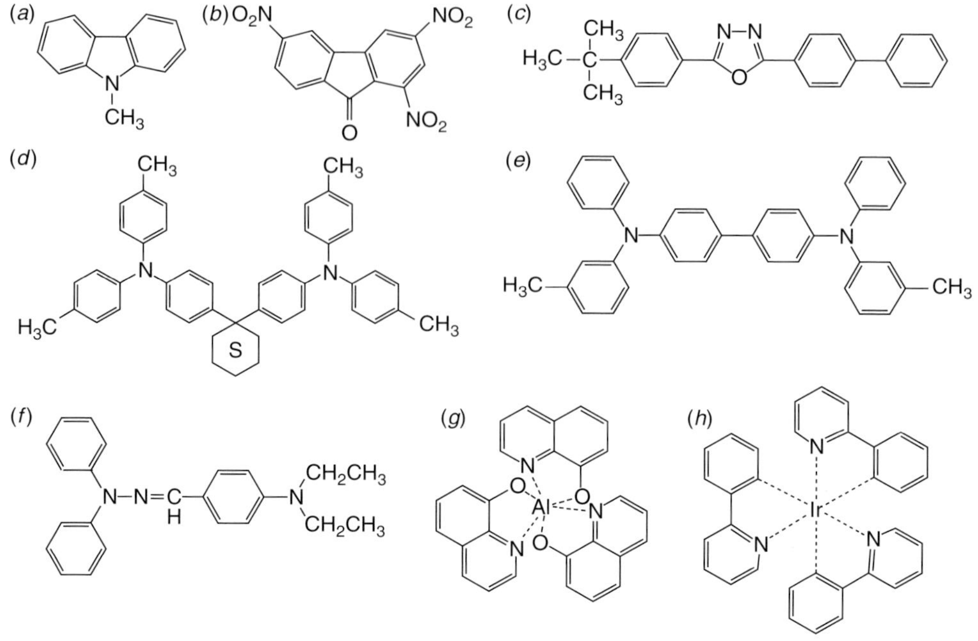

Fig. 8.24 Molecular species that assist carrier motion in molecularly doped polymers.

8.4.1 Carrier mobility

The measurement of carrier mobility uses equipment, shown in Fig. 4.8, to determine the time of flight (TOF) of carriers through the polymer sample. Light from a short-pulse laser is incident on the transparent electrode, typically gold or indium tin oxide. The wavelength of the laser is chosen so that it is strongly absorbed by either the semiconducting polymer or a charge generation layer deposited on the surface, such as copper phthalocyanine. Photo-excitation produces a thin sheet of carriers, typically 50 nm thick, close to the electrode. The carriers then drift through the sample under the influence of an applied field of the appropriate polarity. The quantity of carriers generated is kept small enough to avoid space charge effects, i.e. modification of the field by the localised concentration of carriers. For materials where all the carriers transit the sample with the same mobility a current pulse will be observed. The current decreases sharply as the charge sheet exits the sample, see Fig. 8.25(a). The mobility is then given by Equation (4.32).

Polymeric semiconductors are usually disordered, so that charge transport is dispersive, i.e. there is a spread in the apparent mobility of the carriers. This results from the presence of the distribution of dopant and host energy states

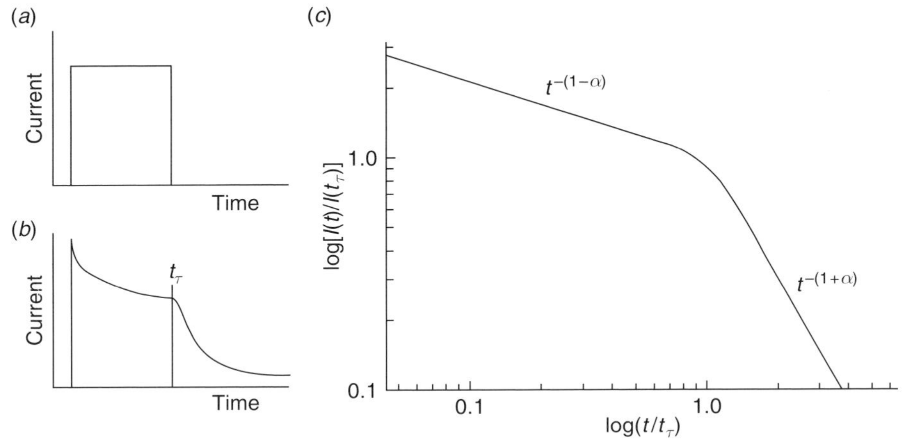

Fig. 8.25 Current transients in a TOF experiment, (a) ideal, (b) dispersive transport and (c) predicted variation according to Scher and Montroll (1975).

within the band gap of the host. Several factors contribute to the variation in mobility:

a. *Dispersive hopping*. As a consequence of the random distribution of the conjugated moieties in space and energy within the polymer matrix the hopping rate is not constant but varies from place to place and hence along the paths of carriers transiting the sample.
b. *Trap-controlled hopping*. Charge carriers fall into low-lying energy levels either in the tail of the Gaussian distribution of states of the dopant or in the polymer. The carriers are trapped and remain immobile until released from the trap by thermal activation. Since the distributions of trap levels are continuous, there is also a distribution of activation energies, resulting in a spread in transit times in addition to that due to dispersive hopping.
c. *Multiple trapping*. Where there is a multiplicity of shallow traps near to the conduction (or valence) band, carriers may also propagate by successive hops from trap to trap. This situation is, however, not encountered in polymeric semiconductors with localised regions of conjugation.
d. *Percolation*. When the concentration of conjugated moieties is high, aggregation and percolation pathways can occur. Then carriers will move primarily by tunnelling between sites in close proximity. This process will be aided by local molecular vibrations, which act to lower the potential barrier between the sites.

Dispersive transport was analysed by Scher and Montroll (1975) for a random array of hopping sites in a regular crystal lattice. The transition rate between sites is an exponential function of both inter-site distance and activation energy. The distribution of transition rates may arise from a variation in separation with fixed activation energy, a variation in activation energy for fixed site separation or a variation in both distance and activation energy. Thus, the results of this model should be applicable to amorphous as well as crystalline samples. The hopping of carriers between sites is described by a probability distribution function:

$$\psi(t) \propto t^{-(1-\alpha)}, \tag{8.38}$$

where $\psi(t)\mathrm{d}t$ is the probability that the carrier will hop in the time interval t to $t + \mathrm{d}t$ and where $0 < \alpha < 1$. Provided the dispersion of transit times is not too great, there will be a distinct *knee* in the current transient at time t_T, the transit time, as shown in Fig. 8.25(b). The variation in current is then:

$$I \propto t^{-(1-\alpha)} \text{ for } t < t_T \text{ and } I \propto t^{-(1+\alpha)} \text{ for } t > t_T. \tag{8.39}$$

This model predicts that the sum of the exponents of the current decay before and after t_T will be 2. As α tends to zero, the temporal distribution $\psi(t)$ broadens and the sharp knee seen in Fig. 8.25(b) becomes much less prominent, since the rate of decay of the current is similar both before and after t_T. t_T will depend on the ratio of the sample thickness to the average displacement of the carrier in the field direction during its random walk through the sample. Use of the formal definition of mobility, Equation (4.2), leads to the result that the mobility has a dependence on electric field and sample thickness, L, of the form:

$$\mu \propto \left(\frac{E}{L}\right)^{\frac{(\alpha-1)}{\alpha}}. \tag{8.40}$$

While the Scher and Montroll model gives an analytical solution, it is not clear that it is a correct basis for the analysis of the behaviour of carriers in polymeric semiconductors. In its original formulation the activation energy was assumed to be constant giving a temperature-independent TOF current transient. This is clearly not the case in polymeric systems, as outlined in the introduction to this section. For many semiconducting polymers the density of sites on which the charge carriers can reside is high, e.g. PVK. This is a situation which cannot be analysed analytically but which can be studied by computer simulation. The starting point for such simulation is to assume that *static diagonal disorder* gives rise to a Gaussian distribution of energies of the conjugated moieties in the density of states of the material. This will be superimposed on a background distribution of defect states in the band gap of the host polymer, which is assumed to be negligible. This is the Gaussian disorder model (GDM), treated by Bässler (1993). The distribution of energy states is taken to be:

$$\rho(\varepsilon) = \frac{1}{\sqrt{2\pi\sigma^2}} \exp\left[-\frac{\varepsilon^2}{2\sigma^2}\right], \tag{8.41}$$

where ε is the energy relative to the centre of the distribution, $\rho(\varepsilon)$ is the density of states per unit energy and volume and σ is the standard deviation of the distribution.

The choice of a Gaussian distribution can be justified by the observation that the optical absorption profiles of well-defined conjugated moieties are Gaussian. Charge transport is treated as a biased random walk amongst the conjugated moieties, which have random site energies described by Equation (8.41). The computer simulations provide results which can be fitted with analytical functions over limited ranges. At low fields the mobility is found to be temperature dependent:

$$\mu(T) = \mu_0 \exp\left[-\left(\tfrac{2}{3}\hat{\sigma}\right)^2\right], \tag{8.42}$$

where $\hat{\sigma}$ is σ/kT. Simulations of the dependence of mobility on applied field E give s-shaped curves on $\log(\mu)$ versus $E^{1/2}$ plots, so that the behaviour is Poole–Frenkel-like ($\log(\mu) \propto \sqrt{E}$) only over a limited range of fields. The calculated TOF current versus time plots are dependent on the choice of distribution width.

In the GDM model it is assumed that the energies of adjacent conjugated sites are uncorrelated. However, the broadening of the energy levels of the conjugated moieties is due to the interaction of charges on the site with the local field, i.e. a charge–dipole interaction. Gartstein and Conwell (1995) have argued that this interaction has sufficient range to give rise to correlations between the energies of adjacent sites. Introducing this into the GDM model extends the region of Poole–Frenkel behaviour to lower fields. Dunlap *et al.* (1996) incorporated this concept into an analytical model for a Gaussian distribution of correlated site energies. Their general expression reduces in the limit of large disorder to:

$$\mu = \mu_0(E) \exp\left[-\left(\frac{\sigma}{kT}\right)^2\right] \exp\left[2\frac{\sigma}{kT}\sqrt{\frac{eEa}{kT}}\right], \tag{8.43}$$

where E is the applied field, a corresponds to the size of the conjugated site and $\mu_0(E)$ is slowly varying with respect to the exponential factors. This gives a temperature dependence similar to that of the GDM, Equation (8.42), and an explicit proportionality between log(μ) and $E^{1/2}$.

At high temperatures the injected carriers effectively maintain a dynamic equilibrium in the Gaussian density of states (DOS) of the conjugated moieties as they move across the sample. In this situation the carrier transport process is independent of sample thickness and the transport is said to be *non-dispersive*. As the sample temperature is reduced, a point will be reached where dynamic equilibrium is not reached during the transit of the carriers across the sample. This occurs at a temperature related to the width of the Gaussian distribution. Below this temperature the transport will be *dispersive* and there is no longer a distinct knee in the TOF transient. At low temperature the current decay can appear to be continuous. Despite this, a change in slope can be detected in log–log plots. A further complication is the possibility that defects and impurities in the sample can act as traps for the mobile carriers. If the traps are energetically shallow, i.e. the depth of the trap level is comparable with thermal energies (kT), the carriers will soon escape again, and the only effect will be to reduce the apparent mobility. The *trap-limited* mobility is given by Equation (4.33) and is temperature dependent, since the release of carriers from the traps is thermally activated, c.f. Equation (4.34). Hence, for semiconductive polymers the carrier mobility determined from the transit time, obtained from double log plots of the TOF transients, is an average rather than a single-valued quantity. Figures 8.26 and 8.27 show examples of TOF data, and derived carrier mobilities are listed in Table 8.2.

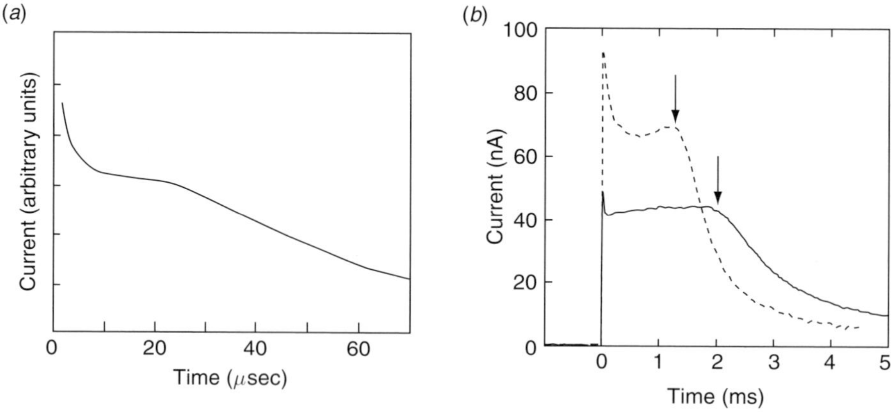

Fig. 8.26 TOF current transients: (a) PC – TAPC (1:3) and (b) PC – TPD (1:1) for photo-generation adjacent to an Au electrode (dashed curve) and injected from a PPV electrode (full curve). Reproduced with permission of the American Institute of Physics from Borsenberger *et al.* (1991) and Antoniadis *et al.* (1993).

Table 8.2 *Mobilities in semiconducting and molecularly doped polymers*

Material	Carrier	Mobility ($m^2V^{-1}s^{-1}$)[a]
PVK	Holes	1.4×10^{-10}
PVK–TNF (1:1.5)	Holes	6.2×10^{-12}
	Electrons	1.2×10^{-11}
Poly(1-vinylpyridine)	Holes	7.5×10^{-11}[b]
Polymer of Fig. 8.23(d)	Holes	4×10^{-10}
PMPS[c]	Holes	1.7×10^{-8}
PC[d]–DEH (1:1)	Holes	10^{-9}
PC–TAPC (1:3)	Holes	6.5×10^{-9}
PC–TPD (4:1)	Holes	1.7×10^{-11}
PC–TPD (1:1)	Holes	2.2×10^{-9}
PC–TPD (1:4)	Holes	1.05×10^{-8}
PS–TTA[e](3:2)	Holes	6×10^{-8}

[a]At room temperature and an applied field of 5×10^7 Vm^{-1}.
[b]Field 2×10^{-5} Vm^{-1}.
[c]Poly(methylphenylsilane).
[d]Bisphenol-A-polycarbonate.
[e]Polystyrene – tri-*p*-tolylamine.

Figure 8.26(a) shows a TOF transient observed for a 9 μm thick sample of bisphenol-A-polycarbonate containing 75 mass% of TAPC subjected to a field of 7×10^7 Vm^{-1} at room temperature. Figure 8.26(b) shows two extremely well-defined TOF signals obtained with a 22 μm thick sample of polycarbonate loaded with TPD at 50 mass%. The dashed curve is for charge generated adjacent to a gold electrode with a 1ns, 337 nm nitrogen laser pulse. The transit time, marked by the arrow, gives a mobility of 1.3×10^{-9} $m^2V^{-1}s^{-1}$ for a field 1.4×10^{-7} Vm^{-1}, in agreement with other data for this system. When a 460 nm wavelength laser pulse is used to inject charge from a poly(*p*-phenylenevinylene), PPV, layer interposed between the PC–TPD and an aluminium electrode, a longer transit time is observed. This is attributed to either local charge accumulation reducing the field across the PC–TPD layer or delayed charge injection from the PPV into the PC–TPD layer. The transients have ideal profiles, because this system is an example of trap-free carrier transport (Stolka *et al.*, 1984). This is observed in polymers containing triphenylamine moieties, e.g. the polymers shown in Fig. 8.23(e) and 8.23(f), TPD, TAPC, DEH, etc. These compounds have relatively low ionisation potentials, e.g. 5.38 eV for TPD and 5.06 eV for the polymer in Fig. 8.23(e), so that impurity levels are likely to lie higher in energy and be inactive as traps. In PVK co-facial dimers act as traps reducing carrier mobility, but the triphenylamines do not form excited state dimers (excimers). Other defects, e.g. chemically related species, are likely to act as shallow traps that lie within the Gaussian DOS of the

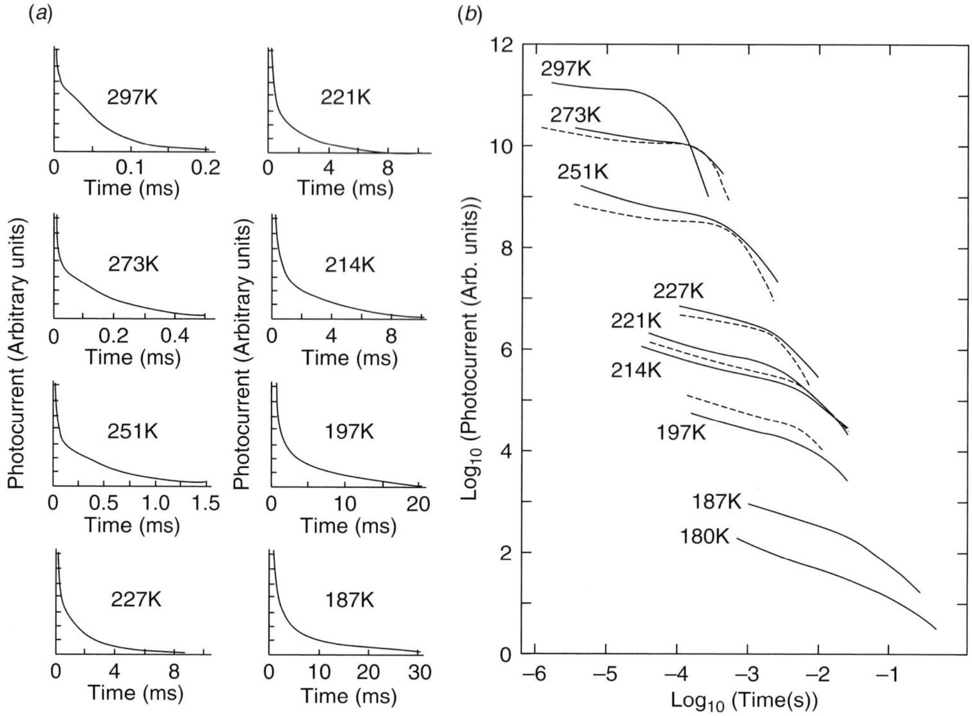

Fig. 8.27 Temperature dependence of the TOF transients for PC–TAPC (1:3) plotted on (a) linear scales and (b) log scales. The dashed curves in (b) are GMT simulations. Reproduced with permission from Borsenberger *et al.* (1993). Copyright 1993 by the American Physical Society.

triphenylamines; typically σ is of the order 0.1 eV in these materials. This set of circumstances accounts for the observation of trap-free transport, despite the structural and chemical disorder present in the samples.

Figure 8.27 illustrates the change in TOF signals accompanying this change in carrier transport from non-dispersive at room temperature to dispersive at low temperature for films of TAPC in polycarbonate. The raw data is shown in Fig. 8.27(a) and is plotted on logarithmic scales in Fig. 8.27(b). This effect is seen in a variety of materials, e.g. DEN in polycarbonate (Borsenberger *et al.*, 1992) and PVK (Pfister and Griffiths, 1978). In the case of PVK the transport at high temperatures is controlled by shallow traps rather than by maintaining dynamic equilibrium in the DOS of the carbazole moieties.

The mobility is concentration dependent, see Table 8.2, since increasing the density of conjugated sites reduces the mean hopping distance. Data for DEH loaded in polycarbonate is shown in Fig. 8.28. The solid line is a fit to the empirical relationship:

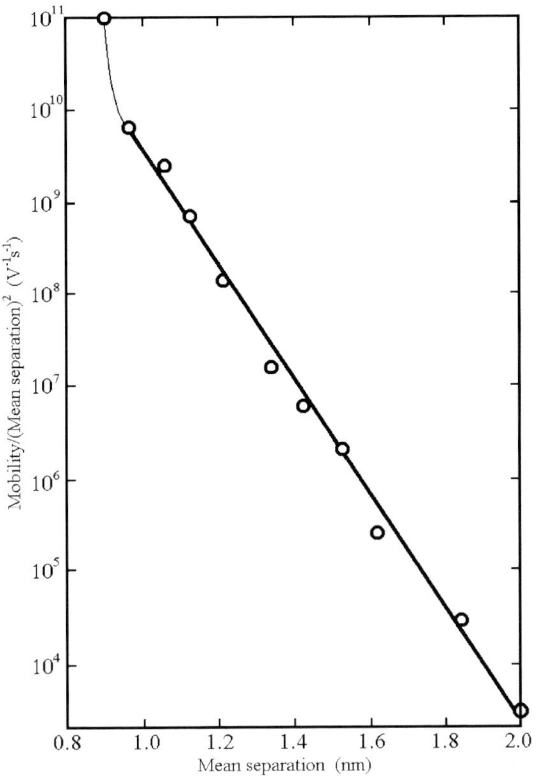

Fig. 8.28 Variation of hole mobility with DEH concentration in DEH-loaded bisphenol-A-polycarbonate. The bold line is Equation (8.44) with r_0 equal to 0.14 nm. Reproduced with permission of the American Institute of Physics from Stolka *et al.* (1984).

$$\mu \propto r^2 \exp\left(-\frac{2r}{r_0}\right), \tag{8.44}$$

where r is the inter-site distance and r_0 is the charge localisation radius on the sites. The exponential factor is a consequence of the distance dependence of wavefunction overlap between sites and the pre-factor is related to the mean square displacement per hop. This equation describes the observed variation at all except the highest concentrations of DEH.

The field dependence of mobility is observed to be Poole–Frenkel-like over a wide range of applied voltage. This has been observed both for pure polymers, e.g. PVK, and molecularly loaded polymers, e.g. DEH and TAPC in polycarbonate. Data for loaded systems are shown in Fig. 8.29. The fit to the DEH data is obtained using the full analytical expression derived by Dunlap *et al.* (1996).

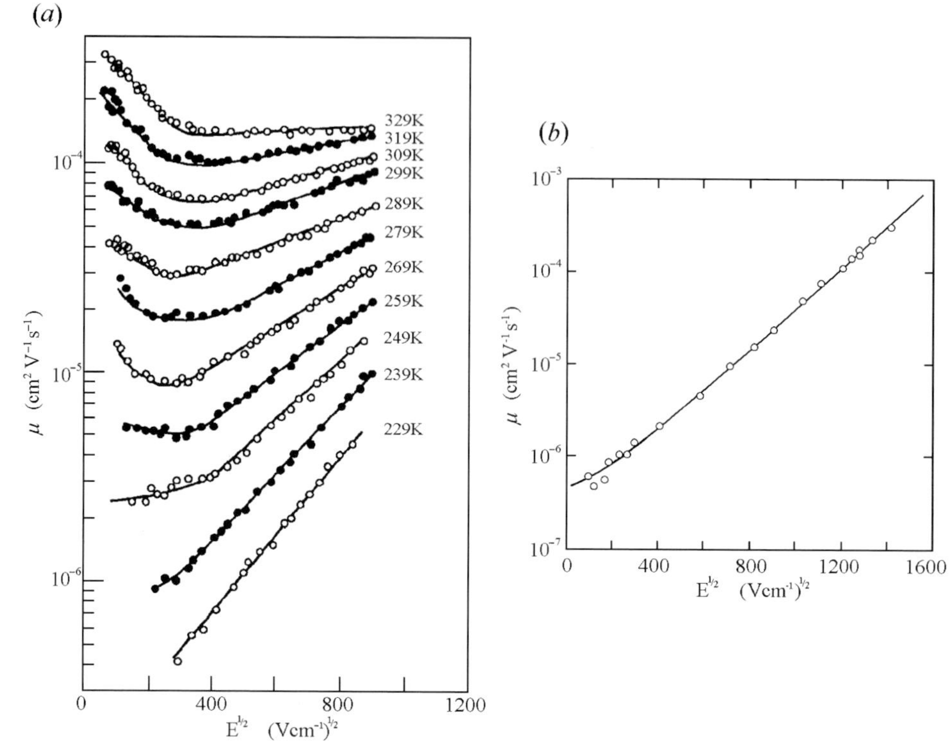

Fig. 8.29 Voltage dependence of hole mobility for (a) PC–TAPC (1:3) for a range of temperatures (reproduced with permission of the American Institute of Physics from Borsenberger *et al.*, 1991) and (b) PC–DEH (1:1) at room temperature (reprinted with permission from Dunlap *et al.*, 1996; copyright 1996 by the American Physical Society.).

The introduction of a second conjugated moiety will generally introduce traps which will reduce the mobility, such as PVK–TNF in Table 8.2, where the addition of TNF gives much lower hole mobility. However, the TNF can act as an electron transport agent, and although the electron mobility in PVK is negligible in PVK–TNF, it is comparable to or greater than the hole mobility. The introduction into tri-*p*-tolylamine doped polystyrene of a 10^{-2} mol fraction of the amines di-*p*-tolyl-*p*-anisylamine, di-*p*-anisyl-*p*-tolylamine and tri-*p*-anisylamine creates traps with depths of 0.07, 0.14 and 0.22 eV, respectively. At room temperature and for an applied field of 7×10^7 Vm^{-1} the hole mobility is reduced marginally and by factors of 6 and 16 by these three additives, respectively (Borsenberger *et al.*, 1998).

8.4.2 Photoconductivity

As discussed in Section 4.2.2, photoconduction in semiconductors is associated with the excitation of free carriers by photons with energy greater than the band gap. This is not the case here, since the species that absorb in the visible region, the conjugated species, are localised. Photo-excitation produces an excited electronic state, an *exciton*, comprising an electron and a hole localised on the same molecular site. However, as pointed out in the earlier discussion, excitons can give rise to a photocurrent, if the charges can be spatially separated, so that they can move independently of one another. This requires that they must be able to move apart by a distance greater than the Coulomb capture radius, i.e. to the point where their Coulomb energy is of the order of the thermal energy and thermal fluctuations will lead to their complete separation. This problem was first analysed in 1938 by Onsager, who showed that the thermalisation length, or Onsager radius, is given by:

$$r_c = \frac{e^2}{4\pi\varepsilon\varepsilon_0 kT}. \tag{8.45}$$

In the presence of an electric field, E, the Coulomb potential barrier in the downfield direction is lowered and escape from *geminate recombination* becomes easier. The quantum efficiency for charge generation is then given, to first order in the field, by:

$$\Phi = \Phi_0 \exp\left(-\frac{r_c}{r_0}\right)\left[1 + \left(\frac{er_c}{2kT}\right)E\right], \tag{8.46}$$

where Φ_0 and r_0 are constants. This can be an efficient process if the Coulomb interaction is screened by the surrounding medium, giving a small Onsager radius. In wide band-gap polymers the relative permittivities are modest and there are no free charges to provide screening. Hence Φ is small, and most excited charges suffer geminate recombination and exist as excitons, which are electrically neutral.

Thus, while *intrinsic single-photon generation* of charge carriers can occur, it is an inefficient process. However, excitons are mobile, i.e. the excitation can migrate between the conjugated moieties in the system. There is, therefore, the possibility that two excitons can come into close proximity and can fuse to give a single exciton of twice the initial energy. The energy excess can then be sufficient to stimulate charge separation. This is a bimolecular process. The exciton may either have spin zero, a *singlet exciton*, or spin one, a *triplet exciton*. In the former the spin of the excited electron is unchanged on excitation, while in the latter the spin changes sign on excitation. Since a change of spin requires a magnetic interaction, it is forbidden for electric dipole interactions, which dominate the interplay between radiation and molecules. Thus the generation of triplet excitons is much less probable than the generation of singlet excitons. The ground state of most molecules is, however, a singlet, and this makes the decay of triplets to the ground state an improbable process. Hence, once created, triplet excitons are long-lived against radiative decay to the ground state and a sizeable population can build up in some materials, where other decay mechanisms do not reduce triplet life times. Thence exciton fusion can occur for singlet–singlet, singlet–triplet and triplet–triplet pairs. The new, high-energy mobile exciton either dissociates at a defect site or is annihilated at an electrode, giving rise to charge injection. While exciton fusion and dissociation in the bulk of the material will generate electron–hole pairs, the decay of an exciton at an electrode will create only carriers of one sign. In addition, it is possible to have *extrinsic single-photon generation* of charge, if molecular species with a high electron affinity and a low ionisation potential are both present in the material. Excitation of a neighbouring pair of dissimilar molecules will then result in a *charge transfer exciton* in which the electron and hole are separated on the neighbouring sites. This reduces the Coulomb interaction and increases the probability that the electron and hole can escape geminate recombination. The single-photon processes are linear in light intensity, while charge generation by exciton fusion depends on the square of the light intensity.

There are other processes for charge generation; these include two-photon absorption and single photon ionisation of singlet and triplet excitons. These processes are proportional to the square and higher powers of the light intensity and usually somewhat weaker than the processes described in the previous paragraph.

Now consider a sample in which photo-generation occurs throughout the sample. In general, the charges are more likely to recombine with charges of opposite sign, that they encounter on their biased random walk through the material, than to transit the sample. The steady-state concentration of charges is then the product of the carrier generation rate per unit volume and time, ρ, multiplied by the recombination time, τ, and the steady-state conductivity is given by Equation (4.1). Assuming that the electrical contacts made to the material are ohmic, i.e. charge is injected into the material at one electrode as it leaves at the other (so that the concentration of charges is constant), the steady-state photocurrent is given by:

$$i_{\mathrm{st}} = e\rho\tau E\mu = e\left(\frac{\Phi I}{L}\right)\tau E\mu, \tag{8.47}$$

where E is the applied field, μ the carrier mobility, Φ the quantum efficiency for charge generation, I the total number of photons absorbed per unit area per unit time and L the thickness of the material. In practice it is difficult to make ohmic contacts to organic materials. More often the electrical contacts are *blocking*, i.e. photo-generated charge can drift out of the material but no charge can be injected from the electrodes. Then, if the recombination time is larger than the transit time, the current is just ΦIe. If the reverse is true, then:

$$i_{\mathrm{bl}} = e\Phi I \frac{\tau}{t_\tau} = e\Phi I\tau\left(\frac{\mu E}{L}\right). \tag{8.48}$$

This is identical to Equation (8.47), reflecting the fact that the majority of photo-generated carriers recombine in the material and only a small number reach the electrodes. The steady state photoconductivity is, therefore, generally dependent on the product $\Phi\tau\mu$. μ can be determined from TOF measurements, while Φ can be measured for thin films by the xerographic discharge technique. The top surface of film deposited on a metal substrate is charged with the same polarity as the majority carriers, typically holes, with a corona discharge. The potential difference across the film is monitored with a capacitive probe. The rate of change in potential is measured when the film is uniformly illuminated with light of a known wavelength and intensity. Provided that charges are not injected from the substrate electrode, the rate of change of potential at times long compared to the carrier transit time is given by the rate of photo-generation of carriers:

$$\frac{\partial V}{\partial t} = -\Phi Ie. \tag{8.49}$$

The process of photo-generation of carriers means that the action spectrum is expected to be similar to that of the conjugated moiety. However, when the absorption coefficient of the material is high, photo-generation will be confined to the surface of the sample. The observed photocurrent can then depend on sample geometry and experimental method, and data must be analysed with care. For example, in the TOF geometry, if all the carriers generated near the front electrode are detected, the integrated signal will be a measure of Φ. But if the photocurrent is measured with surface electrodes on the illuminated surface, and defect states contribute to the photocurrent, the photocurrent can fall as the absorption increases. This occurs if intrinsic photo-generation in the material cannot compensate for the reduction in extrinsic photo-generation as the penetration depth of the light is reduced to 0.1 μm or less.

PVK has been studied as a prototypical semiconductive polymer. The photo-action spectrum mirrors the absorption spectrum of *N*-substituted carbazoles with an onset near 370 nm and maxima at ~340, 300 and 270 nm. At a wavelength of 254 nm the quantum efficiency is $\sim 10^{-3}$ with an applied field of $10^7\,\mathrm{Vm^{-1}}$. The field dependence of the quantum efficiency is described by a second-order Onsager expression, such as Equation (8.46), with $\Phi_0 = 0.11$ and $r_0 = 2.5$ nm. The large value of r_0 rules out exciton annihilation as the source of carriers and suggests the formation of a charge-transfer excited state. Either dissolved oxygen or oxidised carbazole units have been suggested as the electron acceptor present as an impurity in the PVK. The addition of the electron acceptor TNF at a 1:1 molar concentration in PVK results in quantum efficiencies of $\sim 8 \times 10^{-3}$ at $10^7\,\mathrm{Vm^{-1}}$ rising to 0.5 at $9 \times 10^7\,\mathrm{Vm^{-1}}$. Fitting to the Onsager model gives values of $\Phi_0 = 0.23$ and $r_0 = 3.5$ nm. The excitation of charge transfer between the carbazole and TNF moieties is the primary mechanism for photo-carrier generation. As noted above, this increase in quantum efficiency is, however, accompanied by a decrease in the hole mobility. This disadvantage is offset by the fact that the charge transfer excitation gives rise to an additional strong absorption band. The 1:1 PVK–TNF system has an absorption profile that is approximately constant across the visible region and is similar to that of inorganic semiconductors such as selenium. A variety of other polymeric systems containing electron donating and accepting species have been studied, although none have properties that approach those of PVK–TNF.

8.4.3 Space charge effects

The low carrier mobility and the presence of traps in polymeric semiconductors means that when carriers are injected from an electrode, either by a voltage pulse in the dark or by photo-injection, a high charge density can build up adjacent to the electrode. The presence of this space charge distorts the electric field in the material and limits the current flowing through the sample. The occurrence of space-charge-limited currents, SCLCs, is seen in the form of the current–voltage characteristic and time evolution of the dark current induced by a voltage step, see Abkowitz and Pai (1986). An analysis of the current–voltage characteristic for injection of one sign of carrier from an ideal ohmic electrode in the dark (Pope and Swenberg, 1982) indicates four distinct regimes.

a. At low voltages, when little charge is injected, the behaviour is Ohmic, i.e. current is proportional to voltage.
b. As the voltage is increased the space charge accumulates and the current becomes proportional to the square of the applied voltage:

$$I = \frac{9}{8}\mu\varepsilon\varepsilon_0\Theta\frac{V^2}{d^3}, \qquad (8.50)$$

where Θ is a trapping factor, and d is the distance between the electrodes, c.f. Equation (4.35). The trapping factor equals 1 in the absence of traps and is dependent on $\exp(U_T/kT)$, c.f. Equation (4.34), for shallow traps.
c. In the presence of traps, increasing carrier injection will gradually fill the trap levels until they are all filled. At this point further injected charges cannot be trapped and there will be a sudden increase in current at the *trap-filled limit* voltage.
d. Above the trap-filled limit the material will behave as though it were trap free and Equation (8.50) with $\Theta = 1$ will be valid.

In practice, the trap-filled limit is difficult to observe as it is often preceded by electrical breakdown of the sample. The transition from the linear to square law (Child's Law) dependence of current on voltage is usually not sharply defined. Thus samples may display an intermediate power law over a considerable voltage range. This, and the uncertainty of the trapping factor, render the measurement of current–voltage characteristics unsuitable for the determination of carrier mobility.

The voltage-step method is, however, analogous to the TOF method and has been used in studies of semiconductive polymers. The theoretical transient current response is similar to that shown in the upper curve in Fig. 8.30(b). It rises from its initial value to a maximum and then falls away to a steady state

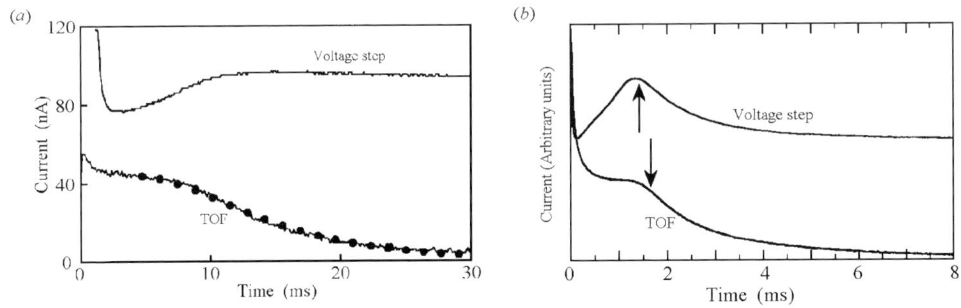

Fig. 8.30 Current transients following the application of a voltage step and for TOF measurements; (a) Vitel-DEH (1:1) (reproduced with permission of the Institute of Physics from Goldie, 1999) and (b) PC – TPD (1:1) (reprinted from Abkowitz *et al.*,1994; copyright 1994, with permission from Elsevier). See the text for experimental details.

SCLC given by Equation (8.50). The analysis of Many and Rakavy (1962) shows that the peak current is approximately 1.2 times the SCLC and that the maximum occurs at approximately 0.8 times the carrier transit time. Goldie (1999) has incorporated the experimentally observed Poole–Frenkel field dependence into this model and finds a range of possible numerical factors for the current maximum of from 1–1.2 times the SCLC and 0.7–0.8 times the transit time. Experimental data come close to these model profiles, see Abkowitz *et al.* (1994), Goldie (1999).

Figure 8.30(a) shows data for DEH dispersed at 50 mass% in a polyester, Vitel, which has the structure:

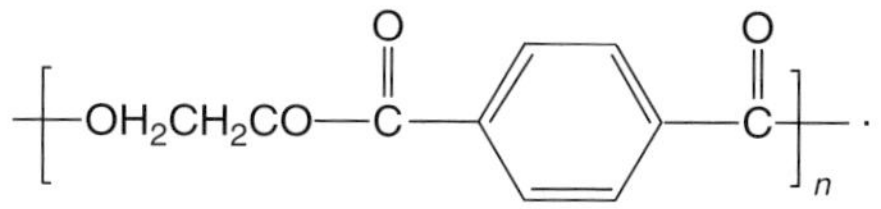

The upper curve shows a current transient produced by a voltage step applied to an unilluminated sample. The applied field was $10^7\,\mathrm{Vm}^{-1}$ for a sample 6.5 μm thick measured at 298 K. The lower curve shows a TOF transient recorded under the same conditions, and the points represent a theoretical fit using an analytical form based on an error function. This function is a reasonable approximation at high temperatures, when the photo-generated carriers rapidly achieve a dynamic equilibrium in the Gaussian DOS of the DEH molecules.

Results for a 20 μm thick sample of polycarbonate containing 50 mass% TPD for a field of $1.5 \times 10^7\,\mathrm{Vm}^{-1}$ at 296 K with charges injected from an indium tin oxide (ITO) electrode coated with a 0.1 μm thick layer of PPV are shown in Fig. 8.30(b). The limiting current is close to the trap-free SCLC, indicating that the PPV-coated ITO acts as an efficient hole-injecting electrode. The lower curve is the TOF transient recorded under identical conditions. The arrow on the lower curve indicates the transit time and that on the upper curve is 0.8 of this value. The step-voltage response is therefore close to the theoretical prediction.

The voltage-step method offers an alternative to the TOF method for determining carrier mobility in semiconductive polymers. Despite the simpler experimental arrangement, it has not been utilised to the same extent as the TOF method. Nevertheless, it is a useful tool for evaluating the nature of the contacts between electrodes and semiconductive polymers. By comparing data from voltage-step measurements with those from TOF measurements in trap-free systems it is possible to distinguish Ohmic contacts from blocking contacts where charge injection requires photo-excitation. The nature of the electrode–polymer interface is important for applications in light-emitting diodes and the photoreceptors used in xerographic copiers and printers (see Section 10.2.1). The former require Ohmic contacts for efficient charge injection while the latter require blocking contacts, since charge injection in the dark is undesirable. Studies of TPD-doped insulating polymers and polymers with integral TPD groups, see Figs. 8.23(e) and 8.23(f), show that for these systems a wide variety of electrodes, e.g. gold, carbon-black polymer composites and conjugated polymers, behave as Ohmic contacts.

8.5 Further reading

Armand (1994) has briefly summarised the history of polymer electrolytes. A more extensive account can be found in Gray (1991). Wakihara and Yamamoto (1998) describe the development of lithium ion batteries. Sahimi (1994) discusses applications of percolation theory. Early work on conductive composites has been covered by Norman (1970). Subsequent edited volumes by Sichel (1982) and Bhattacharya (1986) deal with carbon- and metal-filled materials respectively. Donnet *et al.* (1993) cover the science and technology of carbon blacks including their use in composites. Gul' (1996) presents a detailed account of conductive polymer composites up to the mid-1990s. Borsenberger and Weiss (1998) discuss semiconductive polymers with non-conjugated backbones in the context of xerography. Bässler (1983) reviews transport in these materials.

9

Intrinsically conductive polymers

9.1 Introduction

The essential difference between insulating polymers and those that are electronically conductive is that the former do not contain any unsaturated (conjugated) chemical structures while the latter do. In the preceding chapter we have seen that hopping conduction can occur in polymers with saturated backbones provided there are conjugated groups incorporated into the polymer. We have also seen in Section 4.2.5 that the simple application of band theory to polyacetylene, modelled as an ideal, infinitely-long polymer with a bond-alternated backbone structure, predicts that it will behave as a semiconductor. Few conjugated polymers have structures approaching the ideal of long, fully-extended, defect-free backbones. Most are either partially crystalline or amorphous. In these situations the polymer backbone is distorted and the backbone conjugation is disrupted, so that macroscopic samples contain a distribution of somewhat extended, but separated, conjugated regions. Because of this, charge transport through macroscopic samples will inevitably involve some hopping conduction. Polymers with conjugated backbones have an inherently higher density of conjugated regions with a higher degree of connectivity than polymers with saturated backbones containing only localised conjugated moieties. One consequence of this is the fact that the former can be rendered metallic while the latter are at best semiconductors. In this chapter the science of intrinsically conductive (conjugated) polymers will be reviewed.

While the focus of current interest is on carbon backbone polymers, we should not overlook poly(sulphurnitride), see Section 4.2.6. SN_x is an electrically conductive inorganic polymer that becomes superconducting at low temperature (Labes *et al.*, 1979). SN_x is obtained as a single crystal material by the solid-state polymerisation of S_2N_2. Though S_2N_2 decomposes explosively at room temperature, it is stable at low temperature and the polymerisation occurs at temperatures as low as 77 K. The SN_x polymer chain comprises head-to-tail SN moieties in a *cis*-configuration. The crystals have a metallic lustre, and have electrical conductivity parallel to the polymer chains of $\sim 4 \times 10^5\ \Omega^{-1}m^{-1}$ and a value about 50 times smaller perpendicular to the chains at room temperature. The calculated band structure does not have a gap, but the density of states is low at the Fermi energy. SN_x becomes superconducting at 0.26 K. Although the anisotropy of the conductivity increases to $\sim 10^3$–10^4 above the superconducting transition the inter-chain separation is small with S–N, S–S and N–N distances of 0.326, 0.347 and 0.335 nm, respectively. Thus SN_x is an anisotropic, three-dimensional semi-metal in which the inter-chain contacts suppress the Peierls transition and allow the superconducting transition to occur. Efforts to prepare analogous compounds with similar properties have been unsuccessful, so that, as noted in Section 4.2.6, SN_x appears to be a unique material. Hence, compared to carbon-based polymers, where the diversity possible in the molecular structure allows a vast array of conductive polymers to be synthesised, SN_x is an interesting curiosity rather than the basis for a fruitful avenue of research.

9.2 Conjugated polymers

The report of the doping of polyacetylene (PAc) films to produce metallic levels of conductivity by Shirakawa *et al.* (1977) sparked the interest in electrically conductive polymers that has continued until today. While it was not the first example of a conductive polymer, the increase in conductivity, by a factor greater than 10^7, observed on exposing films of *trans*-PAc to arsenic pentafluoride and iodine, was dramatic, see Fig. 9.1. The impact of this result was immediate, and created an upsurge of interest in conjugated polymers and the possibility of rendering them conductive.

In fact the first reports of semiconductive conjugated polymers appeared in the latter half of the nineteenth century, although it was not then realised that the materials synthesised were polymers. It was not until the 1920s, when the work of Staudinger and others established unambiguously the occurrence of macromolecules, that the true nature of the materials was recognised. PAc, see Fig. 9.1, was originally prepared by heating acetylene. The first studies were those of Berthelot (1866) and Erdman and Köthner (1898). The extensive research into the polymerisation of acetylene undertaken in the early part of the twentieth century was summarised by Mark and Raff (1941). Improved synthetic methods were developed in the 1950s and 1960s by Natta and co-workers using Zeigler catalysts (Natta *et al.*, 1958) and by Luttinger

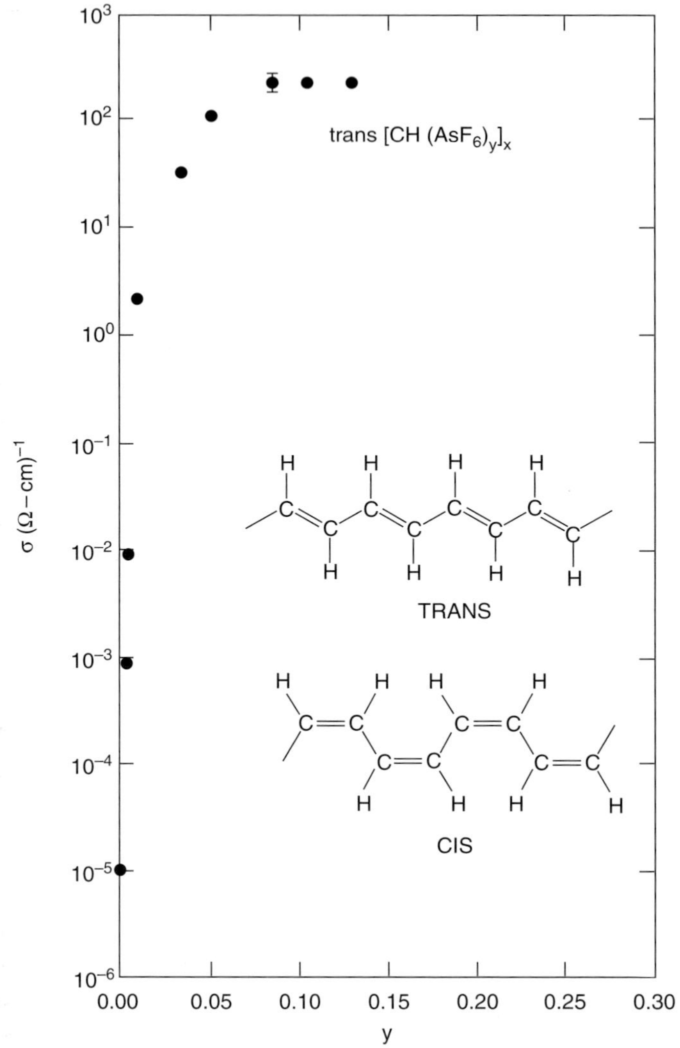

Fig. 9.1 Increase on room temperature conductivity of *trans*-PAc exposed to arsenic pentafluoride. Reprinted with permission from Chiang *et al.*, (1977). Copyright 1977 by the American Physical Society.

(1960). The product of these syntheses was an intractable, poorly-characterised powder that attracted little interest, even though Berets and Smith (1968) showed that the conductivity of a pressed powder sample was increased by oxidising agents and decreased by reducing agents. The crucial step leading to the 1977 papers was the preparation of thin films of fibrous PAc on the surface of very concentrated catalyst solution, Shirakawa and Ikeda (1971) and Ito *et al.* (1974). Parallel to the experimental studies, theoretical models

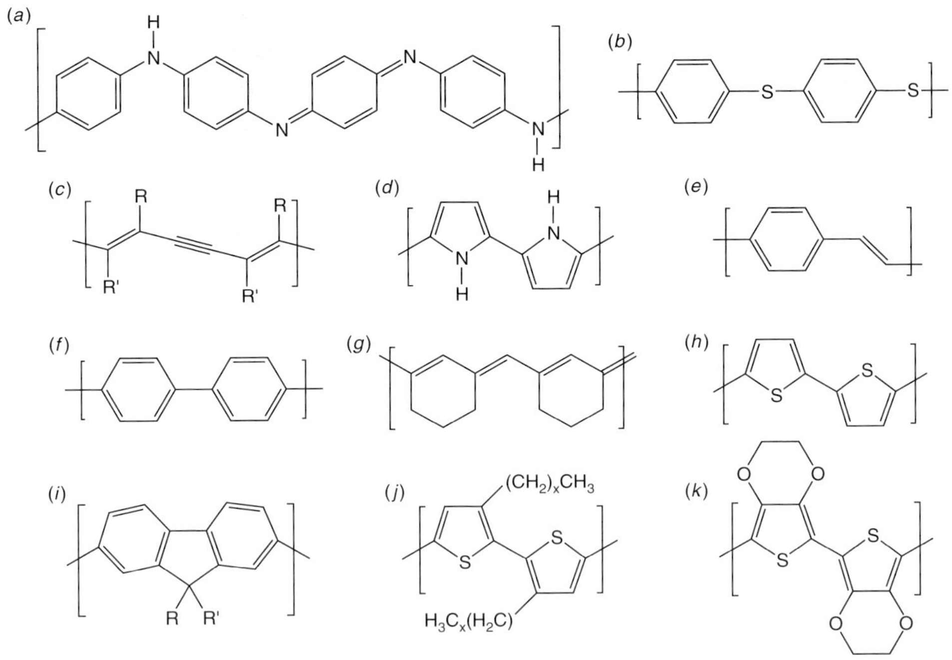

Fig. 9.2 Idealised backbone structures of extensively studied conjugated polymers, R and R′ representing a variety of pendent groups. Note that CH bonds are not shown.

were evolving. Initial work in the 1930s focussing on the length dependence of the electronic states of oligomers was extended in the 1950s, culminating in the work of Pople and Walmsley (1962), previously described in Chapter 4. Pugh (1973) was the first to discuss the excited states in terms of excitons, while Wang and Chiu (1976) considered a mechanism of charge transport along a PAc chain. Thus the emergence of highly conductive PAc was the culmination of the gradual improvement of the starting polymer and the development of theoretical models over an extended period of time.

Three other conjugated polymers have similar extended histories. Letheby (1862) reported the anodic oxidation of aniline, and the properties of the product, aniline black, were studied by others, see for example Mohilner *et al.* (1962). In the 1960s Jozefowicz and co-workers made a thorough study of polyaniline (PAni), establishing the accepted structure of the emeraldine base form shown in Fig. 9.2(a). They reported an electrical conductivity of 330 $\Omega^{-1}m^{-1}$ (Jozefowicz *et al.*, 1965). However, interest in PAni waned and was not revived until the 1980s. Grenvesse (1897) first reported poly(*p*-phenylene-sulphide), Fig. 9.2(b), at the end of the nineteenth century. Lenz and Carrington (1959) improved the synthesis and the resulting polymer has been produced commercially by Philips Petroleum as Ryton® since 1969. Interest in its use as a conductive polymer was explored in the 1980s but has been replaced by other more recently discovered polymers. Bäyer and Landsberg (1882) first noted the colouration of diacetylene crystals, which is now known to be due to the formation in the solid state of polydiacetylenes, Fig. 9.2(c), in 1882. The process attracted attention from the 1930s onwards, but the true nature of the solid-state polymerisation was not proven until 1969 (Wegner, 1969). These polymers remain a unique example of solid-state polymerisation leading to macroscopic polymer single crystals with perfectly-aligned, extended polymer chains. It has not been possible to render them highly conductive, but they are of scientific interest because of their crystallinity.

Polypyrrole (PPy), Fig. 9.2(d), first reported by Angeli in 1916, is another polymer that was found to be conductive before the advent of metallic PAc. Dall'Olio *et al.* (1968) prepared PPy electrochemically and measured a conductivity of 754 $\Omega^{-1}m^{-1}$. Despite this, no further studies appeared until 1979. It is less surprising that other conjugated polymers synthesised during the 1960s attracted less attention. The products of chemical synthesis and of the pyrolysis of saturated backbone polymers were poorly characterised and often intractable semiconducting materials. Although these were considered briefly as alternatives to germanium and silicon, the latter soon achieved a dominant position in the fabrication of electronic devices. However, poly(*p*-phenylene-vinylene) (PPV), Fig. 9.2(e), which was first synthesised at that time (McDonald and Campbell, 1960), has subsequently proved to be of considerable interest. Examples of other polymers that were reinvestigated after 1977 include poly(*p*-phenylene), see Fig. 9.2(f) (Kovacic and Kyriakis, 1962), and poly(1,6-heptadiyne), see Fig. 9.2(g) (Stille and Frey, 1961).

From the late 1970s onwards, efforts to synthesise conjugated polymers rapidly expanded and numerous new materials were prepared. Many of these still proved to be intractable substances that were difficult if not impossible to purify and characterise. The maximum levels of conductivity achieved on doping often fell well short of the metallic range. Such properties meant that the majority of these materials attracted little attention beyond the initial reports, and certainly no commercial interest. An example of the few polymers produced at this time that have been extensively studied subsequently is polythiophene (PTh), Fig. 9.2(h). Although this polymer was also reported in the nineteenth century (Meyer, 1883), the first reliable synthesis appeared in 1980, see McCullough (1998). Another example is polyfluorene, Fig. 9.2(i), which was prepared chemically and electrochemically in 1985, see Rault-Berthelot and Simonet (1986). Much subsequent synthesis has been directed to the inclusion of pendent groups to either enhance solubility, e.g. poly(3-alkylthiophene), see Fig. 9.2(j) (Jen *et al.*, 1986), or to control other physical properties, e.g. poly(3,4-ethylenedioxythiophene) (PEDOT), see Fig. 9.2(k) (Groenendaal *et al.*, 2000 and 2003).

Most research effort, therefore, concentrated on those polymers where it is possible to control morphology, to obtain chemically pure materials and to manipulate them to provide forms, e.g. chain oriented samples and thin films, suitable for basic research and applications. The development of synthetic routes that provide clean, reproducible samples has been critical in the progress achieved in understanding the physics of conjugated polymers. This in turn underpinned the rapid development of technological applications, which appeared on a much shorter timescale than for other, comparable technologies. The level of scientific and technical interest in polymers, such as PAc, PAni, PPy, PThs, PPVs and polyfluorenes, has resulted in the commercial production of them and their precursors.

Although polymer synthesis was outlined briefly in Section 1.2.3, given the importance of synthesis in providing high quality conjugated polymers, as discussed above, this topic will be revisited in the next section.

9.2.1 Preparative methods

As noted in Chapter 1, techniques for the synthesis of saturated polymers such as addition and condensation polymerisation can also be utilised to prepare conjugated polymers. These are rather broad categories where many different catalysts may be used to facilitate addition polymerisation and many different functional groups can be used to form leaving molecules in condensation polymerisation. The control of chemical structure of the polymer and the ease of elimination of catalyst residues and unwanted reaction products are the key issues in determining the particular routes to be employed. It is important to eliminate crosslinks and to control the molecular mass distribution and tacticity of the polymer. Crosslinks and chain defects will disrupt the conjugation along the polymer backbone and have a profound effect on the physical properties of the polymer. The energy of a conjugated chain will be minimised when the conjugated bonds lie in a plane and there is maximum overlap of the π-electrons. The cost in energy of distortion of this planar geometry means that conjugated polymers have rigid backbones making them difficult to either dissolve or melt. Because of this it is difficult either to remove impurities or modify the morphology built into the polymer during synthesis. This problem is illustrated by direct methods of synthesis of PAc. Zeigler–Natta catalyst produces intractable powders (Natta *et al.*, 1958), while the method of Ito *et al.* (1974) produces fibrous thin films. While the latter morphology is better for physical studies, the insolubility of the product means that it cannot be either purified or properly characterised. In particular it is not possible to determine the molecular mass distribution. This problem was addressed initially by the use of a route using soluble, non-conjugated, precursor polymers. These can be purified by solvent extraction and precipitation and their molecular masses determined prior to conversion into a fully-conjugated polymer. Manipulation during the conversion process also provides a means of controlling morphology. Despite these advantages, the products of the conversion process can contaminate the final conjugated polymer. The solution that has been widely adopted is the direct synthesis of soluble forms of conjugated polymers. This can be achieved by the addition of pendent alkyl chains or similar substituents, e.g. Fig. 9.2(j). Rigorous purification of the starting materials will eliminate one source of structural defects on the polymer backbone. The synthetic methodology will control tacticity and molecular mass, and can also reduce or eliminate crosslinking. If necessary, the soluble polymer can be fractionated to provide narrow, defined molecular mass distributions. Solution processing allows films and fibres to be prepared and a degree of control to be exercised over chain orientation.

Electrochemical polymerisation produces films on an electrode surface. Under controlled conditions uniform films up to a few mm thick, which can be removed from the electrode for subsequent study, can be prepared. Physical properties can be modified by choice of the counterions (dopants) included in the film during growth. It is, however, more difficult to control chain structure and crosslinking than in chemical methods. Electrochemically produced polymers are, therefore, less well characterised than the best directly-synthesised polymers. While this is less satisfactory for fundamental investigations, it is of less concern for applications such as battery electrodes, artificial muscles and drug release agents. The two main approaches, direct-synthesis and electrochemical, are described in the following two sections.

Direct synthesis

There are extensive reviews of the synthesis of conjugated polymers, with major compilations by Skotheim (1986), Skotheim *et al.* (1998) and Nalwa (1997). Scherf (1999) describes the synthesis of polymers with backbones containing rigidly-linked, conjugated rings that form two-dimensional ladder- or ribbon-like framework, a subgroup that is growing in importance. The review by Martin *et al.* (2001) is succinct but wide ranging. Rather than attempt to provide an even more compressed account, a few key examples will be described to show how developments in synthesis have resulted in improved materials.

(a) Polyacetylene The first controlled synthesis of acetylene giving a linear polymer product was that of Natta *et al.* (1958) using $Ti(OC_4H_9)Al(C_2H_5)_3$ as a catalyst in an inert solvent. The crystallinity of the insoluble polymer obtained depends on catalyst concentration and reaction temperature. The accidental use of a high catalyst concentration and reaction under quiescent conditions by Ito *et al.* (1974) resulted in the formation of a film of PAc fibrils on the surface of the catalyst solution and all surfaces coated in the solution. Large area films could be prepared by coating the inside of the reaction vessel with catalyst solution and introducing acetylene gas. The reaction proceeds at 195 K when the bulk of the product is *cis*-PAc, typically greater than 70% for quiescent reaction, and close to 100% when acetylene is bubbled through the concentrated catalyst. The polymer can be converted to the *trans*-form by annealing in an inert atmosphere at 473 K. Care must be taken in purifying the acetylene, solvents and inert gases, since the catalyst is sensitive to water and oxygen. Similarly, the preparation of the catalyst requires care to avoid poisoning prior to use, although a trace of oxygen may be beneficial. Despite these precautions the polymer obtained typically contains about 1% oxygen. Other catalysts, notably $Co(NO_3)_2$-$NaBH_4$ introduced by Luttinger (1960), can function in hydrophilic solvents and also tolerate the presence of oxygen. The quality of the polymer obtained is sensitive to the details of the preparation, and this has resulted in conflicting claims about which catalyst system gives the best product. Furthermore, as the products of these syntheses are insoluble, the molecular mass can be determined only by indirect methods and removal of residual catalyst is difficult.

An alternative, two-step process for the production of PAc involves starting from a non-conjugated polymer and converting it into PAc. It was shown in the 1960s that heating poly(vinylchloride) (PVC), see Table 1.1, leads to the loss of HCl and the formation of conjugated sequences along the polymer backbone. Although the PVC could be purified, characterised and manipulated, the thermal elimination is initiated at random on the PVC chain, leading to structural defects, and it is not complete. Edwards and Feast (1980) developed a significantly better route, as illustrated in Fig. 9.3. The precursor polymer is prepared by the ring-opening metathesis polymerisation (ROMP) of a monomer prepared from the reaction of *cyclo*-octatetraene and a substituted alkyne. Tungsten and molybdenum metal-organic compounds are catalysts, which act on the *cyclo*-butene unit of the monomer to produce a pre-polymer with either *cis*- or *trans*-vinylene units. PAc is obtained by the exothermic elimination of either xylene, where R = H, or a substituted xylene, a reaction which occurs slowly at room temperature and rapidly on heating. Elimination can be carried to completion, and the vinylene units present in the

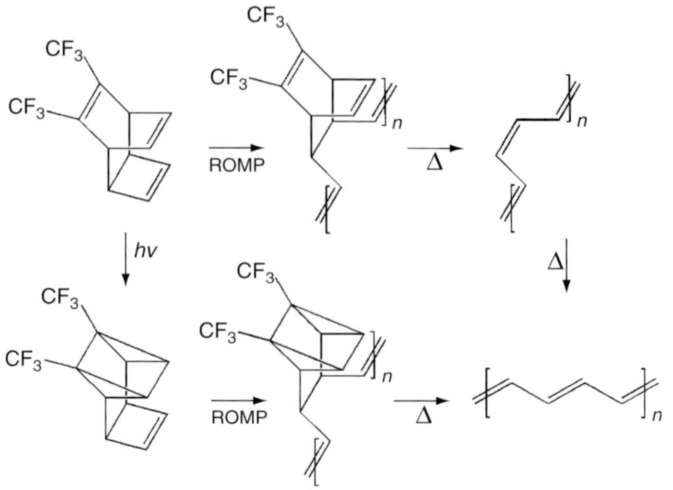

Fig. 9.3 Precursor route to Durham PAc: in the upper route the CF_3 moieties can be replaced by H, $COOCH_3$ or a bridging aromatic ring.

pre-polymer ensure that the chains are free from structural defects. While films of the precursor cast from solution give essentially amorphous PAc, stretching the pre-polymer during elimination results in anisotropic samples with aligned PAc fibres. Thus considerable control can be exercised over the morphology of a polymer derived from a well-characterised pre-polymer. The occurrence of the elimination at room temperature requires care in preparation and manipulation of the pre-polymer at low temperature. An alternative route, where the final elimination occurs above room temperature, involves the photo-isomerisation of the monomer to the form shown in the lower portion of Fig. 9.3, followed by ROMP and thermal elimination. The polymer prepared by these routes has been termed Durham PAc. This polymer facilitated experimental investigations not previously possible into the physics of PAc.

The next major improvement in sample quality was the result of developments in direct synthesis. Oriented samples were obtained by the polymerisation of acetylene with the conventional titanium catalyst on a single crystal substrate and in nematic liquid crystals, see Tsukamoto (1992). Heat-treating the catalyst was found to produce polymer that could be stretched to give oriented samples. This resulted in a higher electrical conductivity along the orientation direction, up to $2\times10^5\ \Omega^{-1}\text{m}^{-1}$ being obtained. Naarman and co-workers heat treated the catalyst at 393 K and carried out the reaction in silicone oil at room temperature (Theophilou *et al.*, 1986). Silicone oil was chosen as it has the same viscosity at room temperature as the usual solvents such as toluene at 195 K. They also claimed that the silicone interacted with the catalyst. The product could be easily stretched and on doping with iodine the conductivity in the alignment direction was $\sim 10^7 \, \Omega^{-1} m^{-1}$. Tsukamoto *et al.* (1990) modified this method by using a high boiling point solvent, e.g. decalin, and heat-treating the catalyst at 473 K. This method also produces a stretchable polymer with a similar conductivity after iodine doping. These improvements have been attributed to an experimentally observed reduction in sp^3 carbon defects (see Section 1.2.1) in the polymer chain, i.e. breaks in the backbone conjugation. Unfortunately, these methods are extremely sensitive to trace impurities, e.g. different batches of silicone oil produce polymer with large differences in conductivity, making it difficult to produce high quality samples consistently.

ROMP has also been used to obtain soluble PAc by polymerisation of *cyclo*-octatetraenes mono-substituted with aryl or alkyl units (Gorman *et al.*, 1993). The product is a PAc with one substituent every four repeat units, sufficient to impart solubility without adverse consequences for backbone conjugation.

There have been major improvements in the quality of PAc samples and consequent enhancement of physical properties. Nevertheless, the instability of PAc under a normal atmosphere prompted the focus of synthetic work to switch to other, more stable, polymers as the emphasis changed from fundamental studies to application at the turn of the century.

(b) Polythiophenes PThs offer simpler synthesis and improved stability relative to PAc. Roncali (1992) and McCullough (1998) reviewed the synthesis of PThs. In the most commonly used methodology di-halogenated thiophene reacts with either magnesium or zinc to form a metal-substituted thiophene, a Grignard reagent, which can be polymerised either *in situ* or after intermediate purification by a number of different nickel catalysts, Fig. 9.4(a). Oxidising agents such as $FeCl_3$ and AsF_5 can also be used to polymerise thiophene, Fig. 9.4(b). The products of these routes are insoluble and have relatively low molecular masses, typically corresponding to less than 50 repeat units. The electrical conductivity on doping is about $10^3 \Omega^{-1} m^{-1}$. The same methods can be utilised with thiophene substituted with a single alkyl chain. Polymers with methyl and ethyl substituents are insoluble, but those with butyl or longer substituents are soluble (Jen *et al.*, 1986). This allows work-up of the conjugated polymer to improve purity and doping in solution to give a homogeneously doped sample. The latter is difficult to achieve with insoluble polymers where dopants must diffuse into a solid sample.

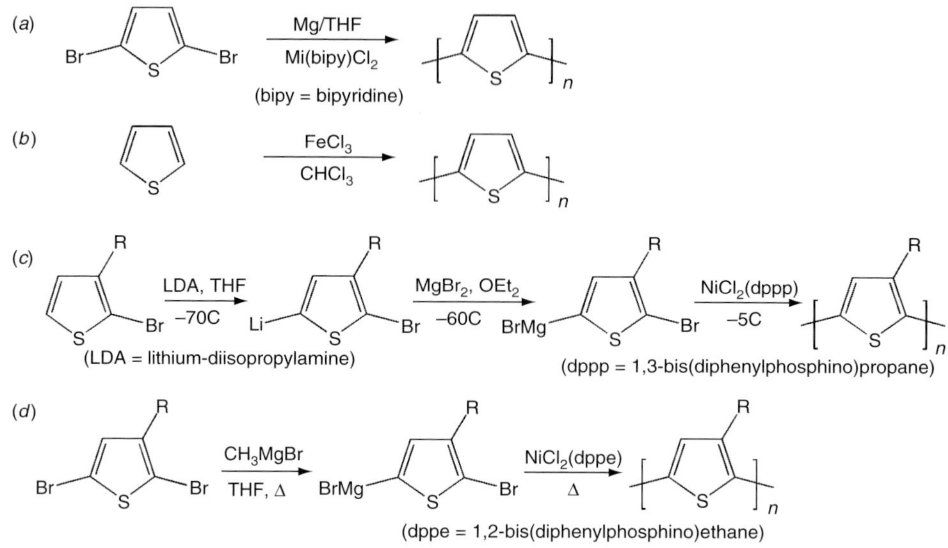

Fig. 9.4 Examples of the synthesis of PTh, (a) and (b), and poly(alkylthiophenes), (c) and (d), where R is an alkyl group.

An important aspect of the synthesis of poly(alkylthiophenes) and other mono-substituted PThs is the regularity of the polymer structure. Ideally, the polymer should have a regio-regular structure, i.e. have only *head-to-tail* (HT) coupling of monomers, see Section 1.2.4. Fortunately, the syntheses automatically favour the production of regio-regular polymer. The polymerisation of 3-(4-octylphenyl)-thiophene by the slow addition of $FeCl_3$ gives a product with 94% HT coupling, although this method produces some crosslinks through the alkyl chains. Polymerisation from a di-halogenated monomer via a Grignard intermediate produces regio-irregular polymer. Regio-regular polymers with an alternation of *head-to-head* (HH) and *tail-to-tail* (TT) units can be formed by polymerising dimers with either an HH or TT structure; providing the monomers couple adjacent to the sulphur site the product must comprise HH–TT repeat units. Polymer containing 98–100% HT units can be prepared by the polymerisation of 2-bromo-5-(bromomanganesio)-thiophene, Fig. 9.4(c). The preparation of the monomer requires cryogenic temperatures and extremely pure starting materials but gives a polymer with a conductivity on doping of $10^5\,\Omega^{-1}\text{m}^{-1}$ compared with $2\times10^3\,\Omega^{-1}\text{m}^{-1}$ for a polymer with a random backbone structure. Alternative methods include the use of specially prepared catalyst and the coupling of a 4:1 mixture of regular 2-bromo-5-(bromomanganesio) and irregular 2-(bromomanganesio)-5-bromo isomers which favours the reaction of the regio-regular monomers, Fig. 9.4(d). The

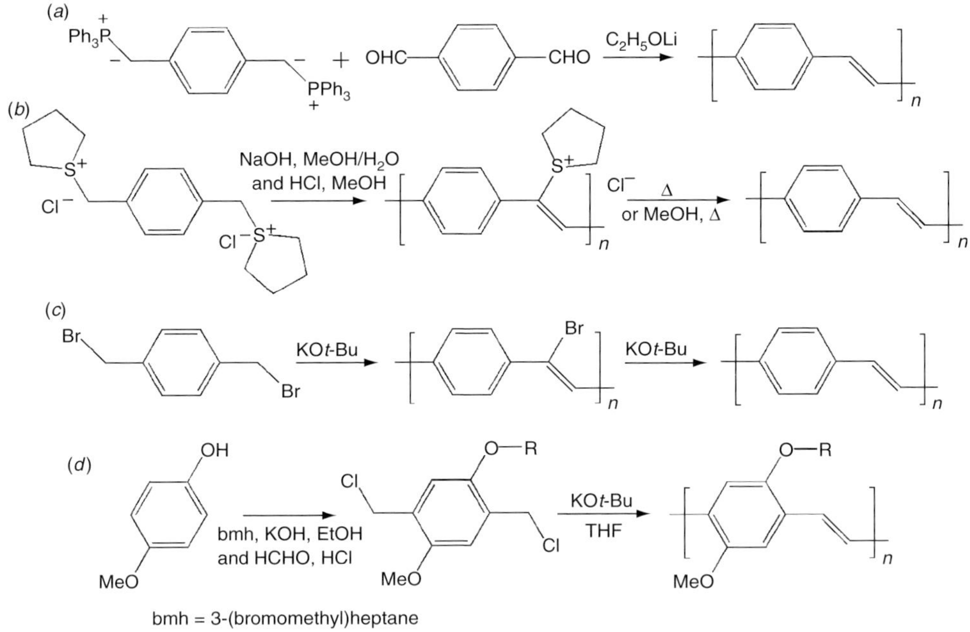

Fig. 9.5 Synthetic routes to PPVs. R is a long chain substituent: in MEH–PPV R is an ethylhexyl branched chain.

latter gives 99% HT polymer with high molecular mass (200–300 repeat units) in 60–70% yield. The improved control over polymer structure and purity is reflected in the increased conductivity after doping.

The introduction of bridging groups on the thiophene ring modifies the physical and chemical properties of the polymers obtained. The energy of the optical absorption is reduced in PEDOT, Fig. 9.2(k), and poly(*iso*thianaphthalene), (PITN) (Wudl *et al.*, 1984), so that in the conductive state thin films are transparent. PEDOT shows high electrochemical stability in the oxidised state and, when combined with poly(styrenesulphonic acid) counter ions, can be processed from aqueous solution.

(c) Poly(p-phenylenevinylenes) Initial efforts to prepare PPV were handicapped by the poor solubility of the polymer. Because of this the products of the coupling of the bisphosphonium salt of *p*-xylene and the dialdehyde, Fig. 9.5(a), and the ROMP of *p-cyclo*phane were low molecular mass oligomers. The search for a precursor route to PPV in the 1980s led to the rediscovery of the method developed by Wessling and Zimmerman at the Dow Chemical Company in 1968. In this method the phenyldisulphonium salt is polymerised to produce a soluble, sulphonium-substituted polymer,

which can be converted to PPV by either heating at 473–573 K in vacuum or treating with methyl alcohol and heating under HCl at 293 K, Fig. 9.5(b). Similar routes have been developed, such as the coupling of halogenated *p*-xylenes, to make a brominated precursor with subsequent debromination, Fig. 9.5(c). Cho (2002) reviewed the chemistry of precursor routes to PPV and related polymers.

The discovery of electroluminescence fuelled interest in PPV, and the precursor route was extensively used to fabricate thin-film light emitting devices (LEDs). Unfortunately, the by-products generated during the conversion of the precursor polymer were difficult to remove and interacted with the electrodes in the device. Subsequent introduction of long-chain alkyl substituents provided soluble polymers. While these can be obtained in high molecular mass by the halogenated precursor route, the direct coupling of di-halogenated disubstituted monomers provides a simpler synthesis. The inclusion of branched alkyl substituents, as in MEH–PPV, Fig. 9.5(d), enhanced solubility further. This route has been used to prepare PPVs with a variety of substituents in the phenyl ring. Introduction of a cyano substituent on the vinylene group can be achieved by the coupling reaction similar to that of Fig. 9.5(a). A dicyano-substituted monomer replaces the bisphosphonium salt and the inclusion of alkyloxy substituents on the rings of both monomers ensures a soluble product. Such substitution increases the electron affinity of the PPV and this is beneficial in the context of LEDs.

While the PPVs produced by these routes are primarily HT, defects in the backbone occur due to HH coupling giving a tolane bis-benzyl moiety which contains a non-conjugated link breaking the backbone conjugation. In the reaction shown in Fig. 9.5(d) polymerisation occurs via an intermediate in which one Cl atom has been removed. A regio-regular polymer requires that the same Cl atom be removed from all the intermediates, otherwise HH linking can occur. Replacement of the –OR group by a 3-alkyloxy substituted aryl group achieves this, leading to $> 99.5\%$ HT coupling. While this does not have any major effect on the physical properties of the polymer, there is a pronounced increase in the stability of the polymer. This reflects the fact that non-conjugated sites are more readily attacked by triplet oxygen generated in LEDs. Covion Organic Semiconductors, Frankfurt, implemented this synthesis on an industrial scale, in order to meet the need for volume production of high quality materials for use in commercial LEDs (Becker *et al.*, 2001).

(d) Polyaniline PAni is readily obtained by oxidative coupling, a fact that accounts for its discovery in the nineteenth century. The chemistry of the polymer is somewhat more complicated than for other conjugated polymers,

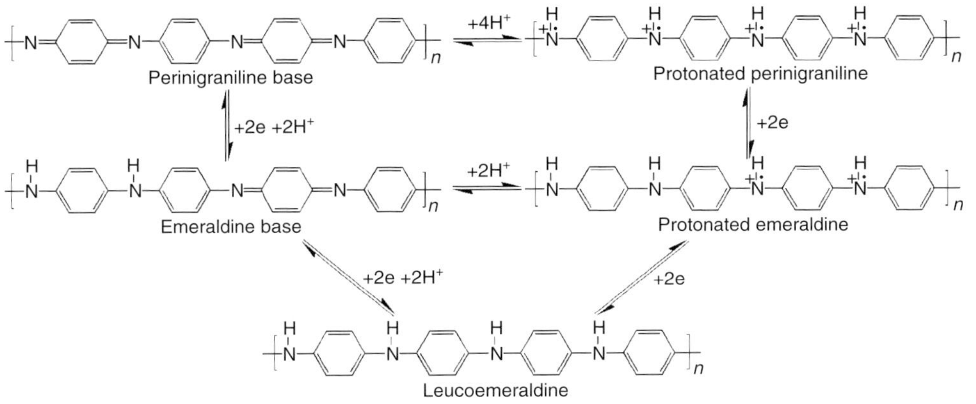

Fig. 9.6 The molecular structures of the main oxidation and protonated states of PAni; the addition of electrons and protons causes the structure to change in the direction of the immediately adjacent arrow.

since it can exist in fully or partially oxidised forms. The chemical structures of the main forms are shown in Fig. 9.6. In the fully reduced form, leucoemeraldine (LM), the nitrogen atoms are amines, while in the fully oxidised form, perinigraniline (PNA), all the nitrogen atoms are imines. The 50% oxidised material, emeraldine (EM), contains equal numbers of amine and imine units, each containing two nitrogen atoms, see Fig. 9.6. There is a series of intermediate oxidation states for defined ratios of the basic units, e.g. a 3:1 ratio of amine to imine units that is termed nigraniline. The partially or fully oxidised forms can be protonated to give the forms on the right of Fig. 9.6. Because of this possibility the non-protonated forms are usually referred to as EM base and PNA base.

When interest was rekindled in the 1980s, the method commonly used to synthesise PAni was the oxidative coupling of aniline with ammonium persulphate in aqueous HCl. This produces partially protonated EM salt which can be deprotonated with ammonia to form EM base. The molecular mass can then be determined in dilute LiCl solution in N-methylpyrrolidone (NMP); typical values correspond to $\sim$160 of the repeat units shown in Fig. 9.6. The reaction shows the characteristics of a *living* polymerisation, since the molecular mass increases if further monomer and oxidant are added to the reaction mixture, when the chain length increases to a maximum of about 240 repeat units. NMR studies of the LM form show about 5% defects in the polymer backbones. When the reaction is performed at 248 K the polymer is produced in nearly 100% yield, with higher molecular mass, typically over 500 repeat units, and greatly reduced defect content (Adams *et al.*, 1996). Other oxidising agents can be employed in the preparation of PAni; Gospodinnova and Terlemezyan (1998) present an extensive review.

The solubility of the electrically conductive form, the protonated EM salt, is poor in common solvents, and so, initially, EM base was used to form films from NMP solution before conversion to the salt with HCl. Attempts to increase solubility by incorporating pendent alkyl chains were counter-productive as such polymers had significantly lower electrical conductivity. Better solubility is obtained for salts produced with sulphonic acids. Thus the camphor sulphonic acid salt dissolves in *m*-cresol and the 2-acrylamido-2-methyl-1-propanesulphonic acid salt dissolves in dichloro-acetic acid. The use of these systems, and polymer synthesised below room temperature, has facilitated the spinning of fibres with oriented polymer chains and conductivity along the fibre as high as $2\times10^5\,\Omega^{-1}m^{-1}$ (Pomfret *et al.*, 1998).

Electrochemical synthesis

The electrochemical oxidation of a wide range of conjugated monomers results in the deposition of an electrically conductive film on the working electrode (Waltman and Bargon, 1986). This method has mostly been used for the polymerisation of heteroatom-containing monomers, e.g. pyrrole, thiophene, aniline, indole, etc., and poly*cyclic* monomers, e.g. azulene, fluorene, pyrene, etc. The example of pyrrole is illustrative of the inherent advantages and problems of electrochemical polymerisation.

Figure 1.6 illustrates the mechanism of the polymerisation of pyrrole. Above the oxidation potential, removal of an electron generates radical cations that couple to form dihydro-dimers. Subsequent deprotonation produces the dimer, which can be oxidised and which may then react with monomer and dimer radical cations to produce higher homologues, the process being repeated until polymer is deposited on the anode. This process can occur because the oxidation potential of the monomer is higher than that of the intermediate oligomers and of the polymer. The polymer film is deposited in an oxidised state, electrical neutrality being maintained by incorporation of counter ions in the film from the electrolyte. The voltammograms of Fig. 9.7 show polymerisation occurring at a potential of about 1 V and the oxidation and reduction peaks of the polymer film growing during successive cycles as the film thickness increases.

Electrochemical polymerisation offers a number of advantages. The apparatus is simple, a cell with working, counter and reference electrodes operated at either constant potential or constant current. Insoluble polymers can either be prepared as films directly on to a variety of electrodes or be incorporated into a porous layer situated on the working electrode. The film can be reduced and

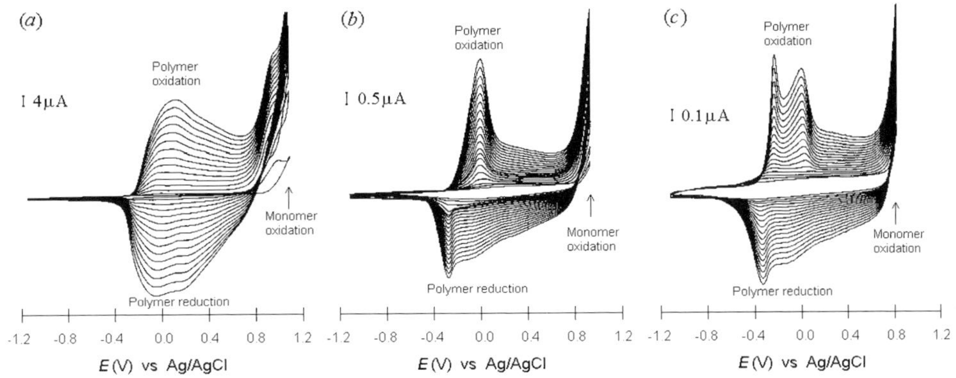

Fig. 9.7 Cyclic voltammograms for the polymerisation of pyrrole with sweeps between (a) −1.12 and +1.13 V, (b) −1.12 and +0.93 V and (c) −1.12 and 0.83 V versus saturated calomel reference electrode; for other details see text. Reprinted from Zhou and Heinze (1999). Copyright 1999, with permission from Elsevier.

oxidised *in situ*, a process which enables the counter ions incorporated during growth to be removed and different counter ions to be inserted. The growth conditions can be easily varied by choice of solvent, working electrolyte, i.e. counter ion, and either polymerisation potential or current. Unfortunately, this flexibility has inherent problems, because the micro- and macro-morphology of the polymer and its properties are strongly dependent on the polymerisation conditions.

The electrochemical polymerisation of pyrrole has been extensively studied since the late 1970s. Despite this, details of the polymerisation process and its effect on the properties of the polymer are still being reported. Choice of electrode material, counter ions and growth conditions are all-important. PPy is deposited on platinum from a tetraethylammonium toluenesulphonate-propylene carbonate electrolyte as a smooth continuous amorphous film that can be grown to millimetres in thickness with an electrical conductivity up to $5\times10^3\,\Omega^{-1}\mathrm{m}^{-1}$, a Young's modulus of 4 GPa, maximum tensile strength of 50 Mpa and good electrochemical stability. 90% of the initial electrochemical activity is retained after more than 1500 oxidation–reduction cycles. Polymerisation below room temperature produces more flexible films that can be stretched to enhance the conductivity in the direction of stretching. The oxidation potential of pyrrole is higher on titanium, iron and aluminium than on platinum. Chemical analysis, Raman spectroscopy, cyclic voltammetry, conductivity measurements and electron microscopy of the films show that there are significant differences in chain structure and morphology. The conductivities are lower and the films are brittle, with inferior mechanical properties, and contain crystalline second phases (Cheung *et al.*, 1988). The use of different counter ions produces similar variations. The oxidation and reduction potentials have separations which vary from 0.27 to 1.09 V and can show multiple peaks. The electronic and vibrational spectra show marked differences, as does the electrochemical stability (Cheung *et al.*, 1990). Similar results have been found for growth from tetrabutyl-ammonium hexafluorophosphate-acetonitrile-1% water electrolyte at 253 K for small changes in oxidation potential, Fig. 9.7 (Zhou and Heinze, 1999). A well-defined polymer with sharp oxidation and reduction peaks can be obtained by growth at very low currents.

These findings reflect the fact that the polymerisation scheme shown in Fig. 1.6 is an idealised oversimplification of a more complex process. The electrochemical reaction may proceed by other than coupling via the carbon atoms adjacent to the nitrogen atom (α–α' linkages); it may also involve the more remote carbons (α–β and β–β linkages), and crosslinks via carbon or nitrogen atoms. Aromatic species are found in polymer produced at very high current densities. Use of oligomers and substituted monomers, where the substituents prevent reaction at the substituted sites, e.g. the β-carbons and the nitrogen, reduce the level of defects in the polymer. The use of substituted monomers modifies polymer properties and introduces new functionality (Higgins, 1997).

Electrochemical polymerisation provides a facile method for the preparation of films of insoluble polymers for both fundamental and applied research. However, it is difficult to ensure that there is a single reaction leading to a unique polymer. The polymers are more often complex, crosslinked materials that may be suitable for practical applications, but are difficult to characterise. Because of the complexities of both the electrochemical polymerisation and the structure and properties of the polymers so produced, these topics will continue to be the subjects of research.

9.3 Theories of electronic properties

After the discovery of the metallic behaviour of PAc (Shirakawa *et al.*, 1977) most of the work on theoretical models concentrated on this polymer. This built on the earlier studies of Longuet-Higgins and Salem (1959) and others who used the Hückel model to demonstrate that the backbone of PAc in its ground state has a bond alternated structure, Fig. 9.8(b), rather than one with equal length C–C bonds, Fig. 9.8(a). In 1979 a Hückel tight binding model was introduced that provided the basis for much of the subsequent discussion of the molecular and electronic structure of PAc (Su, Schrieffer and Heeger, 1979 and 1980). It is now usually referred to as the SSH model. In the adiabatic

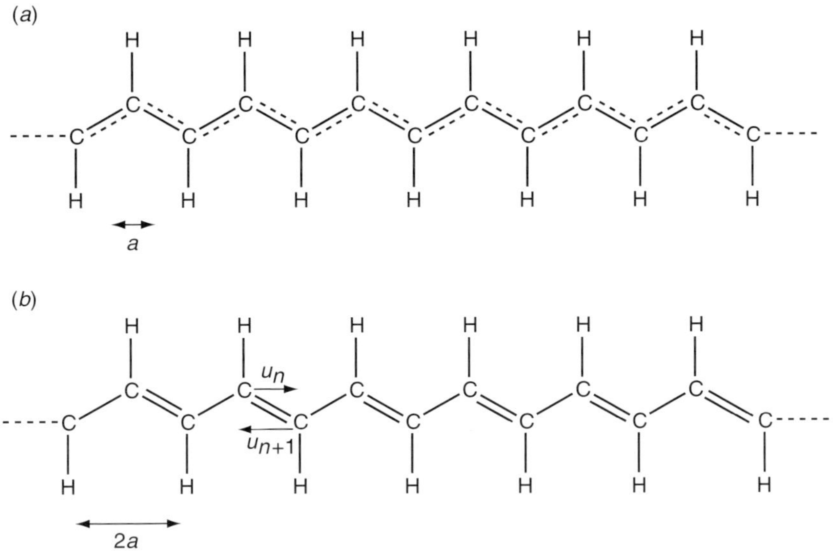

Fig. 9.8 Chemical structure of *trans*-polyacetylene with (a) equal bond lengths and (b) bond alternation.

approximation, i.e. ignoring lattice vibrations, the SSH Hamiltonian contains terms describing the electronic energy of the π-electrons and the elastic energy of the polyacetylene chain:

$$H = H_{\text{elect}} + H_{\text{elast}}, \tag{9.1}$$

where:

$$\begin{aligned} H_{\text{elect}} &= -\sum_{n,s} t_{n+1,n}(c^{+}_{n+1,s}c_{n,s} + c^{+}_{n,s}c_{n+1,s}) \\ H_{\text{elast}} &= \tfrac{1}{2}\sum \kappa(u_{n+1} - u_n)^2. \end{aligned} \tag{9.2}$$

In the electronic Hamiltonian $t_{n+1,n}$ is the transfer integral, i.e. the π-electron wavefunction overlap between nearest neighbour sites in the polymer chain, and is equivalent to the parameter β in Equation (4.20), and $c^{+}_{n+1,s}$ and $c_{n,s}$ are creation and annihilation operators that create an electron of spin s ($\pm 1/2$) on the carbon atom at site $n+1$ and destroy an electron of spin s at the carbon atom on site n, i.e. in effect transfer an electron between adjacent carbon atoms in the polymer chain. The elastic term is just the energy of a spring of force constant κ extended by an amount $(u_{n+1}-u_n)$, where the u are the displacements along the chain axis of the carbon atoms from their positions in the equal bond length structure, as indicated in Fig. 9.8(b). The extent of the overlap of π-electron wavefunction will depend on the separation of nearest neighbour carbon atoms and is approximated by:

$$t_{n+1,n} = t_0 - \alpha(u_{n+1} - u_n), \tag{9.3}$$

where t_0 is the transfer integral for the equal bond length structure and α is a constant.

The electronic states for a given lattice configuration are given by the solution of H_{elect}. In the limit of equal bond lengths, we can use the result already obtained for the one-dimensional, tight binding model, Equation (4.22), to write down the result for the dispersion of the electron band as:

$$E(k) = E_0 - 2t \cos ka. \tag{9.4}$$

As noted in Chapter 4, the Peierls distortion leads to the dimerised structure, Fig. 9.8(b), which can be described by displacements $+u_0$ and $-u_0$ of the nth and $(n+1)$th atoms, where n is an even integer. The transfer integrals for the long and short bonds are then:

$$t_l = t_0 - 2\alpha u_0, \; t_s = t_0 + 2\alpha u_0, \tag{9.5}$$

respectively. Then following Equations (4.20) and (4.21) leads to the result for the k-dependent part of the energy:

$$E(k) = t_0(e^{ika} + e^{-ika}) + 2\alpha u_0(e^{ika} - e^{-ika}) = 2t_0 \cos ka + 4i\alpha u_0 \sin ka.$$

Hence:

$$E^2(k) = 4t_0^2 \cos^2 ka + \Delta^2 \sin^2 ka, \tag{9.6}$$

where $\Delta = 4\alpha u_0$. The bands obtained from this equation shown in Fig. 9.9 are similar to the idealised bands illustrated by the dotted curves in Fig. 4.7. The overall bandwidth of the valence and conduction bands is $4t_0$ and the band gap at $k = \pi/2a$ is 2Δ.

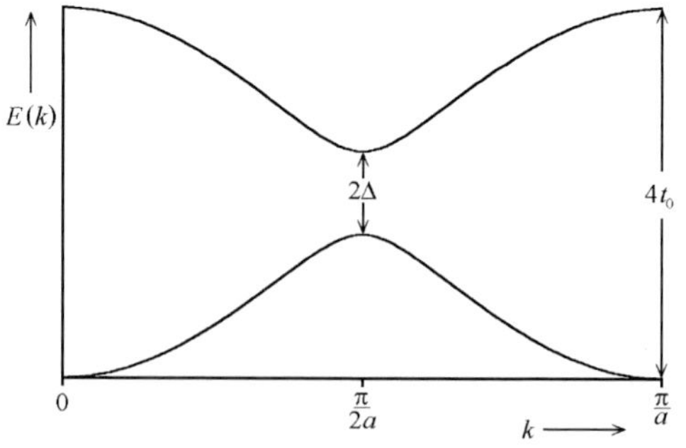

Fig. 9.9 Band structure for a dimerised polyacetylene chain calculated from the SSH Hamiltonian.

The total energy of the polymer is then given as the sum of the elastic and electronic energies. The former is obtained directly from H_{elast} as $2\kappa u^2$ per site, i.e. for a single value of n. The electronic energy is obtained by the product of the energy of the electronics states multiplied by their occupancy. Alternatively, the total electronic energy per site is obtained from the average energy over the first Brillouin zone, i.e. by integrating $E(k)$ from $ka = 0$ to $ka = \pi/2$ and dividing by $\pi/2$. The factor $\cos^2(ka)$ can be eliminated from Equation (9.6) by using the result that $\cos^2(ka) + \sin^2(ka) = 1$, leading to the following expression for $E_0(u)$:

$$E_0(u) = -\frac{4t_0}{\pi}\int_0^{\pi/2} [1 - (1 - z^2)\sin^2 ka]^{1/2} d(ka) + 2\kappa u^2, \tag{9.7}$$

where $z = 2\alpha u/t_0$. The integral is the elliptic integral that for small values of z is approximated by the expression:

$$\xi(1 - z^2) = 1 + \tfrac{1}{2}\left[\ln\left(\tfrac{4}{|z|}\right) - \tfrac{1}{2}\right]z^2 + \cdots \approx 1.$$

The total energy is then given approximately by:

$$E_0(u) \approx -\frac{4t_0}{\pi} + \frac{\kappa t_0^2}{2\alpha^2}z^2. \tag{9.8}$$

Hence for small z and u, the total energy is negative, i.e. the reduction in the electronic energy of the π-electrons is greater than the increase in elastic energy of the σ-electrons, but becomes positive for large values of u. Thus $E_0(u)$ has a minimum at u_0, i.e. at the equilibrium dimerised structure. This is illustrated in the left-hand side of Fig. 9.10. Since the total energy is dependent on u^2 the

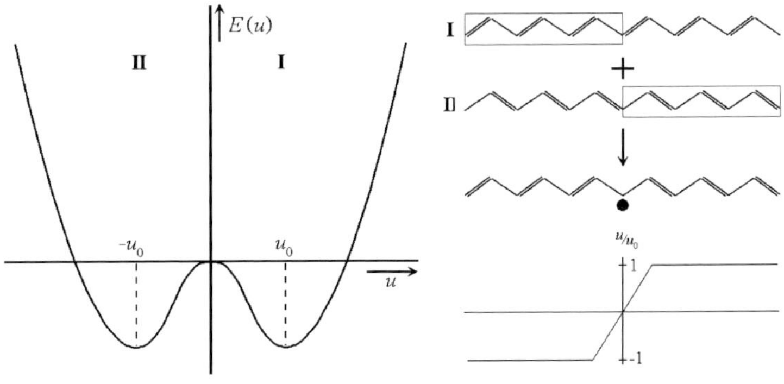

Fig. 9.10 SSH ground state energy of a polyacetylene chain, at left, and the consequent occurrence of a neutral kink (a soliton) in an infinite chain.

diagram is symmetric about $u = 0$ and two energy minima are found at $u = \pm u_0$. This reflects the fact that for an infinitely long chain it does not matter whether the short bond in the dimerised structure is between the nth and $(n+1)$th sites or the nth and $(n-1)$th sites. Thus the energy is not affected by the phase of the bond alternation, i.e. whether $u = -u_0$ or $u = u_0$. The two possible phases of bond alternation gives rise to two alternative ground state structures for the polymer, shown as I and II in Fig. 9.10, that are degenerate in energy. These structures are characterised by a bond alternation parameter, u/u_0, which takes the values ± 1. Note that a bond alternation parameter of zero corresponds to the equal bond length structure. If two portions of a chain have opposite bond alternation parameters a bond alternation defect will occur at the interface where the bond alternation parameter changes sign, as shown on the right-hand side of Fig. 9.10. At this interface where there is a kink in the bond alternation parameter there will be an unpaired π-electron, which is not bound up in the double bonds of either portion of the chain. The polymer chain is electrically neutral but the electron at the interface has an unpaired spin. The π-electrons in the double bonds are spin paired so that the perfect polymer chain is diamagnetic while the chain with the kink has a small but finite paramagnetism. As the energy of the chain on either side of the kink is equal it can move along the chain without any overall change in energy. Thus the unpaired electron should be a mobile species able to move along the polymer chain freely. From the similarity in the theoretical description of the kink to that of solitary water waves it was termed a *soliton*.

If cyclic boundary conditions are used to model the infinite chain, then for an odd number of carbon bonds in the ring a soliton must occur since the first and last bond must be both single bonds or both double bonds. Conversely, if the ring has an even number of bonds, the boundary conditions are satisfied, and a soliton cannot appear on the chain. However, two solitons of opposite spin, a soliton and an anti-soliton, can be created on an even ring, as the change in bond alternation due to the anti-soliton cancels out that due to the soliton. These arguments can be applied to finite length chains where the terminating groups fix the phase of the bond alternation at the chain ends. Thus the ground states of finite length chains with opposite phases of bond alternation at the ends must contain a soliton, while if the phases of bond alternation at the ends are the same the ground state will not contain a soliton.

The *soliton* concept has been central in the discussion of the properties of polyacetylene, and an extensive theoretical framework has been developed. As described in Section 4.2.5, the concept did not originate with SSH, having been discussed earlier by Pople and Walmsley (1962). Their model involved a rapid change in the bond alternation in the vicinity of the soliton, as shown in Figs. 4.15 and 9.10. In the SSH model the soliton has a much larger spatial extent. The lattice displacement and the electronic wavefunction of the soliton have been shown to be:

$$
\begin{aligned}
u_n &= u_0 \tanh\left[\frac{(n-n_0)a}{\xi}\right] \cos n\pi, \\
\psi_n &= \frac{a}{\xi} \operatorname{sech}\left[\frac{(n-n_0)a}{\xi}\right] \cos\frac{n\pi}{2},
\end{aligned}
\tag{9.9}
$$

where the soliton is centred at site n_0, and ξ is a parameter that characterises the spatial extent of the soliton. Numerical calculation yields a value of $7a$ for the parameter ξ. u/u_0 and the electron distribution then have the form shown in Fig. 9.11. The non-bonded electron is no longer associated with a single carbon atom as in Fig. 9.10, but is delocalised over a short section of the chain, Fig. 9.11(c). The wavefunction has the unusual property that it is finite on the even n sites but is zero on the odd n sites along the chain, Fig. 9.11(b).

The gradual changes in bond length associated with the extended soliton mean that displacement of the soliton along the chain by one repeat unit only requires small changes in bond lengths. The activation energy for such motion is then calculated to be of the order of 0.04 meV, i.e. a thermal energy of 1 K.

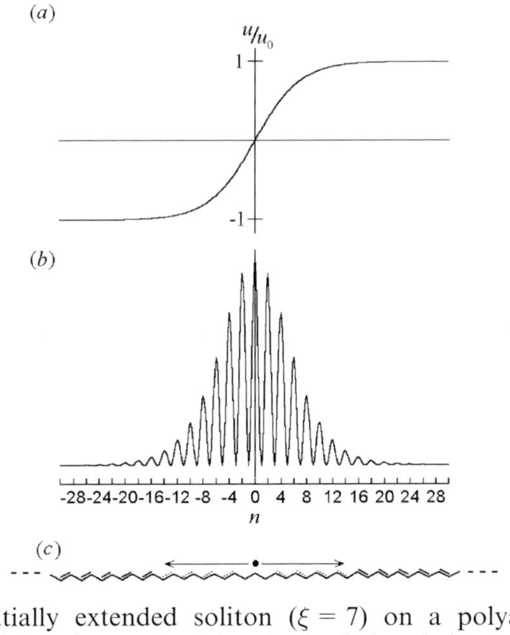

Fig. 9.11 Spatially extended soliton ($\xi = 7$) on a polyacetylene chain centred on the 30th carbon atom, (a) variation in bond alternation parameter, (b) the electronic wavefunction and (c) bonding structure.

Thus the soliton is predicted to be highly mobile on a perfect chain at room temperature. In contrast, the rapid change in bond length associated with the soliton of Fig. 9.10 requires an activation energy for the displacement of the order 40 meV, i.e. a thermal energy of 1000 K, and such a soliton would be immobile at room temperature. The spatial extent of the soliton has less impact on the energy required to create a soliton, or soliton/anti-soliton pair, on an undeformed chain. The creation energy of a single soliton is found by numerical calculation to be 0.6Δ. Experimentally, the energy gap in polyacetylene is 1.4 eV so that the creation energy is $\sim$0.4 eV. This corresponds to a temperature of the order of 10 000 K, so that one cannot therefore expect a thermally excited population of solitons to be present in pure polyacetylene at room temperature; their presence in significant quantities will depend on other external factors.

9.3.1 Doping of conjugated polymers

The electron associated with the soliton is in chemical terms a non-bonded electron that does not participate in either the valence or conduction band states. The soliton is a zero energy state of the chain with an energy level lying at the centre of the energy gap between the band states. The soliton wave function can be decomposed into equal contributions from the states at the edges of the valence and conduction bands. When polyacetylene is *doped* by the addition of strong electron-accepting or -donating chemical species, the creation of charged soliton states will require less energy than the population of band states, and the whole soliton situation is profoundly changed. If an electron is donated to the polymer, the lowest energy level it can occupy is the soliton level, which can incorporate two electrons with opposite spin. Similarly, an acceptor can more easily remove the electron from the soliton level than a valence band state. Thus, in addition to the neutral soliton it is possible to have charged solitons. The different states of occupancy of the soliton level have unusual combinations of spin and charge:

i. the singly occupied, neutral soliton has zero charge but has spin ½,
ii. the doubly occupied soliton has charge $-e$ and spin 0, and
iii. the unoccupied soliton has charge $+e$ and spin 0.

These solitons are illustrated in Fig. 9.12. The mobility of the soliton along the polyacetylene chain means that the charged solitons (ii) and (iii) can behave as spin-less charge carriers. These are distinctly different objects from the electrons and holes found in conventional semiconductors, which carry charge and spin. Thus, while bond-alternated polyacetylene has the band structure of

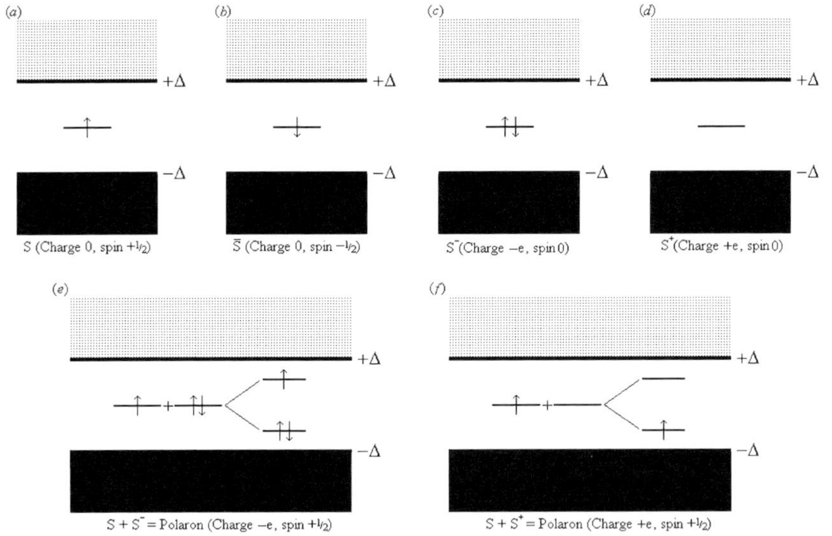

Fig. 9.12 Energy level schemes of solitons and polarons in polyacetylene; (a) soliton, (b) anti-soliton, (c) negative soliton, (d) positive soliton, (e) negative polaron and (f) positive polaron.

a semiconductor, the underlying physics describing its properties is **not** the same as that of an inorganic semiconductor. We should also note that in chemical terms the neutral, negative and positive solitons would be denoted as a radical, radical anion (or carbanion) and radical cation (or carbocation), respectively.

Although added charge can interact with neutral solitons already present in the polyacetylene, if none are available rather than generating a separated soliton/anti-soliton, it forms a lower energy state from a neutral and a charged soliton that are bound to form a polaron. As shown in Fig. 9.12(e) and (d), when a charged and neutral soliton are brought together the interaction of the degenerate energy levels gives rise to two separated levels, just as bonding and anti-bonding levels are formed in a hydrogen atom, Fig. 1.1. The polaron does not affect the phase of the bond alternation and has similar properties to the polaron in conventional semiconductors, i.e. the charge is localised in a region where the lattice is deformed. The polaron is formed by the superposition of the two solitons and can be expressed formally as:

$$u_{\mathrm{p}}(x) = u_0\{1 - \kappa\xi(\tanh\{\kappa(x + x_0)\} - \tanh\{\kappa(x - x_0)\}\}, \qquad (9.10)$$

where κ is the wave vector associated with the polaron and the parameter x_0 is given by the solution of the equation:

$$\tanh 2\kappa x_0 = \kappa\xi. \tag{9.11}$$

The similarity of the deformation of a single soliton and the components of the polaron can be seen by comparing Equations (9.9) and (9.10). The distortion associated with the polaron has the form of a shallow well when $\kappa\xi$ is close to zero and that of a well-separated soliton and anti-soliton as $\kappa\xi$ tends to one, as illustrated in Fig. 9.13. The stable polaron is calculated to have a maximum distortion of $0.41u_0$, which corresponds to the second curve from the top of Fig. 9.13, and a binding energy of 0.07 eV. This is the lowest binding energy state for a single added electron. Comparison with Fig. 9.11 shows that the polaron has a slightly larger spatial extent than that of the soliton.

If two electrons are injected locally on to the polymer chain, the polaron solution is no longer stable. The excess energy of two electrons at the band edge is sufficient to produce a separated soliton and anti-soliton pair, as shown in Fig. 9.13. Model calculations including the lattice dynamics show that this separation into a pair of solitons will occur on a timescale of 10^{-13}s. The same process will result from the photo-excitation of an electron and a hole. This follows since the excess energy of two electrons at the conduction band edge and an electron and a hole at the conduction and valence band edges, relative to two separated soliton states at zero energy, is the same (2Δ). The timescale for the separation of the electron–hole pair into a soliton and an anti-soliton is

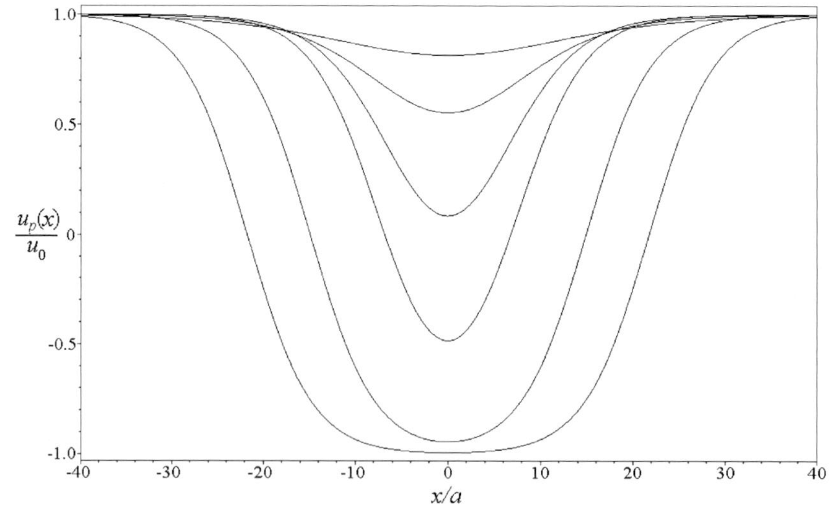

Fig. 9.13 Lattice distortion associated with a polaron on a polyacetylene chain for values of the parameter $\kappa\xi$ varying from 0.42 to 0.999992 from the uppermost to lowest curves respectively.

also predicted to be $\sim 10^{-13}$s. Thus, theoretically solitons can be generated by chemical doping and photo-excitation.

The simple SSH model considers that electron–electron interactions are negligible, i.e. it neglects electron correlation. This is seen in the result quoted in the preceding paragraph where the Coulomb interaction has been neglected. The simple SSH model is said to be charge conjugation invariant since changing the sign of the charge carrier does not affect the result. If the electron–electron interaction is included it results in the appearance of an energy gap even in the un-dimerised polyacetylene chain. In the dimerised chain both the electron–electron and electron–phonon interactions contribute to the energy gap. If the principal contribution still arises from the lattice distortion, the picture presented above is modified and the energy levels of solitons and polarons are no longer symmetrically placed about the middle of the energy gap. Nevertheless, the basic concepts remain valid in this situation.

Using this model, the doping of polyacetylene proceeds by the interaction of dopants with solitons initially present in the polymer, then by polaron formation and at even higher doping levels by soliton formation. The soliton states then broaden into a soliton band that eventually fills the gap, giving a continuum of energy levels and a metallic state.

The situation in most conjugated polymers is different from that encountered on *trans*-polyacetylene, since, although alternative conjugated structures are possible, they are not energetically degenerate. This is neither the case for *cis*-polyacetylene, due to the steric interaction of the pendent hydrogen atoms, nor for poly(*p*-phenylene), where benzenoid and quinonoid structures are possible. The absence of energy degeneracy means that such polymers cannot sustain solitons, since it is energetically unfavourable for an extended section of polymer to be in the higher energy structure. Thus, if a single, soliton-like boundary were to occur, it would be energetically favourable for it to be ejected from the chain, returning the chain to its lower energy form. It is, however, possible for polarons to form on the chain, since the deformation is localised and the distortion of the lower energy structure is relatively small. Some examples are illustrated in Fig. 9.14. Most conductive polymers result from acceptor doping and the formation of positive polarons, since the negatively charged counterparts are chemically unstable. These polarons, or radical-cations, have conventional charge–spin relationships. It is also possible to form bipolarons, or dications, in which the Coulomb repulsion of two positively charged polarons can stabilise a short length of the higher energy polymer structure, see Fig. 9.14. For such polymers heavy doping can be viewed as producing polarons and bipolarons that interact to form bands that eventually fill the intrinsic band gap, leading to a metallic state.

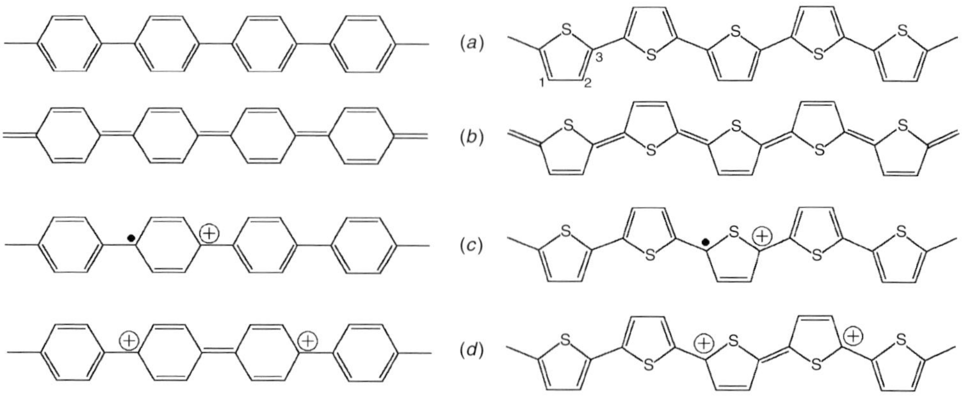

Fig. 9.14 Two alternative bonding structures (a) and (b), polaron (c) and bipolaron (d) for poly(*p*-phenylene) and polythiophene.

The literature contains numerous model calculations of the band structures of ideal conjugated polymer chains with both known and conjectured structures. Versions of most of the quantum mechanical methods described in the introduction to this section have been used. We will not attempt to review all this material but confine our comments to a few specific examples. The VEH method has been applied to a wide range of conjugated polymers (Brédas, 1986). It has been specifically used in an effort to identify conjugated polymers with small or even zero band gaps, i.e. materials that would behave as intrinsic semiconductors or metals without the need for chemical doping. We have seen that for polyacetylene the SSH model predicts a metallic state if the bond lengths are equal, i.e. a combination of equal contributions from the structures I and II shown in Fig. 9.10. The situation is less straightforward for other polymers. In general the equilibrium structure of a conjugated polymer lies somewhere between the limiting aromatic and quinoid forms, (a) and (b) of Fig. 9.14, and can be described as an admixture of the alternative bonding structures. The results of VEH calculations for polythiophene predict that the band gap decreases as the quinoidal contribution increases (Brédas *et al.*, 1986). The theoretical values are as shown in Fig. 9.15 plotted as a function of the difference in bond lengths between the bonds 1–2 and 2–3 in Fig. 9.14. This is a consequence of the fact that the electron distribution in the ground state (HOMO) of the aromatic form is the same as that of the first excited state (LUMO) of the quinoidal form and vice versa, so that the levels will merge for a particular admixture of the alternative bonding structures. Complete closure of the band gap will not occur even under neglect of electron–electron interactions since the crossing of the HOMO and LUMO is forbidden by

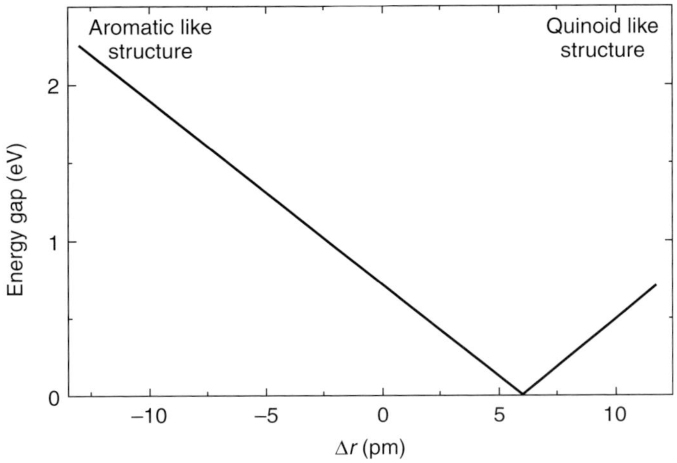

Fig. 9.15 VEH calculated band gap for polythiophene as a function of difference in bond lengths (as defined in the text). Reproduced with permission of the American Institute of Physics from Brédas *et al.* (1986).

symmetry. The reduction of the energy gap is seen in polyisothianaphthalene, see Fig. 9.16(a), where the VEH calculation predicts a reduction in band gap energy from 1.71 eV for polythiophene to 0.54 eV. VEH calculations for substituted poly*iso*thianaphthalenes gave similar band gaps, for instance the dicyano analogue, see Fig. 9.16(b), has a predicted band gap of 0.55 eV. Smaller theoretical band gaps have been calculated for model structures, such as that shown in Fig. 9.16(c), where a value of 0.16 eV is obtained (Toussaint and Brédas, 1995) and a monomer comprised of the moieties shown in Fig. 9.16(c) and (d) (Salzner and Köse, 2002). Although experimentally observed optical and electrochemical energy gaps follow the trends indicated by the theoretical values, they are generally somewhat larger. The inclusion of electron correlation, and a localised first excited state, will inevitably lead to larger values for the calculated band gaps, closer to those observed experimentally.

9.3.2 *Disordered conjugated polymers*

A frequently asked question is *what is the relevance of band structure calculations to real polymers?* For, however sophisticated the calculation is, it is only performed on an ideal structure. The actual polymer materials in question have much more complex structures with a high density of defects, as described in Chapter 1. Two basic models have been used to try to take real structures into account and to model the consequences of defects on single

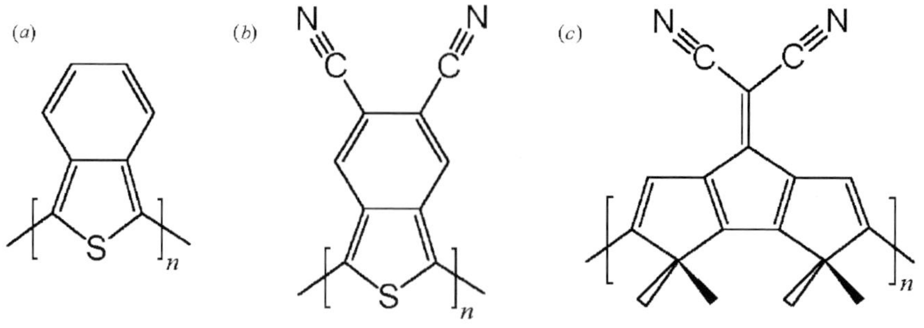

Fig. 9.16 Structures of (a) polyisothianaphthalene, (b) the dicyano substituted analogue and (c) a model small band gap polymer.

conjugated polymer chains. The first is the Kuhn model, in which the chain is considered to consist of short, perfectly straight and flat sections, separated by local chain deformation that *breaks* the conjugation, Fig. 9.17(a). The second is the worm-like chain, in which the inherently stiff, but deformable, polymer chain is considered to undergo gradual changes in structure, Fig. 9.17(b). The Kuhn model is characterised by the length distribution of isolated, perfect-chain sections, while the worm-like chain is characterised by the length over which a similar structure is maintained, the persistence length. The simplest quantum mechanical model that can describe the variation in the energy levels of short conjugated chains is that of an electron in a deep potential well. The analytical solution for the electronic energy levels for an infinitely deep well is:

$$E_n = \frac{\hbar^2}{2m} \cdot \frac{\pi^2 n^2}{L^2} = \frac{h^2}{8m} \cdot \frac{n^2}{L^2}, \tag{9.12}$$

where n is a quantum number labelling the states, which takes values 1, 2, 3, . . . , and L is the length of the potential well. For a short polyacetylene chain, often termed a polyene or more correctly an oligo-ene, if there are N repeat units of length $2a$, each containing two carbon atoms, the energy levels are given by Equation (9.12) with L replaced by $2Na$. There will be a total of $2N$ electrons available to fill the lower levels, and since each level can accommodate two electrons of opposite spin, the first N levels will be occupied. The energy difference between the highest filled level ($n = N$) and the lowest empty level ($n = N + 1$) is then given by:

$$\Delta E = \frac{h^2}{8m} \cdot \frac{1}{4a^2}\left[\frac{(N+1)^2}{N^2} - 1\right] = \frac{h^2}{32ma^2}\left[\frac{2N+1}{N^2}\right] \cong \frac{h^2}{16ma^2} \cdot \frac{1}{N}, \tag{9.13}$$

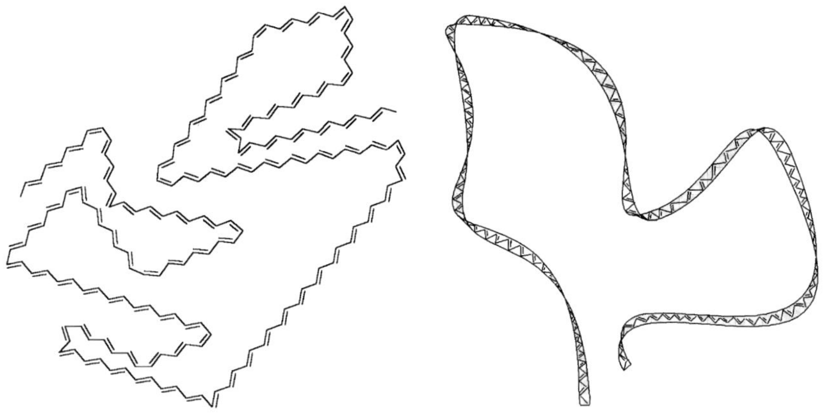

Fig. 9.17 Models for a disordered polyacetylene chain: (a) the Kuhn model and (b) the worm-like chain.

where the final approximation is valid for large values of N. Thus the energy gap goes to zero as N tends to infinity, as expected for a polyacetylene chain with equal bond lengths.

This model can be extended by the use of more realistic potentials, but has been replaced by the use of the quantum mechanical models described in Section 9.3.1. These lead to the same general result that the separation of the highest occupied (HOMO) and the lowest unoccupied (LUMO) molecular orbitals decreases as the length of the oligomers increases. The properties of a conjugated polymer are then determined from the combination of those of an appropriate distribution of perfect chain segments. This result highlights another feature that distinguishes conjugated polymers and conventional semiconductors. In the former the presence of defects that disrupts the overlap of electronic orbitals on adjacent lattice sites produces levels that lie in the band states, since the HOMO and LUMO of short chains have a greater separation than the band edges of an infinite chain. In inorganic semiconductors such local defects produce states that lie in the band gap. This also means that the *longest* chains present will determine the optical absorption of conjugated polymers. The validity of the Kuhn model has been questioned, because the production of the local, large deformations of stiff conjugated chains required to break the conjugation have a high energy cost. This is not the case for the more gradual deformation along the length of the chain in a worm-like chain. Here the clear definition of the *conjugation length* inherent in the Kuhn model is lost. However, in long polymer chains gradual deformations are just as effective in disrupting the conjugation and producing an effective distribution of conjugation lengths. While models describing the properties of worm-like chains have appeared in the literature, many authors continue to use a distribution of *conjugation lengths*, or even talk of a specific conjugation length when discussing the properties of conjugated polymers without providing a definition of what these terms mean. They should always be regarded as *effective* conjugation lengths appropriate to an idealised Kuhn chain, rather than providing a true microscopic picture of the polymer chain morphology.

The presence of a high degree of disorder in most conjugated polymers means that electron transport will be dominated by the inter-chain charge transport rather than the intra-chain charge transport. Quantum mechanical methods have been used to study the inter-chain interaction of idealised chains to provide a basis for understanding intra-chain transport. In semiconductive polymers electrical conduction in bulk samples will be by hopping transport between the polymer chains. This is clearly true for oligomers, where even though they can form crystals, the π-electron systems of the molecules are localised. In this case, rather than using molecular calculations to model some properties of the material, it is possible to use a *quasi-band-structure* description. In this model a discrete number of molecular states are distributed along the dispersion curve of the infinite polymer, see Zojer *et al.* (2000). While this is not useful to describe optical spectra which involve the $k = 0$ states, it is appropriate for the interpretation of electron energy loss spectra (EELs), since this technique probes a large range in k-space. At high doping levels when the polymers become metallic, their behaviour is similar to that of amorphous metals. A model that takes account of the existence of crystalline and amorphous regions in these polymers has been found to give an excellent description of the properties of metallic samples; this will be discussed further in Section 9.4.4.

The theoretical analysis could have started from the premise that the electron–electron interaction (electron correlation) is more important than the electron–phonon interaction. The Hubbard Hamiltonian would then be more appropriate than the SSH one. The Hubbard Hamiltonian has the form:

$$H = H_{\text{elect}} + H_{\text{corr}}, \tag{9.14}$$

where

$$\begin{aligned} H_{\text{elect}} &= -t_0 \sum_{n,s} (c^{+}_{n+1,s} c_{n,s} + c^{+}_{n,s} c_{n+1,s}), \\ H_{\text{corr}} &= U \sum_{n} m_{n,\uparrow} m_{n,\downarrow} \end{aligned} \tag{9.15}$$

and $m_{n,s} = c^{+}_{n,s}c_{n,s}$. The equations are similar to those for the SSH Hamiltonian, Equations (9.1) and (9.2). The creation and annihilation operators are the same as in the SSH Hamiltonian and the transfer integral is a constant (t_0) since the bond lengths are taken to be constant. In applying the Hubbard model it is generally assumed that the interaction between electrons on different sites is screened out, so that only the on-site term given in Equation (9.15) needs to be included. This approximation is reasonable for materials with a reasonable density of free carriers but is questionable in this instance, as discussed below. An analysis similar to that for the SSH Hamiltonian will lead to a band gap, as described in Section 4.2.4. The Hubbard model gives rise to charge density fluctuations or charge density waves, which can be viewed in this context as bond alternation. Thus, although no bond alternation was included in the model it is a property of the model. The use of the Hubbard model rather than the SSH model has been championed but gained less support. In reality neither extreme can be correct, i.e. real polymers will require a model that includes both electron–phonon and electron–electron interactions. This has been known to be true for conjugated molecules for many years, see Salem (1966). A consequence is that, although exact solutions are possible for the SSH and Hubbard models, this is no longer the case when both interactions are included. Theoretical modelling then requires the use of either an approximation to make partial analytical solutions possible or numerical methods for finite size systems.

The vast majority of conjugated polymers have complex morphology with little long-range order. Even within a single polymer chain the regions that adopt an ideal planar, extended structure are on the microscopic scale. And again, in chain-oriented samples, the structure is usually far from perfect and the individual polymer chains have finite length. Carrier motion through such materials is dominated by hopping. Despite these well-known facts, the terminology of the band model, which is appropriate for crystalline materials with good long-range order, is still widely used in the literature, i.e. the highest occupied electronic states are referred to as the *valence band* and the lowest unoccupied electronic states as the *conduction band*. An alternative view, that conjugated polymers should be regarded as disordered molecular solids with electronic wavefunctions having limited spatial extent, has been put forward, but is less widely used. Such a model may also be appropriate for ideal, extended conjugated polymer chains. In order to simplify calculations for theoretical models that include electron–electron interaction, it is assumed that inter-site Coulomb attractions can be neglected, because of screening. However, ideal conjugated polymers will be semiconductors with band gaps of the order of, or greater than, 2eV. As a result of this, the density of thermally excited carriers at room temperature will be extremely small and the screening by free carriers will be negligible. Any screening of Coulomb interactions can, therefore, come only from the polarisation of the polymer chains in the vicinity of the charges. Although the polarisability of conjugated bonds is greater than that of saturated bonds, see Table 2.2, it is not sufficient to screen out inter-site Coulomb interactions completely. This can be seen in crystalline molecular solids such as anthracene, where Coulomb interactions between electrons and holes give rise to two distinct types of exciton. These are Frenkel excitons, where the electron and hole are on the same molecule, and charge-transfer (CT) excitons, where the coupled positive and negative charges reside on separate molecules, see Pope and Swenberg (1982). When the separation of the charges is large, CT excitons are similar to the Wannier–Mott excitons found in inorganic semiconductors. The CT excitons will lie below the conduction band with a binding energy given by:

$$E_{\mathrm{CT}} = I_{\mathrm{g}} - A_{\mathrm{g}} - P(r) - C(r), \tag{9.16}$$

where I_{g} is the molecular ionisation potential, A_{g} is the molecular electron affinity and $P(r)$ and $C(r)$ are the polarisation and Coulomb energies of an electron and hole for a separation r. The size of Wannier–Mott excitons can be very large in inorganic semiconductors with high relative permittivities, as a consequence of the stabilising effect of the polarisation energy. The Coulomb interaction in conjugated polymers will be only partially screened and electrons and holes on a polymer chain will combine to form excitons. A model that has been used to describe excitons on conjugated polymer chains is that of *intermediate* CT excitons, Ladik (1988). These are excitons with a size between that of a Frenkel exciton, where the electron and hole would be on the same repeat unit of the polymer chain, and of a Wannier–Mott exciton, where the electron and hole are separated by a large number of repeat units. The Frenkel exciton model fails because the conjugation along the polymer chain leads to some delocalisation of the wave functions, while that for a Wannier–Mott exciton fails, because the relative permittivity, and hence the polarisation energy, is too small to stabilise large-radius excitons.

The appearance of one or more CT-excitons below the conduction band of the conjugated chain may not appear to be of major importance for the properties of the material. In fact the consequences of the occurrence of excitons are significant. Slater and Shockley (1936) demonstrated that the descriptions of the system by Bloch functions, i.e. the band model, and by localised excitations, i.e. excitons, were related to one another by a unitary transformation. They were also the first to consider the impact of the appearance of a localised state in a one-dimensional system. Their analysis showed that as the localised state split off from the conduction band, it *stole* the oscillator strength from the transition between the valence and conduction bands. Furthermore, the strength of the exciton absorption grew at the expense of the inter-band transition as the exciton moved further away from the conduction band, i.e. as its binding energy increased. This has a pronounced effect in one dimension, since the density of states (DOS) is a maximum at the band edges. The dispersion of the valence and conduction bands shown in Fig. 9.9 occurs only in the chain direction. Parallel chains are assumed to have negligible interaction with one another, so that there is no dispersion of the bands in directions perpendicular to the chain axis. The density of states will be high where the bands are flat, since the states are distributed uniformly along the dispersion curve with respect to the wave vector (k), i.e. the DOS is given by the inverse of the slope of the dispersion curve. In an ideal one-dimensional system the DOS has singularities at the band edges (van Hove singularities), as shown in Fig. 9.18. The occurrence of an exciton in the band gap will sharply reduce the absorption strength associated with the inter-band transition. Thus, as the exciton binding energy increases, it will become the dominant feature in the absorption spectrum, Fig. 9.18. When the effects of thermal perturbation and molecular vibrations are included, the broad exciton absorption will swamp the residual inter-band

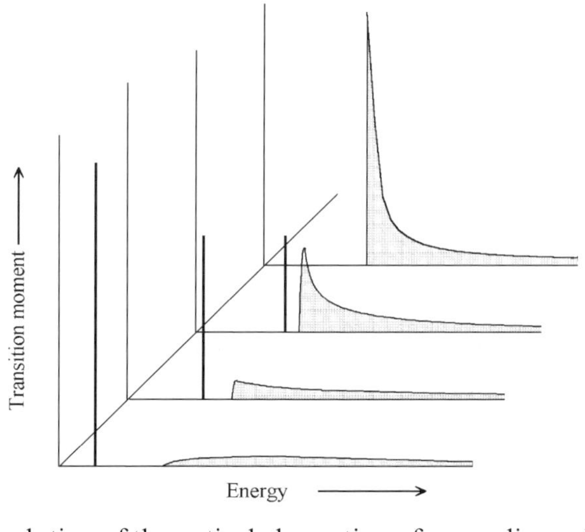

Fig. 9.18 Evolution of the optical absorption of a one-dimensional chain as the binding energy of an exciton located below the conduction band increases. Initially (at the rear) no exciton occurs and the absorption is due to the inter-band transition, shown shaded. This decreases as the transition moment is transferred to the exciton, shown by the solid vertical line.

absorption. Optical excitation will then give rise to a bound electron–hole pair, *not* free carriers, and there will be no directly-excited photocurrent. Even if the excitation occurs well above the onset of the exciton absorption, the generation of carriers is inhibited by geminate recombination. This occurs because, although the electron and hole may be excited into band states, they have insufficient excess energy to overcome their Coulomb interaction. In order to become carriers they must have sufficient energy to move apart to a separation at which the Coulomb energy is comparable with the thermal energy and they can escape from each other. This limit is termed the Coulomb capture radius.

Whether conjugated polymers are best described by a band model, such as the SSH model, or an exciton model, will depend crucially on the relative strengths of the electron–phonon and electron–electron interactions. After the discovery of highly conductive polymers, the band model was widely accepted and applied to the interpretation of experimental data. Gradually since that time, evidence that suggests that an exciton picture is more appropriate has been accumulating. Comparison of experimental results with the models described above has been used to estimate the relative importance of the two types of interaction. This will be discussed in the following sections.

9.4 Physical and chemical properties

Although the upsurge of interest in electrically conductive polymers was triggered by the observation of metallic behaviour in doped samples, subsequent interest in fundamental and applied studies has returned to the properties of the as-prepared, semiconductive polymers. A significant fraction of the literature reports routine measurements made on poorly characterised materials. We now know that even carefully prepared materials can be accidentally doped by residual traces of polymerisation catalyst and oxygen. Thus there are few cases where the data have been obtained on intrinsic material. This situation is similar to the early phase of research on silicon and germanium, where the conductivity was extrinsic due to the presence of trace impurities; once these were eliminated, high-purity, intrinsic materials were found to have rather low conductivity. While exposure to reducing agents can depress the conductivity of conjugated polymers to the levels expected for an intrinsic material, samples are rarely subjected to this treatment. Such treatment compensates for the presence of the species doping the polymer, but it does not remove them and it can also increase the defect density in the sample. Furthermore, the level of conductivity produced by accidental doping is suitable for practical application in semiconducting devices, when it is the properties of the as-prepared polymer that are important. It is necessary to take proper account of the complex morphology of as-prepared polymers when their properties are compared with those predicted by theoretical models. In this context the polydiacetylenes, which can be prepared as high quality, single crystals, can provide model systems with morphology close to that of the idealised linear polymer considered in the theoretical models discussed in the preceding section.

Consideration of Fig. 9.2 shows that most of the polymers contain a sequence of alternating single and double C-C bonds. Exceptions shown are the polydiacetylenes, which have triple C-C bonds in their backbones, and aniline and polyphenylenesulphide, which have a heteroatom in the conjugated chain. A similar situation is found in metal-organic polymers where metal atoms are located in the polymer backbone. The extended, conjugated structure common to all these polymers results in similarities in their properties, and there is a tendency to treat conductive conjugated polymers as a single class of materials. We need to treat this classification with caution, as there can also be marked differences in properties, even within one subset of conjugated polymers, resulting from the different molecular structures and micro- and macroscopic morphologies. The complex behaviour of PAni shown in Fig. 9.6 illustrates this point. The following presentation of the generic properties of conjugated polymers will give examples of the differences that can occur.

The initial application of quantum mechanics to the electronic states of a perfect linear conjugated chain, as in the Hückel model discussed in Section 4.2.5 and above, led to a model of a one-dimensional semiconductor with well-defined valence and conduction bands. This labelling of the electronic states is widespread in the literature. On the other hand, when electron correlation is included, the electronic states are more localised and an exciton description is more appropriate. The disorder present in all but a few exceptional cases inevitably leads to the conclusion that the electronic states must be localised by chain defects. The extent to which the electronic states of conjugated polymers are localised, i.e. deviate from the band model, has been a matter of debate. There is a growing body of experimental and theoretical evidence, discussed in Sections 9.4.2 and 9.4.3 below, that suggests that the exciton description is closer to the truth.

9.4.1 Morphology

While most conjugated polymers display partial crystallinity, see Section 1.2.7, examples are known of extended-chain single crystals and amorphous isotropic solids. The extent of regions in the polymer backbone with uninterrupted overlap of the π-electron wavefunctions will obviously be much greater in an extended chain than in one that is randomly folded. Since physical properties, such as the optical absorption and electrical conductivity, are dependent on the π-electron wavefunctions, they will inevitably be affected by the polymer morphology. Therefore the morphologies encountered in conjugated polymers will be reviewed briefly before discussing their physical properties.

The solid-state polymerisation of di-substituted diacetylene monomers can lead to macroscopic polymer single crystals that retain the perfection of the precursor monomer crystals. While the intrinsic defect density is low, mismatch in the lattice repeat distance between the monomer and polymer means that polymerising chains cannot extend without limit, i.e. a gap opens up between the growing chain end and the adjacent monomer. Measurements of soluble polydiacetylenes indicate chain lengths of several microns (Bloor, 1984). In insoluble polymers this leads to a density of chain ends that is difficult to measure with currently available techniques. However, the chains are free of major conformational defects and are aligned parallel, giving highly anisotropic samples. This chain alignment has been directly observed by high-resolution transmission electron microscopy (Young, 1984).

The X-ray diffraction patterns of PPy films polymerised electrochemically at high current densities, i.e. at rapid growth rates, have weak, extremely broad features, indicating that the material is amorphous (Dyreklev *et al.*, 1996). If the films are grown slowly, sharp diffraction lines occur, indicating the presence of small crystallites, though the pattern is still dominated by the broad amorphous background. This behaviour is characteristic of polymers obtained electrochemically where more than one linkage between monomers is possible. This leads to random crosslinks and defects in the backbone structure; these are more prevalent in rapidly-grown films. Monomers where only a single linkage is possible produce films with a higher crystallinity. Scanning tunnelling microscopy of electrochemically prepared PAni films reveals a granular structure and associated variations in electrical properties indicating partial crystallinity (Yau *et al.*, 1999).

The π-electron bonding in conjugated polymer chains makes them stiff. As a result directly-synthesised and precursor route polymers will tend to aggregate into crystallites. The chain stiffness will also inhibit the occurrence of chain-folded lamella crystals, Fig. 1.11(b), since the formation of tight chain folds is energetically unfavourable. Thus the microstructure of conjugated polymers is best described by the fringed micelle, Fig. 1.11(a), and related models. The difference between non-conjugated and conjugated polymers is illustrated by the morphology of Durham-route precursor polymer and the derived PAc. Films of the precursor polymer are amorphous, but on thermal conversion a crystalline diffraction pattern is observed (White and Bott, 1984). Films that are stretch-aligned during conversion show a sharp electron diffraction pattern indicating a higher degree of crystallinity (Martens *et al.*, 1994). Similar diffraction patterns are observed for Shirakawa PAc. However, the disorder present in these materials has made a precise determination of the symmetry and size of the unit cell difficult. Both orthorhombic and monoclinic unit cells have been assigned to Shirakawa PAc. The most likely structure for Shirakawa PAc is orthorhombic with unit cell dimensions $a = 0.740$ nm, $b = 0.408$ nm, $c = 0.246$ nm, while for the best Durham-route precursor PAc they are $a = 0.725$ nm, $b = 0.419$ nm, $c = 0.243$ nm, all values $\pm$ 0.001 nm. While the precursor route materials are cast as continuous films, chemically synthesised PAcs, including both Shirakawa type and the highly conducting form prepared in silicone oil, have a fibrous morphology. This leads to a model in which the polymer chains are preferentially aligned along the fibre axis with oriented regions interspersed with shorter disordered regions (Ehinger and Roth, 1986). In the as-prepared polymer the fibres are randomly oriented, giving isotropic macroscopic properties and with inter-fibre contacts facilitating charge transport in bulk samples. Stretching then aligns the fibres and reveals the inherent anisotropy of the fibres in the bulk properties.

Highly crystalline films of soluble conjugated polymers can be obtained by casting thin films from solution. Polymer chains in films with thickness of 100 nm or less are aligned parallel to the substrate. Scanning microscopy reveals features in the films having lateral dimensions in the range 20 to 150 nm (Lin *et al.*, 2002). While these have been referred to as aggregates, the quality of the X-ray diffraction patterns obtained from similar films suggests that they are crystallites. This is also likely to be the case for polymers where the average number of repeat units in the chain is hundreds rather than thousands.

Liquid crystalline phases are observed for polymers with rigid backbone segments. Conjugated polymers might, therefore, be expected to behave in a similar manner. LC phases have, however, been observed in only one class of conjugated polymers, the polyfluorenes. Poly(dioctylfluorene) is liquid crystalline between ~170 and ~270 C and poly(di(ethylhexyl)fluorene) is liquid crystalline above 167 C (Grell *et al.*, 1997; Grell and Bradley, 1999). The LC phases are birefringent but the textures observed do not provide an unambiguous identification of the phase. The texture is retained in samples rapidly cooled into a glassy phase, indicating that polymer orientation is retained without crystallisation. Rapid cooling from the LC phase of films on polyimide LC alignment layers results in oriented monodomains.

In view of the diversity in the morphology of conjugated polymers, it is essential that physical measurements be made on samples that are not only chemically pure but where the morphology has been established by thorough characterisation.

9.4.2 Intrinsic optical properties

The optical spectra of conjugated polymers are dominated by an intense absorption due to the electric dipole allowed transition between the highest occupied and the lowest unoccupied electronic states. For solid samples the energy of this transition can lie anywhere between about 0.8 and 4 eV. Data for solid samples are listed in Table 9.1 and for solutions and solid samples in Table 9.2 on page 351.

In many instances the absorption is broad and featureless. Distinct features are observed in materials that have either more ordered polymer chains or an excited state with geometry similar to that of the ground electronic state. Figure 9.19 illustrates this for (a) amorphous samples of Durham-route PAcs, (b) poly(2-methoxy-5-(2′-ethylhexyloxy)-*p*-phenylene vinylene), MEH–PPV, and (c) a rigid poly(*p*-phenylene) ladder polymer, MeL–PPP. The absorption maxima in Fig. 9.19(a) lie between 2.15 and 2.65 eV. The lowest occurs for a low molecular mass sample, where some ordering of the shorter chains is possible. In Shirakawa PAc the degree of crystallinity is higher and the absorption is narrower with maxima at 1.9 and 1.5 eV for random and stretch-aligned samples, respectively. These differences result from the distribution of conjugated sequences that occur in the different samples. In the solid state where the chains are collapsed, the Kuhn model, Section 4.2.5, provides a reasonable model for the amorphous materials. Chemical defects and folds of the polymer chain provide definite breaks in conjugation. This is not, however, the case for spatially extended deformations of the polymer backbone. Thus the extent of conjugation, the *conjugation length*, is generally not precisely delineated, and the distribution of conjugation lengths is merely an experimental parameter descriptive of the form of the absorption spectrum. The spectrum of MEH–PPV is an example of a narrower spectral profile due to the small difference in ground and excited state wavefunctions, but is featureless,

Table 9.1 *Optical absorption maximum, A_{max}, and onset, A_{onset}, for conjugated polymers in the solid state at room temperature*

Polymer	PSQE[a]	PTTP[b]	PITN[c]	PTV[d]	CN–PPV[e]	POFBT[f]	PPV	PPS[g]
A_{max} (eV)	1.41	1.46	1.45	2.35	2.55	2.64	3.03	4.5
A_{onset} (eV)	0.79	0.81	1.0	1.8	2.05	2.1	2.39	3.4

[a]polysquaraine,
[b]poly(thieno(3,4-b)thiophene),
[c]poly(isothianaphthalene),
[d]poly(2,5-thienylenevinylene),
[e]poly(2,5-*bis*(hexyloxy)-1,4-phenylene-(1-cyanovinylene)),
[f]poly(9,9-dioctylfluorene-*co*-benzothiazole),
[g]poly(*p*-phenylenesulphide).

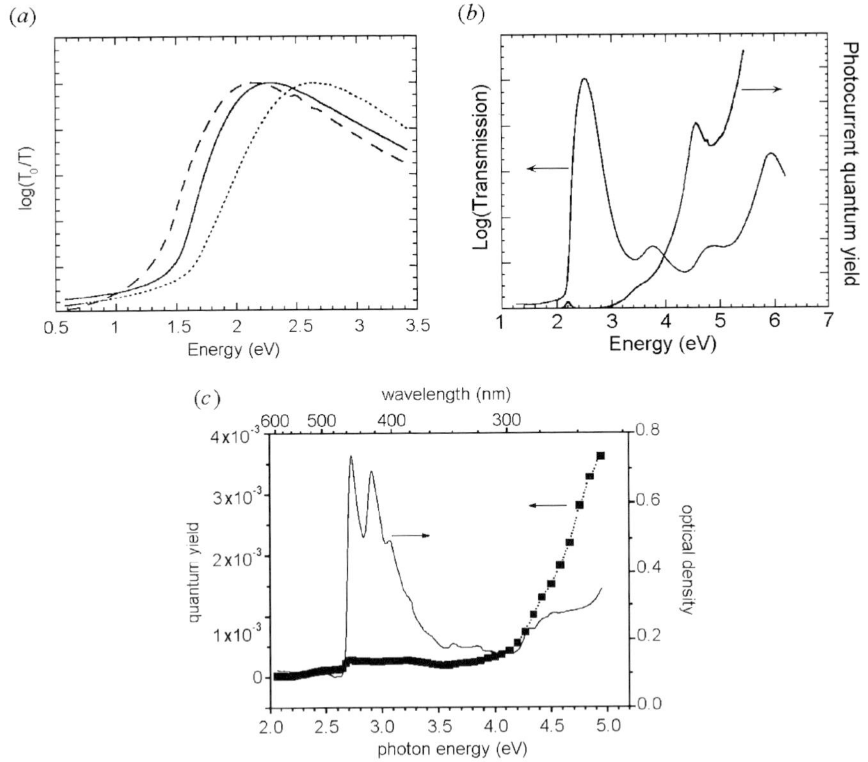

Fig. 9.19 Room temperature optical absorption spectra of (a) three types of Durham PAc normalised to the same maximum absorption (reprinted from Martens *et al.*, 1994; copyright 1994, with permission from Elsevier), (b) MEH–PPV (reprinted with permission from *Nature*, Köhler *et al.*, 1998; copyright 1998 Macmillan Magazines Limited), and (c) a 110nm thick film of MeL–PPP (reprinted from Barth *et al.*, 1998; copyright 1998, with permission from Elsevier).

indicating significant inhomogeneous broadening. MeL–PPP is a rigid rod polymer and has a much sharper absorption profile with a progression of vibronic sidebands separated by ~0.2 eV, an energy characteristic of the vibration frequency of the conjugated bonds in the backbone. Spectroscopic ellipsometry has been used to measure the dielectric properties in the region of the intense optical absorption. Spin coated samples display considerable chain alignment, being birefringent with maximum values of the ordinary and extraordinary components $\varepsilon_0' > 4.5$, $\varepsilon_e' < 2.5$, $\varepsilon_0'' < 2$ and $\varepsilon_e'' < 0.5$ (Zhokhavets *et al.*, 2003).

The perfection of the polymer chains in polydiacetylene crystals results in an absorption spectrum with a narrow electronic transition and sidebands due to vibrations of the backbone double and triple bonds. The absorption is

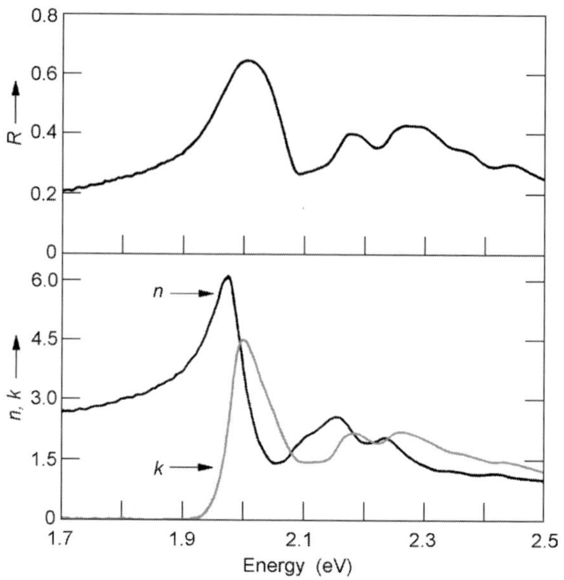

Fig. 9.20 Lower panel, real and imaginary refractive indices of a PDA–TS crystal at room temperature for light polarised parallel to the polymer chains. Upper panel, reflection spectrum recorded under the same conditions. Reprinted with permission of Wiley–VCH from Bloor and Preston (1976).

intrinsically anisotropic. The absorption coefficients of the polydiacetylene with pendent toluene sulphonate groups (PDA–TS) are $9\times10^7\,\mathrm{m}^{-1}$ for light polarised parallel to the polymer chains, and $2\times10^5\,\mathrm{m}^{-1}$ and $< 5\times10^2\,\mathrm{m}^{-1}$ for light polarised perpendicularly to the polymer chains but oriented parallel and perpendicular to the plane of the chains, respectively (Bloor and Preston, 1976). The large absorption coefficient is associated with a strong dispersion of the real part of the refractive index, as discussed in Section 3.7 and illustrated in Fig. 3.19. This behaviour can be seen in the real (n) and imaginary (k) refractive indices of PDA–TS for light polarised parallel to the polymer chains, Fig. 9.20. The reflection coefficient R at a given wavelength depends on n and k at the same wavelength:

$$R = \frac{n^2 + k^2 + 1 - 2n}{n^2 + k^2 + 1 + 2n}. \tag{9.17}$$

The dependence of the reflection coefficient on n and k allows these quantities to be calculated from the reflection spectrum by use of the Kramers–Kronig relations. It follows from Equation (9.17) that an intense absorption gives a high reflectivity, e.g. the reflection spectrum of PDA–TS,

Fig. 9.20, where R has a maximum value of 0.66. This effect accounts for the metallic appearance of PAc and other conjugated polymers.

Are these absorptions in conjugated polymers due to inter-band transitions or are the excited states localised? For many conjugated polymers other measurements point to the latter interpretation. If carriers are excited into a conduction band there should be a measurable photocurrent. The photocurrent in MEH–PPV, plotted in Fig. 9.19(b), rises rapidly for photon energy greater than 3.0 eV compared with absorption onset and maximum at about 2.05 and 2.55 eV, respectively. A similar offset between the optical absorption and the onset of the photo-excitation of charge carriers, detected by electron spin resonance, is observed for MEH–PPV and CN–PPV (Kuroda *et al.*, 2000). The points plotted in Fig. 9.19(c) show the photocurrent generated in the LE–PPP film. The weak photocurrent seen in the vicinity of the optical absorption has been shown to be due to exciton fission and exciton annihilation at an electrode, see Section 4.2.2. Similar data have been reported for other PPVs and PDAs with the onset of photoconduction being higher in energy than that of the optical absorption. Thus, in these cases the main absorption gives rise to electrically neutral species, excitons, and intrinsic photo-generation typically occurs 0.5 to 1 eV above the onset of the optical absorption.

Electro-absorption (EA) spectroscopy, where optical absorption is observed under the application of an electric field to the sample, is another method that can distinguish between localised and inter-band excitations. The electric field produces a Stark shift of allowed optical absorptions and renders forbidden transitions allowed by mixing the wavefunctions of the excited states. Excitons show a quadratic Stark (Kerr) effect with a spectral profile that is the first derivative of the absorption spectrum for localised (Frenkel) excitons and the second derivative for charge transfer excitons, i.e.

$$\Delta A = -\frac{1}{2}\frac{\partial A}{\partial E}(\boldsymbol{F}\cdot\Delta\boldsymbol{\alpha}\cdot\boldsymbol{F}) + \frac{1}{2}\frac{\partial^2 A}{\partial E^2}(\Delta\boldsymbol{\mu}\cdot\boldsymbol{F})^2, \tag{9.18}$$

where A is the absorbance, E the photon energy, $\boldsymbol{F}$ the applied field, and $\Delta\boldsymbol{\alpha}$ and $\Delta\boldsymbol{\mu}$ are the changes in polarisability and dipole moment induced by the field. Electric field mixing of band states, the Franz–Keldysh effect, produces a positive peak below the band gap due to states shifted into the gap and a complementary series of negative peaks above the band gap.

The most striking results have been obtained for electro-reflectance from crystals of a polydiacetylene with carbazole pendent groups (PTS–DCH), Fig. 9.21, where the spectrum comprises a first derivative of the principal absorption and its sidebands and a signal characteristic of an inter-band

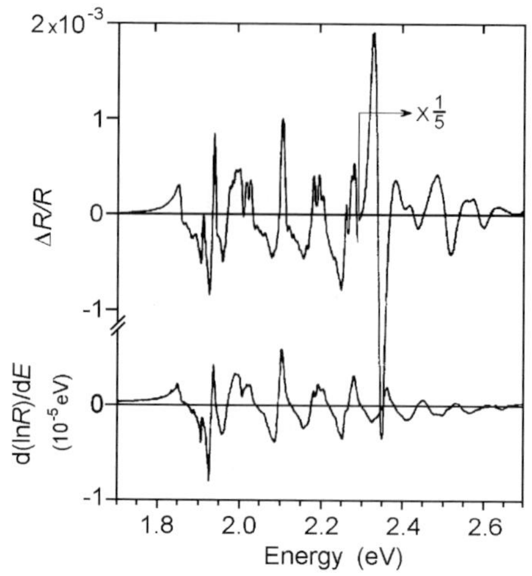

Fig. 9.21 Upper curve: electro-reflectance spectrum of a PDA–DCH crystal at 8 K for light polarised parallel to the polymer chains. (Note the scale change at 2.3 eV.) Lower curve: first derivative of the reflection spectrum. Reprinted with permission from Weiser (1992). Copyright 1992 by the American Physical Society.

transition 0.4 eV higher in energy. No significant feature is discernable in the normal absorption spectrum above the background due to vibrationally excited excitonic states at this energy. The EA signature of the inter-band transition is very intense at low temperatures but is washed out at room temperature. Similar results have been observed for other PDAs (Weiser, 1992). Thus, the localised exciton has *stolen* the oscillator strength from the inter-band transition, as postulated by Slater and Shockley (1936). Such results are not exclusive to PDAs; in the region of the principal absorption the EA spectra of PPV, PThs and MeL–PPP all have profiles similar to first derivates of the absorption spectra.

On the basis of this evidence the principal optical absorption of conjugated polymers can be identified as a transition between a singlet ground state and a localised excited singlet state, an S_0–S_1 transition between a symmetric 1^1A_g ground state and an antisymmetric 1^1B_u excited state.[1] Hence it is similar to the optical transition of a small organic molecule. In this case it should also be possible to observe triplet excitons, i.e. localised states in which the spin of the excited electron and that remaining in the ground state are parallel rather than anti-parallel. The triplet states form a manifold with a lowest state, T_1 or 1^3B_u,

[1] Platt's notation for molecular states is $N\ ^{\text{Multiplicity}}\text{State}_{\text{Symmetry}}$, where N labels the first, second, third, etc., states of the same type, the multiplicity is $(2S + 1)$, A and B states are non-degenerate and degenerate, respectively, and g and u identify symmetric (gerade) and antisymmetric (ungerade) states, respectively.

that, following Hund's rule, is lower in energy than the lowest excited singlet state S_1. This lower energy stems from electron correlation, which keeps the electrons further away from each other than would otherwise be the case, thereby reducing their electrostatic interaction energy. Transitions between S and T states, *intersystem crossing* (ISC), require that the spin state of an electron is reversed, i.e. a magnetic interaction, and S–T transitions are electric dipole forbidden. Absorption can be induced by perturbations that mix S and T states, e.g. the presence of either heavy atoms with free spins or oxygen, which has a triplet ground state. Such species are difficult to incorporate into polymer samples in sufficient concentration to enable S–T absorption to be observed. Lower concentrations can make it possible to observe emission from the T_1 state. As the probability of S_n–T_n ISC remains low, any excited T_n states produced are likely to relax to the T_1 state, and a population will build up in this state. This process can also occur in high purity polymers by ISC due to spin–orbit coupling on the conjugated backbone. However, the T_1 population can decay by T–T annihilation, giving an excited S_n state, and T_1–S_0 transitions that can be either radiative or non-radiative. The observation of long-lived phosphorescent emission is dependent on the competition between the latter two processes. Non-radiative decay occurs by either phonon emission or quenching of the excited state at impurities and defects. The relatively high energy of the vibrational modes of the conjugated polymer chain requires the emission of fewer phonons and increases the probability of non-radiative decay. The mobility of the electronic excitation along the polymer chain also increases the probability of reaching a quenching site. In general the latter process leads to rapid non-radiative decay resulting in very weak, phosphorescent emission. However, at low temperature access to quenching sites is reduced, and delayed fluorescence resulting from T–T annihilation and phosphorescence is observed, e.g. in MeL–PPP, see Table 9.2 (Hertel *et al.*, 2001). Emission spectra of phenylene ethynylene polymers with both conjugated units and platinum atoms in the backbone have similar spectra with a constant T_1–S_1 separation of $\sim$ 0.7eV, despite the fact that the S_1 state varies in energy from 2 to 3.2 eV (Köhler *et al.*, 2002). Analysis of spectroscopic data for PPV thin films leads to a T_1 state energy of 1.55 eV, which is 0.9 eV below the absorption band onset (Österbacka *et al.*, 1999). Phosphorescence, sensitised by the incorporation of platinum porphyrin dopant, has been observed in a phenyl-substituted PPV (Laquai *et al.*, 2003).

The energy of the T_1 state can also be determined for soluble conjugated polymers by the technique of T–T energy transfer (Monkman *et al.*, 2001). When the solvent in a polymer solution is ionised with an intense, pulsed electron beam, charge recombination gives rise to excited solvent molecules in the T_1 and S_1 states in a ratio, determined by spin statistics, of 3:1. The solution also includes an acceptor with a lower T_1 energy than that of the solvent, short-lived excited singlet states and good ISC. Decay of the excited solvent molecules then produces a population of acceptor molecules in the T_1 state. The T_1 state of the polymer will be populated if it lies below that of the acceptor and not otherwise. The presence of the polymer triplet state is detected spectroscopically by the appearance of absorption due to transitions between triplet levels and a reduction in the S_0–S_1 transition as the population of the S_1 state is reduced. Use of several acceptors allows the energy of the T_1 state of the polymer to be determined with an accuracy of between ± 0.05 and ± 0.1 eV. Triplet state energies and T_1–S_1 separations obtained from triplet energy transfer and spectroscopic investigations are given in Table 9.2.

There is a substantive body of evidence that the lowest lying electronic states in conjugated polymers are localised. Furthermore, the large S_1–T_1 state separation of between 0.6 and 1 eV shows that there is strong electron–electron correlation in conjugated polymers. This is contrary to the assumption in the SSH model that correlation can be neglected. A correct theoretical description must include contributions from electron–phonon interaction, i.e. bond alternation, and electron–electron correlation, as discussed in Sariciftci (1997). Most calculations obtain values for the binding energy of the S_1 exciton that are somewhat larger than the experimentally determined values, which fall in the range 0.5 to 0.9 eV. Thus there is an on-going effort to develop reliable theoretical models that incorporate both intrinsic and extrinsic effects (Barford *et al.*, 2002). There is also evidence that in highly ordered polymers, where cofacial stacking of the chains occurs, e.g. the regio-regular PThs, that inter-chain interaction is important (Brown *et al.*, 2003). Overlap of the π–electron orbitals, see Fig. 1.3, on adjacent polymer chains results in an exciton with a wavefunction spread over more than one polymer chain and an absorption lower in energy than that of the intra-chain exciton. This extended exciton is sensitive to the degree of order and is not observed in disordered polymers, where the model of an exciton localised on an individual polymer chain is appropriate. It is now clear that, although the band model was favoured at the time of the discovery of intrinsically conductive polymers, the recent experimental evidence leads inevitably to the conclusion that the exciton model is more appropriate.

Other excited electronic states can be located by direct and induced absorption and by two-photon spectroscopy. Photo-excitation and triplet energy transfer will produce populations mainly in the S_1 and T_1 states, respectively, and thereafter induced-absorption proceeds by transitions from these to higher S_n and T_n states. But photo-excitation will also populate other excited states and

Table 9.2 S_0–S_1, S_0–T_1 *and* T_1–S_1 *separations obtained from absorption and fluorescence spectroscopy and triplet–triplet energy transfer measurements made in benzene solution at room temperature unless otherwise stated (energy in eV). Spectroscopic data are given for the spectral maximum* (A_{max}, Fl_{max}) *and onset* (A_{onset}, Fl_{onset}).

Polymer		PMOT[a]	PPY–RD[b]	PFO[c]	P3OT[d]	MeL–PPP Film 80K	MEH–PPV	PF2/6[e]Glass 80K	PAni[f]
S_0–S_1	A_{max}	3.77	3.35	3.22	2.83	2.72	2.48	3.00	2.00
	A_{onset}	3.22	3.10	3.01	2.42	2.65	2.23	2.90	1.77
S_0–T_1	(T–T transfer)	2.20	2.4	2.30	1.65	2.15[g]	1.30		<0.9
T_1–S_1	A_{max}	1.57	0.95	0.92	1.03		1.18		<1.1
	A_{onset}	1.02	0.70	0.71	0.62		0.93		<0.9
S_0–T_1	Fl_{max}	1.95[h]	2.3[i]			2.08		2.18	
	Fl_{onset}	2.20[h]	1.9[i]						
T_1–S_1	Fl_{max}					0.61[j]		0.75[j]	

[a]poly(3-octyl-4-methylthiophene),
[b]poly(2,5-pyridinediyl),
[c]poly(dioctylfluorene),
[d]poly(3-octylthiophene),
[e]poly(2,7-(9,9-*bis*(2-ethylhexyl)fluorene)),
[f]polyaniline,
[g]benzene solution at room temperature,
[h]glass at 15K,
[i]film at room temperature,
[j]separation of maxima in phosphorescence and delayed fluorescence spectra.

produce separated charge carriers. In consequence, the spectra are complex with overlapping bands. Identification of the excited states involved is aided by other techniques such as the observation of the electron spin resonance spectra of the excited species (Österbacka *et al.*, 1999). Direct and photo-induced absorption only reveals states accessible by allowed electric dipole transitions, so that transitions from S_1, the 1^1B_u state, will be to higher lying N^1A_g states. Therefore, if the first excited 2^1A_g lies below the 1^1B_u level, it cannot be located by these methods. The 1^1A_g to 2^1A_g transition is, however, allowed for two-photon absorption, and this method has been used to determine the energy of the 2^1A_g level. In many instances two singlet and two triplet levels have been located below the band-edge with the 2^1A_g level above the 1^1B_u level. Values observed for the separation of these levels are 0.3 eV in a carbazole-substituted PPV, 0.6 eV in poly(3-octylthiophene) and 0.77 eV in polyfluorene. PAc and the polydiacetylenes are exceptions with the 2^1A_g level below the 1^1B_u level, e.g. ~0.2 eV lower in PAc and 0.16 eV lower in PDA–DCH. This additional information is important in assessing the efficacy of theoretical models and in understanding the non-linear optical properties of conjugated polymers.

Figure 9.19 illustrates the effect of disorder on the absorption spectra of conjugated polymers. Disorder broadens out the pure electronic and vibronic bands seen in Fig. 9.19(c) into a single broad band with the maximum absorption shifted in to the region of the vibronic bands. As noted above, in highly disordered amorphous samples the existence of chain sequences with a range of effective conjugation lengths plays a major role in the spectral broadening. The *particle-in-a-box* model, Section 9.2.2, predicts that the excitation energy is inversely proportional to the size of the box, which is the conjugation length in this instance. Similar dependence is found in theoretical models for excitons (Barford *et al.*, 2002). There are associated variations in chain structure and vibrational frequencies. The latter can be probed by resonance Raman spectroscopy, where use of exciting light at the energy of the electronic level produces a large enhancement of the Raman scattering[2] for vibrations that couple with the electronic level. Tuning the excitation energy through the absorption profile of highly disordered polymers reveals changes in the vibration frequencies commensurate with a distribution of effective conjugation lengths. Broad spectral profiles are, however, also observed in chain-oriented samples, where there are few localised chain defects and the effect of the distribution of effective conjugation lengths should be less pronounced. EA

[2] In Raman scattering the energy of the scattered quanta is either reduced or increased by vibrational energies characteristic of the scattering material. The bands of increased energy are weak since this requires the destruction of a vibrational quantum, and at room temperature the population of the vibrational levels is low.

spectroscopy has been used to investigate the role of disorder in this case. Although measurements on polydiacetylenes in single crystals and films cast from solution show a marked broadening of the optical absorption and shift in the absorption maximum, EA spectra show that there is only a small change in the energy of the S_1 exciton. Thus disorder redistributes transition strength from the pure electronic into the vibronic transitions. The signature of the Franz–Keldysh effect at the band-edge is rapidly obliterated by disorder and replaced by a broad positive signature. This has been attributed to either interaction between the band states and the exciton with transfer of oscillator strength from the exciton, or to domination of the EA response by states localised by disorder at the band edge (Weiser and Horvath, 1998). Spectroscopic studies provide evidence for a distribution of excitonic energy levels even in relatively well-ordered materials such as MeL–PPP. These include the time-dependent narrowing of emission spectra as excitons decay to lower energy levels in the density of states and spectral hole burning. Laser irradiation may change either the chemical or physical state of the absorbing species, resulting in a hole, i.e. reduced absorption over a narrow spectral range, in the absorption profile. These features can be long lived at low temperatures when relaxation of the absorbing species to its original state is slow. Hole burning in MeL–PPP has shown that the intrinsic width of states in the low energy tail of the DOS is ~1meV (Romanovskii *et al.*, 2004).

The optical absorption and photoconduction action spectra of many disordered conjugated polymers are similar to each other, in contrast with the situation found in ordered samples. This would suggest that disorder does indeed mix the exciton and band states and the clear cut distinction between the localised and delocalised states seen in highly ordered materials is lost. Nevertheless, it must be emphasised that, while the spectra of conjugated polymers reflect the disorder present in a particular sample, there is no single relationship between the spectral profile and the level of disorder. Thus samples of different polymers with similar morphology can have distinctly different spectral profiles. This reflects the numerous factors that are involved. Pendent groups affect the relative influence of the electron–phonon interaction and electron–electron correlation. This follows since pendent groups are important in determining the local relative permittivity, and thereby the extent to which electron–electron interactions are screened both within and between chains. Pendent groups can also affect chain rigidity and the extent and nature of the deformation caused by disorder. It is necessary to consider each polymer in its own right rather than to assume that they fit a single overarching model.

The spectral broadening imposed by disorder is clearly seen in the spectra of soluble conjugated polymers. In general, flexible pendent groups are necessary

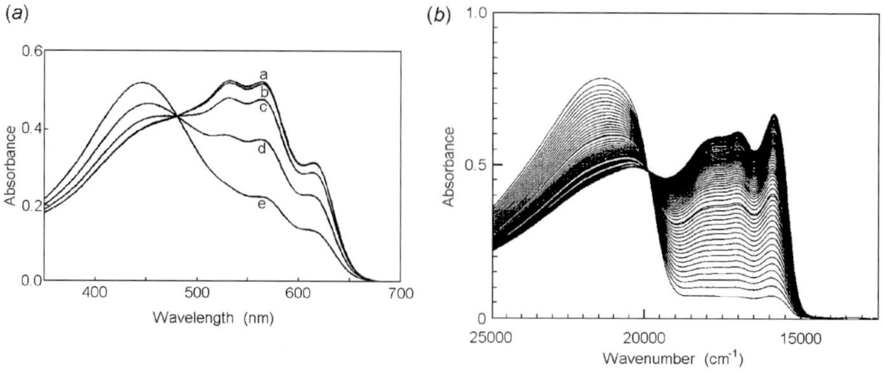

Fig. 9.22 (a) Thermochromism of poly(3-(10-hydroxydecyl)-2,5-thienylene) in N,N′-dimethylpropyleneurea/methanol at (a) 240 K, (b) 248 K, (c) 258 K, (d) 273 K and (e) 290 K (reprinted from Bertinelli *et al.*, 2001; copyright 2001, with permission from Elsevier), and (b) solvatochromism of a soluble polydiacetylene showing spectral evolution for 140 hrs after a small change in solvent composition at 293 K (reproduced with permission of Wiley-VCH from Bloor *et al.*, 1986).

to impart good solubility to a conjugated polymer, dissolution occurring by interaction of the pendent groups with the solvent. The pendent groups are randomised and carry the stiff polymer chain into solution. This model is the basis for the *worm-like* chain model of conjugated polymers in solution. In this situation, unlike that in solid samples, the backbone will be subjected to dynamic disorder. The effect of both static and dynamic disorder is to broaden the absorption profile. Hence polymers dissolved in good solvents at room temperature have broad featureless spectra, see Fig. 9.22. Profound changes in the spectral profile occur when the solvent quality is reduced, e.g. when either the temperature of the solvent is reduced, Fig. 9.22(a), or a poor solvent is added, Fig. 9.22(b). The resulting spectra are indicative of the conversion of the disordered polymer into a more ordered form. Whether this process is due to straightening of individual, isolated chains or to aggregation is a matter of debate. It seems probable that a rapid change in solvent quality in very dilute solutions, where the individual chains are not in contact, causes individual chains to fold back on themselves to form quasi-crystalline nano-particles. At higher concentrations aggregation is more likely to occur, as evidenced by the slow evolution of the more ordered phase, as shown in Fig. 9.22(b). Rapid changes in solvent quality cause residual disorder to be locked into the system of pendent groups, and partially ordered species are observed. The two examples shown in Fig. 9.22 illustrate the effect of gradual changes, which allow the pendent groups to achieve a higher degree of order, resulting in spectra with well defined features. In these examples the occurrence of clear *isobestic* points shows that the disordered polymer is converted into a single ordered species.

As the transition from the S_1 to the S_0 level is an allowed one, radiative decay gives rise to prompt fluorescent emission. When this process follows the optical excitation of a sample, it is termed *photoluminescence*. However, observation of fluorescence also requires that non-radiative decay does not predominate over radiative decay. As noted earlier, quenching of excited states at defects is the principal non-radiative decay channel. Surprisingly, high-quality, single crystal polydiacetylenes are not fluorescent, while fluorescence can be observed in less perfect samples and for many other conjugated polymers. This indicates that the excitons are highly mobile on the perfect chains in the single crystal polymer and readily find quenching sites even though the density of these is low. In less perfect materials the excitons are somewhat less mobile, residing on shorter chain segments, and are less likely to reach quenching sites.

Interest surged in studies of fluorescence after the discovery of *electroluminescence* from PPV (Burroughes *et al.*, 1990), because of its potential for practical application in light emitting devices (LEDs). Electroluminescence is fluorescent emission produced by the recombination of electrons and holes injected into a thin film of conjugated polymer, and will be discussed in the next section. If photoluminescent emission from a polymer is weak, then the electroluminescence is unlikely to be of practical significance, and consequently studies of photoluminescence and photoluminescent *quantum efficiency* have been used as a means of selecting polymers likely to be useful in LEDs.

Fluorescence spectra can often show more structure than absorption spectra. This may be explained by the mobile nature of excitons. On extended chains the exciton wavefunction extends over several repeat units and will be mobile along the chain. Dipole–dipole interactions allow inter-chain transfer to occur, enabling the excitons to move throughout a sample. In the process of diffusion the excitons will migrate to the lowest energy states available, which are those on the longest, most perfect chain segments. This effect occurs in both solid samples and solutions, and this then leads to steady state emission spectra with more sharply defined features, Fig. 9.23. Despite the dynamic disorder in the solution, the latter result, Fig. 9.23(b), shows that more ordered-chain segments persist long enough to act as the source of the structured fluorescent emission. The effect of interactions on individual chains is very well illustrated by the differences in emission from isolated MEH–PPV chains deposited from polar and non-polar solvents (Hollars *et al.*, 2003). The decay of fluorescence intensity of a single expanded chain obtained from a

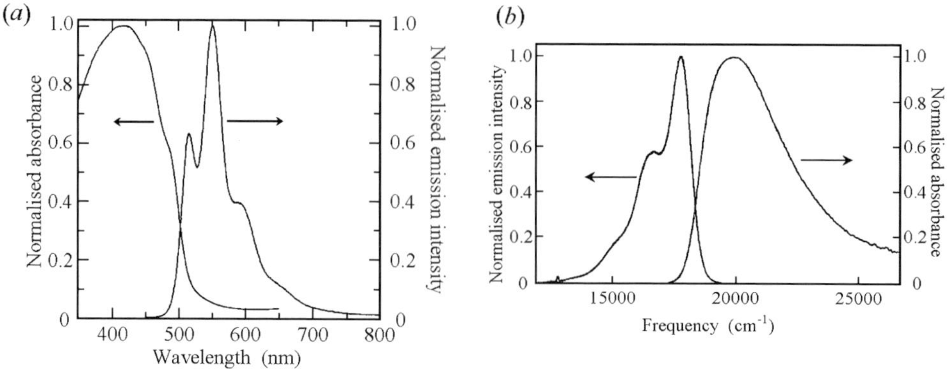

Fig. 9.23 Absorption and steady state emission spectra of (a) a PPV thin film excited at 415 nm (reproduced with permission of Imperial College Press from Samuel *et al.*, 1997), and (b) MEH–PPV solution in chloroform excited at 420 nm (reprinted from Chang *et al.*, 2000; copyright 2000, with permission from Elsevier).

polar solvent indicated emission from many independent luminescent centres. That from a tightly coiled chain deposited from a non-polar solvent gave a fluctuating output, indicating collective behaviour of large numbers of chromophore units with only two or three luminescent centres per molecule. This collective behaviour can be attributed to the lateral interaction between segments of the coiled chain that are in close contact.

The study of the temporal evolution of the photoluminescence and photoinduced absorption spectra, after excitation with short-duration laser pulses, provides a detailed picture of the interaction and decay of the excited states. In particular, such studies can follow the relaxation of the initial excitation to the lowest-lying exciton states on the most-perfect chain segments. This enables us to draw conclusions about the exciton dynamics and diffusion. For example, Fig. 9.24 (Rothberg *et al.*, 1996) shows schematically the mechanism of the decay process deduced in this way for excitation in PPV, prepared by the sulphonium precursor route, and in related polymers. The bulk of the initial excitation resides on chain segments with short effective conjugation length, but relaxes to the lower energy levels on segments with long effective conjugation length. During this process free carriers can be generated; the processes that lead to this will be discussed in detail in the next section. The initial excitons are *hot*, i.e. carry excess vibrational energy, which may occur as self-localised states with coupled vibrational and electronic excitation (Tretiak *et al.*, 2003). These relax to the vibrational ground state, cold excitons, although some hot excitons give rise to carriers on neighbouring chains, bound together by electrostatic interaction. These eventually decay non-radiatively with the

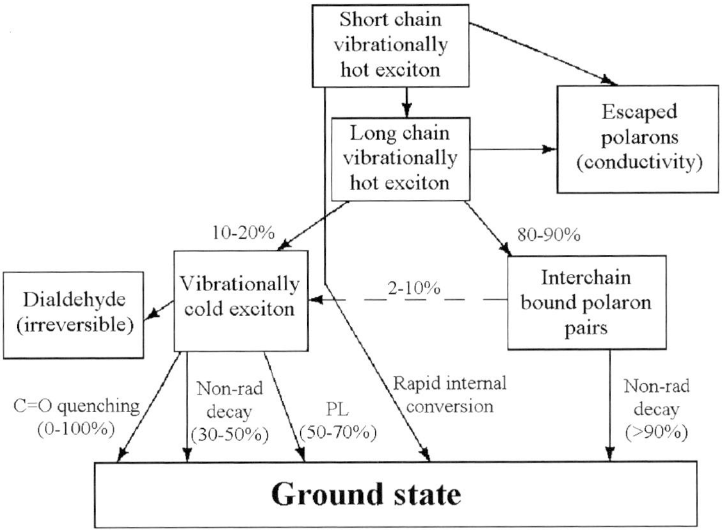

Fig. 9.24 Major pathways for the decay of photo-excited PPV polymers. Reprinted from Rothberg *et al.* (1996). Copyright 1996, with permission from Elsevier.

exception of a few, which recombine to form cold excitons. Radiative decay of the cold excitons is in principle slightly more probable than non-radiative decay, but the cold excitons can be quenched by either irreversible degradation or interaction with carbonyl defects produced by exposure of the polymer to oxygen. Since most polymer samples are partially crystalline, i.e. contain both highly disordered and aligned polymer chains, the process of exciton diffusion involves movement of the excitons from the amorphous to the crystalline regions of the sample. The interaction of the polymer chains is greater in the ordered regions and this reduces the overall photoluminescence quantum efficiency, a significant factor in the development of LEDs (Herz *et al.*, 2001; Sainova *et al.*, 2003).

The model of the non-linear optical properties of conjugated molecules discussed in Section 3.7.2 indicates that the third order non-linearity, i.e. the second hyperpolarisability, is maximised when the π-electrons are delocalised along the entire length of the molecule, which occurs when there is no bond alternation, see Fig. 3.22. This situation is not realised in real conjugated polymers, where there is finite bond alternation. Despite this the model suggests that the largest non-linearity will occur where the polymer chains are extended and free from defects, i.e. the effective conjugation length is maximised. On the basis of this elementary model, polymers with oriented chains, e.g. single crystal polydiacetylenes, should have large third order non-linearity and this is, indeed, found to be the case. Data for third harmonic generation, characterised by the susceptibility coefficient $\chi^3(-3\omega; \omega,\omega,\omega)$ (see Section 3.7.1), shows the predicted general trend, since, taking the wavelength of the one-photon absorption maximum as a measure of the effective conjugation length of the polymer, $\chi^3(-3\omega; \omega, \omega, \omega) \propto \lambda^n_{max}$. The spread of values of $\chi^3(-3\omega; \omega,\omega,\omega)$ is large, falling between bounds of $n = 6$ and n = 11 (Mathy *et al.*, 1996). Thus, this simple approach cannot provide accurate predictions of non-linear optical properties. Nevertheless, the understanding of the nature and ordering of the excited electronic states of conjugated polymers described above provides a better basis for the elucidation of non-linear optical properties. The connection is seen most directly in the derivation of the quadratic electro-optic Kerr effect ($\chi^3(-3\omega; \omega,0,0)$) from the electro-absorption spectrum. Just as n and k can be deduced from absorption and reflection spectra, so Δn and Δk can be obtained from EA spectra, since the leading term of Equation (9.18) enables $\boldsymbol{\Delta\alpha}$ to be evaluated as a function of photon energy. Then the quadratic electro-optic Kerr effect is given by:

$$\chi^3(-\omega; \omega, 0, 0) = \frac{n\Delta n - k\Delta k}{2\pi F^2} + \mathrm{i} \cdot \frac{n\Delta k + k\Delta n}{2\pi F^2}, \tag{9.19}$$

where ω is the phonon angular frequency and F is the applied field. Now we know from its effect on the nature and ordering of electronic states that electron–electron correlation is important in conjugated polymers, so that we expect it to play a part in their non-linear optical properties too. By taking account of both electron–phonon interaction and electron–electron correlation, Meneghetti (1993) went on to obtain excellent agreement between theory and experiment of both the linear and non-linear optical properties of PAc.

As noted in Section 3.7.1, there are several third order non-linear effects including, in addition to those mentioned above, electrical field induced second harmonic generation, optical Kerr effect, four wave mixing and optical phase conjugation, which have distinct dependencies on the electronic excited states. Resonance effects are more complex for second order non-linearity, see Section 3.7.2, since single and multi-photon resonance can occur. There are also a variety of methods available for determining non-linear properties. This makes it difficult to determine non-resonant values of the susceptibility coefficients and to make absolute comparisons between materials from the data available in the literature. Typical values, obtained by third harmonic generation and degenerate four wave mixing, are given in Table 9.3. These illustrate the effects of polymer chain orientation, resonance and of pendent groups that dilute the concentration of the active conjugated chains in the material.

Table 9.3 *Third harmonic generation (THG) and degenerate four wave mixing (DFWM) non-linear optical coefficients of conjugated polymers.*

Polymer	Measurement	Type of response	χ^3 (10^{-12} esu)[a]
Pac	THG oriented polymer	3-photon resonant	17000–27000
PDA–TS	THG single crystal	3-photon resonant	850–1000
PDA–DCH	THG single crystal	3-photon resonant	170
PPV	THG film	?	160
MEH–PPV	THG film	?	20
PDA–TS	DFWM single crystal	Resonant	9000
PDA–TS	DFWM single crystal	Non-resonant	500
PTh	DFWM film	Resonant	400
PTh	DFWM film	Non-resonant	40
PPV	DFWM film	Resonant	400
PDOTPV[b]	Z-scan	Resonant	500

[a]Esu units are still used in the literature: the esu unit for $\chi^{(3)}$ is cm^2 statV^{-2}; the unit in MKS is m^{2}V^{-2} and the conversion factor is:

$$\text{MKS/esu} = \left[4\pi/(c\times10^{-4})^2\right],$$

[b]poly(2′,5′-dioctyl-4,4′,4″-terphenylenevinylene).

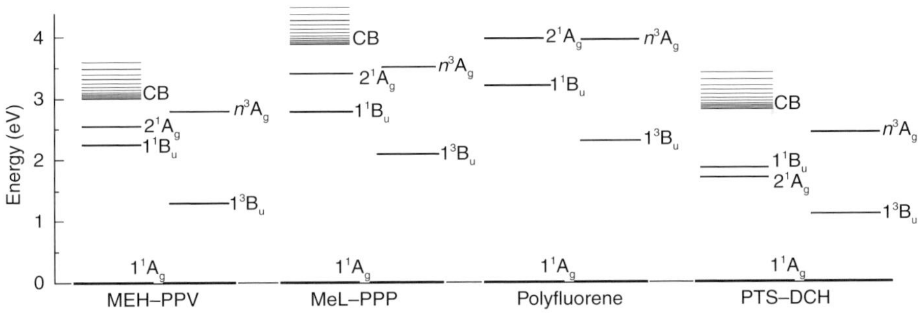

Fig. 9.25 Energy level diagrams for conjugated polymers derived from spectroscopic measurements.

On the basis of the measurements of the optical properties of conjugated polymers described in this section it is possible to construct energy level diagrams for the lowest lying electronic excitations. Some examples of the results obtained are illustrated in Fig. 9.25. The diagram for MeL–PPP is for data obtained with samples of the same morphology; those for the other polymers combine data for samples of different morphology and for related polymers, and are indicative rather than exact.

9.4.3 Intrinsic electronic properties

The experimental evidence presented above shows that the gap between the valence and conduction bands of conjugated polymers is typically a few electron volts, which is comparable with those of silicon and germanium. The energy gap is much larger than thermal energy at room temperature, ~0.025 eV, so that the number of thermally excited carriers will be very small, see Section 4.2.2. Consequently, the intrinsic conductivity of pure materials will be very low. This is observed for silicon and germanium where the addition of small quantities of either acceptors or donors (dopants) is required to provide extrinsic conduction suitable for many device applications. The levels of conduction in many as-prepared conjugated polymers are higher than is expected for intrinsic conduction. As noted in Section 9.2, this is a result of *accidental doping* of the polymers by residual catalyst and either oxidising or reducing impurities introduced during either synthesis or subsequent sample manipulation. Oxygen is the most commonly found of the latter class of impurities. As mentioned earlier, the effect of accidental doping can be neutralised by chemical treatment. Since in most conjugated polymers the accidental dopants are oxidising agents giving rise to *p*-type semiconductors, treatment with a reducing agent will reduce the observed conductivity. The conductivity of Shirakawa PAc is reduced from $10^{-3}\ \Omega^{-1}\mathrm{m}^{-1}$ to $10^{-7}\ \Omega^{-1}\mathrm{m}^{-1}$ by exposure to ammonia. Even the latter value appears to be due to residual doping as values below $10^{-10}\Omega^{-1}\mathrm{m}^{-1}$ have been reported in high purity samples of poly(*p*-phenylene) and PPV. The lowest levels of accidental doping are found in polydiacetylenes prepared as single crystals by solid-state polymerisation. This process does not require chemical catalysts, the chain reaction proceeding by radical species generated either thermally, or by UV or ionising radiation. Any oxygen present in the monomer crystal is likely to combine with the reactive chain ends producing chain breaks, which will decrease rather than increase the sample conductivity. The conductivity of thermally polymerised PDA–TS is below $10^{-10}\Omega^{-1}\mathrm{m}^{-1}$. This order of magnitude appears to be appropriate for the intrinsic conductivity of conjugated polymers at room temperature.

In principle either the removal or the addition of an electron can produce charge carriers in a conjugated polymer. This can be achieved by either chemical doping, which is considered in the next section, or photo-excitation. In these processes, charge neutrality of the sample is maintained by the creation of equal numbers of charges of opposite polarity. Theoretical models of the quasi-particles associated with added charge on conjugated chains have been discussed in Section 9.3. PAc is distinguished from most other conjugated polymers by the degeneracy of its ground state with respect to interchange of bond alternation. This leads to the possibility of neutral solitons in undoped material. The introduction of charges can produce charged solitons and charged polarons, which are formed by the combination of neutral and charged solitons. In polymers where the structures obtained by the alternative arrangements of conjugated bonds are not energetically degenerate, solitons cannot exist and the possibilities for charged species are polarons and bipolarons, see Fig. 9.14. This terminology follows from that adopted for inorganic semiconductors, where free carriers in the conduction and valence band carry with them a local lattice distortion and the *dressed* carrier is termed a polaron, see Section 4.2.2. In conjugated polymers the chain is distorted in the vicinity of the neutral and charged species and, since the polymer chain is not as rigid as the covalently bonded lattice in an inorganic semiconductor, the distortion is larger than that in the latter materials. As noted in Section 9.3.1, these species can also be described in chemical terms, although this is less frequently used in the literature. Neutral solitons are radicals, charged solitons are either carbanions or carbocations, polarons are radical cations or anions, and bipolarons are bications or bianions.

In an ideal, isolated, extended polymer chain the charged solitons, polarons and bipolarons are predicted to have wave functions extending over several repeat units of the polymer chain, see Figs. 9.11 and 9.13. Because of this the energetic barrier to motion of the charged species along the chain will be small. At room temperature the intra-chain motion of the charge carriers will be high, provided that the Coulomb interaction with the counter charge, which may be immobile, is insufficient to prevent the charged species moving away from the site where it was created. The motion of the charged species through macroscopic samples will be impeded by defects in the polymer chains and their finite extent. In order to transit the sample, charge carriers must traverse many different polymer chains distributed in crystalline and amorphous regions and make many inter-chain hops. The carrier motion will be similar to that described in Section 8.4.1 for non-conjugated semiconductive polymers, though with a much higher density of conjugated regions and shorter range hopping. Thus measurements of mobility will generally reflect the inter-chain hopping rather than the intra-chain motion. A variety of techniques were used to obtain the mobilities of carriers listed in Table 9.4.

The time of flight (TOF) method has been described in Section 8.4.1. The FET mobility, μ_{FE}, is obtained from the characteristic of a field effect transistor constructed with the conjugated polymer as the conduction channel; see below. With source–drain voltages larger than the gate voltage the source–drain current, I_{sd}, saturates and is given by:

Table 9.4 *Carrier mobilities observed for conjugated polymers, units* $m^2V^{-1}sec^{-1}$. *Carriers are positive polarons (holes) unless otherwise noted.*

Polymer	Method (for details see text)	Mobility	Zero field mobility
PDA–TS (electrons)	TOF ($E = 4.2 \times 10^6$ Vm^{-1})	4×10^{-4}	
P3HT[a] (96% HT)	FET	5×10^{-6}	
P3HT[a] (81% HT)	FET	2×10^{-8}	
P3BT[b] (94% HT)	TOF ($E = 3 \times 10^5$ Vm^{-1})	1.4×10^{-6}	
P3BT[b] (94% HT)	TOF ($E = 2 \times 10^7$ Vm^{-1})	8×10^{-8}	
EHO–OPPE[c] (el)	TOF ($E = 3 \times 10^6$ Vm^{-1})	2.2×10^{-7}	
EHO–OPPE[c]	TOF ($E = 3 \times 10^6$ Vm^{-1})	1.8×10^{-7}	
MeL–PPP	TOF ($E = 4 \times 10^8$ Vm^{-1})	8×10^{-8}	3.7×10^{-7}
Durham route PAc	FET	$\approx 1 \times 10^{-8}$	
POBTD[d] (electrons)	TOF ($E = 5 \times 10^7$ Vm^{-1})	2×10^{-8}	
POBTD[d] (electrons)	LED		2×10^{-10}
POBTD[d]	LED		2×10^{-11}
P3OT[e]	CT ($E = 2 \times 10^6$ Vm^{-1})	1×10^{-8}	2×10^{-5}
PPV[f]	TOF ($E = 4 \times 10^7$ Vm^{-1})	1.5×10^{-9}	
MDMO–PPV[g]	FET	3×10^{-9}	
MEH–PPV	SCLC		6×10^{-11}

[a]Poly(3-hexyl)thiophene with cited density of head-to-tail units.
[b]Poly(3-butyl)thiophene with cited density of head-to-tail units.
[c]Poly(2,5-dioctyloxy-1,4-diethynyl-phenylene-*alt*-2,5-bis(2′-ethylhexyloxy)-1,4-phenylene).
[d]Poly(2,7-(9,9′-di-*n*-octylfluorene)-3,6-benzothiadiazole).
[e]Poly(3-octyl)thiophene regioregular but HT content not cited.
[f]Precursor route PPV.
[g]Poly(3-(2′-methoxy-5′-octylphenyl)thiophene).

$$I_{sd} = \frac{zC_i}{2L}\mu_{FE}\left(V_g - V_0\right) + \frac{zt\sigma}{L}V_{sd}, \tag{9.20}$$

where z and L are the channel width and length, respectively, t the polymer film thickness, C_i the gate capacitance per unit area, V_g and V_{sd} the gate and source-drain voltages and the second term is the Ohmic current due to the inherent conductivity of the polymer film. In an FET conduction occurs at the interface between the polymer layer and the gate insulator, so that μ_{FE} is not necessarily identical with the bulk mobility, see Section 10.3.1. In a light emitting diode (LED), analysis of the current–voltage and current-emitted power characteristics allows the mobility of both positive and negative carriers to be estimated. The current transient (CT) method is a variation of the voltage step method described in Section 8.4.3, where the voltage is applied as a linear ramp. For conduction through a thin film in the space charge limited current (SCLC) regime, see Section 8.4.3, the zero field mobility, μ_0, can be extracted by substituting a Poole–Frenkel dependence of the mobility on field into Equation (8.50) with $\Theta = 1$, i.e.

$$j \cong \frac{9}{8}\varepsilon\varepsilon_0\mu_0\frac{V^2}{L^3}\exp\left(\sqrt{\frac{V}{F_0 L}}\right), \tag{9.21}$$

where j is the current flowing in a sample of thickness L under an applied voltage V, and F_0 is the field fitting parameter. Comparison with Table 8.2 shows that with the exception of the single crystal polydiacetylene, the mobilities are either similar to, or a little higher than, those of heavily molecularly-doped insulating polymers which also have a high density of conjugated regions. The highest mobility is found in the single crystal polydiacetylene and the regio-regular samples of P3HT, which provide highly crystalline thin films with individual crystallites typically 10 nm across. The crystalline samples provide long defect-free chains and regular chain packing. The latter will have little effect on the polydiacetylenes where the inter-chain separations are large, but it is important in the regio-regular PThs. Here the overlap of π-electron wavefunctions between chains gives rise to both excitons and polarons extended over more than one chain. Inter-chain charge transfer is facilitated, and this leads to a higher mobility in macroscopic samples. Chain packing is influenced by molecular mass, and a strong dependence of mobility on molecular mass has been observed for regio-regular PTh (Kline *et al.*, 2003).

Intra-chain mobility can be estimated using a number of different methods. Electron spin and nuclear magnetic resonance reveal the microscopic motion of species with unpaired spins by the motional averaging out of the hyperfine interaction of the unpaired spin and its environment. Line narrowing has provided evidence for diffusive motion of the unpaired spins on individual polymer chains in *trans*-PAc. These measurements require a significant density of free spins, and so can only be employed for doped materials. A technique that can be used for undoped materials is the measurement of muon spin relaxation – see Blundel *et al.* (2002) and references therein. Positive muons (μ^+) decay into positrons that are emitted in the direction of the muon spin. Observation of the emitted positrons probes the evolution of the muon spin as it interacts with its environment. When muons are deposited in a conjugated polymer, they can extract an electron from the polymer forming muonium ($\mathrm{Mu} = \mu^+ e^-$) bound to the polymer chain, leaving an unpaired spin on the polymer chain. If the unpaired spin can move along the chain, it will interact intermittently with the fixed Mu and give rise to spin relaxation of the Mu that can be analysed to determine whether the diffusion of the spin occurs in one, two or three dimensions, i.e. is either intra- or inter-chain. In other polymers the Mu can, in effect, hydrogenate the polymer and induce a negative polaron on the polymer chain. The relaxation of the muon spin is then affected by the diffusion of the negative polaron. In insulating polymers such as polybutadiene, see Fig. 1.8, the species generated on the polymer chain is, as expected, an immobile radical. For PAc the unpaired spin is localised, i.e. a radical, for *cis*-PAc, but is mobile, i.e. soliton-like, for *trans*-PAc. In electrochemically prepared PPy both mobile and immobile polarons have been observed, reflecting the high defect density in the material. In both PAni and PPV the diffusion of the negative polarons is found to be fast below ~100 K and falls at higher temperatures. Inter-chain diffusion is typically two orders of magnitude smaller, but becomes more important at higher temperatures. Thus thermally induced deformation of the polymer chain scatters the carriers, impeding intra-chain motion. The intra-chain mobility can be derived from the diffusion coefficient (D) using the Einstein–Smoluchowski relation:

$$\mu = \frac{D}{kT/e}. \tag{9.22}$$

This leads to typical values at room temperature of $5 \times 10^{-6}\,\mathrm{m^2V^{-1}sec^{-1}}$ for films of soluble PPV polymers cast from solution. At 30 K a value of 3×10^{-4} $\mathrm{m^2V^{-1}sec^{-1}}$ was obtained for poly(2,3-dibutoxy-1,4-phenylenevinylene). Comparison with ESR data shows that in PAni the intra-chain motion of negative and positive polarons is similar but that negative polarons are more mobile between chains.

Scanning tunnelling microscopy and microwave conductivity measurements have also been utilised in an effort to measure carrier mobility free from the adverse effects of polymer morphology. Current can be injected into very small regions of thin polymer films using the tip of an atomic force microscope (Lin *et al.*, 2002). Current flow has been observed in regions ~20 nm in diameter through MEH–PPV films 30–50 nm thick, corresponding to a volume smaller than that of the polymer crystallites. Applying Equation (8.50) with $\Theta = 1$ leads to a mobility of $7 \times 10^{-7}\,\mathrm{m^2V^{-1}s^{-1}}$ for an applied field of $1.4 \times 10^8\,\mathrm{Vm^{-1}}$. The zero field mobility obtained using Equation (9.21) is $6.2 \times 10^{-9}\,\mathrm{m^2V^{-1}s^{-1}}$, a factor 100 times larger than the value obtained for the bulk, Table 9.4. The relatively low defect density found in single crystal polydiacetylenes offers the prospect of measuring the mobility of carriers on the polymer chain. In principle, measurement of the conductivity at microwave frequencies probes the intra-chain motion of charge carriers, since the rapid reversal of the applied electric field means that the distance traversed by the carriers is less than the length of the polymer chain (Warman *et al.*, 2002). In order to make such measurements a large number of carriers is required. The only method that can be used to achieve this is irradiation with pulses of high-energy electrons, which are deposited uniformly throughout the sample. In solution, provided the solvent has a lower electron affinity than the conjugated polymer chains, electrons are rapidly transferred from solvent radicals to the polymer. However, if the solvent has a higher ionisation potential than that of the polymer, the electron pulse will produce solvent radical cations that will abstract electrons from the polymer, producing positive polarons. In solid samples, particularly of polymers with large pendent groups, the routes for the production of charges on the polymer are less clear-cut, although it is likely that most of the charge deposited will result in charges on the polymer chains. Measurements have been performed on MEH–PPV solution and polydiacetylenes in partially and fully polymerised crystals. In these examples the individual polymer chains have lengths, from a fraction of a micron to a few microns, longer than the mean-free-path distance traversed by the electrons under the influence of the microwave electric field. For PDA–4BCMU the electron pulses initiate polymerisation of the monomer crystal and the mobility is observed to peak at a low level of polymerisation and then fall, illustrating the sensitivity of the mobility to chain defects produced by cumulative electron beam irradiation. Values obtained for the intra-chain mobility are listed in Table 9.5. The exact fraction of the charge deposited in the sample that is transferred to the polydiacetylene chains is not known, but assuming that all the deposited charge is transferred, leads to a lower limit for the carrier mobility. Actual mobility may be as much as 50 times larger than the values listed for PDAs in Table 9.5.

The influence of chain conformation on mobility is clearly seen in the data for PDA–4BCMU. The mobility is highest for isolated, extended linear polymer chains produced at low concentration in the monomer crystal matrix and falls dramatically for the disordered chains in a yellow solution, but is slightly

Table 9.5 *Intra-chain mobility determined from microwave conductivity measurements*

Polymer	Sample	Carrier polarity	Mobility ($m^2V^{-1}s^{-1}$)
PF2/6[a]	Benzene solution	Positive	7.4×10^{-5}
MEH–PPV	$CHCl_3$ solution	Negative	5×10^{-5}
MEH–PPV	Benzene solution	Positive	4.3×10^{-5}
MeL–PPP	Benzene solution	Positive	1.6×10^{-5}
MEH–PPV	Annealed film	Negative/Positive[b]	$> 2.5 \times 10^{-7}$
$(OD)_2$–PPV[c]	Annealed film	Negative/Positive[b]	$> 3.6 \times 10^{-6}$
PDA–TS	Single crystal	Negative/Positive[b]	$> 3.2 \times 10^{-4}$
PDA–4BCMU	Partially polymerised crystal	Negative/Positive[b]	$> 2.7 \times 10^{-4}$
PDA–4BCMU	Yellow solution in dioxane	Negative/Positive[b]	$> 7 \times 10^{-10}$
PDA–4BCMU	Red 'solution' in benzene	Negative/Positive[b]	$> 1.4 \times 10^{-8}$

[a]Poly(9,9′-bis(2-ethyhexyl)fluorene).
[b]Carrier polarity not determined; the mobility is the average for all carriers present.
[c]Poly(di-octadecaoxy-paraphenylenevinylene).

higher for the more extended and hence ordered chains in the red 'solution'. Comparison with Table 9.4 shows that the latter value is comparable with that found in solid samples of other conjugated polymers. This is reasonable since the red 'solution' is better described as a gel or suspension of chains that are not isolated and have local structure similar to that found in solid samples, i.e. inter-chain transfer occurs as well as intra-chain motion. It can also be seen that the intra-chain mobility for carriers on isolated MEH–PPV chains in solution is 10^6 times larger than that obtained for bulk samples and 10^2 times larger than that obtained in annealed solid films and by the AFM measurements described above. Comparable mobilities are observed for polydiacetylene crystals and the solutions of polymers with a rigid conformation, DEH–PF, MEH–PPV and MeL–PPP. Integrated TOF signals for P3HT and PFO films have been analysed to obtain the product of the trapping time of photo-generated carriers, τ_f, and the initial *free-flight* mobility (Juška *et al.*, 2003). Using a value of 100ps for τ_f from fast photocurrent experiments gives an initial mobility of $\sim 2\times10^{-5}\,m^2V^{-1}s^{-1}$. The values obtained for crystalline polydiacetylenes by TOF and microwave conductivity methods, see Tables 9.4 and 9.5, are similar to theoretical estimates for polarons on ideal, but finite, polymer chains. These results suggest that the intra-chain carrier mobility on defect free conjugated chains is in the range from $10^{-3}-10^{-4}\,m^2V^{-1}s^{-1}$.

Examples of photoconduction action spectra showing that the onset of photoconduction occurs above the excitonic absorption have been presented in Fig. 9.19. Although direct photo-excitation of carriers is possible, the strength of the absorption associated with this process is observed to be weak compared to the excitonic absorption. The situation in conjugated polymers should, therefore, be similar to that described in Section 8.4.2, with excitons being the primary photo-excitations and carrier generation occurring by exciton dissociation and annihilation. Such processes are implied in the schematic decay scheme for photo-excited PPV shown in Fig. 9.24. The precise nature of the mechanism for the generation of photo-carriers has been the subject of time resolved photo-induced absorption studies using sub-picosecond duration laser pulses. Despite the weight of evidence that the primary photo-excitations are excitons, summarised in the preceding section, the interpretation of the data from such experiments has been a matter for debate. The question is whether the photo-carriers generated coincident with the laser pulse are the result of either direct on-chain generation of electrons and holes, i.e. inter-band excitation of polarons, or exciton dissociation or annihilation processes. The papers of Moses *et al.* (2001) and Silva *et al.* (2001, 2002) provide examples of the arguments propounded in this debate.

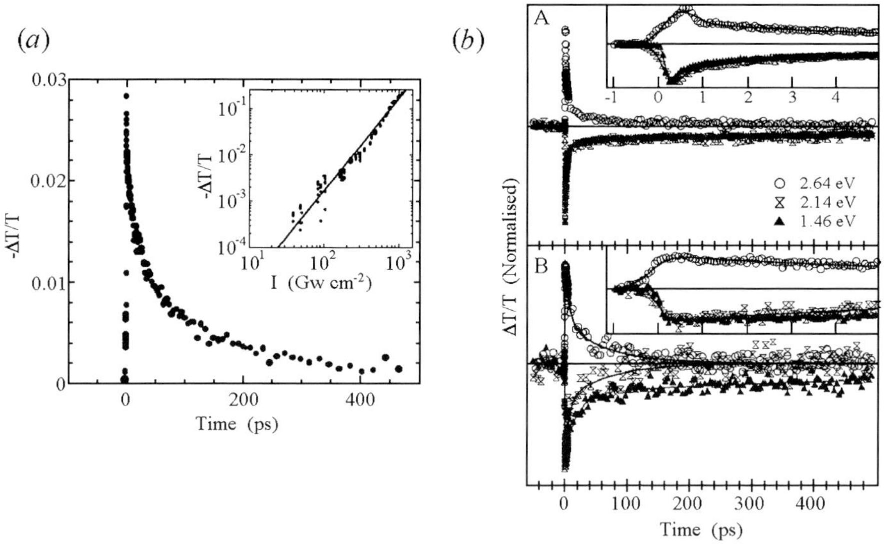

Fig. 9.26 Temporal characteristics of photo-induced absorptions associated with charge carriers generated by short laser pulses. (a) IR absorption in MEH–PPV (reprinted from Moses *et al.*, 2001; copyright 2001, with permission from Elsevier) and (b) optical absorption in a poly(indenofluorene) (reprinted from Silva *et al.*, 2001; copyright 2001, with permission from Elsevier). See text for details.

Moses *et al.* (2001) studied the infra-red active vibrations (IRAV) that occur in both doped and photo-excited conjugated polymers. These are localised vibrational modes of the polymer chain associated with the charged species produced by doping and photo-excitation. The vibrations produce intense infra-red absorptions that are proportional to the number of charge carriers on the polymer chains. Figure 9.26(a) shows the laser-induced IR signal observed for MEH–PPV with very high excitation intensities at photon energy of 1.55 eV. The quadratic intensity dependence of the induced signals, shown in the inset in Fig. 9.26(a), demonstrates that the excitation is via two-photon absorption. Comparison with Fig. 9.24 shows that this excitation is in the vicinity of the two-photon allowed 1^1A_g to 2^1A_g transition as well as the conduction band onset. The growth of the signal is identical with the 100 fs temporal resolution of the equipment used, and is followed by a fast initial drop and then an exponential decay with a time constant of 250 ps. The former is attributed to carriers that recombine with other charges on the same polymer chain, while the longer time decay is of carriers that transfer to other chains after their creation. The fast initial decay is not observed in PPV, where the absence of large pendent groups allows for more efficient inter-chain separation of the photo-excited charges. Because the IRAV grow faster than charges can be created by exciton–exciton annihilation, Moses *et al.* (2001) conclude that they must result from direct one-photon excitation of carriers, i.e. an inter-band transition.

Silva *et al.* (2001) studied the temporal behaviour of the photo-induced spectrum of poly(indenofluorenes) with octyl and 2-ethylhexyl pendent groups between 1.4 and 2.7 eV. Short pulse laser excitation at 3.18 eV was close to the peak of the principal absorption of the polymers, i.e. the 1^1A_g to 1^1B_u transition. The photo-induced spectrum was probed with a pulsed white-light continuum subjected to a variable time delay. The spectrum shows a positive signal above 2.45 eV due to stimulated emission (SE) induced by the probe pulse from the ^1B_u level, Fig. 9.26(b). Below 2.45 eV the photo-induced absorption (PIA) spectrum shows two peaks at ~2.1 and 1.5 eV. The PIA at 2.1 eV is assigned to absorption from the excited ^1B_u state to a higher energy m^1A_g state, while that at 1.5 eV is assigned to charge pairs. The observed growth of the latter signal is identical with the experimental temporal resolution of ~100 fs. Silva *et al.* acknowledge that this is much faster than is possible by exciton–exciton annihilation, which occurs on a ps timescale. At high pump power the SE decay has sub-ps and ~50 ps components, while the PIA has an additional slower ns time constant tail, as shown in the upper part of Fig. 9.26(a). At low pump power, shown in the lower part of Fig. 9.26(b), the sub-ps component is less pronounced in all the signals and the PIA at 2.14 eV is similar to the SE with little evidence of the ns tail, although this is still strong in the PIA at 1.46 eV. The fast decay at high pump power is attributed to the combined effect of the amplified stimulated emission of light propagating in the plane of the thin film sample and exciton–exciton bimolecular annihilation at the high exciton densities produced. A kinetic model for fast charge pair generation during the excitation pulse by two-step one-photon excitation, first to the 1^1B_u level and then to a highly excited exciton, followed by very rapid dissociation, gave a good fit to the experimentally observed power dependence of the 1^1B_u level PIA. This process is observed to be efficient, ~10% quantum efficiency, in the ordered crystalline regions of the polymer.

The fact the charge carriers are generated via excitonic states in conjugated polymers is supported by other studies of both photo-induced optical and IR spectra. Gadermeier and Lanzani (2002) presented an overview and suggested a model that is consistent with the picture of the electronic excited states presented in the previous section, Fig. 9.25. The processes that they suggest occur after photo-excitation, illustrated in Fig. 9.27, are: (1) vibrational relaxation, (2) singlet fission into triplet pairs, (3) inter-chain relaxation, i.e. excitation diffusion, (3′) singlet dissociation into carriers, (4) induced absorption, (5) dissociation of high lying states, (6) radiative and non-radiative decay

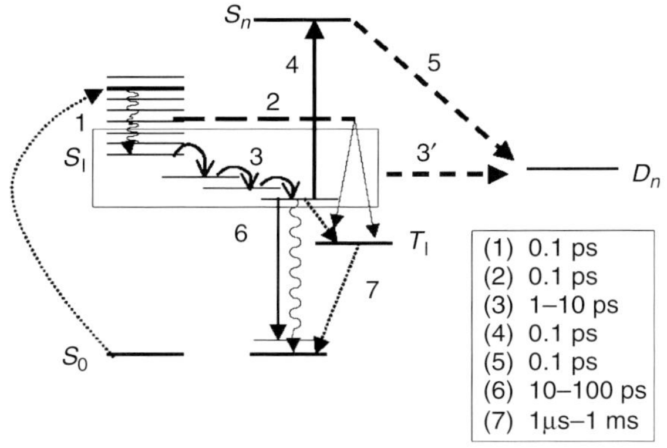

Fig. 9.27 Schematic representation of the decay of photo-excited conjugated polymers. Reproduced with permission of the Institute of Physics from Gadermeier and Lanzani (2002).

of S_1, and (7) decay of T_1. The timescales of these processes are indicated in the diagram. It is clear from the literature that there are a number of possible mechanisms for carrier generation, and while this model is a reasonable basis for understanding the photo-physics of conjugated polymers, there is probably not a single dominant process in all materials. Photoelectron emission has been observed from polyfluorene, PTh and PPV thin films for photon energy less than 4 eV and low light intensity (Wegewijs *et al.*, 2000). The photoelectrons will give rise to a photoconduction signal when surface electrodes are employed. Immersing the sample in a CO_2/SF_6 atmosphere that captures the photoelectrons quenches this signal. In thin PDA–4BCMU crystals disordered by heating, the onset of photoconduction shifts from the vicinity of the band edge, revealed by electro-absorption spectroscopy, to the exciton absorption edge. Möller *et al.* (2001) attribute this to an Auger process in which there is simultaneous excitation of trapped electrons into the conduction band and decay of an excited exciton. These observations would also be consistent with the suggestion made in the previous section that disorder mixes the exciton and conduction band states. Frankevich *et al.* (2002) studied fast charge carrier generation with correlated pairs of femtosecond laser pulses in MeL–PPP and concluded that the initial excitations were intra-chain polaron pairs bound by Coulomb interaction. They argue that electric field assisted dissociation of the initial pairs, followed by random recombination, leads to the release of free polarons. Gulbinas *et al.* (2002), on the basis of fast PIA studies, conclude that free polarons result from the electric field dissociation of both the vibrationally excited and relaxed S_1 exciton. Studies of IRAV in poly(dioctylphenylene-vinylene) (Mizrahi *et al.*, 2003) and the quantum efficiency of carrier generation

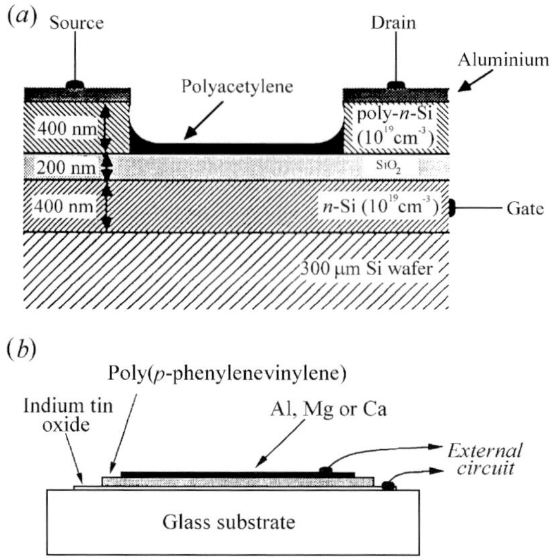

Fig. 9.28 Structures of electronic devices incorporating conjugated polymers, (a) FET incorporating Durham route PAc (reprinted with permission from *Nature*, Burroughes *et al.*, 1988; copyright 1988 Macmillan Magazines Limited) and (b) LED incorporating PPV (reprinted with permission from *Nature*, Friend *et al.*, 1999a; copyright 1988 Macmillan Magazines Limited).

in P3HT and a PFO copolymer (Österbacka *et al.*, 2003) are examples of an increasing body of work supporting the exciton model. On the basis that exciton dissociation occurs at specific charge transfer centres, the rate of carrier generation will be determined by the probability of excitons encountering these centres as they diffuse through the polymer (Arkhipov *et al.*, 2004). Since the jump rate of excitons diffusing in a density of states distribution decreases more rapidly than the exciton density, this model leads to the conclusion that the characteristic time for carrier photogeneration will be shorter than the exciton lifetime. Thus, on the balance of the evidence, the mechanism of fast photo-carrier generation is exciton dissociation.

Processable conjugated polymers can readily be incorporated into electronic devices through the spin-coating process. The devices that have attracted most attention are field effect transistors (FETs) and thin film diodes, the latter primarily as light emitting diodes (LEDs). Typical structures are shown in Fig. 9.28.

The intractability of Shirakawa PAc limited the earliest devices to be studied to diodes fabricated by depositing metal contacts on opposing surfaces of thin films. Gold was used to provide an ohmic contact and low work function metals such as indium were found to form Schottky barriers, see Fig. 7.14. The high bulk resistance of accidentally doped polymers limits the forward current to below 10 Am^{-2}. Useful current densities require the use of doped polymers, as discussed in the next section. The first FETs were made by the electrochemical growth of a conductive film across the gap between closely spaced gold electrodes on an insulating substrate and by subsequent electrochemical reduction of the film. The availability of soluble precursors and conjugated polymers enabled production of FETs by spin coating on to silicon substrates, with preformed gate and drain electrodes of gold or poly-silicon, see Fig. 9.28. The earliest examples used either soluble poly(3-alkylthiophene) or Durham-route precursor PAc: these were accidentally doped and behaved as *p*-type semiconductors. For an *n*-type Si substrate a depletion region forms at the surface of the polymer and the conductance of the channel between the source and the drain electrodes is high. For a positive gate voltage, electrons are attracted to the polymer–gate interface and a thin *n*-type inversion region forms in the polymer; for a negative gate voltage holes are attracted to the interface and a *p*-type accumulation region is formed. The higher carrier density in the polymer results in a higher channel conductivity. For gold electrodes the higher conductivity occurs for negative gate bias, i.e. the accumulation region has more carriers than the inversion region. However, *n*-type source and drain, as in Fig. 9.28, makes Ohmic contact to the *n*-type PAc, giving a higher net conductivity for positive gate voltages. In this device full depletion of the channel occurs at a gate–source voltage, V_{GS}, of $+10\,V$, and maximum channel conductivity of 10^5 times the minimum occurs at V_{GS} $+40\,V$. The development of devices with characteristics suitable for practical application will be described in Section 10.3.1.

The characteristics of the FET described above and Schottky diodes and metal–insulator–semiconductor (MIS) diodes fabricated with Durham-route precursor PAc (Burroughes *et al.*, 1988) show unambiguously the formation of depletion, inversion and accumulation layers in the polymer. The Schottky diodes have characteristics close to those for an ideal device, with no surface states pinning the energy of valence and conduction bands at the interface. Surface states in inorganic semiconductors make the fabrication of reproducible Schottky diodes difficult, since device characteristics are very sensitive to the nature and density of surface states. Furthermore, the formation of depletion, inversion and accumulation layers requires semiconductors with low densities of defects that have energy levels within the semiconductor bandgap, i.e. high purity Si. At first sight the behaviour of the devices made with Durham-route precursor PAc, which has very low crystallinity, is surprising. In fact these experimental findings highlight the drastically different natures of inorganic and polymeric semiconductors. Surface states occur in inorganic semiconductors such as silicon, because the surface contains dangling bonds where the surface disrupts the three-dimensional, covalently-bonded crystal lattice. In comparison, conjugated polymers contain discrete molecules with all chemical bonds fully occupied, so that, provided the surface they are in contact with is clean, there is nothing to form a significant density of surface states. Impurities and lattice defects in inorganic semiconductors tend to inject states into the band gap. In conjugated polymers physical and chemical defects that interrupt the π-bonding give rise to shorter conjugated segments with greater separation between the LUMO and HOMO levels and the continuum states. Thus these defects produce energy levels that do not lie in the band gap and will not affect device behaviour. But this does not mean an absence of trapping levels in conjugated polymers, since these are produced by chemical defects, in particular keto groups, resulting from oxidation, act as electron traps.

Charge storage in inorganic semiconductors occurs as holes in the valence band or electrons in the conduction band. In PAc charge storage occurs as charged solitons, while in other conjugated polymers it occurs as polarons and bipolarons. Since the PAc chains are finite in length and the bonding at the chain ends is fixed, the charged solitons must be injected in pairs, so that the change in bond alternation introduced by a single soliton, Fig. 9.10, is cancelled out by the second. For other conjugated polymers the need to minimise energy constrains the size of the polarons and bipolarons, see Section 9.3.1 and Fig. 9.14. Evidence for the presence of these charged species is provided by the IR spectra induced by voltages applied to the diodes and FETs. These spectra can be observed because the silicon substrates are transparent in this spectral region. The spectra show both broad induced and sharp IRAV absorption bands, similar to those produced by chemical doping, see Section 9.4.4.

Although the physical defects in conjugated polymers do not adversely affect the properties of electronic devices fabricated from them, as noted above, the presence of chemical impurities and dopants must also be taken into account. Chemical impurities or dopants will produce intermediate layers at the interface between the polymer and metal and semiconductor contacts. These additional layers may result from chemistry either at the surface of the inorganic material or in the polymer adjacent to the interface or both. Ion milling has been used in conjunction with analytical scanning electron microscopy to study the structure of such interfaces. Typically, oxide layers are found on the surface of reactive metals and semiconductors, while reaction of the polymer with the electrode materials can produce either insulating, reduced layers or conductive, doped layers. The existence of such accidental insulating layers at the interface has in some instances been beneficial to device performance by allowing charge injection from the electrode, but blocking charge transfer from the polymer to the electrode. The physics of real interfaces has been extensively explored by the use of X-ray and ultra-violet photoelectron spectroscopy (XPS and UPS) (Salaneck *et al.*, 2001). XPS and UPS analyse the photoelectrons ejected from valence band states and can be used to study bulk materials and thin polymer films deposited onto electrically conductive substrates. For bulk samples that are not highly conductive, charge builds up on the sample surface and must be neutralised with a background electron beam, introducing an uncertainty into measured photoelectron energy. This is not a problem for thin films studied by UPS, and, given the interest in thin film devices, this has become the method of choice. Figure. 9.29(a) shows the UPS spectra of a gold substrate and a thin film of poly-(dioctylfluorene) (PFO) on a gold substrate.

The position of the Fermi level, E_F, of the gold relative to the vacuum is determined by its work function, Φ_{Au}. The vacuum level for the polymer film is offset by an amount Δ, if there is a dipole layer at the metal–polymer interface, and the ionisation potential is the sum of the offset of the polymer valence band and vacuum level from the Fermi level of the substrate, $E_F^{VB}$and E_F^{Vac}, respectively, as indicated in Fig. 9.29(a). The maximum photoelectron energy is equal to the photon energy. This enables Δ to be determined from the relative positions of the secondary electron cut-offs for the substrate and the polymer film. Since thermodynamic equilibrium requires that the position of the Fermi level is constant throughout the sample and its position is determined by the metallic substrate, then $E_F^{VB}$and E_F^{Vac}can be found. This information provides a basis for modelling charge injection and removal at the interfaces in polymer

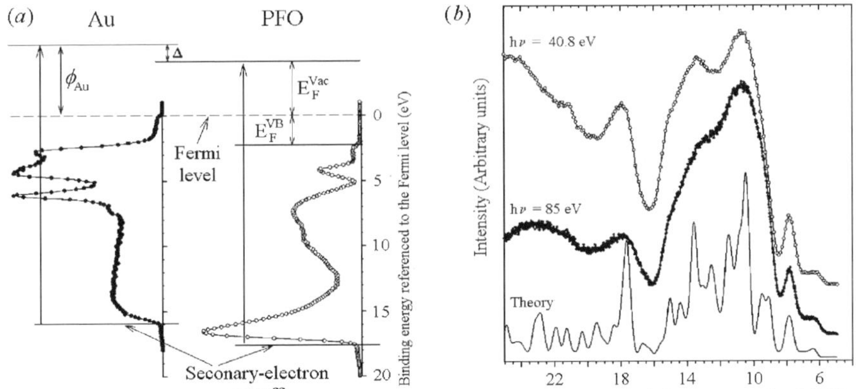

Fig. 9.29 (a) UPS spectra of a gold surface and a thin film of PFO on gold recorded with excitation with photons of energy 21.2 eV, and (b) measured and model valence band density of states for PFO. Reprinted from Salanek *et al.* (2001). Copyright 2001, with permission from Elsevier.

diodes. As for saturated polymers, Section 4.2.5 and Fig. 4.14, the valence band structure obtained by XPS and UPS, particularly with higher energy photons, can be compared with that from model calculations. Although many of the predicted peaks cannot be observed, the general shape of the experimental data mirrors the theoretical density of states. While photoelectron spectroscopy is a valuable tool for studying the density of states of the valence band, the location of the band is subject to some uncertainty, since, despite attempts to eliminate surface charging, the presence of double charge layers may affect the data. XPS and UPS provide no information about the location of the conduction band. In much of the literature the conduction band edge has been located by taking the band gap to be equal to the energy of optical absorption edge. In the light of the results discussed in the previous section, this places the conduction band too low in energy. Since photoelectron and optical spectroscopy involve different electronic excitation and relaxation processes, combining data from these methods can at best give only approximate energy level schemes.

The electro-luminescence from conjugated polymers was discovered accidentally during measurements on thin film polymer diodes (Burroughes *et al.*, 1990). This created great interest in the possibility of practical LEDs and acted as a stimulus for the investigation of charge injection, transport and recombination in thin films, Fig. 9.30. This discovery came at a time of heightened interest in conjugated polymers, but it was not the first observation of electro-luminescence from a polymer diode. Electro-luminescent devices based on PVK were patented in 1974, although the results did not appear in the scientific literature until 1983 (Partridge, 1974, 1983), but it was a discovery ahead of its

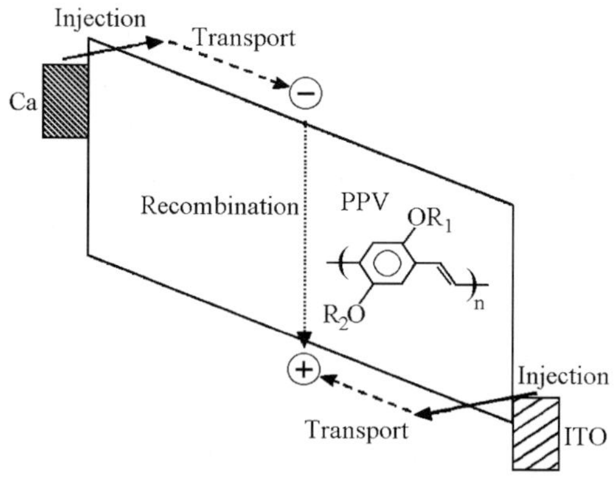

Fig. 9.30 Schematic diagram of the band structure and electronic processes in a PPV LED. Reprinted from Blom and Vissenberg (2000). Copyright 2000, with permission from Elsevier.

time and stimulated little interest. The simplest LEDs consist of a metallic electrode suitable for electron injection, typically Ca or Mg, a thin polymer film and a transparent hole injection electrode, typically indium tin oxide (ITO), Figs. 9.28(b) and 9.30. The properties of LEDs utilising soluble PPVs have been reviewed and analysed by Blom and Vissenberg (2000). They note that it not always possible to determine the relative importance of charge injection and transport from LED characteristics alone. The XPS and UPS studies described in the preceding paragraph and studies of Schottky diodes have been useful in identifying electrode–polymer combinations with either Ohmic or hole- or electron-injecting behaviour. Modelling of device characteristics using established models of SCLCs, Fowler–Nordheim tunnelling, etc., has also been applied. Measurements of carrier mobility, discussed earlier in this section, are compatible with transport dominated by hopping between discrete polymer chains, suggesting that the GDM approach described in Section 8.4.1 should also be applicable to conjugated polymers. Blom and Vissenberg (2000) extend this model by assuming, in addition to a Gaussian disorder of site energies, an inhomogeneous structural order. The structural inhomogeneity affects the local electronic coupling between polymer chains and hence the local carrier mobility. Therefore the carrier mobility fluctuates throughout the polymer. This inhomogeneity is modelled as an alternation of regions of high and low mobility, a physically reasonable approximation for a partially crystalline polymer. Carriers can oscillate within the regions of high mobility before they escape through the adjacent region of low mobility, so that the high mobility regions act as reservoirs of charge. This complex system is thereby reduced to one where carriers hop along a one-dimensional conduction path with charge reservoirs at both ends. This model results in mobility with a dependence on the field E of the form $\mu \propto \exp\sqrt{E}$ over a wide field range, and gives a better fit than the GDM for the mobility of holes in a series of soluble PPVs between 150 and 300 K. The predicted dependence is very similar to that of the correlated GDM of Dunlap *et al.* (1996) described in Section 8.4.1.

Device characteristics are usually analysed using established continuum models (Walker *et al.*, 2002). Charge injection is treated using a conventional band scheme, as shown in Fig. 7.14, and the overall device structure is treated in a similar manner, see Fig. 9.30. In general, the low mobility leads to the result that devices function in a SCLC regime. As was seen in Section 8.4.1, traps play an important role in these models, and the nature of the traps present in thin conjugated polymer films has been a matter of debate. The density of shallow traps is often assumed to decay exponentially into the band gap, and deep traps have been modelled as both discrete and exponential energy distributions. The dependence of the current on voltage for an exponential trap distribution was analysed by Jain *et al.* (2002). Initially the current depends linearly on the voltage, but then increases more rapidly in the SCLC regime with a slope given by V^{α}, where $\alpha = T_C/T$, T_C is the characteristic temperature of the trap distribution and T is the sample temperature. At high applied voltages the current increases more slowly, tending towards a V^2 Child's law dependence. This model fits experimental data, and the deviation from the SCLC regime has been interpreted as the trap-filled limit. However, the model shows that the current approaches, but does not attain, a trap-free Child's law voltage dependence, and that the trap-filled limit is approached asymptotically as the applied voltage goes to infinity. In the SCLC regime the trapped carriers determine the internal field in the sample. The current/voltage characteristic deviates from this when the number of free carriers present is sufficient to affect the internal field. As the number of free carriers increases, they will eventually determine the internal field and the trap-free limit will be reached. Given the high density of traps in a typical sample, this can only be achieved in the limit of an infinite applied field. The distribution of trapped charge in 3–4 μm thick films of poly(2,5-pyridinediyl), MEH–PPV and PFO has been determined by the analysis of the pyroelectric response produced by a thermal wave propagating in a polarised film (Feller *et al.*, 2002). A narrow, positively-charged layer is observed close to the anode with a broader, negative layer at between 0.5 and 1 μm from the electrode. This indicates that injected positive charge carriers are trapped close to the electrode, while negative charges generated in the bulk of the film accumulate in the trapping zone.

There have been surprisingly few direct experimental observations of trap distributions. The thermal release of trapped charge gives rise to thermally stimulated currents (TSC), a method that has been used to study poly(*p*-phenylene), precursor route PPV and soluble PPVs. The capacitance of a Schottky diode is determined by the width of the depletion region, which depends on the charge stored in this region. Application of a forward bias voltage pulse can pull trap levels below the Fermi surface, increasing the stored charge. Deep-level transient spectroscopy (DTLS) utilises the temporal variation in capacitance after a voltage pulse to measure the emptying of trap and interface states. This method has been used to study precursor route PPV. TSC studies indicate the occurrence of traps lying between 0.03 and 0.9 eV into the band gap. Von Malm *et al.* (2002) observed broad trap distributions between 0.2 and 0.6 eV with noticeable peaks and complex shapes. For poly-(2-methoxy-5(3,7-dimethyloctyloxy)-1,4-phenylenevinylene), MDOO–PPV, they found that hole traps were dominant, although electron traps were also observed. Both photo-excitation and forward biasing a diode were used to fill traps. Similar TSC signals were observed for an integrated current that was

$\sim 10^5$ times smaller for optical trap loading than for current loading. This result was rationalised by assuming that the polymer contained percolation paths for hole transport that are relatively free of traps. These paths dominate the transport for low currents and voltages, while at high currents and voltages more uniform charge injection occurs, accessing all existing trap sites. TSC data for precursor route PPV indicates traps with maximum densities over the ranges 0.3 to 0.6 eV and 0.13 to 0.18 eV into the gap. DTLS results for this polymer indicate the existence of a discrete trap level for holes 0.75 eV into the gap, and that carriers decay from this trap into a Gaussian distribution of transport states with a standard deviation of 0.1 eV (Campbell *et al.*, 2000). The density of traps in PPVs with energy depths greater than 0.2 eV increases with oxygen exposure, i.e. are probably associated with defects resulting from degradation of the conjugated polymer.

Recombination of carriers injected into a diode structure requires that carriers pass within a collision-capture radius of one another. For this to be an efficient process in polymer films with thickness $\sim$100 nm there must be a high density of one sign of charge carrier. In many polymers electrons are somewhat less mobile than holes, so that this condition is satisfied. It can also be achieved by device design. Electrodes that inject holes but block electron transfer from the polymer to the electrode have been used to build up a high electron density adjacent to the electrode. Multi-layer devices have been constructed with electron and hole transporting layers that enable high carrier densities to be established at the interface between them. Where the band-edge offset at the interface between the electron and hole transporting layers is less than the exciton binding energy, typically 0.5 eV, excitons are stable at the interface and electro-luminescence is observed. A third emissive layer has also been included, sandwiched between electron and hole transporting layers. The Langevin model is usually used to describe carrier recombination. This assumes that carriers diffuse to the recombination sites, which are not trap sites, and the process is spin-independent. Comparison of fluorescence and electroluminescence spectra shows that both types of emission predominantly occur from the lowest excited singlet exciton energy level. The assumption of spin-independent recombination means that singlet and triplet excitons occur with equal probability, and, since the triplet state is threefold degenerate, only 25% of the carriers will recombine as singlets. Thus most of the injected charge is expected to form triplets that, as seen in the previous section, mostly decay non-radiatively and give rise to only weak emission. This is reflected by the fact that there have been few reports of electro-phosphorescence. Lupton *et al.* (2002) and Reufer *et al.* (2003) describe electro-phosphorescence in MeL–PPP, where ISC was mediated by a 100 ppm palladium impurity. Sinha *et al.* (2003)

Table 9.6 *Measured exciton formation cross-section ratios and maximum fraction of singlet excitons generated during carrier recombination at 20 K unless otherwise noted*

Polymer	PPV	MEH–PPV	MeL–PPP	PFO	PTV[a]	MDOO–PPV[b]
σ_S/σ_T	2.2	2.7	3.4	3.9	4.9	12 ± 6
$\eta_{\text{Rec}}^{\text{Max}}(\%)$	42	47	53	57	62	83 ± 7

[a]Poly(thienylenevinylene),
[b]Dhoot *et al.* (2002), measured at 10 K.

report intrinsic electro-phosphorescence from a polyfluorene. Experiment indicates, however, that a higher proportion of singlet states is produced than suggested by the simple charge recombination model (Köhler and Wilson, 2003). Wohlgenannt *et al.* (2001) have employed PIA and the modulation of the PIA by spin-half magnetic resonance to determine the ratio of the formation cross-sections for singlet and triplet excitons, σ_S/σ_T. The maximum fractional conversion of recombining charge carriers into singlet excitons is then: $\eta_{\text{Rec}}^{\text{Max}} = \sigma_S/(\sigma_S + 3\sigma_T)$. Measured values for these quantities are given in Table 9.6.

While there is some debate about the influence of chemical defects on this data, there are a number of other measurements that give values of σ_S/σ_T that are all significantly higher than the limit for spin-independent recombination. Methods that give theoretical values of σ_S/σ_T similar to those observed experimentally are (1) the inclusion of inter-chain interactions in the calculation of exciton formation by inter-chain charge transfer and (2) calculation of the mixing of singlet and triplet excitons by spin–orbit coupling in PPV as a function of the angle between the plane of the phenyl ring and the plane of the polymer backbone. Although σ_S/σ_T differs from the prediction of the Langevin model, the rate constant for bimolecular recombination of electrons and holes is found to be close to this model both experimentally and theoretically, i.e. the rate-limiting step is the diffusion of the charged carriers towards one another under their mutual Coulomb attraction. This behaviour is expected for hopping conduction and is consistent with the view that carrier transport in conjugated polymers is dominated by hopping.

The picture that has emerged from the studies of the electronic properties of conjugated polymers, described in this and the preceding section, can be summarised as follows. Electron correlation is a major factor that cannot be neglected, as it is in the SSH model of PAc and similar molecular orbital models. The consequence of this is that the lowest excited states of conjugated polymers are excitons. The binding energy of the lowest singlet exciton lies in the range 0.6 to 0.9 eV. The overall energy level schemes of conjugated polymers are as shown in Fig. 9.25. A number of singlet and triplet excitons are observed to lie below the conduction band onset. Photo-excitation leads to a population of excitons, which can either decay radiatively or non-radiatively or dissociate into holes and electrons in the valence and conduction bands. These processes are controlled by the presence in the polymer of sites where one or other of these processes is favoured and the inter- and intra-molecular diffusion of the excitons to these sites. Carriers in the valence and conduction bands form polarons, or solitons, as the structure of the polymer chain is perturbed by the presence of the charged species. In bulk samples, although the charge carriers can be highly mobile on the polymer backbone, charge transport is controlled by hopping of the carriers between polymer chains, not by band-like motion. This results in the low bulk mobilities observed experimentally.

9.4.4 Doped and metallic polymers

It should be noted that the term doping was introduced by analogy with the introduction of dopants into inorganic semiconductors, but the analogy is poor. The level of dopants required to give useful conductivity in inorganic semiconductors is low, usually much less than 1 % by mass. Figure 9.1 shows a much higher level of dopant incorporated into PAc; values of tens of percent are typical for this and other conjugated polymers with metallic conductivity. This is a level typical of the formation of a distinct chemical compound. A better analogy is with the intercalation of graphite, where a new compound is formed by the insertion of layers of the added species between the graphite layers. X-ray analysis shows that this process occurs in the crystalline regions of PAc doped with iodine where the linear I_3^- and I_5^- ions lie between and parallel to the polymer chains. Such a regular arrangement is less likely for bulkier counter ions and impossible in the amorphous regions of partially crystalline polymer samples. Thus a consequence of the polymer morphology is that both the doping and the electrical properties will be heterogeneous.

The nature of the semiconductor obtained by chemical doping will be determined by the oxidation and reduction potentials of the conjugated polymer relative to those of the chemical oxidising and reducing agents. Hence *p*- and *n*-type semiconducting materials can be obtained by suitable choice of oxidising and reducing agents. For example, iodine, arsenic pentafluoride and numerous other compounds are able to oxidise PAc, while it can be reduced with liquid sodium amalgam or sodium naphthalide. Both processes can be achieved electrochemically with the insertion of counter ions to retain charge neutrality in the doped polymer, e.g. for *p*-type PAc $(ClO_4)^-$ and for *n*-type Li^+. The utility of a particular doping process is determined by the ease with which it can be accomplished and the stability of the doped material. In practice most conjugated polymers are more readily oxidised than reduced. The *p*-type materials are susceptible to attack by oxygen, though in favourable cases this can be a rather slow process, e.g. PPy and PAni. *n*-type polymers are more prone to hydrolysis and are generally less chemically stable than *p*-type polymers. Excessive oxidation, over-oxidation and reduction are all liable to be accompanied by chemical degradation and care must be taken to avoid this during the doping process. The doping process requires the diffusion of the dopant species into the polymer and the rate of this process is found to affect the distribution of dopant in the polymer. In general a slow rate of doping gives a more uniform dopant distribution than a fast rate.

In addition to the increase in conductivity when PAc is doped, there are also changes in the magnetic and optical properties. Electron spin resonance (ESR) reveals the presence of unpaired electrons in as-made, accidentally doped PAc. At dopant concentrations less than 0.1 mass% this signal is unaffected, but it decreases rapidly at a doping level of ~1 mass%. No paramagnetism is observable until the doping level reaches 5 mass%, when an increasing Pauli paramagnetism[3] is observed. Comparison of these results with Fig. 9.1 shows that the electrical conductivity increases rapidly at ~1 mass% dopant concentration. These observations were interpreted in terms of the presence of intrinsic chain defects, solitons, in the as-made material. The solitons are electrically neutral, but have an unpaired spin, see Fig. 9.12. On doping, electrons will be either added or removed from the lowest unoccupied or highest occupied energy level, i.e. the soliton level. This will give rise to charged soliton objects, with charge but no spin, that can contribute to the passage of electrical current but cannot be seen by ESR. As the level of doping increases, the solitons will interact to form polarons that carry both charge and spin, resulting in the Pauli paramagnetism. While this model fits well with the data for doping above the 1 mass% level, the constancy of the ESR signal at low doping levels requires that such disparate dopants as iodine and AsF_5 either combine with existing solitons, removing free spins, or create new solitons on defect-free chain segments with an equal probability (Bernier, 1986). The doping process is heterogeneous, reflecting the partially crystalline nature of the PAc (Pekker and Janossy, 1986). Thus it

[3] This is characteristic of a metal and occurs because, in the presence of a magnetic field, the energy of spins parallel to the magnetic field is lowered and that of anti-parallel spins is raised. With a constant Fermi energy this means that anti-parallel spin electrons above E_F will flip their spins and occupy the empty parallel spin states below E_F. There is then a preponderance of parallel spins that renders the metal paramagnetic.

is possible that, at low dopant concentrations, the pre-existing spins in crystalline regions are not affected by the dopants, while in the amorphous regions they are converted directly into polarons.

While the conductivity of conjugated polymers increases on doping, in only a few cases does it exceed the minimum metallic conductivity of about $10^4 \Omega^{-1} m^{-1}$. Thus most behave as disordered semiconductors. Furthermore, few of the highly conductive materials remain metallic at low temperatures, as the conductivity generally falls rapidly to that of an insulator as the temperature is reduced. This is a consequence of the high structural disorder present in most materials. At present only PAc, PAni and PPy have been observed to remain metallic at low temperatures. Isolated experiments suggest that further reduction in the defect content of the regioregular PThs and oriented PPV may result in materials with similar behaviour. In these limited cases the transformation from a low conductivity semiconductor to a metal as the polymers are doped is rapid and can be described as an *insulator-to-metal* transition. Three distinct regimes have been observed to occur during this transition:

- the insulator (semiconductor) regime,
- an intermediate or critical regime, and
- the metallic regime.

We will first consider how doping affects the properties in the semiconductor regime common to all conjugated polymers.

The optical spectrum of PAc changes continuously during doping to 5–10 mass% of dopant, Fig. 9.31. The initial polymer contains a mixture of *trans*- and

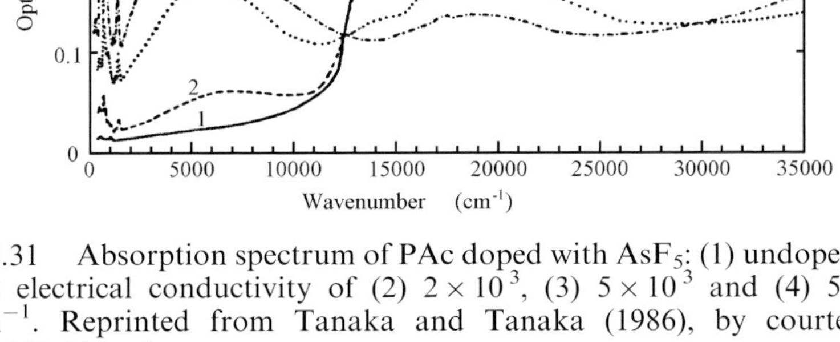

Fig. 9.31 Absorption spectrum of PAc doped with AsF_5: (1) undoped, and to an electrical conductivity of (2) 2×10^3, (3) 5×10^3 and (4) 5×10^4 $\Omega^{-1} m^{-1}$. Reprinted from Tanaka and Tanaka (1986), by courtesy of Marcel Dekker, Inc.

cis-forms, giving the broad shoulder below 16 000 cm^{-1} and the features above 16 000 cm^{-1}, respectively. As doping progresses, the polymer absorptions weaken and a series of absorptions appears at lower energy. Those below 2000 cm^{-1} are the IRAV bands that are also seen in photo-excited, undoped polymer, as described in the previous section. These occur at different frequencies than do the vibrations of the pristine polymer and vary slightly for different materials and dopants. The frequencies and absorption strengths are similar to those predicted for either a charged soliton or a polaron, so that a definite identification of the species involved is not possible from the spectral data (Gussoni *et al.*, 1991). The broad band is attributed to the solitons created by the dopants, since transitions can occur from the valence band to the soliton level for a positive soliton and from the soliton level to the conduction band for a negative soliton, see Fig. 9.12. In the absence of electron correlation the transitions would appear at half the band gap, but correlation shifts the soliton level away from mid-gap. Given that the conduction band lies higher in energy than the optical absorption edges, the position of the doping-induced absorption in Fig. 9.31 lies well below mid-gap, consistent with other data indicating significant electron correlation, see Section 9.3.2.

The other feature of Fig. 9.31 is the occurrence of an isobestic point at ~12 000 cm^{-1}, which indicates that only two species are present, i.e. undoped and doped polymer. This is somewhat surprising given the presence of *cis*- and *trans*-PAc in the sample, but is possible if doping occurs at a similar rate in both isomers, since doping converts the *cis*- to the *trans*-form. Moreover, the data span the region above 1 mass% of dopant, where the ESR indicates direct creation of charged solitons, i.e. there is a single product. Note that the changes in absorption below 0.1 mass% are too small to provide information in the regime where the doping process is less clear.

For other conjugated polymers the species created on doping are expected to be polarons and bipolarons. There are then two levels within the band gap with either one electron in the lower level for the polaron, see Fig. 9.12(f), or two empty levels for the bipolaron. In the latter case there will be two possible absorptions, while for the polaron, the presence of the electron makes an additional absorption possible. Examples of the doping-induced spectra for a PT with the dopant forming part of a group pendent to the thiophene ring, the sodium salt of poly(3-thiopheneethanesulphonate) (PTESNa), and poly(*iso*thianaphthalene) (PITN), are shown in Fig. 9.32(a) and (b), respectively. As doping proceeds, two induced absorption bands are seen. Similar spectra are observed for PT and for soluble, alkyl-substituted PTs as solids and, in the latter case, in solution. The relative strength of the polymer and induced absorptions indicates that the level of doping achieved in the PTESNa

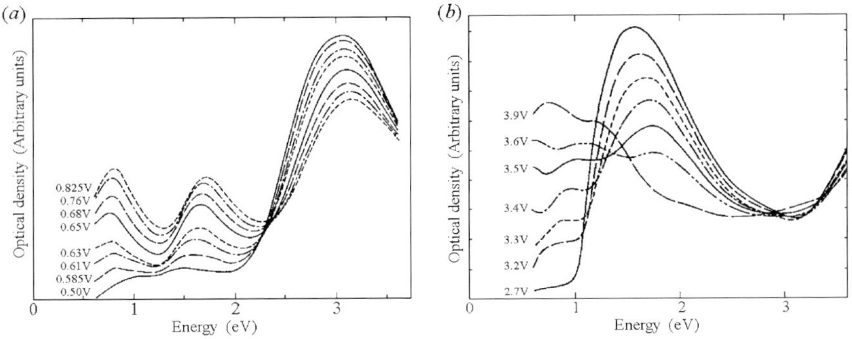

Fig. 9.32 Spectra of (a) PTESNa and (b) PITN during electrochemical doping. The counter-ion in (b) was Cl^- and the voltages indicate the electrode potential. Reprinted with permission from Patil *et al.*, (1988). Copyright 1988 American Chemical Society.

is quite low. The dopant concentration in the PITN sample varies from 0.08 to 12.3 mass%, the maximum achievable. The appearance of two induced absorption bands indicates the formation of bipolarons. The absence of a clear isobestic point suggests that there is some polaron formation, but that these are only a minority species in the doped polymer. During the doping of PPy, three absorption bands appear at low dopant levels, but over most of the range up to doping levels of 33 mass% there are only two induced absorptions. This has been interpreted as the initial formation of polarons that merge to form bipolarons. The changes visible in the spectra shown in Fig. 9.32(a) and (b) are compatible with this model. Comparison of optical spectra recorded during chemical doping, and photo-induced spectra obtained on undoped polymers, also indicates that the charged species produced initially are polarons, despite the presence of the counter-ions in chemically doped materials. In addition to the broad induced absorption bands, sharp IRAV are also observed, similar to those seen in PAc. Both broad and IRAV absorptions are polarised parallel to the polymer chain in oriented samples (Miller *et al.*, 2001).

PAni has a very complex structure and doping behaviour, see Fig. 9.6, and the spectra are sensitive to the polymer morphology, the level of oxidation and degree of protonation. This accounts for the considerable variation in the spectra that have appeared in the literature. The effects are illustrated in Fig. 9.33 for various forms of the protonated salt. These spectra refer to dried films, electrochemically prepared at different electrode potentials, and subject to oxidation by exposure to air. This variation in preparation conditions means that the degrees of oxidation and protonation are not well defined, as evidenced by the pronounced differences in the spectra of the emeraldine prepared at the same electrode potential. The fact that these films contain only short, defect-free segments of polymer chain also affects the spectra. More consistent spectra have been obtained with chemically prepared PAni, which has longer, defect-free chain segments. In N-methylpyrrolidone solution leucoemeraldine has a single absorption peaking at 3.6 eV, emeraldine base has two peaks at 1.94 and 3.75 eV and pernigraniline base has three peaks at 2.3, 3.8 and 4.3 eV. An identical spectrum is observed for spin-cast films of emeraldine base, while similar films of the emeraldine salt have a peak at 3.5 eV and a broad featureless band extending from ~1.8 to < 0.4 eV, the limit of measurement. The absorptions above 3 eV have been assigned to the HOMO–LUMO transition of the polymer chain, on the basis of EA data for emeraldine base (Pomfret *et al.*, 1996), those near 2 eV to self-trapped charge transfer excitons and the lower energy peaks to either localised or delocalised polarons. The short chain segments in electrochemically polymerised samples result in localised polarons, with distinct absorption peaks, as in Fig. 9.33, while the longer segments in chemically polymerised material allow greater delocalisation of the polaron and the broad low energy absorption described above. This is also consistent with the data from muon experiments, described in the previous section, and the higher electrical conductivity found in chemically prepared material.

Jarrett *et al.* (1995) have studied the relation between carrier mobility and conductivity in the semiconductor regime by studying the properties of FETs

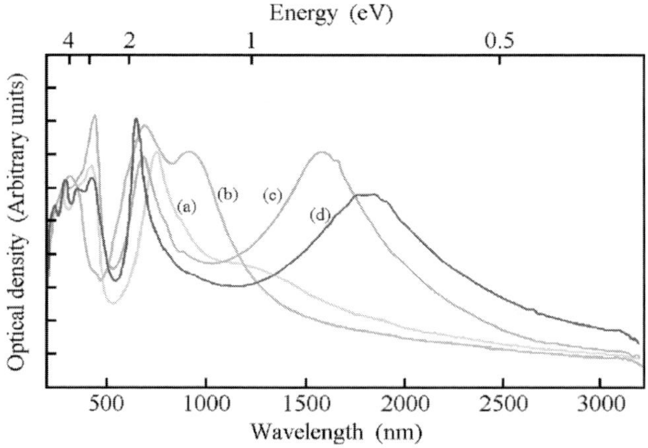

Fig. 9.33 Absorption spectra of dried thin films of electrochemically polymerised PAni on a Pt electrode: (a) yellow-green partially reduced, (b) and (c) green and (d) blue partially oxidised films prepared in water/HCl at 0.72 V and oxidised at –0.2, 0.4 and 0.8 V relative to a saturated calomel electrode, respectively. Reproduced with permission of the Institute of Physics from Monkman *et al.*, (1987).

as a function of doping level of the conduction-channel polymer. μ_{FE} and the conductivity are given by Equation (9.20). Experimentally, the conductivity has a power-law dependence on the concentration of dopants, i.e. $\sigma \propto N^{\theta}$, where N is the number density of dopant species. Combining this experimental observation with Equation (4.1) leads to a power-law dependence of mobility on conductivity, $\mu_{\mathrm{FE}} \propto \sigma^{\varphi}$, where $\varphi = (\theta - 1)/\theta$. This simple model fits the experimental data for poly(alkylthiophenes). Figure 9.34 shows data for poly(β′-dodecacyloxy(-α,α',α',α''-)terthienyl) for a device initially doped to the highest level and subsequently thermally dedoped, and a series of devices with different doping levels. The *thick* straight line is a fit with $\varphi = 0.7$, which is consistent with the experimental value of Θ of 0.35. The results shown in Fig. 9.34 were initially interpreted in terms of a simple model of hopping conductivity. Paasch *et al.* (2002) present a more formal analysis of variable range hopping (VRH), and show that a good fit is obtained for either two- or three-dimensional VRH. A theoretical curve for 3D-VRH for hopping between electronic states with a Bohr radius of about 0.8nm is shown by the *thin* curve in Fig. 9.34. A similar fit is found for 2D-VRH and a Bohr radius > 2 nm, although a 3D model appears more reasonable physically. Thus, while the experimental data do not distinguish clearly between these two models, they do show that carrier motion is dominated by hopping. This is also evident from the temperature dependence of the conductivity of semiconducting conjugated polymers, which is observed to follow a $T^{-\gamma}$ law. The limiting values for γ predicted by 3D- and 1D-VRH are 0.25 and 0.5, respectively. A value of

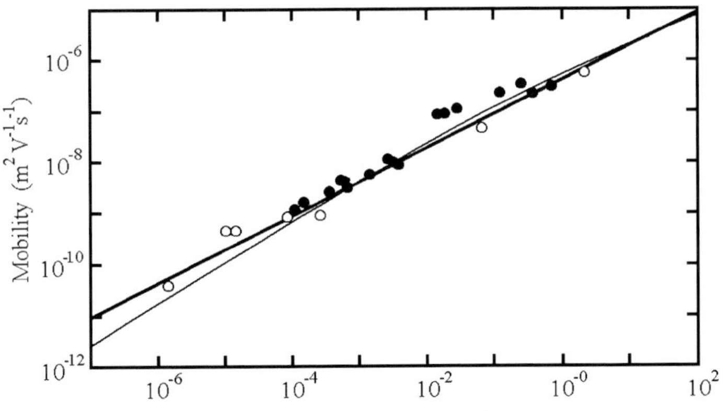

Fig. 9.34 Correlation of electrical conductivity and charge carrier mobility for a doped conjugated polymer: ● thermally dedoped device and o separately doped devices. The fits, thick and thin lines, are described in the text. Reproduced with permission of the American Institute of Physics from Jamett *et al.* (1995).

0.5 is also found for 3D hopping when either electron–electron interaction plays a role or tunnelling occurs between mesoscopic islands that have significant charging energy. Experimentally, γ is found to be in the range 0.3 to 0.5, making it difficult to identify which model is appropriate. Nevertheless, it is clear that in most instances the structural heterogeneity of conjugated polymers and the doping process are dominant in determining their electrical properties.

The nature of the insulator-to-metal transition (IMT) that occurs in a few polymers has been the subject of intensive investigation. Theoretically, the transition has been described in terms of the formation of either a polaron or a bipolaron lattice. As more charged species are created during the doping process, the energy levels in the band gap will broaden and eventually the gap will be filled. This is analogous to the formation of an impurity band in a heavily doped inorganic semiconductor. At some point, rather than the polarons or bipolarons occupying random positions on the polymer chain, it is more favourable for them to form an ordered arrangement, a polaron or bipolaron lattice with a distinct band structure. However, the morphology of even the best samples of the metallic polymers is not good enough for such elegant models to be applicable. The polymers are, at best, partially crystalline and a realistic description of the IMT must take this disorder into account. The two models that have been considered are either an Anderson disorder-induced transition or a percolation threshold in a heterogeneously doped material. The Anderson model requires that the polymers have large, but randomly distributed disorder, so that the disorder is homogeneously distributed throughout the polymer. The disorder produces localised states and a *mobility edge* at the boundary between localised and delocalised states, see Section 4.2.3 and Fig. 4.10. At the IMT the localisation length diverges and the Fermi level crosses the mobility edge, leading to metallic behaviour. The morphology of the partially crystalline polymer forms the basis of the percolation model. The disorder in the polymer is heterogeneous and doping will reflect this. As doping proceeds, some regions within the polymer will be more heavily doped and more conductive than others. The polymer will then behave like the metal-loaded, insulating polymers described in Chapter 8, with a rapid increase in conductivity as the highly conductive regions link up at the percolation threshold.

In normal metals such as copper the conductivity rises as the temperature is reduced from room temperature to about 10 K and then remains constant. The conductivity rises as the phonon scattering of free electrons is reduced as the thermally activated phonons are frozen out at low temperatures. The constant value at very low temperature results from scattering by the static defect concentration of the metal. The literature contains very few examples of conjugated polymers that show a higher conductivity at low temperature than at room temperature (Park *et al.*, 1998, and Tzamalis *et al.*, 2001). In general the conductivity is lower at low temperature than at room temperature, even though it may have increased on cooling to ~200 K, see Figs. 9.35 and 9.36.

The small variation in conductivity shown by many samples is similar to the weak temperature dependence of conductivity displayed by amorphous metals and alloys (Kaiser, 2001a and b). For conjugated polymers the sample with the highest room temperature conductivity will not necessarily have the highest conductivity at low temperature, see Fig. 9.36. In the examples shown in the latter figure, distinctly different techniques were used to prepare the samples of PAni. PAni and d,l-camphorsulphonic acid (HCSA) were mixed by grinding, and samples A and B were cast from an *m*-cresol solution of the mixture, E and F were cast from a mixed *m*-cresol/chloroform solvent, and G and H from a chloroform solution, while C and D were cast from an *m*-cresol solution made by mixing separate solutions of PAni and HCSA. For PAc the normalisation applied in Fig. 9.35 hides the fact that sample A is less conductive than B over most of the temperature range. The room and low temperature conductivities of these samples are $4 \times 10^6\ \Omega^{-1}m^{-1}$ and $7 \times 10^6\ \Omega^{-1}m^{-1}$ for A and $9 \times 10^6\ \Omega^{-1}m^{-1}$ and $6.5 \times 10^6\ \Omega^{-1}m^{-1}$ for B. Thus, the sample with the lower room temperature conductivity has the more metallic temperature dependence. The ratio of the conductivity at room temperature to that at low temperature can be used to give a rough indication whether a sample is in an

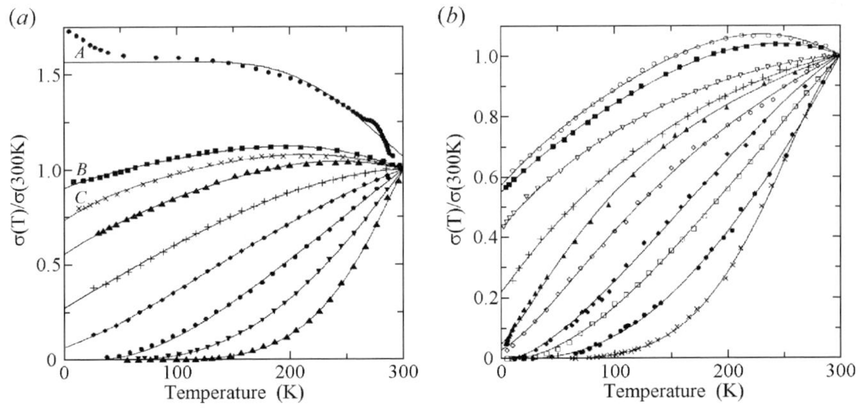

Fig. 9.35 Temperature dependence of the conductivity of doped PAcs. (a) A and B, doped with ClO_4^-; C, doped with $FeCl_3$; the other curves are for $MoCl_5$ doping; and (b) doped with iodine. Reproduced with permission of Wiley-VCH from Kaiser (2001b).

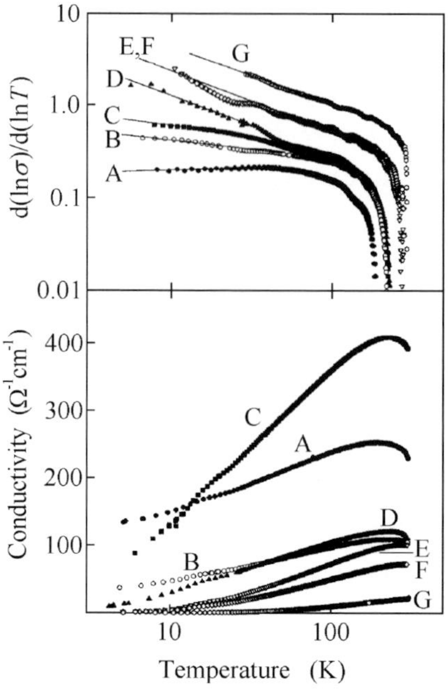

Fig. 9.36 The lower panel shows the temperature dependence of the conductivity of PAni doped with HCSA for samples prepared by different methods; the upper panel displays the reduced activation energy. Reprinted with permission from Kohlman *et al.* (1996). Copyright 1996 by the American Physical Society.

insulating (semiconducting) or a metallic state, but a better indication is given by the reduced activation energy, W:

$$W = \frac{\mathrm{d}\,(\ln \sigma)}{\mathrm{d}\,(\ln T)}. \tag{9.23}$$

In the insulating (semiconducting) regime W has a negative temperature coefficient, in the intermediate (critical) regime it is temperature independent and in the metallic regime it has a positive temperature coefficient (Menon *et al.*, 1998). W is plotted in the upper panel of Fig. 9.36 and shows that sample A, which had a conductivity of $8 \times 10^3\ \Omega^{-1}\mathrm{m}^{-1}$ at 20 mK, is in the metallic regime. The temperature dependences of the conductivity and of W for high molecular mass PAni ($\sim 2 \times 10^5\ \mathrm{g\,mol}^{-1}$) doped with 2-acrylamido-2-methyl-1-propanesulphonic acid (AMPSA) are shown in Fig. 9.37. Although the conductivity is slightly less than that of sample A of Fig. 9.36 at room temperature, it falls less rapidly as the temperature is reduced and W has a larger positive temperature coefficient. Thus these samples are more metallic. Plots of W similar to those in Figs. 9.36 and 9.37 are observed for doped PAc

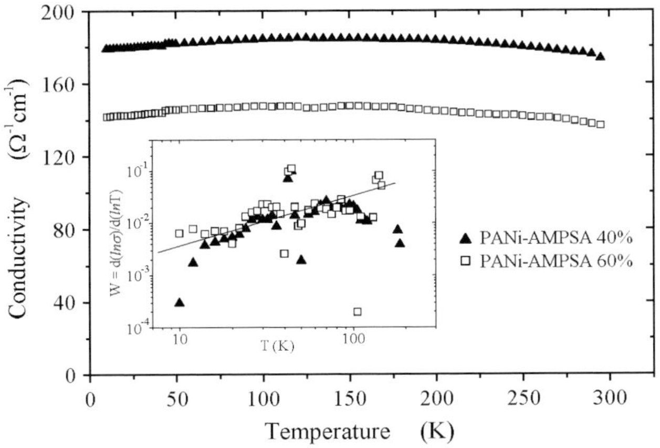

Fig. 9.37 Conductivity and reduced activation energy of high molecular mass PAni doped with AMPSA. Reproduced with permission of the Institute of Physics from Tzamalis *et al.* (2001).

and PPy (Menon *et al.*, 1998 and Kohlman and Epstein, 1998). The difference in temperature coefficient of W across the IMT is small. This feature, together with the comparability with disordered metals and the lack of a distinct correlation between room temperature conductivity and sample behaviour, are indications that disorder plays a dominant role in determining the electrical conductivity of doped conjugated polymers.

Excellent fits can be obtained to the temperature dependence of the conductivity using a heterogeneous structural model, which reflects the experimentally observed morphology of crystalline regions separated by disordered insulating (semiconducting) barriers. The nature of the barrier regions is affected by doping and may be either non-metallic or disordered metallic (Kaiser, 2001b). The sample resistivity can then be written in terms of the fractions of crystalline regions, f_c, non-metallic barrier regions, f_n, and disordered metallic barrier regions, f_d, as:

$$\rho = \frac{1}{\sigma} = f_c \rho_c + \left[\frac{1}{f_n \rho_n} + \frac{1}{f_d \rho_d} \right]^{-1}, \qquad (9.24)$$

where ρ_c, ρ_n and ρ_d are the resistivities of the crystalline, non-metallic and disordered metallic barrier regions, respectively. The resistivity is then given by the following equations.

1. In the insulating (semiconducting) regime:

$$\rho = \frac{1}{\sigma} = f_c \rho_m \exp\left(-\frac{T_m}{T}\right) + f_n \rho_0 \exp\left[\left(\frac{T_0}{T}\right)^{\gamma}\right], \qquad (9.25)$$

where the crystalline regions are considered to behave as quasi-1D metals of resistivity ρ_m, and T_m is a measure of the energy of the phonons that backscatter the charge carriers in these regions. The second term describes VRH or the other hopping mechanisms listed earlier, and ensures that $\sigma \to 0$ as $T \to 0$. ρ_0 and T_0 are model parameters.
2. In the metallic regime:

$$\rho = \frac{1}{\sigma} = f_c \rho_m \exp\left(-\frac{T_m}{T}\right) + \left[\frac{1}{f_n \rho_t \exp(T_t/(T+T_s))} + \frac{1}{f_d \rho_d}\right]^{-1}, \qquad (9.26)$$

where the Sheng fluctuation-induced, tunnelling model, Section 8.3.1, is used to describe carrier transport through the barrier regions, and ρ_t, T_t and T_s are the parameters of this model. ρ_d is taken to be temperature independent as it is dominated by defect scattering.

The lines drawn in Fig. 9.35 are fits to the data points using Equation (9.25) for the samples where $\sigma \to 0$ as $T \to 0$, and using Equation (9.26) for the remainder with the tunnelling term set to zero for trace A. Similar fits are found for PPy, PTs and PPV (Kaiser, 2001a and b). For the most conductive PPy and PMeT samples the conductivity rises below 20 K, although it falls between 300 and 20 K. A similar effect is seen in disordered metals near the IMT and attributed to either electron–electron interaction or weak localisation. The former gives rise to a $T^{1/2}$ dependence for the conductivity, once the temperature independent residual conductivity has been subtracted. A $T^{1/2}$ dependence is observed for both amorphous metals and doped polymers, but, although the magnitude is similar for amorphous metals and PPy, it is much larger for PAni and very much larger for PAc. The heterogeneous model predicts this, if the conductivity of the metallic regions is high and the IMT occurs in the barrier regions, particularly if the barriers are thin. In the PAni samples both increased and decreased conductivity have been observed below 1 K. The former was not affected by a magnetic field while the latter was suppressed (Kohlman and Epstein, 1998); these effects have been attributed to weak localisation.

Other properties that can provide information on the conduction process in doped polymers are the thermopower and the optical conductivity. The thermopower measures the rate at which carriers diffuse in a thermal gradient, a process that is less sensitive to material defects than is conductivity. For iodine-doped PAc the magnitude of the thermopower decreases as the doping level increases. As expected, the thermopower is positive for positive carriers, though, surprisingly, alkali metal-doped PAc also has a positive thermopower, even though the carriers are expected to be electrons. Initially the temperature dependence is weak, but at high doping levels the thermopower has linear temperature dependence right up to room temperature, as does that of highly conductive samples of PAc produced with different dopants, PAni and PPy. The thermopower of normal metals has identical temperature dependence. The magnitude of the thermopower for the polymers is similar to that of metals with low carrier densities. This behaviour suggests that the polymers are metallic at room temperature. However, the plots of the conductivity and reduced activation energy, e.g. Fig. 9.36, show a crossover from metallic behaviour at low temperature to non-metallic behaviour at room temperature. This apparent contradiction can be resolved with the heterogeneous model, since both metallic and non-metallic regions are present, but will contribute differently to the conductivity and the thermopower; both regions contribute to the conductivity but the thermopower is dominated by the contribution of the metallic regions.

A measurement of the optical conductivity of PAni samples over the range 2 meV to 1eV is illustrated in Fig. 9.38. The reflection spectra were obtained using a Fourier transform interferometer and a synchrotron radiation source. The Kramers–Kronig transformation, with appropriate extrapolation of the data to zero and high energy, was used to extract the complex frequency-dependent conductivity and its real part, the optical conductivity. The optical conductivity is related to the imaginary component of the electric susceptibility, since the power dissipated by an optical field of amplitude E is given by:

$$P = \sigma(\omega)\langle E^2 \rangle = \varepsilon'' \omega \varepsilon_0 \langle E^2 \rangle. \tag{9.27}$$

The reflectivity initially falls rapidly and then more slowly as the photon energy increases. The vibrational modes of the polymer give rise to distinctive resonant features between 0.06 and 0.19 eV. In normal metals the optical conductivity is given by the Drude formula:

$$\sigma(\omega) = \frac{\sigma_0}{1 + \omega^2 \tau^2} = \frac{\omega_p^2 \gamma}{4\pi(\omega^2 + \gamma^2)}, \tag{9.28}$$

where σ_0 is the DC conductivity, τ is the carrier relaxation time, $\gamma = 1/\tau$ and ω_p is the unscreened plasma frequency of the free electrons (carriers). This model fits the data above about 0.2 eV but fails at low energy, where the optical conductivity falls rather than increases gradually to the DC limit. This difference arises from the heterogeneity of the polymer materials. At low energy and frequency, conductivity is limited by the non-metallic barriers, while at higher frequency, it is enhanced, since the carrier displacement in the oscillating field is smaller than the average size of the metallic regions. These effects are incorporated in the localisation-modified Drude model (LMDM):

$$\sigma_{\mathrm{LM}}(\omega) = \sigma(\omega)\left[1 - \frac{C}{(k_{\mathrm{F}}\lambda)^2}\left(1 - \left\{\frac{3\omega}{\gamma}\right\}^{1/2}\right)\right], \tag{9.29}$$

where $\sigma(\omega)$ is given by Equation (9.28), C is a constant ≈ 1, K_{F} is the Fermi vector and λ the carrier mean free path. This model gives good fits to the experimental data, as shown in Fig. 9.38.

Unfortunately, there are few direct measurements of the complex conductivity at microwave frequencies, thereby avoiding the need for extrapolation from higher frequency data. Martens *et al.* (2001) present data in the range 8–600 GHz for PPy and PAni that can be described by the Drude model. However, the plasma frequencies are ~7 meV, much lower than the 1 eV calculated from the known carrier density. This indicates that most carriers are localised by the disorder in the samples and that the density of delocalised states at the Fermi level is low. Prigodin and Epstein (2003) note that the low frequency response is provided by a very small fraction of the carriers that are highly mobile and have long scattering times. They develop a model for this

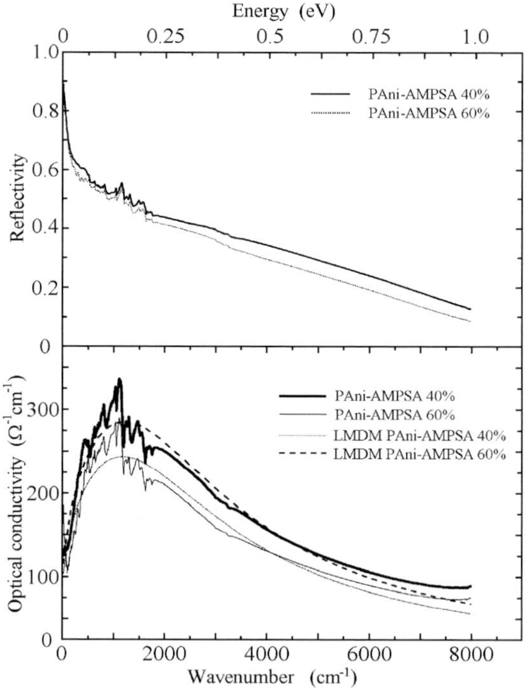

Fig. 9.38 Reflectivity and optical conductivity of PAni doped with AMPSA. A localisation-modified Drude model (LMDM) is used to fit the optical conductivity. Reproduced with permission of the Institute of Physics from Tzamalis *et al.* (2001).

behaviour based on resonant quantum tunnelling between metallic grains, and this provides a good description of the available data.

Although the case is made for the IMT being a homogeneous Anderson transition (Menon *et al.*, 1998), the cumulative weight of the evidence is in favour of the heterogeneous model. Unfortunately, the small number of examples of the IMT currently available limits the gathering of data that could throw further light on the nature of the transition. It is to be hoped that, as improvements are made in methods for the preparation of high molecular mass conjugated polymers with low densities of chemical and physical defects, further examples will be found of polymers with a metallic phase.

As noted in Section 9.2.1, the maximum conductivity is achieved with pure, high molecular mass, oriented polymers. The maximum values are $\sim 10^7\ \Omega^{-1}\mathrm{m}^{-1}$ for PAc (Tsukamoto, 1992) and $2\times 10^5\ \Omega^{-1}\mathrm{m}^{-1}$ for PAni (Pomfret *et al.*, 1998). The specific conductivity, i.e. conductivity per unit mass, is similar to, or even higher than, that of normal metals, since the lower density of the polymers compensates for the lower conductivity. The work functions of the metallic polymers are similar to those of normal metals, e.g. that of PAni is $\sim$4.8 eV and that of PEDOT is $\sim$5 eV, values which are intermediate between those of gold and aluminium. These properties mean that highly conductive conjugated polymers can, in principle, replace normal metals as the electrodes in electronic devices utilising semiconducting conjugated polymers, see Section 10.3.6.

9.5 Further reading

The science of conductive organic materials including polymers is described in an introductory level text that is accessible to all students with a basic scientific background, Roth (1995). The structure and electrical properties of polymers and organic molecular crystals are described in detail by Pope and Swenberg (1982). Theoretical models of conjugated polymers are described by Baeriswyl, Campbell and Mazumdar (1992) and in the compendia by Skotheim (1986) and by Skotheim, Elsenbaumer and Reynolds (1998). The early development of the chemistry and physics of PAc is extensively reported in the monograph by Chien (1984). An overview of work on polydiacetylenes is provided by the compilation edited by Bloor and Chance (1984). The shift in emphasis of research into conjugated polymers over the 1980s and 1990s is illustrated by the contents of the first and second editions of the *Handbook of Conducting Polymers* (Skotheim, 1986, and Skotheim *et al.*, 1998). Another compendium with extensive coverage of conducting polymers and related topics is that of

Nalwa (1997). A shorter summary is provided in the Proceedings of the 81st Nobel Symposium (Salaneck *et al.*, 1993) that led on to the award of the Nobel Prize for the discovery of metallic PAc. Müllen and Wegner (1998) deal with the related area, not discussed here, of conjugated oligomers with chemical structures and physical properties similar to those of conjugated polymers, but with simpler sample morphology.

10

Applications of electro-active and conductive polymers

10.1 Introduction

In this chapter we will present an overview of the applications of polymers that contain conjugated chemical structures. This encompasses the two different classes of materials described in the preceding chapter: polymers with localised conjugated moieties embedded in an otherwise non-conjugated structure and polymers with fully conjugated backbones. The first group also includes the polymers described in Section 3.7.2, where the conjugated group imparts non-linear optical properties rather than electrical conductivity. A number of the applications are well established, e.g. the use of semiconductive non-conjugated polymers in xerography and conjugated polymers as antistatic coatings. There are examples of applications that have been commercialised but proven to be unsuccessful, though they may be rendered viable by further research, e.g. batteries with polyaniline electrodes and multi-element sensors utilising conjugated polymers. In some areas progress is rapid and new applications are emerging from the laboratory into the market place, e.g. light-emitting diodes utilising conjugated polymers. Finally, there are applications that have promise, since materials exist with superior electrical or electro-optic properties to those currently utilised technologically, but they lack the thermal, chemical or temporal stability required for device manufacture and use in an uncontrolled environment, e.g. non-linear optical devices for telecommunications systems. Here it is the long-term potential gains that drive research into materials with the overall property combinations that are necessary for practical applications to be realised.

10.2 Electro-active polymers

10.2.1 Xerography

The photocopiers with which we are all familiar are based on a process known as either electrophotography or xerography. The latter is derived from the

Greek for dry (xeros) and writing (graphos). The process was invented in 1938 and patented in 1940 by Chester F. Carlson and involves several different physical phenomena. At the time the invention was first made these processes were poorly understood and the first demonstration copies were not of the quality we are used to today. Carlson demonstrated his invention to twenty American companies and generated 'an enthusiastic lack of interest'. Nevertheless, development was undertaken initially by the Battelle Memorial Institute and subsequently by the Haloid Company, now known as the Xerox Corporation. The first commercial xerographic copier appeared on the market in 1959 and is today a worldwide multi-billion pound industry.

The xerographic process and equipment are illustrated in Fig. 10.1. The steps illustrated in Fig. 10.1(a) are:

1. Electrical charge is deposited on the surface of the photoreceptor which has a high enough dark resistivity to prevent the charge leaking away.
2. An image of the document to be copied is projected on to the photoreceptor which becomes conductive in the illuminated regions, so that the charge leaks away there. This leaves the photoreceptor with a charge pattern that is a positive copy of the original document.
3. The pattern is developed by coating the charged areas of the photoreceptor with a fine, coloured powder, the toner. The toner particles are first charged by agitated contact with larger carrier particles, which also serve to transport the toner to the surface of the photoreceptor. The materials of the toner and carrier are selected so that their mutual contact charging gives the right sign of charge on the toner for its electrostatic attraction to the remaining charged area on the photoreceptor.
4. The toner neutralises the charge on the photoreceptor, so that the image can be transferred to a charged piece of paper.
5. The paper is heated to fuse the image into the surface of the paper.

In a typical photocopier, Fig. 10.1(b), the photoreceptor is deposited on the surface of a drum that rotates as the image of the document is scanned. Charge is deposited using a corona discharge from a wire held at a high potential close to the surface of the drum (1). The exposure is made either with a scanned document using a light source and projection system (2) or, for digitised images, with a scanned modulated laser (3). The toner and carrier are agitated with rollers and brought into proximity with the exposed photoreceptor (4). The image is then transferred to paper that is charged with a corona discharge (5) and heated (6) to fix the image. The photoreceptor then has to be cleaned to remove any residual toner (7) and then fully discharged either by illumination with an intense light source (8) or by contact with a flexible, grounded conductor (not illustrated).

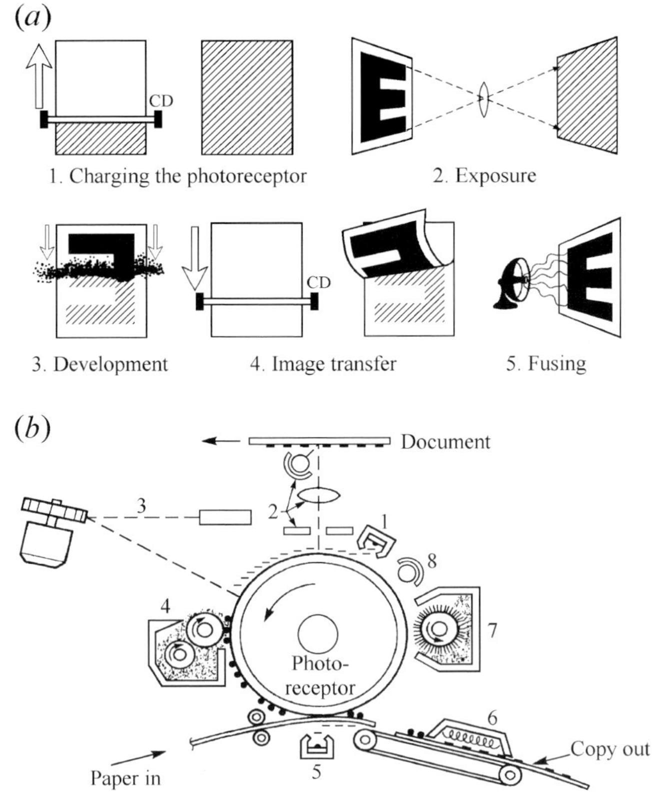

Fig. 10.1 (a) The xerographic process, CD – corona discharge. (b) The components of a typical photocopier; see text for the identification of the component parts. Reprinted with permission from Pai and Springett (1993). Copyright 1993 by the American Physical Society.

The properties required for the photoreceptor are a high dark surface resistivity, photoconductivity when illuminated with either incandescent lamps or semiconductor laser diodes and good abrasion resistance. The latter is needed to ensure an adequate photoreceptor lifetime by minimising damage during the image transfer, cleaning and discharging processes. The first practical copiers utilised photoreceptors fabricated by the vacuum deposition of selenium and selenium alloys. Amorphous silicon has also been employed with deposition by plasma decomposition of volatile silicon and dopant compounds. Nowadays, the majority of commercial systems employ semiconducting polymers. These offer simple fabrication techniques such as dip or spray coating of an aluminium drum and solution coating of metallised polyethylene terephthalate fabric that is used to make photoreceptor belts for high speed copiers. Typically, polymeric photoreceptors have multilayer structures comprising an outer layer of hole-transporting material adjacent to a

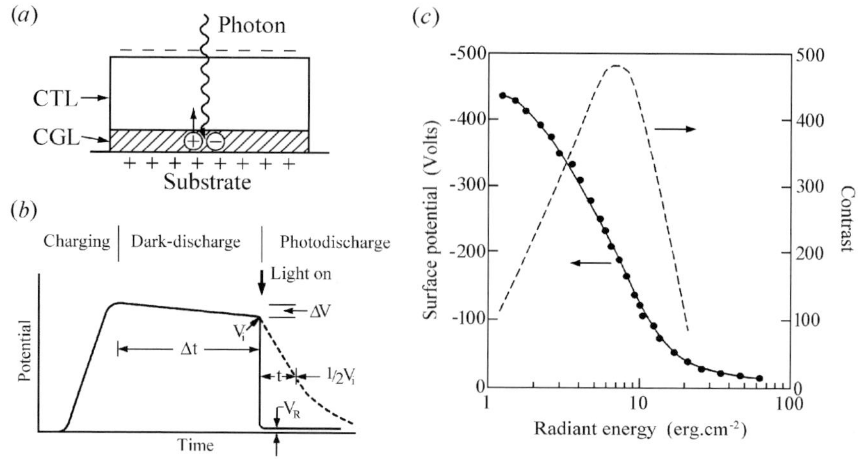

Fig. 10.2 (a) Polymeric photoreceptor, CGL – charge generation layer, CTL – charge transport layer. Reprinted with permission from Pai and Springett (1993). Copyright 1993 by the American Physical Society. (b) Surface potential during a photo-induced-discharge curve measurement. Reprinted with permission from Law (1993). Copyright 1993 American Chemical Society. (c) PIDC and contrast for a dual layer photoreceptor. Reprinted from Borsenberger and Weiss (1998) by courtesy of Marcel Dekker, Inc.

photo-generation layer that is separated from the metal substrate by an undercoat layer that prevents charge injection from the metal. The materials of choice for the charge transport layer are the non-conjugated backbone semiconductive polymers described in Chapter 8. The physical properties that are important in the choice of materials for xerographic photoreceptors, e.g. carrier mobility, and space charge effects, have been discussed in Section 8.4. Much of the underpinning science has involved studies of polymers with pendent conjugated groups such as PVK. However, commercial photoreceptors use molecularly-doped, insulating polymers, avoiding the synthesis of polymers with specific pendent groups. This allows molecules that might be difficult to incorporate as a pendent group to be used, the facile variation of the concentration of the active molecules and incorporation of several different active molecules in a single layer.

Corona charging deposits a uniform distribution of negative charge on the surface of the charge transport layer and a corresponding positive charge is induced in the earthed metallic substrate, Fig. 10.2(a). The undercoat between the charge generation layer and the substrate is chosen to prevent this positive charge being injected from the substrate and neutralising the charge on the upper surface. The charges on the upper surface and the metallic substrate establish an electric field across the photoreceptor. The transport layer must be transparent at wavelengths that generate electrons and holes in the underlying charge photo-generation layer. The holes produced by illumination of this layer migrate through the charge transport layer to neutralise the negative charge on the upper surface. Clearly, in this arrangement, it is important for the transport layer to have high hole mobility. The undercoat, while blocking injection of holes from the substrate, must allow the photo-generated electrons to reach the substrate. During the exposure of the photoreceptor the field created across the photoreceptor by the corona charging will be reduced as the initial charge is neutralised by the photo-generated charge carriers. The discharge rate is determined by measuring the decay of the initial surface potential while the photoreceptor is illuminated, to obtain the so-called photo-induced-discharge curve (PIDC). In this procedure a sample of photoreceptor is corona charged and then moved underneath an electrometer probe and shuttered light source. The variation in surface potential is as shown in Fig. 10.2(b). The sample is kept in the dark for a period Δt and the rate of dark-discharge, $\Delta V/\Delta t$, determined. The residual voltage, V_R, is measured after exposure to an intense light source – see *full* photo-discharge curve in Fig. 10.2(b). The sample is recharged and exposed to monochromatic light of a known intensity to determine the photosensitivity as a function of wavelength – see *dashed* curve in Fig. 10.2(b). Different measures of the photosensitivity appear in the literature. Law (1993) expresses it as the radiant energy that reduces the surface potential to half its initial value, V_i, while Pai and Springett (1993) use:

$$S = \frac{1}{I}\left(\frac{\partial V}{\partial t}\right), \tag{10.1}$$

where I is the radiant energy flux per unit area and time and the rate of decay of the surface potential is usually taken at V_i. The contrast in an image will be highest when the rate at which the surface potential falls is a maximum. Since optical density is related to the logarithm of the light intensity reflected from the image, the contrast is given by the slope of the plot of surface potential against the logarithm of the total radiant energy, Fig. 10.2(c).

The rate at which the electric field decays will be determined by the rate at which free electrons and holes are generated, provided that the transit time of holes through the transport layer is shorter than the exposure time. The decay will be linear if the intensity of illumination is constant, the efficiency of carrier generation is independent of the electric field and no charge is trapped in the charge transport layer. If there are deep traps with long escape times within the layer, then there will be a residual potential, V_R, at the end of the discharge.

A field-dependent, photo-generation rate will lead to non-linear PIDCs with a sigmoidal form, i.e. a slow initial decrease in surface potential followed by a more rapid, approximately linear, fall and final slower decrease giving a drawn-out tail. As discussed in Section 8.4.1, semiconducting polymers have relatively low carrier mobility; so that the transit time can be longer than the exposure time, and the decay of the field will be limited by the carrier mobility. The exposure can then be likened to the light pulse used to generate a charge sheet in time- of-flight (TOF) measurements, Section 8.4.1. For an ideal transit, see Fig. 8.25(a), the PIDC will fall linearly while the charge sheet transits the transport layer, and be constant thereafter, provided that the decrease in potential due to the discharge is less than the initial potential. Actual TOF signals are rarely close to the ideal, as shown in Figs. 8.26, 8.27 and 8.30, with long, slowly decaying tails. Thus it is possible that the carrier transit is not complete before the photoreceptor is developed. This limits copier speed and size and this drives the search for trap-free materials with higher mobility. Examples of compounds developed for use in molecularly doped polymers such as TAPC, TPD and TTA have been described in Section 8.4, see Fig. 8.24 and the accompanying text and Table 8.2. Typical dark-discharge rates are in the range of a few volts to a few tens of volts per second. Photo-discharge rates are primarily determined by the efficiency of generation of free charge carriers in the charge generation layer.

The photo-generation layer usually comprises fine crystals of pigment dispersed in a polymer binder. Pigments for the photo-generation layer must have intense absorption from the near infra-red to the ultra-violet. Examples of materials with suitable characteristics include azo dyes, perylene and phthalocyanines (Law, 1993). The azo dyes have a good photo-response between 500 and 650 nm and are used with white light sources. Perylene has a high, approximately constant, photo-response from 480 to 720 nm, and can be used with white light sources and semiconductor LEDs. Phthalocyanines, such as vanadyl phthalocyanines, have a lower response, but maximum sensitivity occurs near 800 nm making them suitable for use with semiconductor laser diodes. Charge generation in the molecular materials is by excitonic processes, Section 8.4.2, and charge transport involves percolation between grains, Section 8.3.1. The PIDC for a two-layer photoreceptor comprising a granular CGL and a hole transporting CTL is shown in Fig. 10.2(c). The CGL comprises crystallites of a cationic dye, 2,6-diphenyl-4(4-dimethylaminophenyl)thiapyrylium hexafluorophosphate dispersed in bisphenol-A-polycarbonate containing tri-*p*-tolylamine (TTA) to assist hole transport within the CGL. The CTL contains 40% mass TTA dispersed in a polyester matrix (O'Reagan *et al.*, 1996). The discharge was stimulated with a source generating

1 μs duration pulses and the surface potential was measured 1 s after each exposure to allow all the mobile carriers to drift through the CTL. Measurements made with continuous illumination at 680 nm show that the photo-generation efficiency rises from 3% at $10^6\,Vm^{-1}$ to 40% at $7\times10^7\,Vm^{-1}$. The photo-response was high, between 550 and 700 nm.

While there is academic research into the physical processes underlying xerography, the bulk of development activity is conducted in industrial laboratories in the USA and Japan. The commercial sensitivity of this work inevitably means that the data in the open literature is limited.

10.2.2 Organic light emitting diodes and solar cells

While the advent of polymeric light emitting diodes (LEDs) has been the cause of the rapid growth in interest in the photo-physics of conjugated polymers, Sections 9.4.2–3, electroluminescence in organics was first observed in crystalline organic materials (Pope *et al.*, 1963). LEDs utilising small organic molecules were reported in the 1980s (Tang and VanSlyke, 1987). These were the product of research carried out at Kodak Research Laboratories from the late 1970s onwards. Alq_3, see Fig. 8.24(g), was used as the electro-luminescent material in these devices. Although the device performance was good, this work was not commercialised. The topic was pursued in academic and industrial laboratories in Japan, and the Pioneer Corporation was the first to develop commercial displays, incorporated by them in their products from the mid-1990s onwards. The simplest structure of an organic light emitting diode (OLED) is shown in Fig. 10.3(a) and comprises a metal electrode, typically MgAg alloy, vacuum deposited layers of Alq_3 and TPD, see Fig. 8.24(e), and a transparent ITO electrode, see Fig. 9.28(b) which shows a simple polymeric LED. Alq_3 is the emitter and transports electrons injected from the metal to the interface between

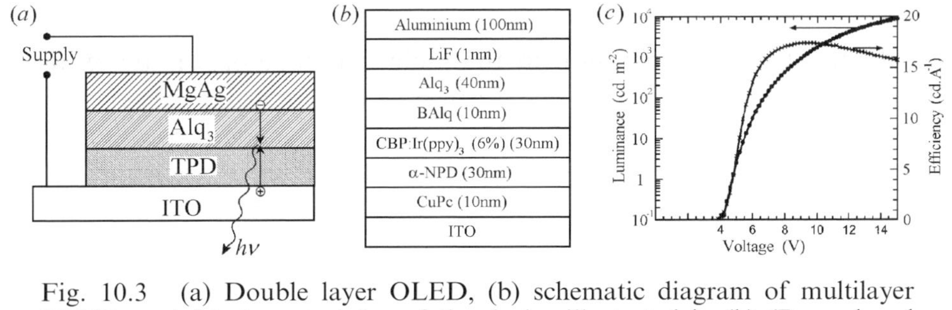

Fig. 10.3 (a) Double layer OLED, (b) schematic diagram of multilayer OLED, and (c) characteristics of the device illustrated in (b) (Reproduced with permission of the American Institute of Physics from Weaver *et al.*, 2002).

the Alq_3 and TPD layers. The TPD transports holes injected from the ITO electrode to the hetero-junction interface where they recombine with electrons giving a broad green emission. The internal field distribution of two layer devices has been determined by electro-absorption spectroscopy (Martin *et al.*, 2002).

Numerous organic materials have been utilised as hole injecting, electron and hole transporting and emitting layers (Hung and Chen, 2002). Electroluminescent moieties have been employed as evaporated layers, as dopants in polymers and as pendent groups in preformed polymers. An important requirement for all these classes of materials is a high glass transition temperature to prevent crystallisation when the device temperature rises during use. Similarly, molecularly-doped materials must be stable against diffusion and aggregation of the molecular component. Cleaning and oxidising the surface of the ITO electrode with either an oxygen plasma or treatment with UV radiation in an ozone atmosphere improves hole injection, although the energy barrier against hole injection remains $\sim$0.5 eV. This is reduced by the introduction of a thin layer of copper phthalocyanine (CuPc), PAni, PEDOT–PSS or hole-transporting materials doped with an oxidising agent. Hole-transporting materials include the molecules illustrated in Fig. 8.24(a)–(f) and the polymers illustrated in Fig. 8.23(a), (e) and (f). The bi-phenyl diamine moiety, see Fig. 8.24(e), is the basis for numerous, related linear and starburst molecules. PVK is widely used as the host polymer for such hole-transporting molecules which enhance the native hole-transporting properties of the matrix. Electron injection requires the use of low work function metals, like Ca or K. Bare metals of this type have poor corrosion resistance, and so their alloys like Mg-Al and Al-Li are widely used in their stead. However, Al with a very thin layer of either Al_2O_3 or LiF provides good electron injection into Alq_3. The insulating layer is stable, thin enough to allow electron tunnelling, and is thought to eliminate defects that form at a direct Al–Alq_3 interface. Alq_3 is the most widely used electron-transporting agent and emitter, although many other similar metal-organic compounds based on metals such as Be, Mg, Ca, Sr, Sc, Y, Cu and Zn can be employed as electron-transporting materials. Other materials that have been investigated for this purpose include fully-fluorinated oligo-phenylenes, thiophene and oxadiazole based molecules and aryl substituted cyclooctatetraenes. Of these, PBD, Fig. 8.24(c), is the most widely used organic electron transporting and hole blocking material.

Considerable control can be exercised over the wavelength of the emitted radiation by using guest–host emitters where a dye with longer wavelength emission is incorporated in the emitter layer. If the absorption band of the dye overlaps the emission of the host then efficient energy transfer can occur from the excited host to the dye by non-radiative Förster transfer. Green and red emission is easily achieved, but blue emission requires special high energy-gap hosts and dyes. Combinations are available that provide the three basic colours for colour displays. Emission has also been observed from *exciplexes*, the excited states where the electron and hole reside on the separate constituents of a molecular complex, e.g. from the complex formed between the pendent carbazole moieties of PVK and TPD. Efficient device operation has been achieved by the use of metal–organic compounds where the emission occurs from the triplet state, rather than from the singlet state. $Ir(ppy)_3$, Fig. 8.24(h), is an example of this class of emitter, which is usually used in a guest–host system with 4,4′,*N*,*N*′-dicarbazole-biphenyl (CBP) as a host. White light emission has been obtained by using an emitting layer of PVK doped with three dyes with overlapping emission bands (Kido *et al.*, 1994). A single layer device with an external quantum efficiency of 2% photons per electron has been fabricated by spin coating of polystyrene doped with Alq_3 and rubrene, which act as electron- and hole-transporting agents, respectively (Jin *et al.*, 2000).

Figure 10.3(b) illustrates the level of complexity encountered in an OLED incorporating many of the features described above. (BAlq is the metal organic compound aluminium(III)bis(2-methyl-8-quinilato)4-phenylphenolate.) The dominant emission in this device is from the $Ir(ppy)_3$. The turn-on voltage is 4 V and the peak efficiency is just over 16 cd A^{-1} between 8 and 10 V.[1] This device was fabricated on a polyethylene terephthalate substrate coated with a multilayer polyacrylate-Al_2O_3 barrier film to limit diffusion of oxygen and water vapour into the device. Both oxygen and water degrade device performance by oxidising the Al electrode and reacting with the organic components. A hermetic seal was provided by an epoxy-bonded glass backing with a CaO getter to remove residual water and by-product from the cure of the epoxy seal. The light output of the device decayed to 50% of its original value of 600 cd m^{-2} in 3800 hr compared to 10 700 hr for a glass-encapsulated device. At a reduced output of 100 cd m^{-2} the lifetime is estimated to be about 16 000 hr. The degradation of OLEDs leads to the appearance of dark spots. These occur at defects in the Al cathode that allow water and oxygen to penetrate to the metal organic surface where either reduction of the water or oxidation results in separation of the electrode and emitting layer (Schaer *et al.*, 2001). However, it has been shown that the use of hydrophobic coating allows organic FETs to function with the conduction channel exposed to water (Someya *et al.*, 2002).

[1] The unit of luminous intensity (candela, cd) is defined as the luminous intensity of (1/600 000) m^2 of a black body at a temperature of 2040 K and a pressure of 101 325 Pa.

Glass encapsulated devices using more common emitters show small decreases in output intensity over 10 000 hr and life times in excess of 50 000 hr have been reported. Flexible devices with lifetimes greater than 5000 hr have been fabricated on plastic substrates (Kwong *et al.*, 2003). The technology has reached the point where small displays are being incorporated in commercial products and large-size, full-colour, flat-panel displays that offer an alternative technology to the currently available liquid crystal displays have been demonstrated (Forrest, 2003).

The process of recombination of electrons and holes at a hetero-junction to emit photons may be inverted: the absorption of photons to generate electrons and holes at a hetero-junction is the photovoltaic effect. The early studies at Kodak showed that the device structure shown in Fig. 10.3(a) could also be used as a photovoltaic cell, commonly referred to as a solar cell (Tang, 1986). Investigation of the use of electron- and hole-transporting materials for solar cells has attracted less attention than work on OLEDs. Bach *et al.* (1998) have used the amorphous hole-transporting material, 2,2′,7,7′-tetrakis(*N*,*N*-di-*p*-methoxyphenyl-amine)9,9′-spirobifluorene to make a solid state version of a photo-electrochemical cell (Grätzel, 2001). Although devices have been made with doped pentacene single crystals and vacuum deposited organic compounds (Brabec *et al.*, 2001a and Nelson, 2002), most progress has been achieved with conjugated polymers, see Section 10.3.3.

10.2.3 Non-linear optics

Optical modulators and switches

Many of the non-linear optical effects described in Section 3.7.2 have the potential for utilisation in devices. The most extensively explored device applications are second harmonic generators, modulators and switches. Second harmonic generation was an initial focus with the aim of generating blue radiation from long-wavelength semiconductor LEDs and lasers for use in compact disc technology. This work was overtaken by the advent of blue semiconductor sources, and research effort subsequently refocussed on optical modulators and switches utilising a non-linear refractive index. Emphasis is placed on devices which may be integrated with other fibre-optic components and couplers for use in the all-important area of telecommunications and computer networks. Here the insertion loss of the devices must be small, and therefore absorption and scattering losses must be minimised. This, in turn, requires non-linear optical materials that are both transparent at the operating wavelength and homogeneous, i.e. either single crystals or glasses. The difficulty of preparing single crystals of organic compounds a few microns thick by

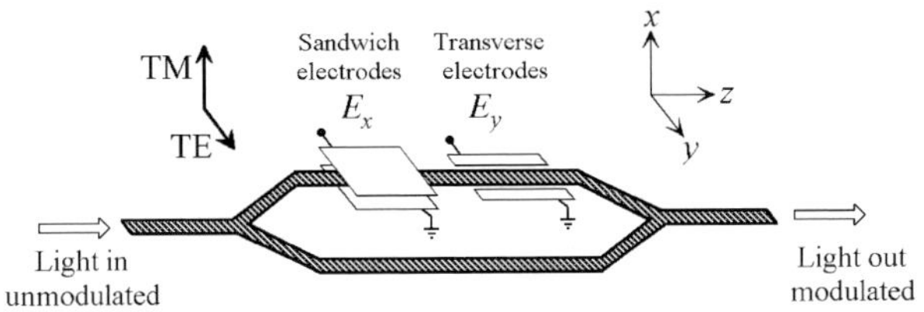

Fig. 10.4 Schematic diagram of an integrated optical Mach–Zehnder interferometer modulator. Reprinted from Donval *et al.* (2000). Copyright 2000, with permission from Elsevier.

several centimetres across, with the necessary orientation for accessing an appropriate component of the non-linear tensor, makes their use unattractive. In contrast, the techniques of spin coating and lithographic patterning are readily applied to the fabrication of thin-film, integrated optical structures utilising amorphous polymers. Light will be guided in a polymer layer provided its refractive index is higher than that of the adjacent layers. Confined optical waveguides, rib waveguides, can be formed by either etching the active polymer layer to leave a narrow, thick region or coating a substrate in which a channel has been previously etched. The latter type requires that the polymer layer does not conform to the shape of the substrate but acts as a planarising layer. The layer on either side of the rib is too thin for light to propagate in it, so that the light is confined to the thicker rib waveguide.

Electric field poling is required to induce polar order in an amorphous polymer, see Section 3.7.2. Light propagates in an optical waveguide in TM or TE modes with the electric vector either normal or parallel to the plane of the waveguide, see Fig. 10.4. In order to maximise the interaction of any chromophore with the incoming light, the chromophores must be aligned either normal or transverse to the plane of the film for the TM and TE modes, respectively. Normal alignment is achieved either by applying a voltage across sandwich electrodes placed on the surfaces of the polymer thin film, or by producing a surface charge on the exposed surface of the polymer with a corona discharge. The latter has the advantage that electrical breakdown, which can occur at high poling fields with fixed electrodes, is avoided. Transverse alignment is achieved with electrodes formed on the substrate before it is coated with the polymer. This reduces breakdown that can occur at the edges of electrodes on the upper surface of the polymer film. Buffer layers are used to separate the waveguide from the electrodes, see Fig. 10.5(a), since metal electrodes in direct contact with the waveguide rapidly attenuate the guided light.

Examples of polymers incorporating non-linear optical moieties in their chemical structure have been described in Section 3.7.2 and shown in Fig. 3.24,

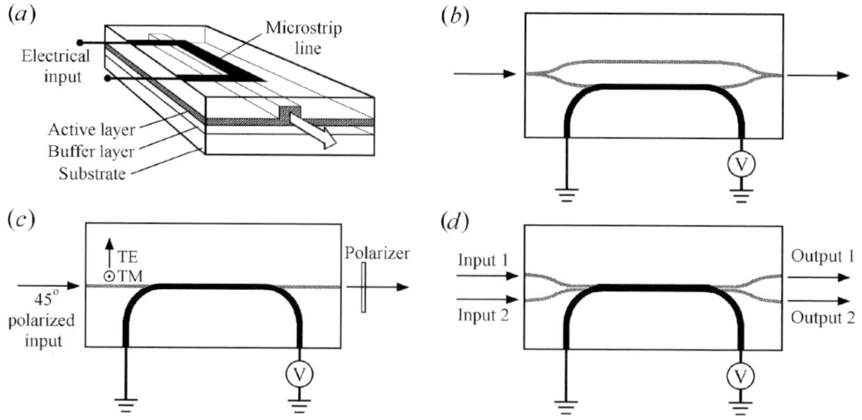

Fig. 10.5 High frequency optical modulators and switches: (a) construction of a rib wave guide with microwave strip line upper electrode, and plan views of (b) Mach–Zehnder, (c) birefringent, and (d) bi-directional coupler modulators. Reproduced with permission of Wiley-VCH from Dalton *et al.* (1995).

but much exploratory work has been done with guest–host systems where non-linear molecules are dispersed in an insulating, transparent polymer matrix. Despite the disadvantage that the polar order induced by electric field poling relaxes too rapidly for practical application in this kind of system, there are a number of advantages: (a) a high transparency host can be used, e.g. PMMA or polycarbonate, (b) the concentration of the non-linear moiety can be readily varied, (c) molecules not readily chemically bound to a polymer can be employed. The emphasis of materials research is to produce polymers that incorporate highly non-linear moieties and that are amorphous and have glass transition temperatures high enough to ensure that the poled polymer is stable over many years at the highest temperatures to be encountered in use (Man and Yoon, 1992). As noted in Section 3.7.2, excellent progress has been achieved towards this goal.

The structures of some typical polymeric integrated optical devices are illustrated in Figs. 10.4 and 10.5. In the widely-used Mach–Zehnder interferometer, Figs. 10.4 and 10.5(b), guided light splits evenly between two branches of equal length, so that the light recombines constructively with undiminished intensity in the output waveguide. Electrodes placed on one arm allow the refractive index, and hence the phase of the light emerging from that arm, to be changed so that the output intensity is reduced. Application of an oscillating voltage results in modulation of the output. Sandwich electrodes will modulate TM modes of the guide and transverse electrodes the TE modes. Polarisation-insensitive modulation can be achieved either by a combination of sandwich and transverse electrodes or by using an array of electrodes to produce non-uniform poling that interacts equally with TE and TM modes (Donval *et al.*, 2000). In the birefringent modulator, Fig. 10.5(c), TE and TM modes of equal intensity are launched into a single-mode waveguide and sandwich electrodes are used to alter the phase of the TM mode, so that elliptically polarised light emerges from the guide to be analysed with a polariser. Directional couplers, Fig. 10.5(d), utilise the fact that the overlap of the field distribution of light propagating in closely spaced parallel waveguides results in energy transfer between them. This is analogous to the transfer of energy between coupled pendulums. Thus light incident on one input can exit from either output, depending on the length of the coupler in relation to the length over which energy transfer takes place. Light can be switched between the output guides by the application of an electric field to alter either the phase in one guide or the coupling between the two guides. The coupler can be used either as a modulator or as a switch. The latter is a key component of an optical communication network, where it is necessary to switch light from one of many inputs to one of many outputs without any cross-talk.

High modulation rates are desirable since they allow greater throughput on a single optical fibre, thereby avoiding the need to use multiple fibres for high data rates. However, for modulation frequencies in the GHz regime, propagation of the signal in the electrode structure also becomes important (the electrodes themselves act as waveguides for the electrical signal). It is then preferable to use a travelling wave structure, a microstrip line, for the electrode, as shown in Fig. 10.5(a). This facilitates matching of the electrode structure and the high frequency source. The modulation signal then travels in the electrode parallel to the light in the optical waveguide, and, for optimum coupling, the electrical and optical signals should remain in phase over the length of the electrode. The propagation velocities of the electrical signal and the light are primarily determined by the permittivity of the guiding medium at the modulation and optical frequencies, respectively. If the difference in the permittivities is large, the distance over which the electrical signal and the light are in phase will be short. This problem is encountered in crystalline, ferroelectric non-linear optical materials such as lithium niobate, which have a very high permittivity at microwave frequencies. Non-linear optical polymers have a distinct advantage, since the difference in permittivities at the two frequencies is much lower. Thus coupling of the electrical with the optical signals is possible at much higher frequencies in polymer devices. A further advantage is that, while the non-linearity in materials like lithium niobate derives from displacement of the atomic nuclei, in organic non-linear materials it is due to a displacement of the electron clouds and is inherently faster.

The operating frequency of polymer integrated optical modulators has been increasing steadily from a few tens of GHz to over 100 GHz while operating voltages have been falling to below 1 V (Dalton *et al.*, 1995 and Dalton, 2002). Lee *et al.* (2002) describe a Mach–Zehnder modulator utilising a guest–host non-linear polymer with a bandwidth of 150–200 GHz and detectable modulation of the light at 1.6 THz. As they note, these modulation rates are higher than the anticipated bandwidth requirements for the foreseeable future. While the temporal stability of poling for crosslinked polymer matrices with high glass transition temperatures is sufficient for commercial application, the photolability of the non-linear chromophore means that device lifetimes fall below that required for telecommunication system components, typically ten years. Absorbed oxygen is a major cause of the problem at the telecommunications wavelength of 1.3 μm because an excited triplet energy state, which is highly chemically reactive, is generated by absorption of radiation in this region. This is less of a problem at 1.5 μm, although the losses in the polymer waveguides are higher at this wavelength. Solution of the chemical stability problem requires rigorous exclusion of oxygen during production and the elimination of subsequent permeation of atmospheric oxygen into the device. An additional inherent problem is that harmonics of the guided radiation will be generated in the non-linear polymer, and the high-energy photons this produces can degrade the chromophore even in the absence of oxygen. A device such as a modulator can be in continuous use, so although the effect may be small, over extended periods it can degrade the non-linear polymer. Improvement in photostability is, therefore, a major research goal. The use of a modulator structure in which the non-linear polymer is in contact with a stable, passive polymer waveguide can reduce the problem. In the absence of an applied field the refractive index of the active polymer is made to be less than that of the passive polymer. Radiation is confined to the passive guide and the active polymer is subject to only the low intensity evanescent tail of the guided mode. When a voltage is applied, the refractive index of the active polymer is increased until the mode is drawn into the active polymer in the region under the electrode. Thus the intensity of the radiation in the active polymer is high for only half the time that the modulator is in use. Commercial exploitation of non-linear polymer devices has been explored by a number of companies, and 40 GHz modulators are available from Pacific Wave Industries Inc. With further progress in materials and device design, practical devices with higher bandwidths are likely to appear on the market.

Photorefractive materials

In materials that possess both photoconductive and electro-optic responses, the photo-generation and trapping of charge carriers produces internal fields that modify the refractive index. A sinusoidally varying intensity caused by the interference of two, crossed, coherent beams of light of wavelength Λ_g will generate charges in the regions of high intensity, Fig. 10.6(a). One sign of carrier will then drift into the regions of low charge density, typically the holes, as they are generally more mobile than the electrons, Fig. 10.6(b). Eventually the mobile carriers are trapped and the electrons and holes are spatially separated, giving rise to a periodically varying charge density, ρ_{sc}, Fig. 10.6(c). This distribution will not necessarily be coincident with the initial intensity distribution, particularly if an external field has been applied to assist both the photo-generation, which is generally field dependent, and charge separation. The internal fields E_{sc} induce a periodic variation in the refractive index of the material Δn, which is shifted by a phase angle ϕ from the initial intensity distribution, Fig. 10.6(d). Patterns written in this manner will persist until the separated charges recombine. This can occur on timescales from nanoseconds to years depending on the material. The pattern can therefore be either dynamic or effectively static. The pattern in materials with long lifetimes is erased by uniform illumination. Multiple patterns can be recorded with writing beams having different orientations. Uses of photorefractive materials include holographic recording for data storage and holographic interferometry, image processing, and pattern recognition.

Photorefractive materials are characterised by wave-mixing experiments, either two-beam coupling or degenerate four-wave mixing, DFWM. In two-beam coupling a refractive index grating is written with overlapping coherent beams, Fig. 10.6. The phase shift of the grating ϕ results in a transfer of energy from one of the incident beams to the other, i.e. it causes more light to be diffracted into one beam than into the other. The optical gain coefficient is then given by:

$$\Gamma = \frac{1}{l}\ln\left(\frac{1+\alpha}{1-\beta\alpha}\right), \quad \text{with } \alpha = \frac{\Delta I_1(l)}{I_1(l)} \text{ and } \beta = \frac{I_1(I_2=0)}{I_2(I_1=0)}, \tag{10.2}$$

where I_1 is the intensity of the amplified beam, I_2 is that of the attenuated beam and l is the path length of the amplified beam in the photorefractive material. The experiment can be performed with either *s*- or *p*-polarised beams. Since photorefractive materials must absorb radiation, there is competition between this absorption and the optical gain. Ideally, the optical gain coefficient should exceed the absorption coefficient. In DFWM a grating is established by two *s*-polarised beams and is probed by a *p*-polarised beam propagating counter to one of the writing beams, and the diffracted intensity propagating counter to the second write beam is measured. The internal diffraction efficiency is given by:

$$\eta_{int} = \frac{I_p(diff)}{I_p(diff) + I_p(tran)}, \tag{10.3}$$

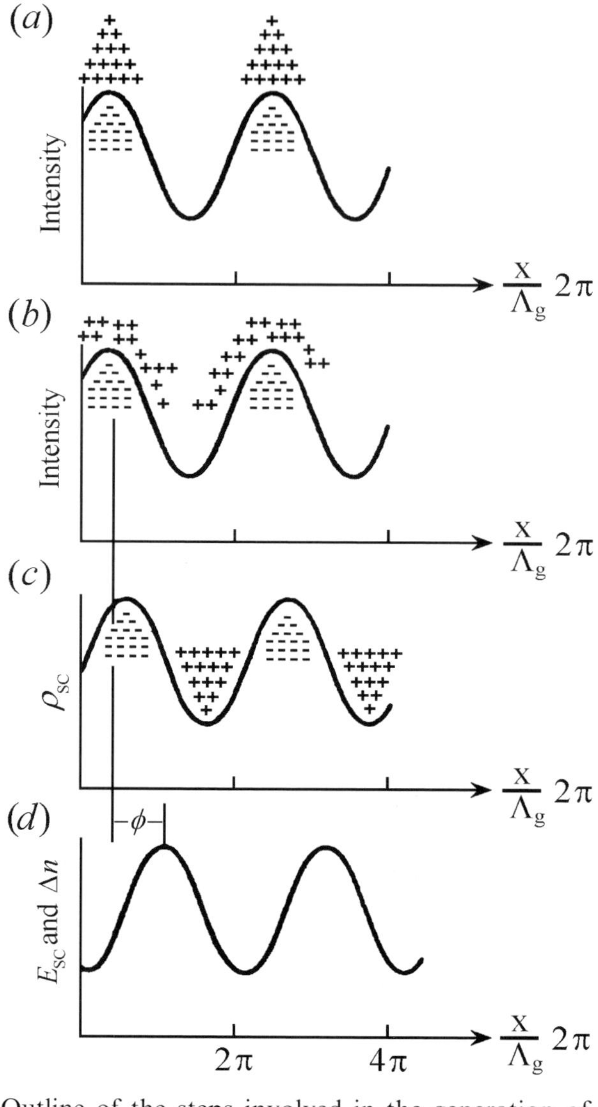

Fig. 10.6 Outline of the steps involved in the generation of a refractive index grating in a photorefractive material. Reprinted with permission from Yu *et al.* (1996). Copyright 1996 American Chemical Society.

where $I_p(diff)$ and $I_p(tran)$ are the diffracted and transmitted intensities of the probe beam.

While polymeric photorefractive materials can be prepared by adding separate molecules for photo-generation, charge transport and non-linear optical response into an inert polymer, most systems studied use polymers where one or more of these components are covalently bonded to the polymer. Polymer

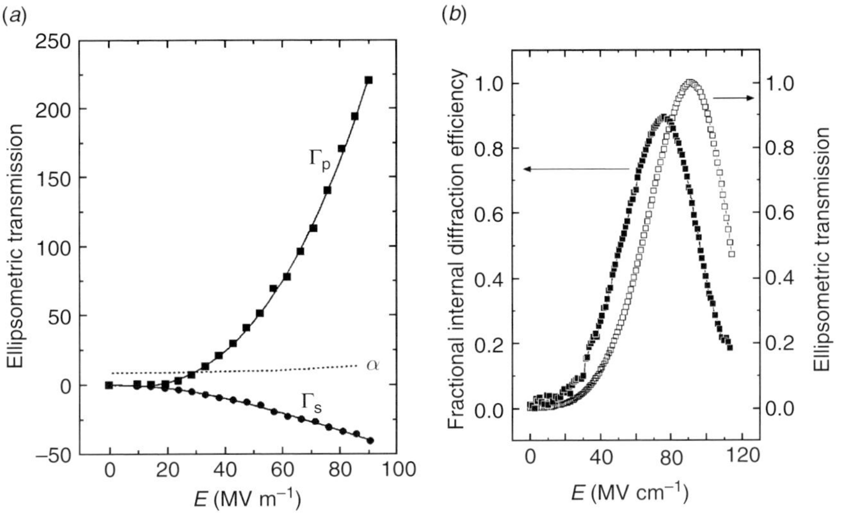

Fig. 10.7 Field dependence of (a) the optical gain and linear absorption of a DMNPAA, PVK, ECZ, TNF composite (reprinted with permission from *Nature*, Meerholz *et al.*, 1994; copyright 1994 Macmillan Magazines Limited), and (b) internal diffraction efficiency and ellipsometric transmission of a DMNPAA, PVK, ECZ composite (reproduced with permission of Wiley-VCH from Bittner *et al.*, 1999).

hosts should have a low glass transition temperature T_g, as this allows the non-linear moieties to reorient parallel to the internal fields, giving a larger photo-refractive effect. PVK has been widely used, but has a T_g of over 200 C and requires the addition of a plasticising agent, typically *N*-ethylcarbazole (ECZ), to reduce T_g. TNF (Fig. 8.24(b)), and C_{60} are used as sensitizers to improve photo-generation efficiency. The host acts as the hole-transporting medium, and a variety of non-linear molecules, similar to those illustrated in Fig. 3.23, have been incorporated to provide the electro-optic response. Alternatively, a host polymer with pendent non-linear groups, see Fig. 3.24(a), can be used with a hole-transporting agent such as DEH, Fig. 8.24(f). In this instance the pendent substituent acts as both the photo-generation and non-linear optical component of the composite. In all these materials the photo-generation efficiency and the charge separation are enhanced by the application of an electric field. This is illustrated in Fig. 10.7(a), which shows the increase in optical gain as a function of applied field for a material comprising the non-linear chromophore DMNPAA (2,5-dimethyl-(4-*p*-nitrophenylazo)-anisole), PVK, ECZ and TNF at 50, 33, 16 and 1% by mass, respectively. Figure 10.7(b) shows the variation of internal diffraction efficiency for a DMPAA–PVK–ECZ composite with 40, 50 and 10% by mass, respectively.

The field for maximum η increases linearly as $(T_g - T)$ for materials with a range of compositions and T_gs, except at high concentrations of the non-linear chromophore where the field has a minimum value just below T_g. This is due to the T_g dependence of either the poling efficiency or the magnitude of the internal field. The birefringence produced by the electro-optical effect was measured independently using an ellipsometric method which gives maximum transmission when the electro-optic effect is a maximum, Fig. 10.7(b). The field for maximum birefringence, which is a measure of the poling efficiency of the material, increases as T_g is increased irrespective of the composition of the material. Thus the minimum in η at high chromophore loadings must be due to variations in the internal field which are primarily due to changes in the recombination rate as a function of T_g.

The need to apply an electric field to obtain good photorefractive performance limits the utility of polymeric materials. The fields required are large, see Fig. 10.7, which limits the polymer thickness to the range 100–200 μm. The field is applied perpendicular to the plane of the film and will also orient the polar non-linear molecules in this direction. Light normally incident on the film will not access the induced non-linearity, because the electric vector is normal to the orientation direction. Thus the film must be tilted with respect to the incident beams. These drawbacks mean that polymeric photorefractive materials remain the subject of research and further improvements. In particular, the strength of the poling field, needed to achieve the desired level of non-linearity, will have to be reduced before significant applications emerge.

10.3 Intrinsically conductive polymers

In all applications, in addition to the property directly relevant to the application in question, the polymer must be able to withstand the conditions encountered in use. Thus mechanical properties must be such that the integrity of the polymer is retained when in use. This also means that the polymer must possess thermal stability over the expected operating temperature range, which may include low as well as high temperatures. Chemical and physical stability must be sufficient to ensure a lifetime adequate for the application. Consideration of these requirements has come to the fore as conjugated polymers have moved from being the subject of fundamental research to materials employed in an increasingly wide range of applications.

Conjugated polymers are very susceptible to degradation when exposed to atmospheric gases, causing the conductivity to vary with time. Exposure of as-prepared conjugated polymers to oxygen can lead to an initial increase in conductivity as the oxygen acts as a dopant. Subsequently, the conductivity falls as oxidation destroys the backbone conjugation. The conductivity of deliberately-doped samples is also observed to fall on exposure to either oxygen or a normal atmosphere. The rate of decrease initially reflects the diffusion of the reactive gas into the bulk of the polymer and, once the material is saturated, first order reaction kinetics. In a variety of different polymers it has been shown that oxidation and hydrolysis occur at reactive chain defects. Reaction products include carbonyl, peroxy and hydroxyl moieties. The decline in conductivity of the most stable conjugated polymers in a normal atmosphere is slow enough that they can be utilised for some applications without the use of protective barrier coatings or encapsulation. Nevertheless, electronic applications require rigorous exclusion of oxygen and water and special methods to prevent their ingress are necessary.

Photo-oxidation can also be a problem. In bulk samples the high absorption coefficient of the polymer at actinic wavelengths means that photo-oxidation occurs only on the surface of the sample. The initial lustre of some PDA crystals is lost on exposure to sunlight, but the degradation products can be removed by dissolution and the initial lustre recovered. Notably, no photo-oxidation is observed in PDA–DCH crystals, because the packing of the carbazole moieties is so dense that oxygen cannot come into contact with the conjugated backbone. This situation is exceptional, and photo-oxidation is a potential problem for the spin-coated films utilised in polymeric light emitting diodes unless oxygen and actinic radiation are excluded from the active region of the device.

Attempts to dope conjugated polymers by ion beam implantation revealed that this process resulted in extensive damage with samples subjected to high dose having a carbonised surface layer. Although early studies indicated that polyacetylene was resistant to γ-ray irradiation, recent ESCA studies of PEDOT show that electrons with energy as low as 3 eV cause significant chemical modification. Fortunately, most applications do not involve exposure of the polymer to ionising radiation.

It is frequently stated in the literature that conjugated polymers have poor mechanical properties and that it is necessary to incorporate them in a composite with a conventional polymer to achieve useful properties. While such work is worthwhile, see Section 10.3.6, the justification for it is incorrect. The π-bonding in conjugated polymer chains means that the backbones are *stronger* than those of non-conjugated polymers; poor mechanical properties stem rather from defects in the sample morphology. For example, electrochemically prepared polymers often consist of grains grown at different nucleation sites with weak interfacial contacts (Cheung *et al.*, 1990). Excellent mechanical properties can be obtained when care is taken to eliminate such defects, e.g. continuous, tough films of polypyrrole up to 1mm thick with properties comparable to those of epoxy resins can be prepared under appropriate conditions (Bloor *et al.*, 1985). Similarly, carefully prepared fibres of high molecular mass polyacetylene and polyaniline have excellent mechanical properties. In many applications the conjugated polymer is employed as a thin film on a supporting substrate, rendering mechanical properties less important. Nevertheless, film integrity requires good adhesion to the substrate and sufficient strength to withstand stresses caused by differences in the thermal expansion of film and substrate during thermal cycling.

10.3.1 Soft electronics

Soft electronics is a term that has been coined to distinguish electronic devices fabricated with organic materials and polymers from those based on inorganic semiconductors. Since the carrier mobility is typically smaller in semiconductive polymers than in inorganic semiconductors, soft electronic devices have a lower maximum operating frequency. Also, the long-term stability of the best semiconductive polymers is inferior to that of their inorganic counterparts. For these reasons soft electronics cannot currently compete with existing devices as components for high-speed electronic systems. But soft electronic devices can compete on cost for applications where high speed and long lifetime are not essential, because spin- and dip-coating are low-energy, low-cost processes that can be used to produce devices on large, inexpensive polymer or glass substrates. Typical low-cost, but large-volume, applications include externally-powered devices for smart and credit cards and electronic barcodes and security tags on merchandise.

The basic device structures, and their utilisation to study the fundamental properties of semiconductive polymers, have been described in Section 9.4.3. Here it was noted that high quality Schottky diodes could be fabricated with semiconductive polymers, because the low temperature processing gives low densities of surface states that normally degrade the performance of inorganic semiconductor Schottky diodes. Device performance is affected by chemical modification of the interface between the polymer and the inorganic contact. This means that good devices require the use of either noble metals or functionalised silicon substrates. For example, Vermier *et al.* (1999) used a pyrrole-terminated alkyl-lithium reagent to attach alkyl pyrrole moieties to the surface of an *n*-type silicon substrate. The attached groups were used as the nuclei for the growth of chlorate-doped, polypyrrole films. The resulting diodes sustained a current density of 50 $\mathrm{A\,m^{-2}}$ at a forward bias of 0.7 V and had an

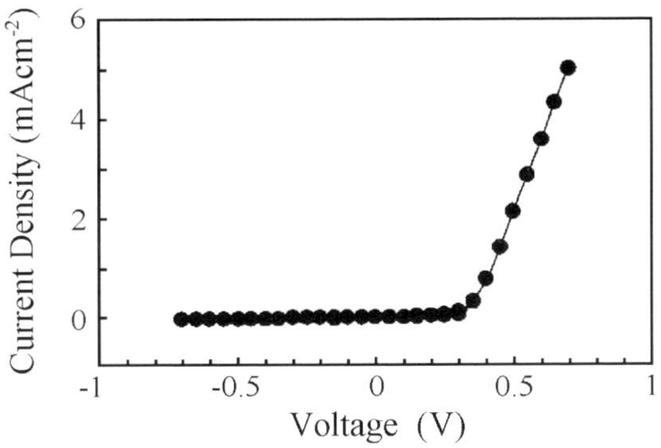

Fig. 10.8 Current–voltage characteristics of a polypyrrole/*n*-silicon diode. Reproduced with permission of the American Institute of Physics from Vermier *et al.* (1999).

ideality (or quality) factor[2] of 1.5, better than that of inorganic metal-semiconductor junctions. The current-voltage characteristic of this device is shown in Fig. 10.8.

Field effect transistors (FETs), also described as thin film transistors (TFTs), have been the main focus of research into soft electronic devices. The earliest devices, which were used to construct logic circuits and to evaluate the performance of different polymers as the conduction channel, had structures similar to that illustrated in Fig. 9.28(a) (Brown *et al.*, 1995 and 1997). In the devices used for materials evaluation the substrate was heavily-doped n^{++}-silicon. It served as the gate electrode and was insulated from the conduction channel by either a thermally grown oxide layer or a chemical vapour deposited silicon nitride on a thin oxide layer. Gold source and drain electrodes were laid down prior to a final coating with a protective polymer. The materials studies revealed the correlation of carrier mobility and conductivity, and hence doping level, discussed in Section 9.3.2 and illustrated in Fig. 9.34. This means that in the design of a device there is trade-off to be made between maximum operating frequency, which is dependent on carrier mobility, and current modulation, which will be small if the resistance of the conduction channel is low when the device is turned off. Although the relationship between mobility and conductivity, $\mu_{FE} \propto \sigma^{\varphi}$ with $\varphi \cong 0.7$, holds for different polymers, the mobility at a given conductivity can vary by a factor of up to 10^2 (Brown *et al.*, 1997). Thus, the performance of the FET can be optimised by

[2] The current in a forward biased diode is given by:

$$J = J_0 \exp(eV_F/nkT),$$

where J_0 is the saturation current density and n is the ideality factor.

the selection of a polymer with the highest mobility at a low doping level. This resulted in the choice of precursor route polythenienylenevinylene (PTV) to make logic gates and circuits.

The devices used to make the demonstrator devices were fabricated on a lightly doped silicon substrate with a thick thermal oxide layer. Phosphorus-doped polysilicon was deposited on this substrate and etched to form the gate electrodes. A further insulating layer of either silicon nitride or thermal oxide separates the gate from the gold source and drain electrodes and the PTV active layer. The PTV was produced from a spin-coated layer of precursor polymer by heat treatment under N_2-HCl atmosphere. Several transistors were fabricated on a single substrate to make inverter and NOR logic gates. Ring oscillators running at between 10 and 50 Hz were realised with these units. Similar devices running at a frequency ten times higher were fabricated from the conjugated oligomer, pentacene. The maximum frequency was a factor of ten smaller than anticipated on the basis of carrier transit times calculated from the maximum experimental carrier mobility observed in organic semiconductors. The PTV devices were more stable than those made with pentacene, device performance being unaffected by six months' storage in air. The source current versus gate bias characteristic displays slight hysteresis that is reduced when the device is in vacuum. Applying a gate bias of -20 V for 20 minutes results in a significant increase in channel conductance at zero gate bias that is reduced by the application of a gate bias of $+20$ V. This is attributed to the trapping of positive charge under the gate electrode in the active polymer. The rate of charge trapping was shown to be dependent on the sheet charge density induced in the conduction channel but independent of the source–drain current. While these effects can affect device performance, the effects will be minimal when the gate bias is either being actively switched or is applied for durations less than one minute.

The development of soft electronic devices has seen the replacement of the inorganic materials shown in Fig. 9.28(a) by polymers. The structure of an FET fabricated entirely from polymers is shown in Fig. 10.9(a). A layer of PAni doped with camphorsulphonic acid is spin coated from *m*-cresol solution on to the polyimide substrate. A photoinitiator, hydrocyclohexylphenyl-ketone, incorporated in this film enables the film to be reduced to the insulating leucoemeraldine form on exposure to short wavelength ultra-violet irradiation. The photoinitiator is sublimed out of the unirradiated areas by heating to 110 C. The surface resistivities of the exposed and unexposed regions are 10^3 and $10^{14}\Omega$, while the reduction in thickness of the 0.2 μm thick conducting layer is less than 50 nm. This enables source and drain electrodes (S and D, Fig. 10.9(a)) and conduction tracks to be patterned in the PAni layer and the

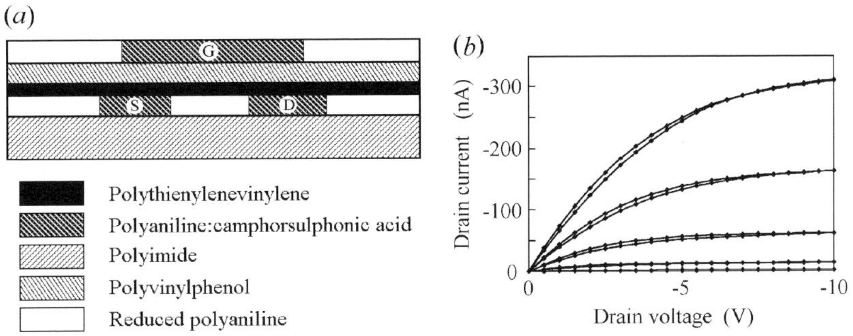

Fig. 10.9 Structure (a) and drain current–voltage characteristic (b) for an all-polymer FET. Reproduced with permission of the American Institute of Physics from Drury *et al.* (1998).

50 nm thick PTV layer to be added without further treatment. The PTV is formed from a precursor polymer, as described above. A 250 nm thick insulating layer of polyvinylphenol is then spun on before the top layer of PAni, in which the gate electrodes (G) are formed, is deposited. Typical devices have conduction channels 2 μm long and 1 mm wide, which give device characteristics as shown in Fig. 10.9(b). Carrier mobility in the PTV is $3\times10^{-8}\,m^2V^{-1}s^{-1}$ and the devices operate at frequencies between 40 and 200 Hz. Circuits using these devices require contacts between the source and drain electrodes of one device and the gate of the next device. This is achieved by driving a sharp pin through contact pads in the top and bottom layers, resulting in the formation of contacts between the upper and lower PAni layers. A 15-bit code generator was fabricated from 326 TFTs and operated successfully. Performance was unaffected by deformation of the flexible multi-layer structure. A variety of other logic circuits were demonstrated with this technology. While the operating speeds of these devices seem low, they are adequate for the operation of displays. This has been demonstrated in a 64 by 64 pixel polymer dispersed liquid crystal display (PDLC)[3] driven by a matrix of 4096 PTV TFTs (Huitema *et al.*, 2002). Figure 10.10 shows a flexible E-ink[4] display with soft electronics driving circuitry.

3 PDLCs comprise LC droplets dispersed in a polymer matrix. Random director orientation leads to the scattering of light due to refractive index mismatch between LC and polymer. A field applied across the film orients the LC so that the refractive indices match and the scattering is reduced, rendering the film transparent.

4 E-ink consists of microcapsules of oppositely charged black and white particles in a fluid matrix. The microcapsules are dispersed in a polymer film with an upper transparent electrode and an electrode array on the lower surface. The white particles are attracted to the visible surface of the display by an applied field of one polarity; the black particles are attracted to the surface when the polarity of the field is reversed.

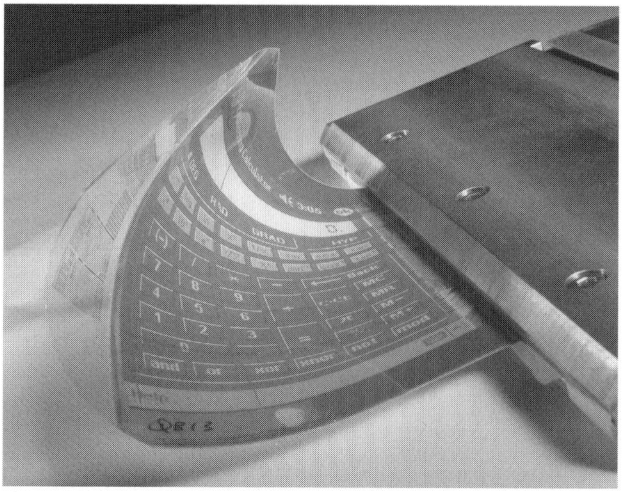

Fig. 10.10 E-ink display with soft electronics active matrix driving circuits. Reproduced with permission of Philips Research.

FETs have been fabricated with regioregular polyhexylthiophene (P3HT) using the device structure shown in Fig. 9.28(a) (Sirringhaus *et al.*, 1998, and Sandberg *et al.*, 2002). P3HT is solution processable, allowing direct spin- and dip-coating of the active layer. As the fraction of head-to-tail repeat units increases, films with a lower level of disorder and more regular packing of the polymer chains are obtained. Structural analysis shows that the plane of the polymer backbones is orthogonal to the substrate. This results in face-to-face contact of the thiophene groups in adjacent chains enhancing inter-chain transfer of charge carriers. As a result the field effect mobility increases as the fraction of head-to-tail units increases, up to $\sim 10^{-5}\,m^2V^{-1}s^{-1}$, as shown in Fig. 10.11(a). Higher mobilities are observed for dip-coated than spin-coated films; solvent evaporation is slower in the former process, allowing the polymer chains to adopt a more ordered structure. FETs fabricated with spin-coated P3HT films 50–70 nm thick have an ON/OFF current ratio of more than 10^6 and maximum drain currents of 10–40 μA (Sirringhaus *et al.*, 1998). These transistors have been used to drive polymer LED arrays. The characteristics are also sufficient to drive a quarter VGA (320 rows × 240 columns) PDLC display at a 50 Hz frame rate. Despite these advantages the devices require rather large gate and drain voltages, typically 20–80 V. Thus, while they can be used as display drivers, they are not currently suitable for use in externally powered devices such as smart cards. FETs with dip-coated P3HT films 20 nm thick have slightly different characteristics, as shown in Fig. 10.11(b). Similar characteristics are obtained for films prepared by repeated dipping, although these are not of uniform thickness.

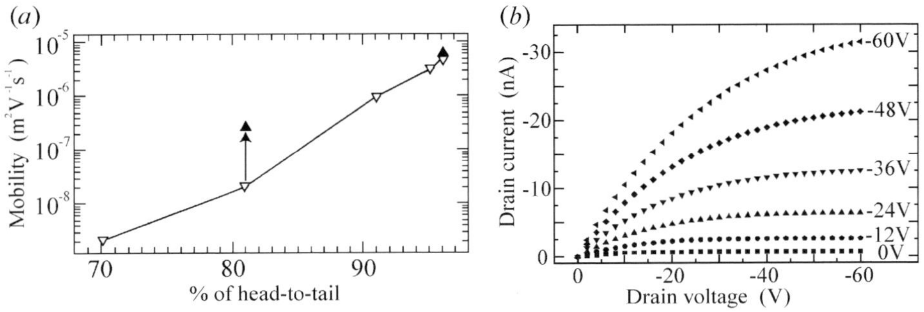

Fig. 10.11 (a) Variation in mobility with stereoregularity of spin-coated, ▽, and dip-coated, ▲, P3HT films (reprinted with permission from *Nature*, Sirringhaus *et al.*, 1999; copyright 1999 Macmillan Magazines Limited), and (b) drain current/voltage characteristic of an FET with a dip-coated P3HT conduction channel gate voltages are shown on the right (reprinted with permission from Sandberg *et al.*, 2002; copyright 2002 American Chemical Society).

Sirringhaus *et al.* (2000) have described the production of devices by a combination of lithography, ink-jet printing and spin coating. Source and drain electrode patterns are lithographically defined in a thin polyimide film on a glass substrate. Next, ink jet printing is used to deposit a commercially available aqueous dispersion of PEDOT–PSS (Baytron-P, Bayer AG) on to the exposed glass. An active layer of poly(9,9-dioctylfluorene-co-bithiophene) (PF8T2) is then spin-coated on to the electrode array followed by an insulating layer of polyvinylphenol. A PEDOT–PSS gate electrode is finally ink jet printed in air. The PF8T2 is utilised rather that P3HT as it has greater air stability and tolerance of impurities that could be incorporated from the printed gate electrode. Devices have conduction channels with lengths down to 5 μm and widths of 3 mm and have drain currents ~5 nA at drain and gate voltages of −20 V. Circuits capable of switching at 250 Hz were fabricated with these devices.

The use of polymers with a low relative permittivity, see Section 2.7, as the FET gate insulator has been found to result in higher carrier mobility and improved device performance (Veres *et al.*, 2003). The authors developed stable semiconductive polymers, poly(triarylamines), and used them to fabricate FETs with silicon dioxide and poly(methylmethacrylate) gate insulators. Values of the carrier mobilities measured in the FETs were about 10 times smaller than those determined by TOF experiments. A dramatic increase in FET mobility was observed when a low permittivity fluoropolymer ($\varepsilon = 2.1$) was employed as gate insulator. The FET mobilities for devices with gate insulators with relative permittivities in the range 2–18 are shown in Fig. 10.12(a). The devices were made with two different poly(triarylamines)

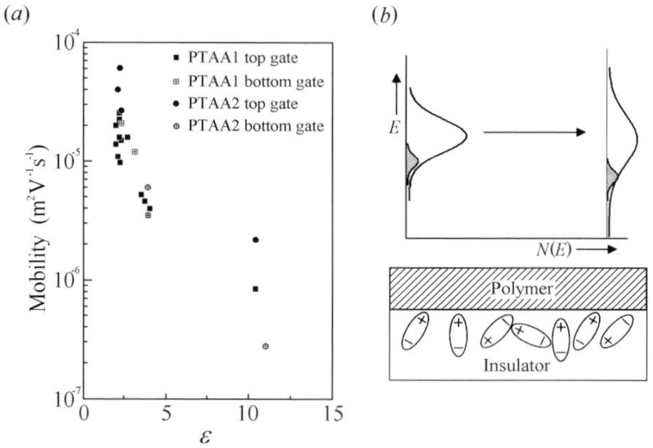

Fig. 10.12 (a) Dependence of the FET mobility on the relative permittivity of the gate insulator. (b) Random dipoles in the gate insulator at the interface with the conductive polymer, shown schematically at the bottom, broaden the DOS in which the carriers, shown shaded, move. Reproduced with permission of Wiley-VCH from Veres *et al.* (2003).

and with either the conductive channel deposited on the preformed gate, bottom gate, or vice versa, top gate. The increase in mobility is greatest for insulating layers with $\varepsilon < 2.5$. Assuming that the GDM model, Section 8.4.1, describes the carrier motion in the conduction channel, these results indicate that the density of states distribution is broadened at the interface between the semiconducting and insulating polymers, Fig. 10.12(b). Carriers moving in a lower density of states in the tail of the broad DOS, at the right in Fig. 10.12(b), will be less mobile that those moving in a narrower DOS in a bulk sample, at the left in Fig. 10.12(b). The broadening of the DOS by charge–dipole interaction between carriers and randomly oriented dipoles in the insulator is a likely explanation of the adverse effect of more polar insulating dielectrics.

All the devices utilising polymers described above are unipolar, i.e. use one sign of charge carrier, and are almost exclusively FETs, although other geometries have been investigated, e.g. a triode with a PAni layer incorporated between two MEH-PPV films (Yang and Heeger, 1994). One reason for this is that the polymers stable enough not to require processing *in vacuo* are all *p*-type semiconductors. In order to make bipolar devices, analogous to conventional *pnp*-silicon transistors, compatible *p*- and *n*-type polymers would be required. Although *n*-type polymers are known, typically having nitrogen-containing aromatic rings such as poly(pyridine-2,5-diyl), these are unstable towards oxygen. Bounds can be placed on the redox potential of *n*-type polymers for stability against degradation by water and oxygen (de Leeuw *et al.*, 1997). *n*-type polymers with oxidation potentials above about 0.5 V would be stable against hydrolysis and oxidation. Conversely, *p*-type polymers with oxidation potentials below about 0.5 V are stable in a normal atmosphere. Even allowing an uncertainty of 0.5 V, these limits mean that polymeric *pn*-junctions would be thermodynamically unstable. This follows since a stable *pn*-interface requires the potential of the *n*-type polymer to be less than that of the *p*-type polymer, the reverse of the situation required to ensure that the polymers are stable in air. The fabrication of polymeric *pn*-devices would require the use of air-unstable *n*-type polymer, i.e. fabrication in an inert atmosphere, the inclusion of stabilisers that scavenge oxygen and a high quality hermetic package. Likely device performance would not seem to be sufficient to justify the additional processing costs compared with polymeric FETs. Similarly, although small molecules such as tetracene have been used in organic FETs, either single crystals or very high quality thin films are required to realise high field effect mobility (de Boer *et al.*, 2003), rendering them less attractive for mass production. Meanwhile, the performance of polymeric FETs continues to improve to the extent that the appearance of soft electronic devices in the low-end applications described above is a matter of commercial rather than technical judgment.

While the development of soft electronic devices has focussed on improvements in material and device stability, it has also been shown that the instability of conjugated polymers can be utilised in device manufacture. Möller *et al.* (2003) report the fabrication of write-once read-many-times memory using the degradation of PEDOT. PEDOT shows good stability at low current densities, but becomes non-conducting when subjected to a high current density. The memory comprises thin-film silicon diodes deposited on a flexible metal foil substrate coated with PEDOT and an upper set of electrodes. Orthogonal upper and lower electrodes connecting rows of diodes allow cells to be turned off by the application of $\sim$10V across the cell. Cells that remain conducting can be interrogated indefinitely at lower voltage.

10.3.2 Light emitting diodes

The intensive investigations that followed the discovery of electroluminescent emission from conjugated polymers (Burroughes *et al.*, 1990) simultaneously provided a better understanding of the nature of the electronic excited states and carrier transport in these materials, as described in the previous chapter. In parallel with these fundamental studies, applied research grew dramatically, fuelled by the recognition of the commercial potential for polymer LEDs and emissive displays. A major advantage, which was immediately apparent, of such devices over inorganic semiconductor LEDs is the facile production of large-area emitters by spin coating. This provides the basis for the development of both displays and lighting. Rapid progress was achieved in the development of displays in the 1990s. The appearance of commercial OLEDs (Section 10.2.2) provided an added impetus to this, and the similarity in the device structures allowed for synergistic development. While the potential remains for the use of large area polymer emitters in surface mounted illumination systems, it has received less attention than displays where the financial returns are likely to be greater.

Fundamental studies established that simple single-layer diodes, as shown in Fig. 9.28(b), can exhibit reasonable internal quantum efficiency η_{int}, as given by:

$$\eta_{\mathrm{int}} = \gamma \eta_{\mathrm{Rec}} q, \tag{10.4}$$

where γ is the ratio of the number of excitons formed by the recombination of electrons and holes in the diode to the number of electrons flowing in the external circuit, η_{Rec} is the fraction of singlet excitons formed, see Section 9.4.3, and q is the efficiency of radiative decay of the singlet excitons.[5] A higher η_{int} than is presently achievable is required for practical devices. A balanced flow of electrons and holes in the diode will maximise γ, but this is difficult to achieve in single-layer devices with direct contact to the electrodes, as illustrated in Fig. 9.30. Greater control of charge injection is possible in multi-layer structures with either separate electron and hole transporting layers, see Fig. 10.3(a), or layers at the interface between the electrode and polymer to modify charge injection, as described for OLEDs in Section 10.2.2, or both, see Fig. 10.3(b). The literature is full of numerous examples of the synthesis of polymers designed to provide improved hole and electron transport and of multi-layer device structures. Illustrative examples are described below.

a. An example is shown in Fig. 10.13(a) of a single poly(2,7-[9,9′-di-*n*-octyl-fluorene]-*alt*-3,6-benzothiadiazole) (PF8BT, Fig. 10.16(a)) layer diode with an ITO electrode modified to enhance hole injection and block electron extraction (Reynolds *et al.*, 2002). PF8BT has relatively high electron mobility and a high ionisation potential of 5.9 eV which gives a barrier for hole injection from an ITO electrode of 1 eV. A layer of MoS_2 was deposited on the ITO and the surface oxidised in an oxygen plasma to form a MoS_2/MoO_3 bi-layer 3–7 nm thick. The MoS_2 acts as a high work function metal and the insulating MoO_3 layer blocks electron extraction. The energy level diagram for this device is shown in Fig. 10.13(b). Facile hole injection into the HOMO states of the PF8BT occurs from the valence band states of the MoS_2

[5] The efficiency of OLEDs can be defined and measured in a variety of ways. Forrest *et al.* (2003) have discussed these and recommended a standard approach to enable accurate comparison to made between devices originating from different sources.

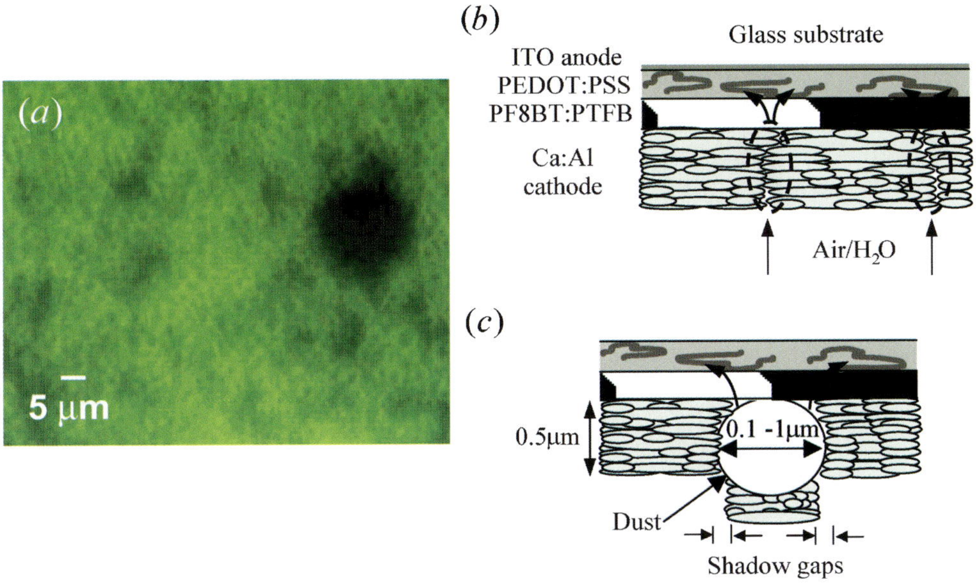

Fig. 10.14

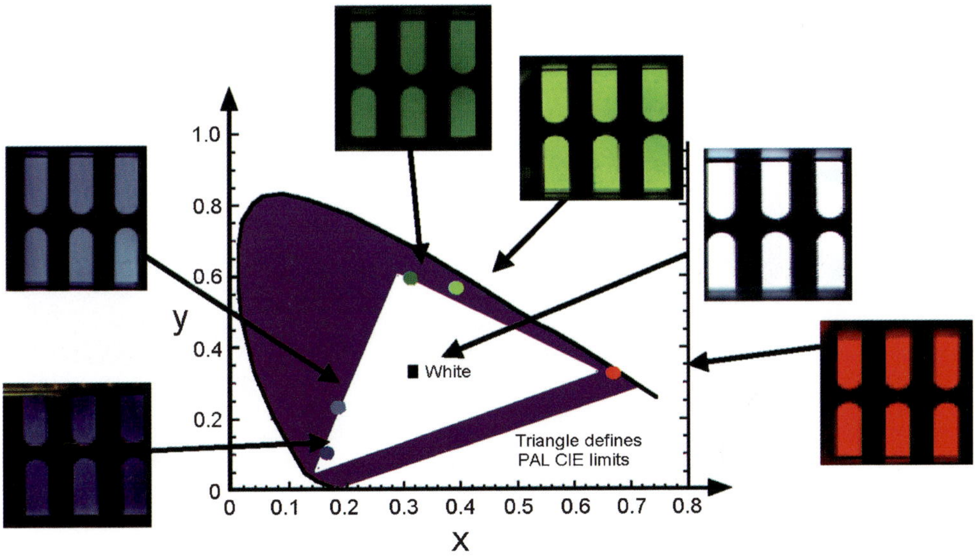

Fig. 10.15

Fig. 10.17

Fig. 10.18

Fig. 10.19

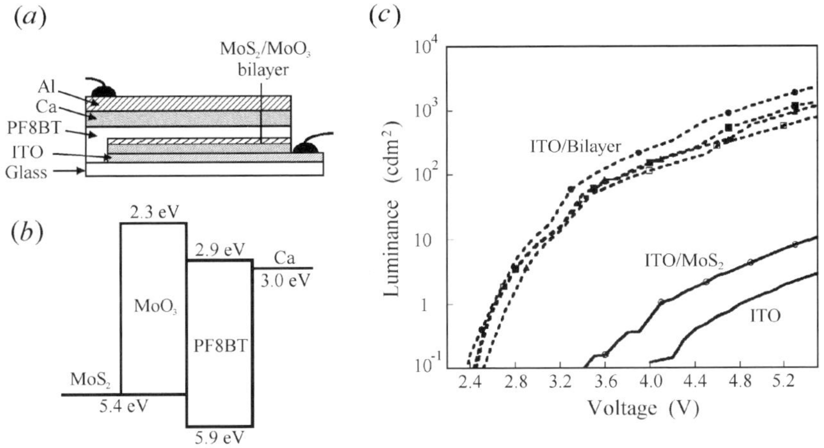

Fig. 10.13 Polymer LED with modified electrodes: (a) structure, (b) energy level scheme, and (c) device characteristics, the curves for the ITO with the bilayer are for different oxidation times. Reproduced with permission of the American Institute of Physics from Reynolds *et al.* (2002).

and MoO_3 which lie at about 5.4 eV below the vacuum level. The barrier to electron injection from the Ca electrode into the PF8TB LUMO states is low, ~0.1 eV, but the conduction band of the MoO_3 lies ~0.6eV higher than the LUMO states of the PF8BT, providing a barrier to electron extraction. PF8BT emits in the blue at 3 eV, ~415 nm, and emission characteristics for untreated and treated ITO electrodes are shown in Fig. 10.13(c). The power efficiency is increased from 0.003 lumW^{-1} at 5.5 V for the ITO electrode to 0.4 lumW^{-1} at 4 V for the modified electrodes.[6]

b. A number of conjugated polymers have been utilised to reduce the barrier for hole injection into the emissive layer. PEDOT–PSS provides an electrode with a work function of –5.2 eV giving improved hole injection, higher device efficiency and lower turn-on voltages (Kiebooms *et al.*, 1999). PEDOT has been compared with emeraldine base PAni protonated with different organic acids (Higgins *et al.*, 2001). Diodes with an MEH–PPV emissive layer and hole transport layers protonated with polystyrene sulphonic acid gave efficiencies up to 50% greater than a diode with a PEDOT–PSS hole transport layer. However, PAni was found to be less effective than PEDOT when the emissive layer was MeL–PPP. Similar comparisons have been made for devices with a poly(9,9′-dioctylfluorene) (PFO) emissive layer and PEDOT and poly(4,4′-dimethoxy-bithiophene) (PDBT) modified ITO electrodes (Gross *et al.*, 2000). The PDBT was deposited electrochemically

[6] The unit of luminous flux (lumen, lum) is defined as the energy emitted in one second into a cone of solid angle one steradian by a point source of intensity one candela.

and highly doped layers were found to reduce the turn-on voltage relative to ITO and PEDOT-coated ITO.

c. PPV, which has been extensively used in single layer devices, Figs. 9.28(b) and 9.30, is a hole-transporting material. Jang *et al.* (2000) describe the synthesis of poly(2-[{2,3,4,5,6-pentafluorobenzene}-2,3,5,6-tetrafluorophenyl]-1,4-phenylenevinylene) (PPFPV). The polymer is soluble in solvents such as chloroform and xylene and the electron-withdrawing fluorinated pendent groups are expected to assist electron injection. Diodes with the structures ITO/PPV/Al, ITO/ PPFPV/Al and ITO/PPV/PPFPV/Al were constructed and their characteristics compared. The emission from the diodes containing PPFPV peaked at 520 nm compared with 510 nm for the PPV diode. Although the turn-on voltage for emission of the PPFPV diode was double that of the PPV diode, its quantum efficiency was 64 times higher. The turn-on voltage of the PPV/PPFPV diode was 40% higher than that of the PPV diode, but its quantum efficiency was 380 times higher. While these devices were not optimised, the results illustrate the greater efficiency possible for diodes with separate electron and hole transporting layers.
d. Diodes have been fabricated with layers to enhance both hole and electron injection and transport on either side of an emissive layer. Poly(2-{4-biphenylyl}-5-phenyl-1,3,4-oxadiazole) (PBPPO) has been incorporated as an electron transport layer in a diode with the structure ITO/PEDOT/MEH–PPV/PBPPO/Al (Wang *et al.*, 2001). The PBPPO layer also blocks hole extraction so that injected electrons and holes are confined to the MEH–PPV emissive layer. Diodes made without the PBPPO layer had external quantum efficiency of $\sim$ 0.01% and maximum brightness of $\sim$15 cd m^{-2}, while diodes incorporating the PBPPO layer had values of 0.26% and 800 cd m^{-2}.

Enhanced generation of singlet excitons is controlled by polymer structure, since morphology is a dominant factor in spin–orbit coupling and inter-chain interactions, Section 9.4.3 and Table 9.6. Radiative emission from the singlet states is influenced by both the structure of the diode and the polymer structure. Since in most conjugated polymers the electron mobility is somewhat smaller than the hole mobility, see Table 9.4, electron–hole recombination will generally occur close to the metallic electron-injecting electrode. The proximity of the metal has a strong influence on the emission characteristics. Energy can be lost by coupling to plasmon[7] modes in the metal at distances below 30 nm. Radiative decay is also affected by standing waves that are set up by the interaction of the emitted electromagnetic radiation and the metal

[7] Plasmons are collective excitations of the electron cloud in a metal.

surface. This reduces the radiative emission rate for excitons created at the nodes and increases it near the anti-nodes of the standing wave pattern. Damping by plasmon modes is reduced in two-layer diodes relative to that in single-layer diodes since the excitons are created at the hetero-junction between the hole and electron transporting materials, which is separated from the metal electrode by the electron-transporting layer. Emission can be further enhanced if the layer thickness is adjusted to place the active region at an anti-node of the standing wave pattern. The photoluminescence of conjugated polymers is generally lower in the solid state that in solution. This is a result of the easier transfer of excitons between chains in the solid, enabling the excitons to migrate to regions where the chains are aggregated, thereby favouring non-radiative decay, see Section 9.4.1. In principle, the inclusion of bulky pendent groups will reduce inter-chain exciton transfer and the formation of aggregates. Alkyl chains are not particularly useful in this context, as demonstrated by the example of P3HT discussed in the previous section, where close contact of the chains occurs. Bulky conjugated pendent groups can act as trap sites for carriers, leading to reduced carrier mobility and poor charge balance in the emissive layer, nullifying any beneficial effects. The potential gains in device efficiency, which could be achieved if inter-chain interaction and packing is disrupted, have been demonstrated with rotaxanes comprising polyparaphenylene, polydiphenylenevinylene (PDV) and PFO oligomers threaded through cyclodextrin molecules that are retained by bulky terminal groups. The luminance of simple single-layer diodes with ITO and Ca electrodes with a PDV rotaxane layer was twice that of a diode with a PDV oligomer emissive layer (Cacialli *et al.*, 2002).

In planar diodes employing electron and hole transporting layers, as shown in Fig. 10.3(a), the electrons and holes recombine close to the hetero-junction and the area of the junction is the same as that of the front face of the device. Junction area is effectively increased if the hetero-junction is dispersed throughout the volume of the LED by blending immiscible electron- and hole-transporting polymers (Moons, 2002). The blend forms an interpenetrating network of the two polymers with large interfacial area. Carriers injected at the electrodes will reach the dispersed hetero-junction provided that the network contains percolation pathways. When there is a large difference in the HOMO–LUMO energy gap, excitons created in the high gap material can readily transfer across the hetero-junction, improving device efficiency. Blends of PPV and CN–PPV have been used to make LEDs using this effect. Diodes fabricated by combining this approach with tailored electron- and hole-injecting electrodes are highly efficient (Kim *et al.*, 2002). The device structure is ITO/PEDOT–PSS (50 nm)/PF8BT–PTFB (80 nm)/Ca (5 nm)/Al (400 nm), as shown in Fig. 10.14(b). The device is encapsulated with a glass cover slip sealed with

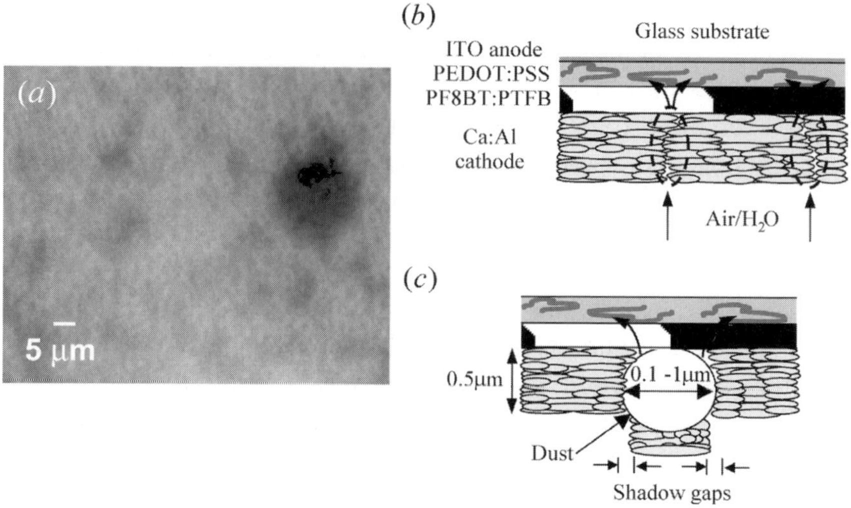

Fig. 10.14 Non-emissive defects in a polymer LED: visual appearance (a), gas penetration at grain boundaries (b) and dust particles (c) in the cathode. Reproduced with permission of Wiley-VCH from Kim *et al.* (2002).

epoxy resin. PF8BT–PTFB is a 1:1 by mass mixture of PF8BT and a hole-transporting polyfluorene, poly(2,7-(9,9′-di-*n*-octylfluorene)-*alt*-(1,4-phenylene-((4-secbutylphenyl)amino)-1,4-phenylene)), Fig. 10.16(b). PEDOT–PSS enhances hole injection from the ITO electrode and transports these to the blend layer, where they propagate in the PTFB. Similarly, the Ca layer ensures efficient electron injection into the PF8BT component. The diodes emit yellow–green light, have a low threshold voltage with a current density $>100\,\mathrm{mAcm^{-2}}$ at 3 V and an initial brightness of $\sim$10 000 cd $\mathrm{m^{-2}}$ at a current density of 100 $\mathrm{mAcm^{-2}}$. The power efficiency is 20 $\mathrm{lmW^{-1}}$ and the lifetime is >5000 hr at a light output of 100 cd $\mathrm{m^{-2}}$. PF8BT–PTFB devices can have turn-on thresholds below 2 V and brightness of 100 cd $\mathrm{m^{-2}}$ at 2.1 V. This has been attributed to the fact that the band-edge offsets are close to the exciton binding energy, i.e. the system is close to the limit for the formation of stable excitons at the interface between the two polymers. In this situation *exciplexes* can be formed, with electrons and holes residing on the polymers on either side of the interface. The interfacial exciplex has a strong charge transfer character and is stabilised by the deformation of the polymer chain structures in the vicinity of the charges. Charge recombination and emission can now occur by charge transfer across the interface without the need for the formation of an exciton on a single polymer chain. The exciplex emission is red-shifted relative to the single polymer exciton emission. LED emission is observed to have contributions from PF8BT/PTFB exciplexes and PF8BT excitons. Polyfluorenes now form an important class of materials for LEDs. Enhanced performance has been reported for PFOs end-capped with amine groups (Miteva *et al.*, 2001). This is attributed to holes trapped at the end groups recombining radiatively with electrons on the PFO backbone. This radiative channel competes effectively with the non-radiative decay of electrons and holes at chain aggregates and excimer forming sites.

Long operational lifetime is crucial for commercial devices. While the lifetime of devices was low at first, it has increased steadily and lifetimes in excess of 15 000 hrs have been reported (Blom *et al.*, 2000). Device failure has two principal causes, the degradation of conjugated polymers by oxygen and water (noted in the discussion of soft electronics, Section 10.3.1) and disintegration of the device structure. The search for polymers stable enough for solution processing, by either spin coating or ink jet printing, in air has been wide ranging. It has resulted in the commercial availability of substituted PPVs from Covion (Becker *et al.*, 2000) with improved stability due to the elimination of the tolane bis-benzyl chain defects that provide sites for the initiation of degradation, see Section 9.2.1, and polyfluorenes from Dow (Bernius *et al.*, 2000). Additionally, the development of improved materials and device structures has, as described above, resulted in more efficient devices with lower threshold voltage and lower current density for useful light outputs. This means that devices are subjected to less stress during use. The principal effect of failure of device integrity is the appearance of dark spots which eventually cease to emit, Fig. 10.14(a). These are attributed to the penetration of water into the diode through defects or gaps in the upper metal electrode, Fig. 10.14(b) and (c) (Kim *et al.*, 2002). In the diodes studied by Kim *et al.* (2002) spectroscopic analysis shows that the PEDOT–PSS layer is dedoped within the dark spots but that there is no detectable change in the emitting polymer blend. The degradation is considered to be by electrochemical action where water penetrates the diode. Normally, transfer of electrons from the metal electrode to the PEDOT layer is prevented by the built-in potential difference, but decomposition of water also generates ions that can either neutralise the built-in potential difference or diffuse into the PEDOT layer. In either case the PEDOT will be reduced. This electrochemical process differs from the mechanical failure observed for OLEDs (Schaer *et al.*, 2001). Similar processes may occur in polymer LEDs depending on the structure of the diode. However, in either case, elimination of defects in the cathode will result in improved device lifetime.

While there are markets for monochrome displays, that for colour displays is much larger. This has driven the development of polymers emitting in the blue, green and red (Kraft *et al.*, 1998, and Shim and Jin, 2002). Figure 10.15 illustrates the colours required for a full colour display plotted on a CIE diagram, together with some examples of polymer diodes that emit close to

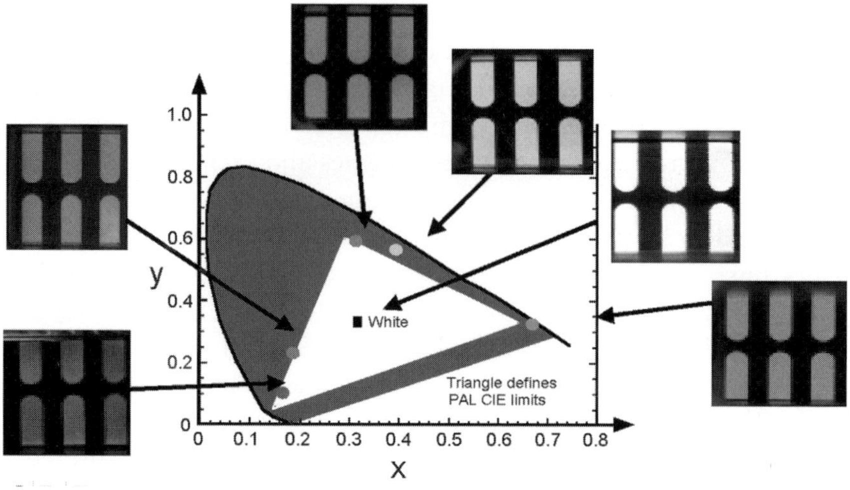

Fig. 10.15 Examples of polymer diodes with CIE co-ordinates close to those required for a full PAL colour display, which lie at the corners of the white triangle. Reproduced with permission of the Institute of Physics from Friend *et al.* (1999b).

the desired chromaticity, as developed by CDT. Green emitting polymers are readily available. The principal approach for blue emitters has been to introduce short conjugated sequences into the polymer chain. Longer conjugation sequences are required for red emitters. The non-radiative (Förster) transfer of energy from the excitons on the conjugated polymer chain to red emitting species, e.g. porphyrins or rare earth compounds, incorporated as either dopants or pendent groups, has been studied. Most success has been achieved by the modification of known efficient emitters by incorporating either different pendent groups or conjugated groups in the polymer backbone. Dow developed polyfluorenes with a wide range of emission wavelengths, whilst Covion synthesised modified PPVs and introduced spiro-polyfluorenes. Crosslinkable spiro-polymers have been utilised in prototype three-colour devices and provide materials suitable for the solution processing of colour displays (Müller *et al.*, 2003). The basic structure of these polymers is shown in Fig. 10.16(c). The 50% of backbone segments labelled Ar are a mixture of the moieties shown in Fig. 10.16(d) to (i). The ratios required for crosslinkable polymers with blue, green and red emission are (i) blue, 5:2:3::d:e:i, (ii) green, 5:2:3::d:e:f, and (iii) red, 5:2:2:1::d:e:g:h. A mixture of 1:4::e:i gives blue emission but is not cross-linkable.

A number of companies are actively developing and commercialising monochrome and colour displays. Examples are shown in Figs. 10.17–19. Flexible displays can be fabricated by the use of flexible glass or polymer substrates,

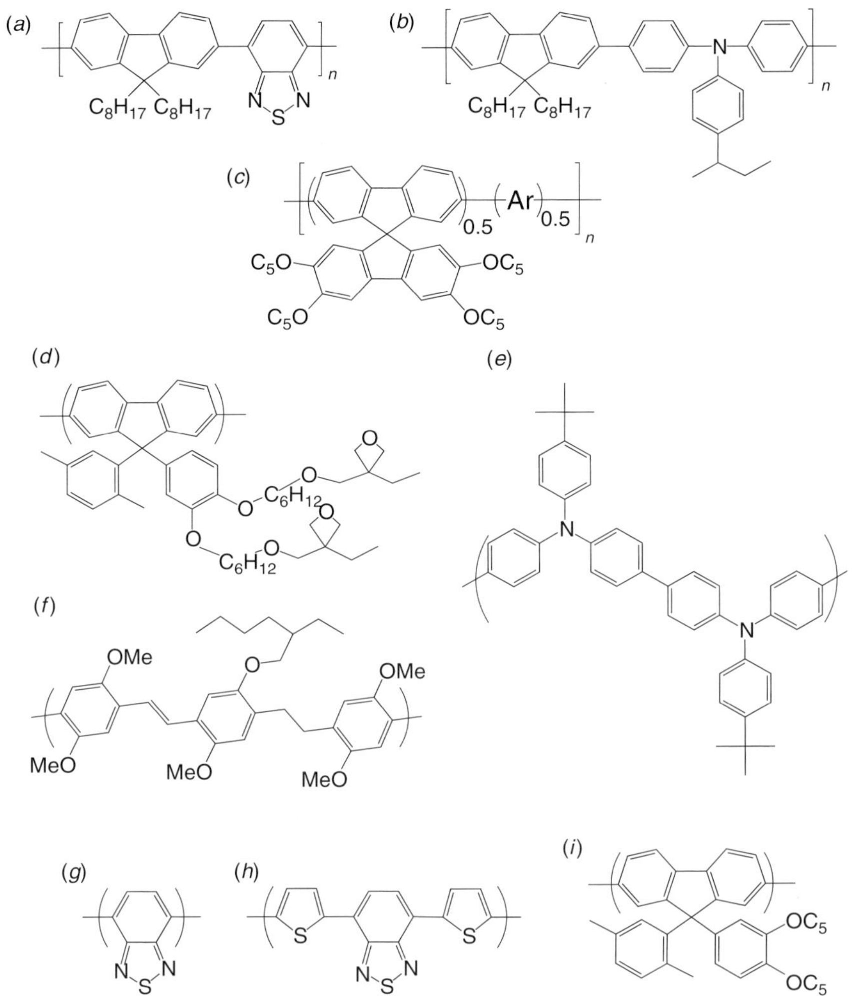

Fig. 10.16 Chemical structures of polymers with tailored transport and emission characteristics.

Fig. 10.17. Philips incorporated a monochrome display in a shaver that came on to the market in September 2002, Fig. 10.18. OSRAM Opto Semiconductors introduced evaluation kits for a 128 by 64 pixel monochrome Pictiva OLED display in 2003, Fig. 10.19. Larger colour, video rate displays for use in portable systems are under development, Fig. 10.18. Polymer emissive displays appear to be well placed to make a considerable impact on the display market.

Polymer LEDs offer further interesting properties. Orientation of the polymer chains in the plane of the diode results in polarised emission since, like the absorption, the principal component of the emission is polarised parallel to the

Fig. 10.17 Left: a 96 by 64 pixel passive matrix display fabricated with Covion polymer. Right: a flexible polymer LED. Reproduced with permission of Covion Organic Semiconductors GmbH.

Fig. 10.18 Polymer displays developed by Philips. Left: a colour display showing the pixelation, upper right: a small colour display, and lower right: shaver incorporating a monochrome polymer display. Reproduced with permission of Philips Research.

polymer backbones (Grell and Bradley, 1999, and Sainova *et al.*, 2002). The mono-domain orientation of the LC phase of PFOs has been used to obtain polarisation ratios as high as 15 (Grell *et al.*, 1999 and Miteva *et al.*, 2001). All polymer displays are made possible by the use of soft electronic driving circuits (Sirringhaus *et al.*, 1998). Placing the emitting layer in a resonant micro-cavity limits the bandwidth of the emitted radiation. A Bragg mirror, alternating layers of low and high refractive index, is used in conjunction with the metallic

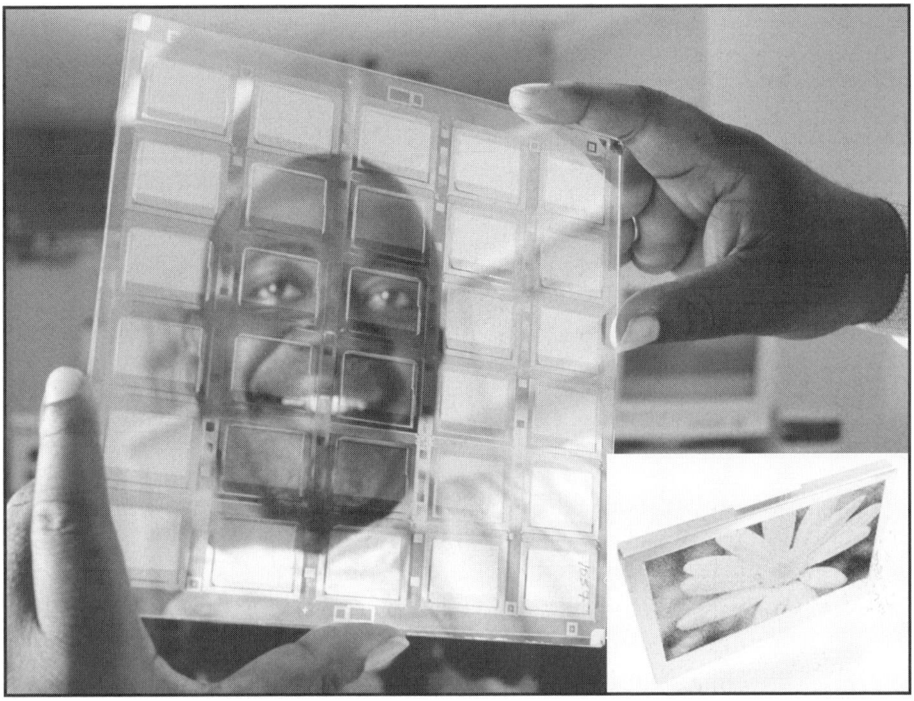

Fig. 10.19 Pictiva polymer OLED displays developed by OSRAM Opto Semiconductors. Main picture: a batch of 128 by 64 pixel monochrome displays. Insert: an operating display. Reproduced with permission of OSRAM Opto Semiconductors, Inc.

cathode to define a cavity in which the allowed modes enhance emission at the Bragg wavelength, while emission at other wavelengths is suppressed. Alternatively, the emitting layer can be used as an optical waveguide with Bragg gratings forming a resonant structure. The use of resonant cavities to adjust emission characteristics to match those required for colour displays was proposed, but was overtaken by the development of the new polymers described above. These structures are still of interest as the basis for electrically pumped polymer lasers (Tessler, 1999, and Scherf *et al.*, 2001). Optically pumped amplified stimulated emission (ASE) and laser emission have been observed from conjugated polymers. ASE is observed as line narrowed emission from a sample that is subjected to intense pumping but is not in a resonant cavity, Fig. 10.20. ASE has been observed in derivatives of PPV, polyfluorene and PPP, and blends of PPV derivatives. High optical gain has been observed in PFOs (Xia *et al.*, 2003), while orientation of the LC phase further enhances the ASE (Heliotis *et al.*, 2003). Laser emission, with a line-width of a few nanometres, has been seen in optically pumped solutions in macroscopic cavities and in solid films in microcavities and waveguides. On the basis of

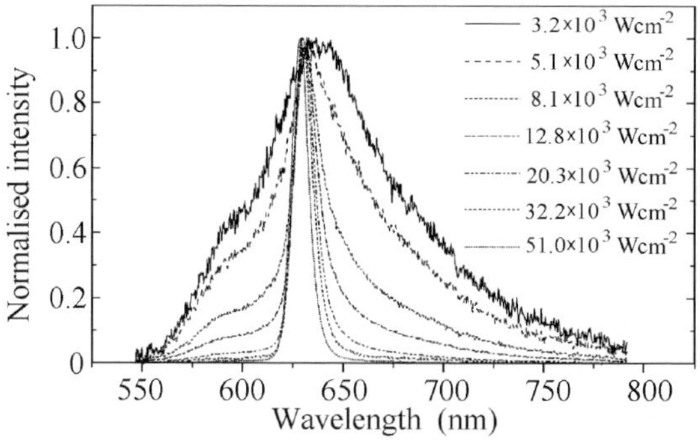

Fig. 10.20 Narrowing of the photoemission of MEH–PPV, due to ASE of light guided in a thin film, as the pump laser power is increased. Reprinted from Gupta *et al.* (2002). Copyright 2002, with permission from Elsevier.

the results for optically pumped devices and the properties of inorganic semiconductor laser diodes, the current density needed to achieve laser action in a polymer diode is estimated to be a few thousand amperes per square centimetre. Currents of this magnitude can be reached by pulsed operation of polymer LEDs, but laser emission was not observed. This can be attributed to the intrinsic differences in the behaviour of inorganic and polymeric semiconductors. The carriers injected into the polymer exist as polarons. These cause absorption that counteracts the optical gain due to the excitons created by carrier recombination. The low mobility of carriers in polymers means that very thin diodes are required to obtain nanosecond response times. High current densities also require thin diodes. The metal electrodes are then close to the emitting layer, quenching emission and strongly attenuating light propagating in the plane of the diode. This means that waveguide structures have unacceptably high optical loss. Furthermore, at high excitation densities, non-radiative decay of singlet excitons by singlet–triplet annihilation, mediated by long range dipole–dipole interaction, becomes more significant. Progress has been made, but these difficulties remain to be surmounted before an electrically pumped polymer laser is realised.

10.3.3 Photovoltaic devices

The photovoltaic effect in conjugated polymers and its application to solar cells was investigated in the late 1970s and early 1980s (Kanicki, 1986). Cells fabricated with PAc had relatively short lifetimes which resulted in the proposal that they could perhaps be commercialised as disposable power sources. The development of commercial solar cells has, however, been slow and now lags behind that of LEDs, despite the much more recent discovery of electroluminescence. The principal reason for this is the relatively poor power conversion efficiency of conjugated polymer photovoltaic cells compared to that of other technologies such as polysilicon solar cells and photo-electrochemical cells (Grätzel, 2001). In silicon solar cells, light absorbed at a hetero-junction or Schottky barrier directly produces electrons and holes in the conduction and valence bands. In photo-electrochemical cells, *n*-type TiO_2 coated with a sensitising dye is in contact with liquid electrolyte; light absorbed by the dye results in efficient charge separation at the TiO_2/electrolyte interface. Here conduction in the TiO_2 is in band states. This is not the case in conjugated polymers, where conduction entails more localised states. The physics of the photo-generation of charge carriers has been described in Sections 9.4.2 and 9.4.3. Examples of photocurrent action spectra for MEH–PPV and MeL–PPP are illustrated in Figs. 9.19(b) and (c). The photoconduction onset, indicating the position of the conduction band, lies well above the excitonic state that gives rise to the intense optical absorption, see Fig. 9.25. Carrier generation below this onset is then by exciton dissociation and annihilation, Sections 8.4.2 and 9.4.3. These indirect processes are much less efficient than direct inter-band excitation of charge carriers.

In a hetero-junction cell, see Fig. 10.3, excitons, created by the absorption of photons close to the hetero-junction, will dissociate with charge transfer between the polymers if the exciton binding energy is smaller than the offsets in HOMO and LUMO energy levels at the junction. The Coulomb attraction between the separated charges means that there is competition between geminate recombination of the electron and hole and dissociation into free carriers. Barker *et al.* (2003) have developed a numerical model for the prediction of the properties of bi-layer hetero-junction cells that incorporates charge generation, injection, drift, diffusion and recombination and the effect of space charges in the cell. This model predicts that the short-circuit quantum efficiency is dependent on the competition between recombination and dissociation of bound charges at the hetero-junction. Hence improved efficiency of hetero-junction cells will require the development of systems where dissociation is the dominant process. Methods that have already been used to improve cell efficiency are either to increase the area of the hetero-junction or to incorporate molecular additives to enhance carrier generation (Brabec *et al.*, 2001a).

Just as described for LEDs in the preceding section, the area of the hetero-junction in solar cells with the planar structure shown in Fig. 10.3(a) is defined by the dimensions of the cell. Similarly, this area is dramatically increased by the use of blends of immiscible electron- and hole-transporting polymers (Moons, 2002). Carriers generated at the dispersed interface between the two polymers will be able to diffuse to the electrodes if the network contains

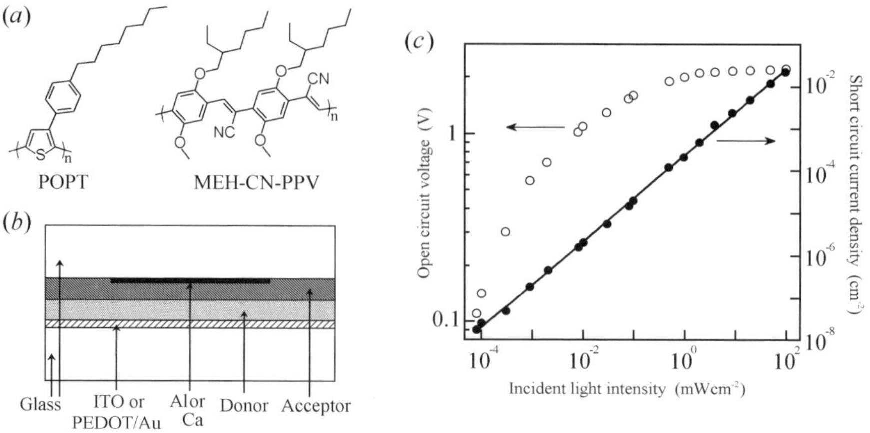

Fig. 10.21 (a) Electron donating and accepting polymers, POPT and MEH–CN–PPV, (b) hetero-junction solar cell, and (c) the light intensity dependence of the cell current and voltage. Reprinted with permission from *Nature*, Granström *et al.* (1998). Copyright 1998 Macmillan Magazines Limited.

percolation pathways. Solar cell performance will be optimised if the length scale of the network is similar to the exciton diffusion length, so that the excitons produced by photons absorbed in the bulk can reach the hetero-junction and dissociate. The destruction of excitons by efficient charge separation results in a quenching of photo-luminescence which can be used to identify blends suitable for use in solar cells (Halls *et al.*, 1995). However, the blend with the greatest reduction in photoluminescence may not give the most efficient photodiode, since this requires that the transport of the separated charges through the blend is also optimal, which may not be the case. Strong quenching of photoluminescence has been observed in blends of POPT and MEH–CN–PPV, see Fig. 10.21(a). These polymers were used to fabricate two-layer cells with an acceptor-rich layer and a donor-rich layer, as shown in Fig. 10.21(b) (Granström *et al.*, 1998). The layers are blends of POPT and MEH–CN–PPV in the ratios 1:19 and 19:1 for the acceptor and donor layers, respectively. The short-circuit photovoltaic quantum efficiency has a maximum value of 29% at ~480 nm. The response falls from this peak value to ~1% at 800 nm and the power conversion efficiency under a simulated solar spectrum is 1.9%. Performance of solar cells is also characterised by the *fill factor*, defined as the ratio of the maximum power $(IV)_{max}$ that can be extracted divided by the product of the maximum, short-circuit, current I_{sc} and the maximum, open-circuit, voltage V_{oc}. For the above two-layer cell Fig. 10.21(c) shows the dependence of I_{sc} and V_{oc} on light intensity. Over the range illustrated the fill factor lies in the range 0.3–0.34.

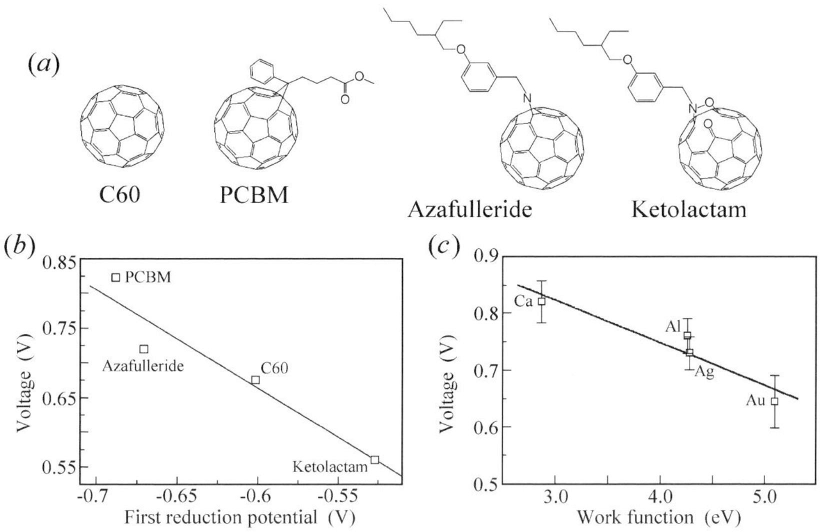

Fig. 10.22 (a) Fullerenes with different reduction potentials, (b) dependence of solar cell potential on fullerene reduction potential, and (c) cathode work function. Reproduced with permission of Wiley-VCH from Brabec *et al.* (2001b).

Most conjugated polymers behave as *p*-type semiconductors and act as donors when photo-excited. Although *n*-type polymers can be used as acceptors, e.g. MEH–CN–PPV and poly(pyridopyrazinevinylene) (Zhang *et al.*, 2003), as discussed in Section 10.3.1, they are less stable than *p*-type polymers. Stable *molecular* acceptors have been utilised, with the fullerenes proving to be particularly useful as acceptors when combined with PPVs. Both planar hetero-junctions, with separate fullerene and PPV layers, and bulk hetero-junctions, with the fullerene dispersed in a PPV matrix, have been employed (Brabec *et al.*, 2001a). The fullerenes are not strong enough acceptors to promote charge transfer in the absence of photo-excitation, but greatly enhance exciton dissociation. By analogy with chemical doping, this process has been termed *photo-doping*. The acceptor strength (first reduction potential) can be varied by covalently bonding different pendent groups to the basic C_{60} molecule. Figure 10.22(a) illustrates fullerenes that have been used to study the effect of acceptor strength on the performance of PPV solar cells (Brabec *et al.*, 2001b). These have first reduction potentials between –0.53 V and –0.69 V. Photovoltaic cells were fabricated using a PEDOT–PSS hole-transport layer on an ITO anode, a layer of poly(2-methoxy-5-(3′,7′-dimethyloctyloxy)-1,4-phenylenevinylene) (MDMO–PPV) doped with the fullerene in a 1:1 molar ratio and a cathode of either Ca, Ag, Al, or Au. V_{oc} was found to depend on the acceptor strength of the fullerene more strongly than on the work function of the cathode, Fig. 10.22(b) and (c). The weak dependence on

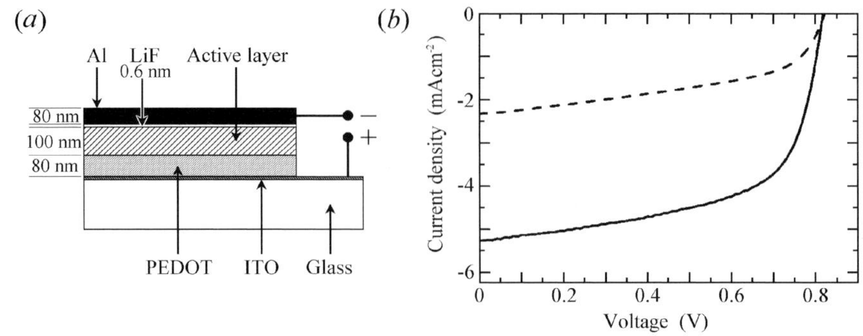

Fig. 10.23 (a) Solar cell and (b) cell characteristics for active layers of MDMO–PPV and PCBM prepared from toluene (full line) and chlorobenzene (dashed line) solutions. Reproduced with permission of the American Institute of Physics from Shaheen *et al.* (2001).

cathode work function is attributed to the presence of charged interface states between the active layer and the cathode, resulting in Fermi level pinning of the cathode and the fullerene. This leaves the energy difference in the HOMO of the donor and the LUMO of the acceptor as the main determinant of V_{oc}. Charge transfer has also been demonstrated between a conjugated polymer and compound semiconductor nanoparticles (Pientka *et al.*, 2004), but the performance of these materials in devices was not evaluated.

A further important factor is the morphology of the PPV-fullerene blend (Brabec *et al.*, 2001a). Solar cells with the structure illustrated in Fig. 10.23(a) were fabricated with an active layer comprising a 1:4 mass ratio blend of MDMO–PPV and PCBM (see Fig. 10.22(a)) that was formed by spin-coating from solution in either toluene or chlorobenzene. Atomic force microscope images show that the former produces films with large variations in thickness and features with dimensions $\sim$0.5 μm in size. The latter films show a more uniform thickness with finer features, typically 0.1 μm in size. V_{oc} is independent of solvent, but I_{sc} is much larger for the film prepared from chlorobenzene, Fig. 10.23(b). Since the absorption spectra of the two active layers are very similar, this difference in I_{sc} is attributed to improved carrier transport in the film cast from chlorobenzene. The finer features observed in the microscope images suggest a more intimate mixing of the fullerene with the PPV which would be expected to lead to the observed result. The incorporation of the LiF layer at the cathode, see Section 10.2.2, improves the fill factor of the cell. The internal quantum efficiency was >85% at 460 nm with a measured power efficiency of 9.5% under monochromatic illumination at 488 nm. The power conversion efficiency for a simulated solar source is 2.5% with a fill factor of 0.61.

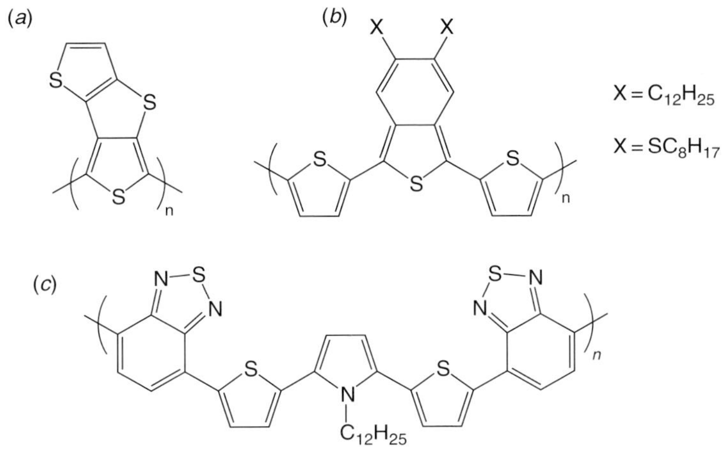

Fig. 10.24 Structures of polymers absorbing beyond 600 nm: (a) poly(dithieno(3,4-b:2′,3′-d)thiophene), (b) thiophene/isothianaphthalene copolymers, and (c) a thienyl-pyrrole/benzothiazole copolymer.

The reduced power efficiency in these cells reflects the low absorption at wavelengths longer than 600 nm. Further improvements in efficiency require the use of polymers with higher absorption at longer wavelengths. Polymers with absorption maxima at wavelengths longer than 600 nm are shown in Fig. 10.24. While the longest absorption occurs for poly(dithieno(3,4-b:2′,3′-d)thiophene), Fig. 10.24(a), this polymer is obtained electrochemically and is less useful than the solution processable copolymers containing either thiophene and isothianaphthalene or pyrrole and benzothiadiazole moieties, Fig. 10.24(b)–(c). Cells with active layers of the latter polymers blended with fullerene acceptors have been reported with power efficiency ~1% under simulated solar radiation (Brabec *et al.*, 2002).

Polythiophenes and MEH–PPV have been used as hole-transporting layers to replace the electrolyte in cells with TiO_2 electrodes (Grätzel cells). Bulk hetero-junction devices have been made by the hydrolysis in air of titanium(IV) *iso*propoxide, co-deposited in MDMO–PPV spin-coated films, to form TiO_2 nanoparticles (van Hal *et al.*, 2003). While reasonable peak performance under monochromatic illumination has been reported, the performance as a solar cell does not match that of the photo-electrochemical cells.

While progress towards more efficient polymer solar cells has been steady and directions for future work are mapped out, there are few reports of cell performance under continuous illumination. Kroon *et al.* (2002) studied the response of encapsulated MDMO–PPV cells, similar to that illustrated in Fig. 10.23(a), subjected to thermal stress and continuous illumination. A 30–50% decrease in efficiency was observed for cells stored in the dark at

40 C, while performance was almost completely degraded under simulated solar illumination at 50 C. Thus much remains to be done to realise cells with performance, under continuous solar irradiation, comparable to those of either silicon solar cells or Grätzel photo-electrochemical cells which give >10% power efficiency and a fill factor >0.7.

While solar cells have been the focus of most research effort, arrays of photosensitive polymer devices for the recording and processing of images have also been reported (Heeger *et al.*, 1995 and Pede *et al.*, 1998).

10.3.4 Sensors

The earliest investigations of the use of conjugated polymers for chemical sensing were based on the changes in properties that occur during chemical and electrochemical doping and dedoping. These provide a means of detecting compounds and ions with redox activity, i.e. those capable of oxidising or reducing the polymer. Catalysis may also occur, and although conductive polymers are poor chemical catalysts, they can mediate electron transfer with entrapped enzymes. The simplest sensors monitor the change in resistivity on exposure to an active chemical of a polymer film formed between interdigitated electrodes (chemo-resistors). The polymer can be used as the working electrode in an electrochemical cell with the effect of analytes detected either potentiometrically, or amperometrically, or voltammetrically. Potentiometry measures the open-circuit potential of the polymer electrode with respect to a reference electrode. The potential is sensitive to changes in dopant (anion) concentration in the film, ion exchange with ions other than the dopant in the electrolyte, and redox reactions in the polymer film and at the underlying metal electrode. While this method is simple, the sensitivity is greatly reduced if the polymer is either reduced, becoming insulating and rendering the cell open circuit, or highly conductive, when the potential is almost independent of anion concentration. Pinholes in the film expose the underlying metal electrode so that it, and not the polymer, determines the electrode potential. Amperometry determines the changes in the oxidation/reduction current at a constant electrode potential. This reflects the activity of the polymer in either the redox reaction or electro-catalysis. In voltammetry the electrode potential is swept cyclically between variable limits and the oxidation and reduction peaks associated with an analyte observed. The sensitivity of the polymer electrode potential to its environment is also utilised in chemically sensitive FETs (ChemFETs). These are insulated-gate devices in which the polymer gate electrode is exposed to either liquid or gas containing an analyte. The analyte brings about changes in the gate electrode potential, and this in turn modulates the drain current. Since the earliest conductive polymers studied were intractable, studies of sensors utilised electrochemically prepared polymers. These could be either directly deposited on the working electrode in an electrochemical sensor or, since they are produced as oxidised, conducting films, grown as films bridging the gap between electrode arrays for chemo-resistors. Hence, much of the early work utilised PPy films. The advent of soluble polymers and colloidal suspensions has allowed the more direct coating of electrodes for electrochemical sensors and chemo-resistors.

The properties of conductive polymers are influenced by the dopants used to render them conductive. This has been utilised in the development of arrays of resistive sensors for use in electronic noses, see Section 8.3.1 (Bailey and Persaud, 2000). Different dopants provide the differences in response to volatile organic compounds (VOCs) necessary to construct a detector array providing a unique *fingerprint* for a particular analyte. The use of different polymers and chemically modified variants of a single polymer allow the production of chemo-resistors with a wide range of responses. PPy has been a polymer of choice for this application as it is relatively stable under ambient conditions. Many VOCs do not act as dopants but can still produce large changes in resistance. In some instances this response may be due to changes in structural order of the polymer. In lightly doped polymers the conductivity increases if the relative permittivity of the VOC is greater than that of the polymer, and decreases if the reverse is true, since the barriers to carrier hopping are reduced if the relative permittivity of the medium is increased and vice versa (Vercelli *et al.*, 2002). Figure 10.25(a) shows an example of a micro-fabricated PPy chemo-resistor, with an integral heater for constant temperature operation. The response to ethanol of a single electrochemically prepared film is shown in Fig. 10.25(b) and the responses of three films with different counter ions and one produced by the direct polymerisation of pyrrole with $FeCl_3$ are shown in Fig. 10.25(c) (Fang *et al.*, 2002). Although *electronic nose* systems utilising conductive polymer arrays have appeared on the market these have not been a commercial success. This is partly due to the initial high cost of electronic nose technology and partly due to genuine problems encountered in use. Response times tend to be slow, see Fig. 10.25(b), because the analyte only diffuses slowly into the dense polymer films, the resistors are sensitive to humidity and temperature and there are gradual changes in chemo-resistor characteristics on prolonged exposure to the ambient atmosphere. The use of other polymers, such as PAni (Nicolas-Debarnot and Poncin-Epaillard, 2003) and PTs (Gallazzi *et al.*, 2003), is being explored in an effort to surmount these difficulties. The stability of PEDOT makes it suitable for use in biosensors, and PEDOT–PSS has been

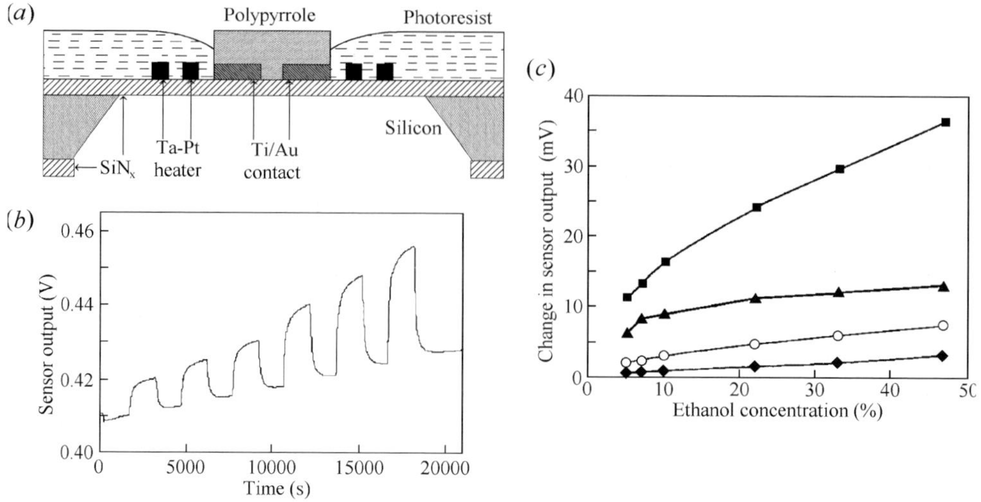

Fig. 10.25 (a) PPy chemo-resistor fabricated on a silicon nitride membrane, (b) typical response to increasing concentrations of ethanol, and (c) response to ethanol in air of relative humidity 15% with counter ions: ■, 1-butane sulphonate, ○, methyl phosphonic acid, ◆, decane sulphonate, and ▲, pyrrole oxidised with $FeCl_3$. Reprinted from Fang *et al.* (2002). Copyright 2002, with permission from Elsevier.

used to make hydrogels for the incorporation of enzymes (Åsberg and Inganäs, 2003).

Work on biosensors has extensively explored the use of conductive polymers as a matrix for enzymes, antibodies and whole cells (Gerard *et al.*, 2002). The biological entity can be added to the electrolyte used to prepare the film of conductive polymer, and incorporated either as an ion within the growing film or as an adsorbed species on its surface. Alternatively, pendent groups providing binding sites for the biological moiety can be incorporated into the polymer. Some authors maintain that the conductive polymer mediates the redox cycle of enzymes by providing a pathway for the transfer of electrons to the underlying electrode, while others conclude that chemical mediators diffusing in the polymer provide the necessary electron transfer. Many enzymes appear to be more stable, and function better in PPy than in other hosts. Glucose, fructose, D-alanine, urea, glutamate and cholesterol are some of the substances detected amperometrically with enzymes immobilised in PPy matrices. PAni has also been employed in conductivity detectors for haemoglobin and triglycerides. The incorporation of antibodies at specific covalent binding sites can provide a more bioactive surface with better reproducibility than incorporation by other methods. Despite extensive research on biosensors based on conductive polymers, few have been commercialised, because their advantages relative to those of other types of sensors, using less expensive supporting materials, are insufficient to outweigh the additional cost.

There are reports of conjugated polymer sensors using optical detection. The colour of PDAs is sensitive to their environment (Batchelder, 1988), so that if either the backbone is deformed or the relative permittivity of the surrounding medium changes, a shift in the absorption maximum will be observed. In compressed PDA monolayers, formed at the air–water interface by the UV polymerisation of diynoic acids, the chains are extended and the films appear blue. Binding sites for biological species can be incorporated in the monolayer either as pendent groups or embedded molecular species. Binding of bulky biomacromolecules at these sites disrupts the order of polymer chain order producing a blue to red colour change. Examples include the detection of virii and bacteria (Charych *et al.*, 1993 and Zhang *et al.*, 2002). Though specific and sensitive, this effect is irreversible. Fluorescence has also been utilised for the detection of biological molecules with water-soluble conjugated polymers. The methyl viologen di-positive ion associates with poly((2-methoxy-5-propyloxysulphonate)phenylenevinylene), MPS–PPV, in water, quenching the fluorescence emission (Chen *et al.*, 1999). Methyl viologen functionalised with biotin also quenches the MPS–PPV fluorescence. The quenching is reversed when avidin is introduced; this binds to the biotin and causes the methyl viologen to separate from the polymer, allowing it to fluoresce. Gaylord *et al.* (2002) have studied the interaction of the water-soluble polyfluorene, poly(9,9-bis(5′-N,N,N-trimethylammonium-hexyl)-fluorenephenylene), with peptide nucleic acids (PNAs) and DNA. The PNAs are labelled with a fluorescent dye chosen so that light absorbed by the polyfluorene is channelled to the dye by Förster energy transfer. If the DNA is complementary to the PNA, an anionic hybrid is formed which binds with the cationic polymer giving a strongly fluorescent complex. If the DNA is not complementary to the PNA, a hybrid is not formed, and the DNA complexes with the polymer; this complex does not associate with the PNA and the fluorescence remains weak. The fluorescence of the DNA–PNA–polyfluorene complex is 25 times larger than that of the isolated PNA and complementary DNA can be detected at 10 pM with a commercial fluorimeter.

10.3.5 Electrochemical applications

There are a number of other applications, in addition to chemical sensing, described in the previous section, that stem from the electrochemical activity of conjugated polymers. The incorporation and release of ions (dopants) during oxidation and reduction provides a way of storing energy and charge, and this has led to battery and capacitor developments. The same principle also provides a potential mechanism for controlled drug release, electromechanical actuators, electrochromic displays and windows with variable transmission.

(a) Batteries and capacitors The action of a battery where one electrode is a metal and the other a polymer, e.g. PPy, PT or PAni, can be described by the equations:

$$x(P^+I^-) \leftrightarrows xP + xI^- - xe^-, \text{ and } xM \leftrightarrows xM^+ + xe^-, \tag{10.5}$$

where P is the polymer, M is the metal, M^+ and I^- are monovalent cations and anions in the electrolyte, and the arrows to the left and right represent the charging and discharging of the cell, respectively. In principle, it is possible to replace the metal electrode with an n-type conducting polymer, e.g. PAc, that can incorporate cations. The low stability of n-type polymers means, however, that this is not a practical proposition. Alternatively, cations can be incorporated into p-type polymers by the use of polymeric anions, e.g. polystyrenesulphonate and polyvinylphosphate, that are fixed in the polymer. During reduction mobile cations must be incorporated to retain charge neutrality:

$$x(P^+Þ^-) \leftrightarrows xP + xÞ^- + xC^+ - xe^-, \tag{10.6}$$

where C^+ is a mobile monovalent cation and $Þ^-$ is the polymeric anion.

Most work has dealt with batteries with lithium cathodes and either PPy or PAni anodes, although PAc, PPV, polycarbazole and polyimidobenzyl have all been utilised (Arbizzani *et al.*, 1997). Cells with both liquid and polymeric electrolytes, Section 8.2.3, have been developed. Li-polymer batteries deliver voltages of 3–4 V. Discharging characteristics are good with a gradual decrease in cell voltage during discharge, providing that large currents are not drawn. Excellent lifetimes have been reported with battery capacity only slightly reduced by more than a thousand charge–discharge cycles. Care must be taken not to reduce the polymer electrode fully, as it then becomes insulating and the battery ceases to operate. This restriction on the extent of discharge also limits the useful storage capacity of the battery. Even when partially reduced the increased electrode resistance will limit the current that can be drawn from the battery. The electrolytes in these batteries must be extremely pure, since impurities give rise to self-discharge, i.e. an impurity oxidised at the polymer can diffuse to the anode, where it is reduced before diffusing back to the cathode and repeating the cycle. Bridgestone-Seiko produced commercial Li-PAni batteries with a $LiBF_4$ electrolyte in the late 1980s but these were withdrawn from the market after five years. At the same time Varta-BASF developed Li-PPy batteries but these were not commercialised. The problems of self-discharge, low current drain and energy storage capacity mean that Li-polymer batteries are not competitive with Ni-Cd and other established batteries. Whether the regulation of the use of heavy metals, to limit their release into the environment, will revive interest in the use of conductive polymers in batteries remains to be seen.

Small electrochemical (electrolytic) capacitors have been available for many years to provide low voltage capacitors with a capacitance that is much larger than that which can be achieved with dielectric capacitors of comparable size. The Matsushita Corporation produces electrolytic capacitors containing PPy (Panasonic SP-Cap) with Al and C electrodes, the former coated with an oxide layer, separated by a layer of PPy. The capacitor functions by the setting up of an electrochemical double layer (Helmholtz layer) at the Al-PPy interface. Similar cells with Ta electrodes have recently been introduced to the market. These capacitors have lifetimes of 1000–5000 hrs. *Supercapacitors* are electrochemical capacitors with energy storage intermediate between that of dielectric capacitors and batteries (Kötz and Carlen, 2000). These have been developed to provide back-up power during temporary interruption of the normal power source, e.g. during the replacement of batteries in portable electronic systems, and to provide faster, higher-current discharge than is possible with batteries. In principle, supercapacitors can be regarded as a form of battery, although there are distinct differences in structure, particularly for conductive polymers, that set them apart. Supercapacitors may have polymer electrodes made from either (a) similar *p*-type polymers, (b) dissimilar *p*-type polymers or (c) *p*-type and *n*-type polymers. In (a) and (b) the electrodes will initially be in equilibrium and both partially doped, and become unequally doped during charging. The operating voltage for structure (a) is $\sim$1 V while that for (b) will be higher depending on the oxidation potentials of the two polymers. Structure (c) has the highest potential working voltage and charge storage capacity but suffers from the poorer stability of *n*-type polymers. The US Department of the Environment programme for the development of supercapacitors for use in electrically powered vehicles set targets of 100 000 cycles with a minimum energy storage density of 5 $\mathrm{Whkg^{-1}}$. The working voltages and energy storage densities for type (a) PPy, type (b) PPy/poly(2,2-bithiophene) and type (c) PPy/poly(3-(4-fluorophenyl)thiophene) supercapacitors with polymer-coated carbon paper electrodes are (a) 1 V, 11 $\mathrm{Whkg^{-1}}$, (b) 1.5 V, 27 $\mathrm{Whkg^{-1}}$, and (c) 3.5 V, 39 $\mathrm{Whkg^{-1}}$, respectively. Note that these energy storage densities are per kilogram of the active electrode material while the DOE target is per kilogram of supercapacitor. The performance of the type (a) cells was, after an initial reduction, retained over 25 000 operating cycles. Cells utilising PAni have a larger initial drop but suffer no further loss of capacity over 100 000 cycles (Arbizzani *et al.*, 1997). These devices suffer from the self-discharge problem encountered with batteries, and further development is needed before practical devices are available.

(b) Electromechanical actuators The motion of ions in and out of a conjugated polymer during oxidation and reduction will be accompanied by changes in the dimensions of the polymer sample. These changes will depend on the size of ions employed (and their solvation cells) and the structure of the polymer, i.e. whether it is a high density film or a low density gel, the degree of crystallinity and any preference for orientation of the polymer chains. Values for dimension changes reported in the literature vary widely. Measured expansion of PPy and PAni films parallel to the film surface varies between ~0.5% and a few percent while estimates of volume change vary between ~7% for PAc to 35–120% for PPy and PAni. Although oxidation is normally expected to result in swelling, and reduction to result in contraction, swelling has been observed during the first reduction of electro-polymerised films. The complexity of the swelling behaviour is illustrated by that of electro-polymerised PPy doped with dodecylbenzenesulphonate counter ions, as revealed by atomic force microscopy during the cyclic voltammetry of films 1–15μm thick (Smela and Gadegaard, 1999). Figure 10.26(a) and (b) shows how the current flow and height change for the first and third cycles of a film 0.8μm thick as the potential is swept from 0 to −1 V and back to 0 V at 5 mVs^{-1}. During the first reduction at −0.53 V there is an unexpectedly large current peak and a very large increase in film thickness. That this is not accompanied by any in-plane deformation is shown by the absence of any bending of a bimorph formed with an inactive substrate, Fig. 10.26(c). Subsequent changes in current and thickness are smaller and reproducible, but the dimensional changes normal to, and in the plane of, the film are in the opposite sense. Other measurements of the in-plane dimensional changes of films and the expansion of gels show the expected swelling and de-swelling behaviour, but cannot on their own be used as a basis for determining the actual changes in volume.

The poor mechanical properties often found in extension restrict the application of electro-polymerised films and gels. Therefore many of the devices described in the literature are bimorphs with a stronger supporting layer (Otero, 2000). The presence of an impermeable backing will cause bending, as the diffusion of counter ions in and out of the polymer film will create a larger dimensional change at the surface of the film than in the bulk. The speed of response is limited by ion diffusion and falls in the range of a few seconds to a few minutes. Fibres with aligned polymer chains are intrinsically strong, and

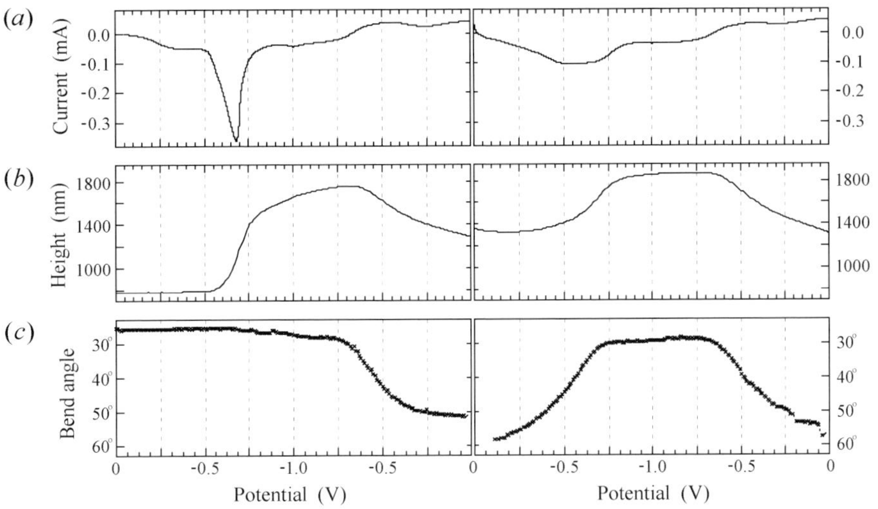

Fig. 10.26 (a) Current flow and (b) film thickness during the cyclic voltammetry of a PPy film compared with (c) bending angle of a PPy bimorph. Reproduced with permission of Wiley-VCH from Smela and Gadegaard (1999).

actuators have been made from PAni fibres spun from solution and drawn at elevated temperature. An all-polymer actuator is fabricated by coating a PAni fibre with a layer of a solid polymer electrolyte and an outer electrode formed by the oxidative polymerisation of pyrrole in ferric salt solution (Mazzoldi *et al.*, 2000). The reduced fibres have an elastic modulus of 1.5 GPa, and although the isometric stress is larger than that of muscle, the strain is only 0.2% at an energy density of 0.01 kWm^{-3}. The deformation of bimorphs of PPy on a gold foil can be precisely controlled and has been used to manipulate micro-fabricated structures (Smela *et al.*, 1995). A bimorph combining two PPy layers arranged so that one expands while the other contracts can move objects 1000 times its own weight (Otero and Cortés, 2003). Strain of 5% has been demonstrated in PPy tubular actuators 20–60 mm long and 125 μm diameter contacted with internal helical electrodes (Ding *et al.*, 2003). These results indicate that actuators made from conductive polymers are likely to find application in small-scale mechanical devices. Smela (2003) has presented a more extensive survey of work on conjugated polymer actuators.

(c) Electrochromism The changes in the optical absorption of conjugated polymers that occur during oxidation and reduction, see Figs. 9.31, 9.32 and 9.33, give rise to distinct changes in colour. A variety of colours are observed, and some examples are listed in Table 10.1. In general two distinct colours occur. PAni is more versatile since it can exist in a variety of forms depending

Table 10.1 *Visual colours of polymer thin films in the reduced and oxidised states*

	Colour	
Polymer	Reduced	Oxidised
PPy	Yellow	Blue-violet
P3MT	Purple	Pale blue
PITN		Colourless
PEDOT	Colourless (Very pale blue)	Dark blue
Poly(2,2′-bithiophene)	Orange-red	Blue

on its state of protonation, see Fig. 9.6. Perinigraniline base is violet in colour, protonated perinigraniline and emeraldine base are blue, protonated emeraldine is green and leucoemeraldine is pale yellow. An electrochemically-cycled film of PAni appears successively yellow, green, blue and dark purple as it is switched from the reduced to the oxidised state. The perceived colour is dependent on film thickness. Thin films in a weakly coloured state can appear colourless while thick films in the oxidised state can appear black. Argun *et al.* (2003) have reported a flexible electrochromic device which is comprised of two transparent electrodes of PEDOT–PSS deposited on insulating polymer films containing a sandwich of cathodically and anodically colouring conjugated polymer films separated by a polymer gel electrolyte.

Initially investigated as a display technology, electrochromism has been overtaken by electroluminescence for this application. Potential applications that are now envisaged are the control of the intensity of transmitted sunlight by *smart windows*, of light reflected from automobile rear-view mirrors, and of the brightness of cathode ray tube displays (Somani and Radhakrishnan, 2002). Many of the early studies utilised PAni, either on its own or in combination with the mixed valence iron cyanide Prussian blue, although PPy and a range of PTs were also investigated. More recent work on displays and smart windows uses commercially available PEDOT. The combination of conductive polymers with solid polymer electrolytes has enabled flexible electrochromic devices to be fabricated. The large increase in absorption at wavelengths beyond 1 μm, see Figs. 9.31 and 9.38, means that conductive polymer electrochromic devices are effective across the infra-red region. Controlled transmission at microwave frequency has been demonstrated with a PAni-polymer electrolyte-Ag particle composite. Electrochromic devices utilising conductive polymers offer an interesting range of properties, but their long-term stability under exposure to sunlight and repetitive cycling remains to be established. There is also competition from commercially available electrochromic smart windows (Schott and Flabeg) and automatically dimming rear-view mirrors (Gentex) based on WO_3.

(d) Controlled ion movement The flow of ions in and out of a conductive polymer film in an electrolyte may be controlled by the potential applied to the film, so that the film can be used as an *ion gate*. Since many drugs are ionic this phenomenon can be utilised in the controlled release of drugs. The biocompatibility of PPy, see Section 10.3.7(b) below, and its hydrophobicity in the reduced state, make it suitable for this purpose (Kontturi *et al.*, 1998). Anionic drugs incorporated during the electro-polymerisation of PPy include glutamate, dopamine, salicylate, naproxen and nicoside. Chlorpromazine and other cation drugs have been incorporated into PPy containing the polymeric anion melanin, see Equation (10.6). In saline solution a small amount of drug is released by exchange of anions with Cl^-. With the exception of nicoside, controlled release of the drug was achieved by the application of potentials below 1 V. The quantity of drug released was small, indicating that the method is likely to be restricted to use with drugs with a high therapeutic activity.

PPy films containing polymeric counter ions can be used to control the transport of metal ions through the film (Davey *et al.*, 2001). A solution, containing Na, K, Ca and Mg cations, is separated from pure water by an electro-polymerised film which in the oxidised state is impermeable to the cations. Applying a square wave potential that switches the films from the oxidised to the reduced state incorporates cations from the solution and releases some of them into the water, so that the film is rendered permeable. Permeation rates scale with cation size. This process is distinct from the permeation of neutral species described in Section 10.3.7(c).

10.3.6 Conductive coatings and composites

Although the specific conductivity of PAc can be greater than that of copper, see Section 9.2.1(a), the use of polymers as the core in power cables, which was proposed in the years immediately after the discovery of conductive polymers, has never been realised. The poor stability of PAc renders it unsuitable for an application where lifetimes of many decades are necessary. PAni is more stable and can be spun into strong fibres with high conductivity, Section 9.2.1(d), but it is significantly less conductive than copper. While the low density of the conductive polymers was seen as an advantage, the specific heat is also lower than that of a normal metal, so that any imperfection giving a locally lower conductivity is likely to lead to thermal run-away and failure of the polymer conductor, and, of course, the insulation too. Conductive polymers were also considered for applications that are less demanding of conductivity: the layers applied to the conductors in high voltage cables to reduce the electrical stress at the metal-insulator interface (see Section 6.7.1), and electromagnetic screens in coaxial cables. Now the market for stress control materials for high voltage power cables, a large and important one, is at present met by carbon black-polyethylene composites. Any conductive polymer replacement for this material needs to be as compatible with the extrusion process used to manufacture the cables and to be as low in cost. While it is possible that solution processing of conductive polymers could be employed, it would require modification of the manufacturing process. Furthermore, the necessary organic solvents are increasingly subject to legislation, restricting their use, and this would present added complications. Hence, for this, and a number of other applications, conductive polymers are required that either can be processed from an aqueous medium or are thermoplastics. As a result, considerable effort has been devoted to the development of conductive polymer composites, blends and colloids (De Paoli, 1997, Anand *et al.*, 1998, Armes, 1996 and Wessling, 1998).

As noted in the introductory paragraphs to Section 10.3, the reason often cited for work on composites and blends is the perceived poor mechanical properties of conductive polymers. But, also as noted there, conjugated polymers are inherently stronger than saturated polymers, their usual low strength resulting from the presence of morphological defects. The full potential for high strength conjugated polymers would be realised if these defects were eliminated, see also Section 9.2.2. But perhaps even more significantly, composites and blends can confer other properties not readily available in conductive polymers alone, one of the most important being thermoplastic behaviour, albeit this comes at the expense of conductivity. The loss in conductivity is acceptable in many applications such as the stress control layers for high voltage cables, screening of electromagnetic interference (EMI), antistatic coatings, and microwave absorption. Composites can be produced by several different processes: (a) electrochemical or chemical polymerisation of the conductive polymer in the host matrix, (b) deposition of the conductive polymer from solution either into a host or by co-precipitation with the second component, and (c) mechanical mixing. Electrochemical preparation simply requires the positioning of a porous matrix on the surface of the working electrode, so that the conductive polymer grows through the pores in the matrix. Any porous solid that does not dissolve in the electrolyte can be employed, e.g. porous glass and ceramics, polymer gels, paper, woven materials, etc. The introduction of direct chemical routes and the use of soluble polymers, methods that are more suitable for large-scale production, have, however, led to a decline in interest in the use of electrochemistry. Synthetic methods and solution processing can yield blends with a spectrum of structures, from composites having phase-separated components with dimensions of microns through to mesoscopic mixtures and interpenetrating networks on the molecular scale. The use of stabilising agents, typically surfactants, during synthesis can also yield colloidal suspensions of conductive polymers, which can be used to produce either composites or coatings.

In an electrochemically prepared conductive composite, a continuous polymer network is automatically formed through the matrix, ensuring that conductivity is achieved even for low fractional content of conductive polymer. In contrast, composites comprising a conductive and a non-conductive polymer prepared by other methods will possess a percolation threshold below which the composite is insulating, Section 8.3. The conductive polymer fraction at the percolation threshold is dependent on the morphology of the composite. Composites formed from particles of conductive and insulating polymers have percolation thresholds above 10%. Composites made from insulating polymer particles coated with a conductive polymer have percolation thresholds between 0.5 and 2%. The lowest percolation thresholds, well below 1%, are found in composites made by the gelation of conductive and insulating polymer mixtures from a common solvent. Initially it was thought that a fibrous network of the insulating polymer was formed, which was subsequently coated with the conductive polymer. Further study has indicated, however, that the composite is a phase-separated, bicontinuous-interpenetrating network. Conductivity values on the plateau above the percolation threshold typically fall in the range 1–1000 Sm^{-1}. These values are adequate for the dissipation of static charges, and one of the first applications of a conductive polymer was as a conductive brush to remove residual charge from the photoreceptor in a xerographic copier, Section 10.2.1. The colour associated with conductive polymers limits their use as antistatic agents in products such as carpets where the tint they produce is unacceptable. They have been used where colour is not important, e.g. in antistatic films and bags for the protective storage of integrated circuits. The high optical conductivity of highly doped polymers, e.g. PAni, Fig. 9.38, means that they are suitable for use as absorbers and EMI shields at microwave frequencies. Useful absorption has been demonstrated in the range 10–100 GHz for PAni composites with insulating polymers. At lower frequency, composites with either metal or magnetic fillers in PAni are effective. Despite this, the low cost of composites containing carbon black or metal powders makes them difficult to replace in these applications.

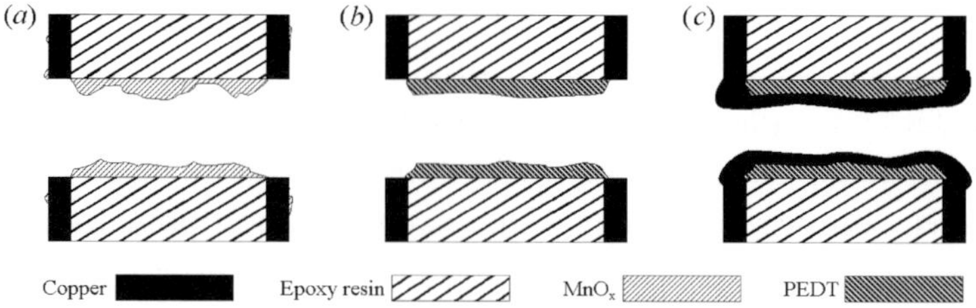

Fig. 10.27 Plating of via using PEDT: (a) chemically deposited MnO_x, and (b) and (c) electrochemical deposition of PEDT and copper in the via.

During the polymerisation of pyrrole in aqueous ferric salt solution, solid objects immersed in the reaction mixture are coated with a dense film of PPy. Reaction intermediates adsorbed on to the surface of the immersed object nucleate separate growth sites that coalesce to form a continuous film. This approach has been used to coat textiles with both PPy and, under carefully controlled conditions, PAni (Kuhn and Child, 1998). PPy deposits on most fabrics with only small variations in conductivity, with the exception of clean glass fibres where silanes must be used to promote the adhesion of the PPy to the glass. A single immersion of a fabric gives surface resistivities of 20–200Ω, and repeated immersion gives values as low as 5Ω. There is a slow increase in surface resistivity with time when the fabric is exposed to air, but the stability is adequate for applications such as static dissipation, EMI shielding and microwave absorption. Milliken & Company has manufactured textiles for these applications under the trade mark Contex®. Research continues into the development of textiles for microwave absorption, driven by the need to screen radiation from mobile telephones and other high frequency electronic devices in domestic use. An alternative method for the chemical deposition of conductive polymers is to coat the object with oxidant and then expose it to the monomer. This method was tried for coating textiles and impregnating paper, but has found successful application in the important process of plating through holes (vias) drilled in printed circuit boards to form an electrical connection between circuits on opposite sides of the boards (Schultze *et al.*, 1999). Initially the process was developed with PPy, but subsequently with polyethylenedioxythiophene (PEDT). The process is illustrated in Fig. 10.27. Exposure to $KMnO_4$ passivates the copper conductors but oxidises the epoxy resin in the vias and coats it with a non-stoichiometric manganese oxide. During electrochemical polymerisation of PEDT from acidic solution, anions from the electrolyte are incorporated in the polymer layer, while the H^+ ions reduce the oxide releasing Mn^{2+} ions into solution. This eliminates the oxide layer, and leaves a conductive PEDT film ~200 nm thick on which copper can be electroplated. Blasberg Oberflachentechnik market both PPy and PEDT processes.

The requirement on health, safety and environmental grounds for industrial processes employing aqueous media rather than organic solvents has prevented the commercial application of many soluble polymers. An alternative approach mentioned above is to develop water-borne colloids that can be used to deposit thin conductive films (Armes, 1996 and Wessling, 1998). PPy and PAni have been most extensively studied. Water soluble oligomers and polymers with pendent aniline and thiophene groups can be grafted on to the surface of PAni and PPy particles, respectively, to give colloids with polymer particles ~100 nm in diameter. PAni colloids are marketed under the trade name Ormecon® by Zipperling Kessler for use in corrosion protection, EMI shielding and as conductive tracks in soft electronic devices, Section 10.3.1. These PAni colloids have also been used to coat polymer films used as the diaphragm in electrostatic loudspeakers marketed by FINAL. Bayer-Industrie has developed PT and PEDOT colloids, their small particle size allowing the polymers to be deposited by inkjet printing and to be utilised in soft electronics, polymeric LEDs and solar cells, Sections 10.3.1–3. Agfa-Gevaert, a Bayer daughter company, produces conductive films using the PT colloids under the trade name Orgacon®. Thin transparent films suitable for the discharge of static electricity can be deposited on polyester substrates and are used as antistatic layers in Agfa photographic films. Conductive films made utilising PEDOT colloid are also being produced. A variety of methods can be employed for the micro- and nano-structuring of these and other conductive polymer films (Schultze *et al.*, 1999). Both inkjet printing and lithography are used to fabricate polymer electronic devices. Photoinitiators, as described for PAni in Section 10.3.1, have also been used with PEDOT–PSS. In this case azide photoinitiators crosslink the PEDOT–PSS with 365 nm radiation before dissolving unirradiated areas. Examples of the structures fabricated are shown in Fig. 10.28.

10.3.7 Other applications

(a) Corrosion control The economic cost of the corrosion of metals is high and there is a commensurate effort to develop methods to inhibit corrosion. Passivation with a stable oxide layer is possible in only a few cases. Impermeable coatings lack permanency and can therefore only slow corrosion but not eliminate it. Active corrosion protection requires a reduction in the rate of oxidation of the metal. Available methods are chemical inhibition, cathodic protection and anodic protection. Electrically conductive polymers have been shown to be capable of providing all of these protection mechanisms. While most research has focussed on the protection of iron and steels, the protection of aluminium, copper, silver, zinc, titanium, magnesium and their

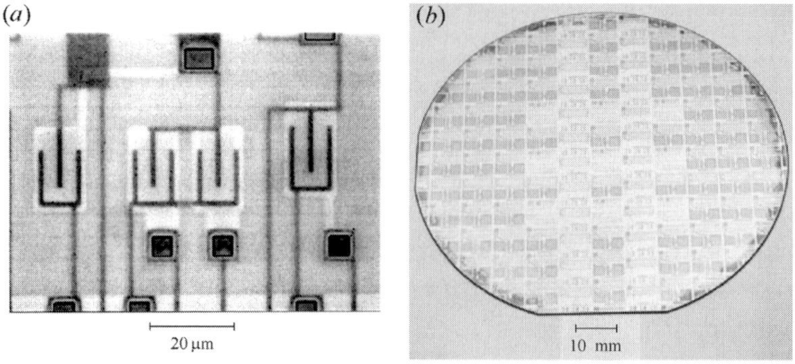

Fig. 10.28 (a) Lithographically defined PEDOT inter-digitated electrodes (dark lines) and (b) 4 inch wafer with soft electronic logic circuits and test structures with PEDOT interconnects. Reproduced with permission of the American Institute of Physics from Touwslager *et al.* (2002).

alloys has also been studied (Lu *et al.*, 1998, and Tallman *et al.*, 2002). PPy, PAni and PT, together with related substituted polymers, have been most widely studied. The principal corrosion mechanism in the presence of water is electrochemical, and electrochemical methods are generally employed to characterise the effectiveness of anticorrosion coatings. Since the most useful polymers can be grown electrochemically, most test samples are prepared by direct electro-polymerisation onto the metal under study. In some instances this is not possible due to the formation of a passive layer, although prior treatment of the surface can eliminate this problem. A simpler method is to cast films from either solution or suspension. In addition to protecting coated metal, conductive polymers reduce corrosion where the metal is exposed. While exact details of the mechanisms of protection are still being studied, it appears that the oxidative power of the partially oxidised polymer coating maintains the metal in a passive regime. In addition, as the polymer is conductive, it can form a galvanic couple with the metal, so that *it* is corroded rather than the metal. The conductivity of the polymer is also thought to be important in relocating cathodic reactions, e.g. oxygen reduction, from the metal interface to the surface of the polymer coating exposed to the atmosphere. One inherent shortcoming is the high permeability of the films, allowing ingress of water and ions, and leading to failure when films are immersed in water or subjected to salt spray. The difficulty is associated with the high ionic content of the polymer and is therefore difficult to solve altogether. Bonding to the metal surface can be a problem, particularly for films cast from solution or suspension. Kendig *et al.* (2003) have made use of the high ionic content of doped PAni films to use them as carriers of corrosion inhibitors. The PAni film acts as a *smart* coating which release the anionic inhibitor when damaged. Conductive polymer coatings can greatly reduce rates of corrosion in metals, but commercial exploitation requires that progress be made in solving the outstanding problems.

(b) Biocompatible materials The possibility that PPy would be biocompatible was suggested by the similarity of its molecular structure to that of naturally occurring compounds such as the bilirubins. This was borne out by the incorporation of enzymes, antibodies and living cells into PPy with retention of their biological activity, Section 10.3.4. Proteins and DNA bind to PPy, so that *in vivo* PPy is coated with a layer of biomacromolecules. PPy also possesses useful bioactivity. Langer and co-workers (Wong *et al.*, 1994, and Schmidt *et al.*, 1997) observed the non-invasive control of the growth of mammalian cells and neurites on PPy substrates in the presence of an electrical field. In the light of these results electro-polymerised PPy coating of titanium surgical implants has been demonstrated with the aim of improving biocompatibility and promoting osteointegration (De Giglio *et al.*, 2001). Other studies used PPy-coated fabrics *in vivo*, although *in vitro* appreciable conductivity is retained for only about a week. Biodegradable conductive polymers are being investigated for biomedical applications such as the provision of a temporary matrix for cell attachment and the *in vivo* application of electrical signals to promote tissue regeneration (Rivers *et al.*, 2002).

(c) Permeable membranes The permeability of conductive polymer films, seen as a problem in the development of anticorrosion coatings, offers an opportunity for their use as membranes for the separation and purification of gases and liquids. The fibrillar morphology of many conjugated polymers, e.g. PAc, Section 9.4.1, is unsuitable for this application, because the gaps between the fibres allow ready passage of all the constituents of gas and liquid mixtures. Most suitable materials are those that approach full packing density, but retain amorphous regions with sufficient free volume to allow molecules to permeate through the film. An average pore size less than 5 nm is necessary for gas separation, reverse osmosis and *pervaporation*. Pervaporation occurs when a liquid in contact with a membrane permeates through it as a gas. PAni has proved to be the most suitable polymer in this context (Conklin *et al.*, 1998). PAni, emeraldine base, films are prepared by casting from N-methyl-2-pyrrolidinone solution. The permeability of these films is much higher for helium and hydrogen than for larger molecules such as nitrogen and methane. The permeability falls for all gases when the film is acid doped, and rises and falls with successive un-doping and doping, although after the first doping the

permeability for oxygen compared to that for nitrogen shows a steady increase. Permeability is also dependent on the acid used to dope the film. Selective gas permeation also occurs in electro-polymerised PPy and di-methoxy-PPV cast from suspension. Liquid mixtures can be separated if pervaporation occurs for one component. Pervaporation has been observed for PAni, PPy and methyl substituted PPy and PT. Doped PAni provides good separation of water from organic acid/water mixtures. Practical application of the permeability of conductive polymer films requires the development of robust, pinhole-free films with areas large enough to achieve high fluxes of the permeating species.

(d) Xerography Conjugated semiconducting polymers display high photo-generation efficiency and high carrier mobility properties that render them suitable for use as single-layer photoreceptors for xerography. Despite this potential there are few examples of work in this area in the open literature. Photo-induced decay curves have been recorded for the ladder polymer MeL–PPP (Pan *et al.*, 2000). The photo-generation efficiency rises from 0.3% at 2×10^6 Vm^{-1} to 25% at 1.6×10^8 Vm^{-1}. MeL–PPP also displays low dark decay of surface potential and low residual surface potential when discharged. Although the spectral response is narrow, see Fig. 9.19(c), the photosensitivity is constant between 380 and 460 nm and has a value about 2.5% that of typical two-layer photoreceptors, described in Section 10.2.1. This response is suitable for use with blue semiconductor laser diodes which offer compact, faster and higher-resolution copiers.

(e) Non-linear optics Conjugated polymers posses a large second hyperpolarisability, Section 9.4.2, and have been explored as the basis for non-linear optical devices. Of the several third order non-linear effects that occur, see Section 3.7.1, the intensity-dependent refractive index, the optical Kerr effect, has been the most widely explored as a basis for devices (Luther-Davies and Samoc, 1997). This emphasis reflects the fact that all the devices described in Section 10.2.3 will function if the control of waveguide refractive index is effected by light rather than an electric field. This offers the prospect of all-optical systems that can be switched by a remotely generated control beam. The non-linear coefficients will be large when close to resonance with the electronic levels of the polymer. These situations are more numerous for third order non-linearity than for second order non-linearity, see Section 3.7.1. The simultaneous occurrence of linear and non-linear absorption hinders the realisation of practical devices, since the control beam can be attenuated before it produces a useful change in refractive

index. The utility of a material for optical switching can be assessed by the figures of merit:

$$W = \frac{n_2 I_0}{\alpha\lambda}, \text{ and } T = \frac{\beta\lambda}{n_2}, \tag{10.7}$$

where n_2 is the intensity-dependent refractive index, Equation (3.84), I_0 is the light intensity, and α and β are the linear and non-linear absorption coefficients at wavelength λ. Ideally $W \gg 1$ and $T \ll 1$.

In the 1990s extensive measurements of non-linear coefficients and non-linear absorption for single crystals and solution-cast films of PDAs revealed promising figures of merit in the near-resonance regime, just below the linear absorption peak and above the main two-photon absorption. Non-linear directional couplers were fabricated using both thin single crystals and solution-cast films, and optical switching was observed. Optical bistability was observed in Fabry–Perot and waveguide interferometers. The optical power inputs required were, however, too high for use in fibre-optic networks. More recent studies revealed effects due to fifth and seventh order non-linearity. In the latter case this gives rise to a strong four-photon absorption extending over a wide wavelength range, e.g. for PDA–PTS from 1.46 to 2.12 μm. Some increase in non-linear coefficients has been achieved by the introduction of pendent groups with electronic interaction with the conjugated backbone. So far these increases are insufficient to overcome the adverse effects of non-linear absorption. Attention has turned to the investigation of non-linear Bragg reflectors fabricated in solution-cast waveguides, Section 10.3.2. MEH–PPV and related PPV are both promising in this context, but practical devices remain to be demonstrated (Bader *et al.*, 2002).

Degenerate four-wave-mixing (DFWM, Section 10.2.3) in conjugated polymers has also been explored. Its use in optical image correlation has been demonstrated using Durham-route PAc and similar polymers. The very large optical non-linearity of PAc, Table 9.3, and the fast electronic response allow correlation to be performed with sub-picosecond laser pulses. The poor stability of PAc prevented this laboratory demonstration being turned into a practical device.

(f) Photorefractive materials Mecher *et al.* (1999) explored conjugated polymers as matrices for photorefractive materials, comparing the properties of composites based on PPV with those of the established PVK composites, see Section 10.2.3. The host polymer was poly(1,4-phenylene-1, 2-di(4-benzyloxyphenyl)vinylene) with DMNPAA and the related

3-methoxy-(4-*p*-nitrophenylazo)anisole, MNAA, as non-linear components, diphenylphthalate, DPP, as plasticiser and a derivatised C_{60} as sensitiser. This was compared with PVK–ECZ with the same chromophores and either the same C_{60} compound or TNF as sensitiser. The PPV composite gave better steady-state photorefractive performance than the PVK materials, due to easier orientation of the chromophores and a higher trap density giving a larger internal field. Despite the higher hole mobility in the PPV polymer, the response times were, however, similar to those of the PVK materials. Poorer photo-generation efficiency in PPV composites accounted for this. Thus the anticipated improvement in performance was not realised.

(g) Miscellaneous Numerous other applications have been considered in the scientific literature but have attracted little attention. Some examples are as follows.

a. The morphology of conductive polymers grown electrochemically is sensitive to the growth conditions. Long filaments can be produced, and in the early 1990s this was proposed as a method of making electrical interconnects for small-scale electronics. The development of the lithography of tractable conductive polymers, Section 10.3.6, supplanted this approach.
b. The stretch alignment of blends of conductive polymers with polymers such as polyethylene and polyvinyl alcohol produces films with anisotropic optical absorption that can be used as polarisers in the infra-red between 2.5 and 25 μm. Although the properties of these polarisers are comparable to those of wire-grid polarisers, they do not appear to have been developed commercially.
c. The formation of strong joints by microwave welding of thermoplastics coated with a conductive polymer powder has been demonstrated.
d. The use of the thermochromism of conjugated polymer films has been proposed for use in reversible thermal recording.
e. Conjugated polymers have high refractive indices just below the first absorption band. Lower refractive indices are obtained for composites containing a powder with a low refractive index. Alternating layers of conjugated polymer and composite have been used to form Bragg gratings in microcavity LEDs. Although this technique is applicable to the fabrication of other optical elements, such developments do not seem to have been pursued.

10.4 Further reading

The tract by Borsenberger and Weiss (1998) is the most complete account available of organic photoreceptors for xerography. Hung and Chen (2002) provide a comprehensive account of low molar mass molecular materials for organic light emitting devices and device fabrication and properties. Dalton (2002) presents an overview of the materials and device technology for polymeric non-linear optical integrated devices. Zilker (2000 and 2002) reviews the design and physics of organic photorefractive materials, while Kippelen and Peyghambarian (2003) review polymer photorefractive materials and their applications. Reviews of the applications of conductive polymers are found in the handbooks edited by Skotheim (1986), Skotheim *et al.* (1998) and Nalwa (1997). Preezant *et al.* (2002) present an overview of soft electronics from an engineering standpoint. The edited volumes by Sinar (2003) and Brabec *et al.* (2003) provide overviews of research on organic light-emitting diodes and organic photovoltaic devices, respectively. Akcelrud (2003) reviews the chemistry and physics of electroluminescent polymers.

References

Abkowitz, M. & Pai, D. M. (1986) *Philos. Mag.*, **B53**, 193.

Abkowitz, M. A., Antoniadis, H., Facci, J. S., Hsieh, B. R. & Stolka, M. (1994) *Synth. Metals*, **67**, 187.

Adams, P. N., Laughlin, P. J., Monkman, A. P. & Kenwright, A. M. (1996) *Polymer*, **37**, 3411.

Akcelrud, L. (2003) *Prog. Polym. Sci.*, **28**, 875.

Alberti, G. & Casciola, M. (2001) *Solid State Ionics*, **145**, 3.

Alquié, C., Dreyfus, G. & Lewiner, J. (1981) *Phys. Rev. Lett.*, **47**, 1483.

Alvarez, F., Alegría, A. & Colmenero, J. (1999) *Phys. Rev.*, **44**, 7306.

Anand, J., Palaniappan, S. & Sathyanarayana, D. N. (1998) *Prog. Polym. Sci.*, **23**, 993.

Anderson, P. W. (1958) *Phys. Rev.*, **109**, 1492.

André, J.-M., Delhalle, J. & Brédas, J.-L. (1991) *Quantum Chemistry Aided Design of Organic Polymers*, Singapore, World Scientific.

Angelli, A. (1916) *Gazz. Chim. Ital.*, **46**, 279.

Antoniadis, H., Hsieh, B. R., Abkowitz, M. A. & Stolka, M. (1993) *Appl. Phys. Lett.*, **62**, 3167.

Arbizzani, C., Mastragostino, M. & Scrosati, B. (1997) in *Handbook of Organic Conductive Molecules and Polymers* (Nalwa, H. S., ed.), Chichester, John Wiley & Sons, **4**, 595.

Argun, A. A., Cirpan, A. & Reynolds, J. R. (2003) *Adv. Mater.*, **15**, 1338.

Arkhipov, V. I., Emelianova, E. V. & Bässler, H. (2004) *Chem. Phys. Lett.*, **383**, 166.

Armand, M. B. (1994) *Solid State Ionics*, **69**, 309.

Armand, M. B., Chabagno, J. M. & Dudot, M. J. (1979) in *Fast Ion Transport in Solids* (Vashishta, J. N., Mundy, J. N. & Shenoy, G. K., eds.), New York, North-Holland.

Armes, S. P. (1996) *Curr. Opin. Coll. & Interface Sci.*, **1**, 214.

Arridge, R. G. C. (1967) *Brit. J. Appl. Phys.*, **18**, 1311.

Åsberg, P. & Inganäs, O. (2003) *Biosen. & Bioelectron.*, **19**, 199.

Astin, A. V. (1936) *J. Res. Nat. Bur. Stds.*, **21**, 425.

Bach, U., Lupo, D., Comte, P., Moser, J. E., Weissörtel, F., Salbeck, J., Spreitzer, H. & Grätzel, M. (1998) *Nature*, **395**, 583.

Bader, M. A., Marowsky, G., Bahtiar, A., Koynov, K., Bubeck, C., Tillmann, H., Hörhold, H.-H. & Pereira, S. (2002) *J. Opt. Soc. Am. B*, **19**, 2250.

Baeriswyl, D., Campbell, D. K. & Mazumdar, S. (1992) in *Conjugated and Conducting Polymers* (Kiess, H., ed.), Springer Series in Solid State Sciences, Berlin, Springer-Verlag, **102**, 7.

Bailey, R. A. & Persaud, K. C. (2000) in *Polymer Sensors and Actuators* (Osada, Y. & De Rossi, D. E., eds.), Berlin, Springer Verlag, p. 149.
Bardeen, J., Cooper, L. N. & Schrieffer, J. R. (1957) *Phys. Rev.*, **106**, 162; **108**, 1175.
Barford, W., Burshill, R. J. & Smith, R. W. (2002) *Phys. Rev. B*, **66**, 115205.
Barker, J. A., Ramsdale, C. M. & Greenham, N. C. (2003) *Phys. Rev. B*, **67**, 075205.
Barth, S., Bässler, H., Scherf, U. & Müllen, K. (1998) *Chem. Phys. Lett.*, **288**, 147.
Bässler, H. (1993) *Phys. Stat. Sol. (B)*, **175**, 15.
Batchelder, D. N. (1988) *Contemp. Phys.*, **29**, 3.
Bates, T. W., Ivin, K. J. & Williams, G. (1967) *Trans. Faraday Soc.*, **63**, 1964.
Battacharya, S. K., ed. (1986) *Metal Filled Polymers*, New York, Marcel Dekker.
Baur, M. E. & Stockmayer, W. H. (1965) *J. Chem. Phys.*, **43**, 4319.
Bauser, H., Klöpffer, W. & Rabenhorst, H. (1971) in *Advances in Static Electricity* (de Geest, W., ed.), Brussels, Auxilia, 1, 2.
Bäyer, A. & Landsberg, L. (1882) *Ber. Deutsch Chem. Ges.*, **15**, 57.
Bechgaard, K., Jacobsen, C. S., Mortensen, K., Pedersen, M. J. & Thorup, N. (1980) *Solid State Commun.*, **33**, 1119.
Becker, H., Spreitzer, H., Kreuder, W., Kluge, E., Schenk, H., Parker, I. & Cao, Y. (2000) *Adv. Mater.*, **12**, 42.
Becker, H., Spreitzer, H., Kreuder, W., Kluge, E., Vestweber, H., Schenk, H. & Treacher, K. (2001) *Synth. Metals*, **122**, 105.
Berets, D. J. & Smith, D. S. (1968) *Trans. Faraday Soc.*, **64**, 823.
Bernier, P. (1986) in *Handbook of Conducting Polymers* (Skotheim, T. A., ed.), New York, Marcel Dekker, **2**, 1099.
Bernius, M. T., Inbasekaran, M., O'Brien. J. & Wu, W. (2000) *Adv. Mater.*, **12**, 1737.
Bertein, H. (1973) *J. Phys. D*, **6**, 1910.
Berthelot, M. P. E. (1866) *Ann. Chim. Phys.*, **9**, 402.
Bertinelli, F., Costa-Bizzari, P., Dalla-Casa, C. & Lanzi, M. (2001) *Synth. Metals*, **122**, 267.
Billing, J. W. & Groves, D. J. (1974) *Proc. Instn. Electr. Engrs.*, **121**, 1451.
Billings, M. J., Smith, A. & Wilkins, R. (1967) *IEEE Trans. Electr. Insul.*, **EI-2**, 131.
Billmeyer, F. W. (1971) *Textbook of Polymer Science*, New York, Wiley.
Bittner, R., Däubler, T. K., Neher, D. & Meerholz, K. (1999) *Adv. Mater.*, **11**, 123.
Blom, P. W. M. & Vissenberg, M. C. J. M. (2000) *Mater. Sci. & Eng. R*, **27**, 53.
Blom, P. W. M., Bernsten, A. J. M., Liedenbaum, C. T. H. F., Schoo, H. F. M., Croonen, Y. & van de Weijer, P. (2000) *J. Mater. Sci.: Mater. in Electron.*, **11**, 105.
Bloor, D. (1976) *Chem. Phys. Lett.*, **40**, 323.
Bloor, D. (1982) in *Developments in Crystalline Polymers-1* (Bassett, D. C., ed.), Barking, Applied Science Publishers, p. 151.
Bloor, D. (1984) in *Polydiacetylenes: Synthesis, Structure, and Electronic Properties* (Bloor, D. & Chance, R. R., eds.), Dordrecht, Martinus Nijhoff, p.1.
Bloor, D. & Chance, R. R. (eds.) (1984) *Polydiacetylenes: Synthesis, Structure, and Electronic Properties*, Dordrecht, Martinus Nijhoff.
Bloor, D. & Preston, F. H. (1976) *Phys. Stat. Sol. (A)*, **37**, 427.
Bloor, D., Ando, D. J., Obhi, J. S., Mann, S. & Worboys, M. R. (1986) *Macromol. Chem. Rapid Commun.*, **7**, 665.
Bloor, D., Hercliffe, R. D., Galiotis, C. G. & Young, R. J. (1985) in *Electronic Properties of Polymers and Related Compounds* (Kuzmany, H., Mehring, M. & Roth, S., eds.), Springer Series in Solid State Science, Berlin, Springer, **63**, 179.
Blundell, S. J., Pratt, F. L., Marshall, I. M., Steer, C. A., Hayes, W., Husmann, A., Fischmeister, C., Martin, R. E. & Holmes, A. B. (2002) *J. Phys.: Condens. Matter*, **14**, 9987.

Blythe, A. R. (1975b) *J. Electrostatics*, **1**, 101.
Blythe, A. R. (1984) *Polymer Testing*, **4**, 195.
Blythe, A. R. & Jeffs, G. M. (1969) *J. Macromol. Sci., Phys. B*, **3**, 141.
Blythe, A. R., Briggs, D., Kendall, C. R., Rance, D. G. & Zichy, V. J. I. (1978) *Polymer*, **19**, 1273.
Borsenberger, P. M. & Weiss, D. S. (1998) *Organic Photoreceptors for Xerography*, Opt. Eng. Series, **59**, New York, Marcel Dekker.
Borsenberger, P. M., Gruenbaum, W. T., Wolf, U. & Bässler, H. (1998) *Chem. Phys.*, **234**, 277.
Borsenberger, P. M., Pautmeier, L. & Bässler, H. (1991) *J. Chem. Phys.*, **94**, 5447.
Borsenberger, P. M., Pautmeier, L. & Bässler, H. (1992) *Phys. Rev., B*, **46**, 12145.
Borsenberger, P. M., Richert, R. & Bässler, H. (1993) *Phys. Rev., B*, **47**, 4289.
Bosshard, Ch., Sutter, K., Prêtre, Ph., Hullinger, J., Flörsheimer, M., Kaatz, P. & Günter, P. (1995) *Organic Nonlinear Optical Materials*, Amsterdam, Gordon and Breach.
Böttcher, C. J. F. (1973) *Theory of Electric Polarization:* Vol. 1, rev. van Belle, O. C., Bordewijk, P. & Rip. A., Amsterdam, Elsevier.
Böttcher, C. J. F. (1978) *Theory of Electric Polarization:* Vol. 2, rev. Böttcher, C. J. F. & Bordewijk, P., Amsterdam, Elsevier.
Boyd, R. H. & Phillips, P. J. (1993) *The Science of Polymer Molecules*, Cambridge, Cambridge University Press.
Boyd, R. W. (2003) *Nonlinear Optics*, 2nd ed., Boston, Academic Press.
Brabec, C. J., Cravino, A., Meissner, D., Sariciftci, N. S., Fromherz, T., Rispens, M. T., Sanchez, L. & Hummelen, J. C. (2001b) *Adv. Func. Mater.*, **11**, 374.
Brabec, C. J., Dyakonov, V., Parisi, J. & Sariciftci, N. S., eds. (2003) *Organic Photovoltaics*, Springer Series in Materials Science, **60**, Berlin, Springer.
Brabec, C. J., Sariciftci, N. S. & Hummelen, J. C. (2001a) *Adv. Func. Mater.*, **11**, 15.
Brabec, C. J., Winder, C., Sariciftci, N. S., Hummelen, J. C., Dhanabalan, A., van Hal, P. A. & Janssen, R. A. J. (2002) *Adv. Func. Mater.*, **12**, 709.
Brédas, J.-L. (1986) in *Handbook of Conducting Polymers* (Skotheim, T. A., ed.), New York, Marcel Dekker, **2**, 859.
Brédas, J.-L., Heeger, A. J. & Wudl, F. (1986) *J. Chem. Phys.*, **85**, 4673.
Broadbent, S. R. & Hamersley, J. M. (1957) *Proc. Camb. Phil. Soc.*, **53**, 629.
Brown, A. R., Jarrett, C. P., de Leeuw, D. M. & Matters, M. (1997) *Synth. Metals*, **88**, 37.
Brown, A. R., Pomp, A., Hart, C. M. & de Leeuw, D. M. (1995) *Science*, **270**, 972.
Brown, P. J., Thomas, D. S., Köhler, A., Wilson, J. S., Kim J.-S., Ramsdale, C. M., Sirringhaus, H. & Friend, R. H. (2003) *Phys. Rev. B*, **67**, 064203.
Brumfiel, G. (2002) *Nature*, **419**, 419.
Brydson, J. A (1999) *Plastics Materials*, 7th edn., Oxford, Butterworth-Heinemann.
Bueche, F. (1962) *Physical Properties of Polymers*, New York, Interscience.
Bueche, F. (1973) *J. Appl. Phys.*, **44**, 532.
Bulgin, D. (1945) *Trans. Inst. Radio Engrs.*, **21**, 188.
Bulinski, A. T., Crine, J.-P., Noirhomme, B., Densley, R. J. & Bamji, S. (1998) *IEEE Trans. on Dielectrics & Electrical Insulation*, **5**, 558.
Bur, A. J. & Roberts, D. E. (1969) *J. Chem. Phys.*, **51**, 406.
Burroughes, J. H., Bradley, D. D. C., Brown, A. R., Marks, R. N., Mackay, K., Friend, R. H., Burns, P. L. & Holmes, A. B. (1990) *Nature*, **347**, 539.
Burroughes, J. H., Jones, C. A. & Friend, R. H. (1988) *Nature*, **335**, 137.
Butcher, P. N. & Cotter, D. (1990) *The Elements of Nonlinear Optics*, Cambridge, Cambridge University Press.

Cacialli, F., Wilson, J. S., Michels, J. J., Daniel, C., Silva, C., Friend, R. H., Severin, N., Samori, P., Rabe, J. P., O'Connell, M. J., Taylor, P. N. & Anderson, H. L. (2002) *Nature Mater.*, **1**, 160.

Campbell, A. J., Bradley, D. D. C., Werner, E. & Brütting, W. (2000) *Synth. Metals*, **111–112**, 273.

Carlson, C. F. (1940) *U. S. Patent* 2,221,776.

Castle, G. S. P. (1997) *J. Electrostatics*, **40 & 41**, 13.

Chang, R., Hsu, J. H., Fann, W. S., Liang, K. K., Chang, C. H., Hayashi, M., Yu, J., Lin, S. H., Chang, E. C., Chuang, K. R. & Chen, S. A. (2000) *Chem. Phys. Lett.*, **317**, 142.

Charych, D. H., Nagy, J. O., Spevak, W. & Bednarskit, M. D. (1993) *Science*, **261**, 585.

Chen, L., McBranch, D. W., Wang, H.-L., Hegeson, R., Wudl, F. & Whitten, D. G. (1999) *Proc. Nat. Acad. Sci. USA*, **96**, 12287.

Cheung, K. M., Bloor, D. & Stevens, G. C. (1988) *Polymer*, **29**, 1709.

Cheung, K. M., Bloor, D., & Stevens, G. C. (1990) *J. Mater. Sci.*, **25**, 3814.

Chiang, C. K., Fincher, C. R. Jr., Park, Y. W., Heeger, A. J., Shirakawa, H., Louis, E. J., Gau, S. C. & MacDiarmid, A. G. (1977) *Phys. Rev. Lett.*, **39**, 1098.

Chien, J. C. W. (1984) *Polyacetylene: Chemistry, Physics and Material Science*, Orlando, Academic Press.

Cho, B. R. (2002) *Prog. Polym. Sci.*, **27**, 307.

Cole, R. H. & Cole, K. S. (1941) *J. Chem. Phys.*, **9**, 341.

Conklin, J. A., Su, T. M., Huang, S.-C. & Kaner, R. B. (1998) in *Handbook of Conducting Polymers, 2nd Edn.*, (Skotheim, T. A., Elsenbaumer, R. L. & Reynolds, J. R., eds.), New York, Marcel Dekker, p. 945.

Copple, C., Hartree, D. R., Porter, A. & Tyson, H. (1939) *Proc. Instn. Electr. Engrs.*, **85**, 56.

Crine, J.-P. (1998) *IEEE Trans. on Dielectrics & Electrical Insulation*, **5**, 681.

Cross, J. A. (1987) *Electrostatics: Principles, Problems and Applications*, Bristol, Adam Hilger.

Dall'Olio, A., Dascola, G., Varacca, V. & Bocchi, V. (1968) *C. R. Hebd. Seances Acad. Sci. Ser. C*, **267**, 433.

Dalton, L. (2002) in *Polymers for Photonic Applications I* (Lee, K.-S., ed.), *Advances in Polymer Science*, **158**, Berlin, Springer Verlag.

Dalton, L. R., Harper, A. W., Wu, B., Ghosn, R., Laquindanum, J., Liang, Z., Hubbel, A. & Xu, C. (1995) *Adv. Mater.*, **7**, 519.

Davey, J. M., Ralph, S. F., Too, C. O., Wallace, G. C. & Partridge, A. C. (2001) *React. & Func. Polym.*, **49**, 87.

Davidson, D. W. & Cole, R. H. (1950) *J. Chem. Phys.*, **18**, 1417.

Davies, D. K. (1969) *J. Phys. D*, **2**, 1533.

Davies, D. K. (1970) in *Advances in Static Electricity* (de Geest, W., ed.), Brussels, Auxilia, **1**, 10.

Davies, G. R. & Ward, I. M. (1969) *J. Polym. Sci. B*, **7**, 353.

Davies, G. R., Yamwong, T. & Voice, A. M. (2002) *J. Appl. Phys.*, **91**, 1472.

Davis, D., Gutfreund, H. & Little, W. A. (1976) *Phys. Rev. B*, **13**, 4766.

Dawson, J. C. & Adkins, C. J. (1996) *J. Phys.: Condens. Matter*, **8**, 8321.

de Boer, R. W. I., Morpurgo, A. F. & Klapwijk, T. M. (2003) *Appl. Phys. Lett.*, **83**, 4345.

de Gennes, P. G. (1971) *J. Chem. Phys.*, **55**, 572.

De Giglio, E., Guascito, M. R., Sabbatini, L. & Zambonin, G. (2001) *Biomater.*, **22**, 2609.

de Leeuw, D. M., Simenon, M. M. J., Brown, A. R. & Einerhand, R. E. F. (1997) *Synth. Metals*, **87**, 53.

De Paoli, M.-A. (1997) in *Handbook of Organic Conductive Molecules and Polymers* (Nalwa, H. S., ed.), Chichester, John Wiley & Sons, **2**, 773.

Debye, P. (1929) *Polar Molecules*, Chemical Catalog Co., reprinted in NewYork by Dover Publications.

Devreux, F., Döry, I., Mihály, L., Pekker, S., Jánossy, A. & Kertész, M. (1981) *J. Polym. Sci: Polym. Phys. Ed.,* **19**, 743.

Dhoot, A. S., Ginger, D. S., Beljonne, D., Shuai, Z. & Greenham, N. C. (2002) *Chem. Phys. Lett.*, **360**, 195.

Diaz, A., Fenzel-Alexander, D., Wollmann, D. & Barker, J. A. (1992) *Langmuir*, **8**, 2698.

Ding, J., Liu, L., Spinks, G. M., Zhou, D., Wallace, G. G. & Gillespie, J. (2003) *Synth. Metals*, **138**, 391.

Dissado, L. A. & Fothergill, J. P. (1992) *Electrical Degradation and Breakdown in Polymers*, IEE Materials and Devices Series **9**, London, Peter Peregrinus Ltd.

Doi, M. & Edwards, S. F. (1986) *The Theory of Polymer Dynamics,* Oxford, Oxford University Press.

Doleman, B. J. & Lewis, N. S. (2001) *Sensors & Actuators B*, **72**, 41.

Donnet, J. B., Bansai, R. & Wang, M. (1993) *Carbon Black Science and Technology*, New York, Marcel Dekker.

Donval, A., Toussaere, E., Hierle, R. M. & Zyss, J. (2000) *Synth. Metals*, **115**, 21.

Doyle, M., Lewittes, M. E., Roelofs, M. G. & Perusich, S. A. (2001) *J. Phys. Chem. B*, **105**, 9387.

Drury, C. J., Mutsaers, C. M. J., Hart, C. M., Matters, M. & de Leeuw, D. M. (1998) *Appl. Phys. Lett.*, **73**, 108.

Dunlap, D. H., Parris, P. E. & Kenkre, V. M. (1996) *Phys. Rev. Lett.*, **77**, 542.

Dyreklev, P., Granström, M., Inganäs, O., Gunaratne, L. M. W. K., Senadeera, G. K. R., Skaarup, S. & West, K. (1996) *Polymer*, **37**, 2609.

Edwards, J. H. & Feast, W. J. (1980) *Polymer*, **21**, 595.

Ehinger, K. & Roth, S. (1986) *Philos. Mag. B*, **53**, 301.

El-Tantawy, F., Kamada, K. & Ohnabe, H. (2002) *Materials Lett.*, **56**, 112.

Erdman, H. & Köthner, P. (1898) *Zeit. Anorg. Chem.,* **18**, 48.

Fang, Q., Chetwynd, D. G., Covington, J. A., Toh, C.-S. & Gardner, J. W. (2002) *Sensors & Actuators B*, **84**, 66.

Feller, F., Geschke, D. & Monkman, A. P. (2002) *Polym. Internat.*, **51**, 1184.

Fenton, D. E., Parker, J. M. & Wright, P. V. (1973) *Polymer*, **14**, 589.

Flandin, L., Chang, A., Nazarenko, S., Hiltner, A. & Baer, E. (2000) *J. Appl. Polym. Sci.*, **76**, 894.

Forrest, S. R. (2003) *Org. Electron.*, **4**, 45.

Forrest, S. R., Bradley, D. D. C. & Thompson, M. E. (2003) *Adv. Mater.*, **15**, 1043.

Frankevich, E., Müller, J. G. & Lemmer, U. (2002) *Chem. Phys.*, **285**, 13.

Friend, R. H., Gymer, R. W., Holmes, A. B., Burroughes, J. H., Marks, R. N., Taliani, C., Bradley, D. D. C., Dos Santos, D. A., Brédas, J. L., Löglund, M. & Salaneck, W. R. (1999a) *Nature*, **397**, 121.

Friend, R., Burroughes, J. and Shimoda, T. (1999b) *Physics World*, June 1999, 35.

Fröhlich, H. (1947) *Proc. Roy. Soc. A*, **188**, 521.

Fröhlich, H. (1949) *Theory of Dielectrics*, Oxford, Oxford University Press.

Fu, Y., Liu, J. & Willander, M.(1999) *Int. J. Adhesion & Adhesives*, **19**, 281.

Fuoss, R. M. (1941) *J. Amer. Chem. Soc.*, **63**, 378.

Gadermaier, C. & Lanzani, G. (2002) *J. Phys.: Condens. Matter*, **14**, 9785.

Gallazzi, M. C., Tassoni, L., Bertarelli, C., Pioggia G., Di Francesco, F. & Montoneri, E. (2003) *Sensors & Actuators B*, **88**, 178.

Gartstein, Y. N. & Conwell, E. M. (1995) *Chem. Phys. Lett.*, **245**, 351.
Gaylord, B. S., Heeger, A. J. & Bazan, G. C. (2002) *Proc. Nat. Acad. Sci. USA*, **99**, 10954.
Gerard, M., Chaubey, A. & Malhotra, B. D. (2002) *Biosens. & Bioelect.*, **17**, 345.
Glasstone, S., Laidler, K. J. & Eyring, H. (1941) *The Theory of Rate Processes*, New York, McGraw- Hill.
Goldie, D. M. (1999) *J. Phys. D: Appl. Phys.*, **32**, 3058.
Gorman, C. B., Ginsburg, E. J. & Grubbs, R. H. (1993) *J. Amer. Chem. Soc.*, **115**, 1397.
Gospodinova, N. & Terlemezyan, L. (1998) *Prog. Polym. Sci.*, **23**, 1443.
Granström, M., Petritsch, K., Arias, A. C., Lux, A., Andersson, M. R. & Friend, R. H. (1998) *Nature*, **395**, 257.
Grätzel, M. (2001) *Pure & Appl. Chem.*, **73**, 459.
Gray, F. M. (1991) *Polymer Electrolytes*, Weinheim, VCH.
Greene, R. L., Street, G. B. & Suter, L. J. (1975) *Phys. Rev. Lett.*, **34**, 577.
Grell, M. & Bradley, D. D. C. (1999) *Adv. Mater.*, **11**, 895.
Grell, M., Bradley, D. D. C., Inbasekaran, M. & Woo, E. P. (1997) *Adv. Mater.*, **9**, 798.
Grell, M., Knoll, W., Lupo, D., Meisel, A., Miteva, T., Neher, D., Nothofer, H.-G., Scherf, U. & Yasuda, A. (1999) *Adv. Mater.*, **11**, 671.
Grenvesse, P. (1897) *Bull. Soc. Chim. France.*, **17**, 599.
Griffiths, C. L., Freestone, J. & Hampton R. N. (1998) *IEEE Int. Symp. on Electrical Insulation*, IEEE, Arlington, USA, p. 578.
Groenendaal, L., Jonas, F., Freitag, D., Pielartzik, H. & Reynolds, J. R. (2000) *Adv. Mater.*, **12**, 481.
Groenendaal, L., Zotti, G., Aubert, P.-H., Waybright, S. M. & Reynolds, J. R. (2003) *Adv. Mater.*, **15**, 855.
Grosberg, A. Y. & Khokhlov, A. R. (1997) *Giant Molecules*, San Diego, Academic Press.
Gross, B. (1949) *J. Chem. Phys.*, **17**, 866.
Gross, M., Müller, D. C., Nothofer, H.-G., Scherf, U., Neher, D., Bräuchle, C. & Meerholz, K. (2000) *Nature*, **405**, 661.
Grunlan, J. C., Gerberich, W. W. & Francis, L. F. (2001) *J. Appl. Poym. Sci.*, **80**, 692.
Gul', V. E. (1996) *Structure and Properties of Conducting Polymer Composites (New Concepts in Polymer Science)*, Utrecht, VSP.
Gulbinas, V., Zaushitsyn, Y., Sunström, V., Hertel, D., Bässler, H. & Yartsev, A. (2002) *Phys. Rev. Lett.*, **89**, 107401.
Gupta, R., Park, J. Y., Srdanov, V. I. & Heeger, A. J. (2002) *Synth. Metals*, **132**, 105.
Gussoni, M., Castiglioni, C. & Zerbi, G. (1991) in *Spectroscopy of Advanced Materials* (Clark, R. J. H. & Hester, R. E., eds.), Chichester, John Wiley & Sons, p. 251.
Halleck, M. C. (1956) *Trans. Amer. Inst. Electr. Engrs.*, **75**, 211.
Halls, J. J. M., Walsh, C. A., Greenham, N. C., Marseglia, E. A., Friend, R. H., Moratti, S. C., & Holmes, A. B. (1995) *Nature*, **376**, 498.
Hamon, B. V. (1952) *Proc. Instn. Electr. Engrs.*, **99**, 151.
Hamon, B. V. (1953) *Austral. J. Phys.*, **6**, 304.
Han, D. G. & Choi, G. M. (1998) *Solid State Ionics*, **106**, 71.
Harper, W. R. (1967) *Contact and Frictional Electrification*, Oxford, Oxford University Press.
Hartshorn, L. & Ward, W. H. (1936) *J. Instn. Electr. Engrs.*, **79**, 597.
Havriliak, S. & Havriliak, S. J. (1997) *Dielectric and Mechanical Relaxation in Materials – Analysis, Interpretation and Applications to Polymers*, Munich, Hanser.
Havriliak, S. & Negami, S. (1967) *Polymer*, **8**, 161.
Hearle, J. W. S. (1957) *J. Textile Inst.*, **48**, 40.

Hedvig, P. (1977) *Dielectric Spectroscopy of Polymers*, Bristol, Adam Hilger.

Hedvig, P. (1984) *IEEE Trans. Electr. Ins.*, **EI-19**, 371

Heeger, A. J., Heeger, D. J., Langan, J. & Yang, Y. (1995) *Science*, **270**, 1642.

Heliotis, G., Xia., R., Whitehead, K. S., Turnbull, G. A., Samuel, I. D. W. & Bradley, D. D. C. (2003) *Synth. Metals*, **139**, 727.

Henry, P. S. H., Livesey, R. G. & Wood, A. M. (1967) *J. Textile Inst.*, **58**, 55.

Hertel, D., Romanovskii, Yu. V., Schweitzer, B., Scherf, U. & Bässler, H. (2001) *Macromol. Symp.*, **175**, 141.

Herz, L. M., Silva, C., Phillips, R. T., Setayesh, S. & Millen, K. (2001) *Chem. Phys. Lett.*, **347**, 318.

Higgins, R. W. T., Zaidi, N. A. & Monkman, A. P. (2001) *Adv. Func. Mater.*, **11**, 407.

Higgins, S. J. (1997) *Chem. Soc. Revs.*, **26**, 247.

Hill, N., Vaughan, W. E., Price, A. H. & Davies, M. (1969) *Dielectric Properties and Molecular Behaviour*, London, van Nostrand.

Hinchcliffe, A. & Munn, R. W. (1985) *Molecular Electromagnetism*, Chichester, J. Wiley & Sons.

Hindermann-Bischoff, M. & Ehrburger-Dolle, F. (2001) *Carbon*, **39**, 375.

Hirata, S., Head-Gordon, M. & Bartlett, R. J. (1999) *J. Chem. Phys.*, **111**, 10774.

Hirsch, J. E. & Scalapino, D. J., (1985) *Phys. Rev. B*, **32**, 117.

Hoffman, J. D., Williams, G. & Passaglia, E. (1966) *J. Polym. Sci. C*, **14**, 173.

Hollars, C. W., Lane, S. M. & Huser, T. (2003) *Chem. Phys. Lett.*, **370**, 393.

Holm, R. (1946) *Electric Contacts*, Stockholm, Hugo Gebers.

Hougham, G., Tesoro, G. & Viehbeck, A. (1996) *Macromols.*, **29**, 3453.

Hougham, G., Zhang, Q. & Jean, Y. C. (1996) *Pol. Preprints*, **39**, 871.

Hubbard, J. (1964) *Proc. Roy. Soc. (London) A*, **281**, 401.

Huitema. H. E. A., Gelink, G. H., van der Puten, J. B. P. H., Kuijk, K. E., Hart, K. M., Cantatore, E. & de Leeuw, D. M. (2002) *Adv. Mater.*, **14**, 1201.

Hung, L. S. & Chen, C. H. (2002) *Mater. Sci. & Eng. R*, **39**, 143.

Hyde, P. J. (1970) *Proc. Instn. Electr. Engrs.*, **117**, 1891.

Ishida, Y. (1960) *Kolloid Z.*, **168**, 29.

Ito, K., Shirakawa, H. & Ikeda, S. (1974) *J. Polym. Sci. Polym. Chem. Ed.*, **12**, 11.

Jäger, K.-M., McQueen, D. H., Tchmutin, I. A., Ryvkina, N. G. & Klüppel, M. (2001) *J. Phys. D: Appl. Phys.*, **34**, 2699.

Jain, S. C., Kapoor, A. K., Geens, W., Poortmans, J. & Mertens, R. (2002) *J. Appl. Phys.*, **92**, 3752.

Jang, M.-S., Song, S.-Y. & Shim, H.-K. (2000) *Polymer*, **41**, 5675.

Jarrett, C. P., Friend, R. H., Brown, A. R. & de Leeuw, D. M. (1995) *J. Appl. Phys.*, **77**, 6289.

Jen, K. J., Miller, G. G. & Elsenbaumer, R. L. (1986) *J. Chem. Soc. Chem. Commun.*, 1346.

Jeszka, J. K., Ulański, J. & Kryszewski, M. (1981) *Nature*, **289**, 390.

Jeszka, J. K., Tracz, A., Sroczyńska, A., Kryszewski, M., Yamochi, H., Horiuchi, S., Saito, G. & Ulański, J. (1999) *Synth. Metals*, **106**, 75.

Jin, Y. D., Yang, J. P., Heremans, P. L., Van der Auweraer, M., Rousseau, E., Geise, H. J. & Borghs, G. (2000) *Chem. Phys. Lett.*, **320**, 387.

Johnson, C. F. & Cole, R. H. (1951) *J. Amer. Chem. Soc.*, **73**, 4536

Jonscher, A. K. (1977) *Nature*, **267**, 673.

Jozefowicz, M., Yu, L. T., Belorgey, G., Perichon, J. & Buvet, R. (1965) *J. Polym. Sci.: C*, **16**, 2943.

Juška, G., Genevičius, K., Österbacka, R., Arlauskas, K., Kreouzis, T., Bradley, D. D. C. & Stubb, H. (2003) *Phys. Rev. B*, **67**, 081201.

Kaiser, A. B. (2001a) *Rep. Prog. Phys.*, **64**, 1.

Kaiser, A. B. (2001b) *Adv. Mater.*, **13**, 927.

Kanicki, J. (1986) in *Handbook of Conducting Polymers* (Skotheim, T. A., ed.), New York, Marcel Dekker, **1**, 543.

Kayzar, F. & Swalen, J. D. (1996) *Organic Thin Films for Waveguiding Nonlinear Optics*, Amsterdam, Gordon and Breach.

Kendig, M., Hon, M. & Warren, L. (2003) *Prog. Org. Coatings*, **47**, 183.

Kepler, R. G. (1960) *Phys. Rev.*, **119**, 1226.

Kido, J., Hongawa, K., Okuyama, K. & Nagai, K. (1994) *Appl. Phys. Lett.*, **64**, 815.

Kiebooms, R., Aleshin, A., Hutchinson, K., Wudl, F. & Heeger, A. (1999) *Synth. Metals*, **101**, 436.

Kim, J.-S., Ho. P. K. H., Murphy, C. E., Baynes, N. & Friend, R. H. (2002) *Adv. Mater.*, **14**, 206.

Kippelen, B. & Peyghambarian, N. (2003) in *Adv. in Polym Sci.: Polymers for Photonic Applications II* (Lee, K.-S., ed.), Berlin, Springer Verlag, **161**, 87.

Kirkwood, J. G. (1939) *J. Chem. Phys.*, **7**, 911.

Kittel, C. (1996) *Introduction to Solid State Physics*, 7th edn., New York, Wiley.

Kline, R. J., McGehee, M. D., Kadnikova, E. N., Liu, J. & Fréchet, J. M. J. (2003) *Adv. Mater.*, **15**, 1519.

Köhler, A. & Wilson, J. (2003) *Org. Electron.*, **4**, 179.

Köhler, A., dos Santos, D. A., Beljonne, D., Shuai, Z., Brédas, J.-L., Holmes, A. B., Kraus, A., Müllen, K. & Friend, R. H. (1998) *Nature*, **392**, 903.

Köhler, A., Wilson, J. S., Friend, R. H., Al-Suti, M. K., Khan, M. S., Gerhard, A. & Bässler, H. (2002) *J. Chem. Phys.,* **116**, 9457.

Kohlman, R. S. & Epstein, A. J. (1998) in *Handbook of Conducting Polymers*, 2nd edn., (Skotheim, T. A., Elsenbaumer, R. L. & Reynolds, J. R., eds.), New York, Marcel Dekker, p. 85.

Kohlman, R. S., Joo, J., Min, Y. G., MacDiarmid, A. G. & Epstein, A. J. (1996) *Phys. Rev. Lett.*, **77**, 2766.

Kohlrausch, F. (1863) *Pogg. Ann. Phys.*, **119**, 352.

Kontturi, K., Pentti, P. & Sundholm, G. (1998) *J. Electroan. Chem.*, **453**, 231.

Kosaki, M., Sugiyama, K. & Ieda, M. (1971) *J. Appl. Phys.*, **42**, 3388; **31**, 1598.

Kötz, R. & Carlen, M. (2000) *Electrochem. Acta*, **45**, 2483.

Kovacic, P. & Kyriakis, A. (1962) *Tetrahedron Lett.,* 467.

Kraft, A., Grimsdale, A. C. & Holmes, A. B. (1998) *Angew. Chem. Int. Ed.*, **37**, 402.

Kremer, F. & Schönhals, A. (2002) *Broadband Dielectric Spectroscopy*, Berlin, Springer Verlag.

Kreuer, K. D., Fuchs, A., Ise, M., Spaeth, M. & Maier, J. (1998) *J. Electrochem. Soc.*, **43**, 1281.

Kreuger, F. H. (1964) *Discharge Detection in High-Voltage Equipment*, London, Heywood.

Kroon, J. M., Wienk, M. M., Verhees, W. J. H. & Hummelen, J. C. (2002) *Thin Solid Films*, **403–404**, 223.

Kuhn, H. H. & Child, A. D. (1998) in *Handbook of Conducting Polymers*, 2nd edn. (Skotheim, T. A., Elsenbaumer, R. L. & Reynolds, J. R., eds.), New York, Marcel Dekker, p. 993.

Kuhn, W. & Moser, P. (1963) *J. Polym. Sci. A*, **1**, 151.

Kuroda, S., Marumoto, K., Ito, H., Greenham, N. C., Friend, R. H., Shimoi, Y. & Abe, S. (2000) *Chem. Phys. Lett.,* **325**, 183.

Kurtz, S. K. & Perry, T. T. (1968) *J. Appl. Phys.*, **39**, 3798.

Kuzyk, M. G. & Dirk, C. W., eds. (1998) *Characterization Techniques and Tabulations for Organic Nonlinear Optical Materials*, New York, Marcel Dekker.

Kwong, R. C., Weaver, M. S., Lu, M.-H. M., Tung, Y.-J., Chwang, A. B., Zhou, T. X., Hack, M. & Brown, J. J. (2003) *Org. Electron.*, **4**, 155.

Labes, M. M., Love, P. & Nichols, L. F. (1979) *Chem. Rev.*, **79**, 1.

Ladik, J. L. (1988) *Quantum Theory of Polymers as Solids*, New York, Plenum Press.

Landauer, R. (1978) in *Electrical Transport and Optical Properties of Inhomogeneous Media* (Garland, J. C. & Tanner, D. B., eds.), AIP Conf. Proc., New York, p. 2.

Laquai, F., Im, C., Kadashchuk, A. & Bässler, H. (2003) *Chem. Phys. Lett.*, **375**, 286.

Lauter, U., Meyer, W. H. & Wegner, G. (1997) *Macromols.*, **30**, 2092.

Law, K.-Y. (1993) *Chem. Rev.*, **93**, 449.

Lawson, W. G. (1966) *Proc. Instn. Electr. Engrs.*, **113**, 197.

Lee, M., Katz, H. E., Erben, C., Gill, D. M., Gopalan, P., Heber, J. D. & McGee, D. J. (2002) *Science*, **298**, 1401.

Lenz, R. W. & Carrington, W. K. (1959) *J. Polym. Sci.*, **41**, 333.

Letheby, H. (1862) *J. Chem. Soc.*, **15**, 161.

Li, Q., Imanishi, N., Hirano, A., Takeda, Y. & Yamamoto, O. (2002) *J. Power Sources*, **110**, 38.

Lin, H.-N., Lin, H.-L., Wang, S.-S., Yu, L.-S., Perng, G.-Y., Chen, S.-A. & Chen, S.-H. (2002) *Appl. Phys. Lett.*, **81**, 2572.

Little, W. A. (1964) *Phys. Rev. A*, **134**, 1416.

Lonergan, M. C., Severin, E. J., Doleman, B. J., Beaber, S. A., Grubbs, R. H. & Lewis, N. S. (1996) *Chem. Mater.*, **8**, 2298.

Longuet-Higgins, H. C. & Salem, L. (1959) *Proc. Roy. Soc. (London)*, **A251**, 172.

Lu, D., Chen, G., Perry, J. W. & Goddard, W. A. III (1994) *J. Amer. Chem. Soc.*, **116**, 10679.

Lu, S. Y. & Lin, C. Y. (2002) *Appl. Phys. A*, **74**, 675.

Lu, W.-K., Basak, S. & Elsenbaumer, R. L. (1998) in *Handbook of Conducting Polymers*, 2nd edn. (Skotheim, T. A., Elsenbaumer, R. L. & Reynolds, J. R., eds.), New York, Marcel Dekker, p. 881.

Lupton, J. M., Pogantsch, A., Piok, T., List, E. J. W., Patil, S. & Scherf, U. (2002) *Phys. Rev. Lett.*, **89**, 167401.

Lussey, D. L. (1998) World Patent WO. 98/33193.

Lustiger, A. & Markham, R. L. (1983) *Polymer*, **24**, 1647.

Luther-Davies, B. & Samoc, M. (1997) *Curr. Opin. Sol. St. & Mater. Sci.*, **2**, 213.

Luttinger, L. B. (1960) *Chem. Ind. (London)*, 1135.

Lux, F. (1993) *J. Mater. Sci.*, **28**, 285.

McLachlan, D. S. (2000) *J. Electroceram.*, **5**, 93.

McCrum, N. G., Read, B. E. & Williams, G. (1967) *Anelastic and Dielectric Effects in Polymeric Solids*, New York, Wiley.

McCubbin, W. L. & Gurney, I. D. C. (1965) *J. Chem. Phys.*, **43**, 983.

McCullough, R. D. (1998) *Adv. Mater.*, **10**, 93.

McDonald, R. N. & Campbell, T. W. (1960) *J. Amer. Chem. Soc.*, **82**, 4669.

McKeown, J. J. (1965) *Proc. Instn. Electr. Engrs.*, **112**, 824.

Maeno, T. & Fukunaga, K. (1996) *IEEE Trans. on Dielectr. & Electr. Ins.*, **3**, 784.

Maeno, T., Futami, T., Kushibe, H. & Takada, T. (1989) *J Appl. Phys.*, **65**, 1147.

Man, H.-T. & Yoon, H. N. (1992) *Adv. Mater.*, **4**, 159.

Mansfield, M. & Boyd, R. H. (1978) *J. Pol. Sci: Pol. Phys.*, **16**, 1227.

Many, A. & Rakavy, G. (1962) *Phys. Rev.*, **126**, 1980.

Mark, H. F. & Raff, R. A. V. (1941) *High Polymeric Reactions, Their Theory and Practice; High Polymers Vol. III,* New York, Interscience Publs., p. 298.

Martens, J. H. F., Pichler, K., Marseglia, E. A., Friend, R. H., Cramail, H., Khosravi, E., Parker, D. & Feast, W. J. (1994) *Polymer*, **35**, 403.

Martens, J. H. F., Reedijk, J. A., Brom, H. B., de Leeuw, D. M. & Menon, R. (2001) *Phys. Rev., B*, **63**, 073203.

Martin, R. E., Peckham, T. J. & Holmes, A. B. (2001) in *Encyclopedia of Materials Science and Technology* (Buscow, K. H. J., Cahn, R. W., Flemings, M. C., Ilscher, B., Kramer, E. J. & Mahajan, S., eds.), Amsterdam, Elsevier, **2**, 1509.

Martin, S. J., Godschalx, J. P., Mills, M. E., Shaffer II, E. O. & Townsend, P. H. (2000) *Adv. Mater.*, **12**, 1769.

Martin, S. J., Verschoor, G. L. B., Webster, M. A. & Walker, A. B. (2002) *Org. Electron.*, **3**, 129.

Mathy, A., Ueberhofen, K., Schenk, R., Gregorius, R., Garay, R., Müllen, K. & Bubeck, C. (1996) *Phys. Rev. B*, **53**, 4367.

Maxwell, J. C. (1892) *Electricity and Magnetism*, Oxford, Oxford University Press, p. 452.

Mazzoldi, A., Della Santa, A. & De Rossi, D. (2000) in *Polymer Sensors and Actuators* (Osada, Y. & De Rossi, D. E., eds.), Berlin, Springer Verlag, p. 207.

Meakins, J. R. (1962) *Progress in Dielectrics*, **4** (Birks, J. B. & Hart. J., eds.), London, Heywood.

Mecher, E., Bräuchle, C., Hörhold, H. H., Hummelen, J. C. & Meerholz, K. (1999) *Phys. Chem. Chem. Phys.*, **1**, 1749.

Meerholz, K., Volodin, B. L., Sandalphon, Kippelen, B. & Peyghambarian, N. (1994) *Nature*, **371**, 497.

Meneghetti, M. (1993) *Phys. Rev., B*, **47**, 13151.

Menon, R., Yoon, C. O., Moses, D. & Heeger, A. J. (1998) in *Handbook of Conducting Polymers*, 2nd edn. (Skotheim, T. A., Elsenbaumer, R. L. & Reynolds, J. R., eds.), New York, Marcel Dekker, p. 27.

Meyer, V. (1883) *Chem. Ber.*, **16**, 1465.

Meyer, W. H. (1998) *Adv. Mater.*, **10**, 439.

Meyers, F., Marder, S. R., Pierce, B. M. & Brédas, J. L. (1994) *J. Amer. Chem. Soc.*, **116**, 10703.

Miller, E. K., Brabec, C. J., Neugebauer, H., Heeger, A. J. & Sariciftci, N. S. (2001) *Chem. Phys. Lett.*, **335**, 23.

Miteva, T., Meisel, A., Knoll, W., Nothofer, H. G., Scherf, U., Müller, D. C., Meerholz, K., Yasuda, A. & Neher, D. (2001) *Adv. Mater.*, **13**, 565.

Miyamae, T., Hasegawa, S., Yoshimura, D., Ishii, H., Ueno, N. & Seki, K. (2000) *J. Chem. Phys.*, **112**, 3333.

Mizrahi, U., Ehrenfreund, E., Gershoni, D. & Vardeny, Z. V. (2003) *Polymer*, **44**, 691.

Miyamoto, T. & Shibayama, K. (1973) *J. Appl. Phys.*, **44**, 5372.

Mohilner, D. M., Adams, R. N. & Argersinger, W. J. Jr. (1962) *J. Amer. Chem. Soc.*, **84**, 3618.

Möller, S., Weiser, G. & Lapersonne-Meyer, C. (2001) *Synth. Metals*, **116**, 23.

Möller, S., Perlov, C., Jackson, W., Taussig, C. & Forrest, S. R. (2003) *Nature*, **426**, 166.

Monkman, A. P., Bloor. D., Stevens, G. C. & Stevens, J. C. H. (1987) *J. Phys. D: Appl. Phys.*, **20**, 1337.

Monkman, A. P., Burrows, H. D., Hartwell, L. J., Horsburgh, L. E., Hamblett, I. & Navaratnam, S. (2001) *Phys. Rev. Lett.*, **86**, 1358.

Montgomery, H. C. (1971) *J. Appl. Phys.*, **42**, 2971.

Moon, P. & Spencer, D. E. (1961) *Field Theory for Engineers*, London, van Nostrand.
Moons, H. (2002) *J. Phys.: Condens. Matter*, **14**, 12235.
Morse, C. T. (1974) *J. Phys. E, Sci. Instr.*, **7**, 657.
Moses, D., Dogariu, A. & Heeger, A. J. (2001) *Synth. Metals*, **116**, 19.
Mott, N. F., (1949) *Proc. Phys. Soc., A*, **62**, 416.
Mott, N. F. (1968) *J. Non-cryst. Solids*, **1**, 1.
Mott, N. F. (1987) *Conduction in Non-crystalline Solids*, Oxford, Oxford University Press.
Mott, N. F. & Davis, E. A. (1971) *Electronic Processors in Non-crystalline Materials*, Oxford, Clarendon Press.
Mott, N. F. & Gurney, R. W. (1948) *Electronic Processes in Ionic Crystals*, 2nd edn, Oxford, Oxford University Press.
Mrozowski, S. (1959) *Proceedings of the 3rd Conference on Carbon*, New York, Pergamon, p. 495.
Müllen, K. & Wegner, G., eds. (1998) *Electronic Materials: The Oligomer Approach*, Weinheim, Wiley-VCH.
Müller, C. D., Falcou, A., Reckefuss, N., Rojahn, M., Wiederhirn, V., Rudati, P., Frohne, H., Nuyken, O., Becker, H. & Meerholz, K. (2003) *Nature*, **421**, 829.
Nalwa, H. S., ed. (1997) *Handbook of Organic Conductive Molecules and Polymers*, Chichester, John Wiley & Sons.
Narkis, M., Ram, A. & Stein, Z. (1981) *Polymer Eng. & Sci.*, **21**, 1049.
National Physical Laboratory (2003) *Characterisation of Dielectric Materials at RF and Microwave Frequencies*, London, Institute of Measurement & Control.
Natta, G., Mazzanti, G. & Corradini, P. (1958) *Atti. Acad. Naz. Lincei Cl. Sci. Fis. Mater. Nat. Rend.*, **8**, 25.
Nelson, J. (2002) *Curr. Opin. in Sol. St. & Mater. Sci.*, **6**, 87.
Nicholson, J. W. (1997) *The Chemistry of Polymers*, 2nd edn., London, Royal Society of Chemistry.
Nicolas-Debarnot, D. & Poncin-Epaillard, F. (2003) *Anal. Chim. Acta*, **475**, 1.
Norman, R. (1970) *Conductive Rubbers and Plastics*, New York, Elsevier.
North, A. M. (1972) *Chem. Soc. Rev.*, **1**, 49.
North, A. M. & Phillips, P. J. (1968) *Chem. Comm.*, 1340.
O'Dwyer, J. J. (1973) *The Theory of Electrical Conduction and Breakdown in Solid Dielectrics*, Oxford, Oxford University Press.
O'Reagan, M. B., Borsenberger, P. M., Magin, E. H. & Zubil, T. (1996) *J. Imaging Sci. Technol.*, **40**, 1.
Onari, S. (1969) *J. Phys. Soc. Jap.*, **26**, 500.
Onsager, L. (1938) *Phys. Rev.*, **54**, 554.
Orczyk, M. E. (1990) *Chem. Phys.*, **142**, 485.
Österbacka, R., Wohlgenannt, M., Chinn, D. & Vardeny, Z. V. (1999) *Phys. Rev., B*, **60**, R11253.
Österbacka, R., Genevičius, K., Pivrikas, A., Juška, G., Arlauskas, K., Kreouzis, T., Bradley, D. D. C. & Stubb, H. (2003) *Synth. Metals*, **139**, 811.
Otero, T. F. (2000) in *Polymer Sensors and Actuators* (Osada, Y. & De Rossi, D. E., eds.), Berlin, Springer Verlag, p. 295.
Otero, T. F. & Cortés, M. T. (2003) *Adv. Mater.*, **15**, 279.
Paasch, G., Linder, T. & Scheinert, S. (2002) *Synth. Metals*, **132**, 97.
Pai, P. M. & Springett, B. E. (1993) *Rev. Mod. Phys.*, **65**, 163.
Pan, J., Scherf, U., Schreiber, A., Bilke, R. & Haarer, D. (2000) *Synth. Metals*, **115**, 79.
Park, Y. W., Choi, E. S. & Suh, D. S. (1998) *Synth. Metals*, **96**, 81.

Partridge, R. H. (1974) *UK Patent* 74/44704.
Partridge, R. H. (1983) *Polymer*, **24**, 733, 739 & 748.
Patil, A. O., Heeger, A. J. & Wudl, F. (1988) *Chem. Rev.*, **88**, 183.
Pede, D., Smela, E., Johansson, T., Johansson, M. & Inganas, O. (1998) *Adv. Mater.*, **10**, 233.
Peierls, R. E. (1955) *Quantum Theory of Solids*, Oxford, Oxford University Press.
Pekker, S. & Janossy, A. (1986) in *Handbook of Conducting Polymers* (Skotheim, T. A., ed.), New York, Marcel Dekker, **1**, 45.
Perlman, M. M. & Meunier, J. L. (1965) *J. Appl. Phys.*, **36**, 420.
Pfister, G. & Griffiths, C. H. (1978) *Phys. Rev. Lett.*, **40**, 659.
Pientka, M., Dyakonov, V., Meissner, D., Rogach, A., Talapin, D., Weller, H., Lutsen, L. & Vanderzande, D. (2004) *Nanotechnology*, **15**, 163.
Pomfret, S. J., Adams, P. N., Comfort, N. P. & Monkman, A. P. (1998) *Adv. Mater.*, **10**, 1351.
Pomfret, S. J., Rebout, E. & Monkman, A. P. (1996) *Synth. Metals*, **76**, 19.
Pope, M., Kallmann, H. P. & Magnante, P. (1963) *J. Chem. Phys.*, **38**, 2042.
Pope, M. & Swenberg, C. E. (1982) *Electronic Processes in Organic Crystals*, Monographs on the Physics and Chemistry of Materials, **39**, Oxford, Oxford University Press.
Pople, J. A. & Walmsley, S. H. (1962) *Mol. Phys.*, **5**, 15.
Prasad, P. N. & Williams, D. J. (1991) *Introduction to Nonlinear Optical Effects in Molecules and Polymers*, New York, John Wiley & Sons.
Preezant, Y., Roichman, Y. & Tessler, N. (2002) *J. Phys.: Condens. Matter*, **14**, 9913.
Prigodin, V. N. & Epstein, A. J. (2003) *Physica B*, **338**, 310.
Probst, N. & Grivei, E. (2002) *Carbon*, **40**, 201.
Pugh, D. (1973) *Mol. Phys.*, **26**, 1297.
Rault-Berthelot, J. & Simonet, J. (1986) *New J. Chem.*, **10**, 169.
Read, B. E. (1965) *Trans. Faraday Soc.*, **61**, 2140.
Reddish, W. (1950) *Trans. Faraday Soc.*, **46**, 459.
Reddish, W., Bishop, A., Buckingham, K. A. & Hyde, P. J. (1971) *Proc. Instn. Electr. Engrs.*, **118**, 255.
Reufer, M., Schindler, F., Patil, S., Scherf, U. & Lupton, J. M. (2003) *Chem. Phys. Lett.*, **381**, 60.
Reynolds, K. J., Barker, J. A., Greenham, N. C., Friend, R. H. & Frey, G. L. (2002) *J. Appl. Phys.*, **92**, 7556.
Riande, E. & Saiz, E. (1992) *Dipole Moments and Birefringence of Polymers*, NJ, Prentice Hall.
Righi, M. C., Scandolo, S., Serra, S., Iarlori, S., Tosatti, E. & Santoro, G. (2001) *Phys. Rev. Lett.*, **87**, 76802.
Ritsko, J. J. & Bigelow, R. W. (1978) *J. Chem. Phys.*, **69**, 4162.
Ritsko, J. J., Brillson, L. J., Bigelow, R. W. & Fabish, T. J. (1978) *J. Chem. Phys.*, **69**, 3931.
Rivers, T. J., Hudson, T. W. & Schmidt, C. E. (2002) *Adv. Func. Mater.*, **12**, 33.
Roberts, S. & von Hippel, A. (1946) *J. Appl. Phys.*, **17**, 610.
Romanovskii, Y. V., Bässler, H. & Scherf, U. (2004) *Chem. Phys. Lett.*, **383**, 89.
Roncali, J. (1992) *Chem. Rev.*, **92**, 711.
Roth, S. (1995) *One Dimensional Metals*, Weinheim, VCH.
Rothberg, L. J., Yan, M., Papadimitrakopoulos, F., Galvin, M. E., Kwock, E. W. & Miller, T. M. (1996) *Synth. Metals*, **80**, 41.
Rouse, P. E. (1953) *J. Chem. Phys.*, **21**, 1272.

Sahimi, M. (1994) *Applications of Percolation Theory*, London, Taylor & Francis.
Sainova, D., Neher, D., Dobruchowska, E., Luszczynska, B., Glowacki, I., Ulanski, J., Nothofer, H.-G. & Scherf, U. (2003) *Chem. Phys. Lett.*, **371**, 15.
Sainova, D., Zen, A., Nothofer, H.-G., Asawapirom, U., Scherf, U., Hagen. R., Bieringer, T., Kostromine, S. & Neher, D. (2002) *Adv. Func. Mater.*, **12**, 49.
Salaneck, W. R., Löglund, M., Fahlman, M., Greczynski, G. & Kugler, T. (2001) *Mater. Sci. & Eng., R*, **34**, 121.
Salaneck, W. R., Lundström, L. & Rånby, B., eds. (1993) *Conjugated Polymers and Related Materials*, Oxford, Oxford University Press.
Salem, L. (1966) *The Molecular Orbital Theory of Conjugated Systems*, New York, W. A. Benjamin.
Salzner, U. & Köse, M. E. (2002) *J. Phys. Chem., B*, **106**, 9221.
Samuel, I. D. W., Rumbles, G. & Friend, R. H. (1997) in *Primary Photoexcitations in Conjugated Polymers: Molecular Exciton Versus Semiconductor Band Model* (Sariciftci, N. S., ed.), Singapore, World Scientific, p. 140.
Sandberg, H. G. O., Frey, G. L., Shkunov, M. N., Sirringhaus, H., Friend, R. H., Nielsen, M. M. & Kumpf, K. (2002) *Langmuir*, **18**, 10176.
Sanden, B. (1998) in *Space Charge in Solid Dielectrics* (Fothergill, J. C. & Dissado, L. A., eds.), London, The Dielectrics Society, p. 225.
Sariciftci, N. S., ed. (1997) *Primary Photoexcitations in Conjugated Polymers: Molecular Exciton Versus Semiconductor Band Model*, Singapore, World Scientific.
Scarisbrick, R. M. (1973) *J. Phys., D*, **6**, 2098.
Schaer, M., Nüesch, F., Berner, D., Leo, W. & Zuppirol, L. (2001) *Adv. Func. Mater.*, **11**, 116.
Schatzki, T. F. (1962) *J. Polym. Sci.*, **57**, 496.
Scher, H. & Montroll, E. W. (1975) *Phys. Rev., B*, **12**, 2455.
Scherf, U. (1999) *J. Mater. Chem.*, **9**, 1853.
Scherf, U., Riechel, S., Lemmer, U. & Mahrt, R. F. (2001) *Curr. Opin. Sol. St. & Mater. Sci.*, **5**, 143.
Schmidt, C. E., Shastri, V. R., Vacanti, J. P. & Langer, R. (1997) *Proc. Nat. Acad. Sci. USA*, **94**, 8948.
Schön, J. H., Dodabalapur, A., Bao, Z., Kloc. C., Schneker, O. & Batlogg, B. (2001) *Nature*, **410**, 189.
Schultze, J. W., Morgenstern, T., Schattka, D. & Winkels, S. (1999) *Electrochim. Acta*, **44**, 1847.
Seki, K., Ueno, N., Karlsson, U. O., Engelhardt, R. & Koch, E.-E. (1986) *Chem. Phys.*, **105**, 247.
Serra, S., Tosatti, E., Iarlori, S., Scandolo, S. & Santoro, G. (2000) *Phys. Rev., B*, **62**, 4389.
Sessler, G. M. & West, J. (1970) *Appl. Phys. Lett.*, **17**, 507.
Sessler, G. M. & West, J. (1971) *Annual Report of the Conference on Electrical Insulation and Dielectric Phenomena*, Washington DC, National Academy of Sciences, p. 8.
Shaheen, S. E., Brabec, C. J., Sariciftci, N. S., Padinger, F., Fromherz, T. & Hummelen, J. C. (2001) *Appl. Phys. Lett.*, **78**, 841.
Sheng, P. (1980) *Phys. Rev., B*, **21**, 2180.
Shim, H.-K. & Jin, J.-I. (2002) in *Adv. in Polym. Sci: Polymers for Photonic Applications* I (Lee, K.-S., ed.), Berlin, Springer Verlag, **158**, 193.
Shimizu, N. & Laurent, C. (1998) *IEEE Trans. on Dielectrics & Electrical Insulation*, **5**, 651.

Shirakawa, H. & Ikeda, S. (1971) *Polym. J.*, **2**, 231.
Shirakawa, H., Louis, E. J., MacDiarmid, A. G., Chiang, C. K. & Heeger, A. J. (1977) *Chem. Commun.*, 578.
Sichel, E. K., ed. (1982) *Carbon Black-Polymer Composites (Plastics in Engineering 3)*, New York, Marcel Dekker.
Sillars, R. W. (1937) *J. Instn. Electr. Engrs*, **80**, 378.
Sillars, R. W. (1973) *Electrical Insulating Materials and their Applications, IEE Monograph Series*, **14**, Stevenage, Peter Reveginus Ltd.
Silva, C., Stevens, M. A., Russell, D. M., Setayesh, S., Müllen, K. & Friend, R. H. (2001) *Synth. Metals*, **116**, 9.
Silva, C., Russell, D. M., Dhoot, A. S., Hertz, L. M., Daniel, C., Greenham, N. C., Arias, A. C., Setayesh, S., Müllen, K. & Friend, R. H. (2002) *J. Phys.: Condens. Matter*, **14**, 9803.
Sinar, J., ed. (2003) *Organic Light-Emitting Diodes*, Berlin, Springer.
Sinha, S., Rothe, C., Günther, R., Scherf, U. & Monkman, A. P. (2003) *Phys. Rev. Lett.*, **90**, 127402.
Sioncke, S., Verbiest, T. & Persoons, A. (2003) *Mater. Sci. & Eng. R*, **42**, 115.
Sirringhaus, H., Tessler, N. & Friend, R. H. (1998) *Science*, **280**, 1741.
Sirringhaus, H., Brown, P. J., Friend, R. H., Nielsen, M. M., Bechgaard, K., Langeveld-Voss, B. M. W., Spiering, A. J. H., Janssen, R. A. J. Meijer, E. W., Herwig, P. & de Leeuw, D. M. (1999) *Nature*, **401**, 685.
Sirringhaus, H., Kawase, T., Friend, R. H., Shimoda, T., Inbasekeran, M., Wu, W., & Woo, P. E. (2000) *Science*, **290**, 212.
Skotheim, T. A., ed. (1986) *Handbook of Conducting Polymers*, New York, Marcel Dekker.
Skotheim, T. A., Elsenbaumer, R. L. & Reynolds, J. R., eds. (1998) *Handbook of Conducting Polymers*, 2nd edn., New York, Marcel Dekker.
Slater, J. C. & Shockley, W. (1936) *Phys. Rev.*, **50**, 705.
Smela, E. (2003) *Adv. Mater.*, **15**, 481.
Smela, E. & Gadegaard, N. (1999) *Adv. Mater.*, **11**, 953.
Smela, E., Inganäs, O. & Lundström, I. (1995) *Science*, **268**, 1735.
Smith, J. W. (1955) *Electric Dipole Moments*, London, Butterworths.
Smyth, C. P. (1955) *Dielectric Behaviour and Structure*, New York, McGraw-Hill.
Somani, P. R. & Radhakrishnan, S. (2002) *Mater. Chem. & Phys.*, **77**, 117.
Someya, T., Dodabalapur, A., Gelperin, A., Katz, H. E. & Bao, Z. (2002) *Langmuir*, **18**, 5299.
Stark, K. H. & Garton, C. G. (1955) *Nature*, **176**, 1225.
Stille, J. K. & Frey, D. A. (1961) *J. Amer. Chem. Soc.*, **83**, 1697.
Stolka, M., Yanus, J. F. & Pai, D. M. (1984) *J. Phys. Chem.*, **88**, 4707.
Su, W.-P., Schrieffer, J. R. & Heeger, A. J. (1979) *Phys. Rev. Lett.*, **42**, 1698.
Su, W.-P., Schrieffer, J. R. & Heeger, A. J. (1980) *Phys. Rev., B*, **28**, 2099.
Suggett, A. (1972) in *Dielectric and Related Molecular Processes*, **1**, London, Chemical Society, p. 100.
Süle, P., Kurth, S. & Van Doren, V. (2000) *J. Chem. Phys.*, **112**, 7355.
Sun, J.-Q. & Bartlett, R. J. (1996) *Phys. Rev. Lett.*, **77**, 3669.
Sutherland, R. L. (2003) *Handbook of Nonlinear Optics*, 2nd edn., New York, Marcel Dekker.
Szent-Györgyi, A. (1941) *Nature*, **148**, 157.
Tada, Y., Sato, M., Takeno, N., Kameshima, T., Nakacho, Y. & Shigekara, K. (1994) *Macromol. Chem. Phys.*, **195**, 571.

Takada, T. & Sakai, T. (1983) *IEEE Trans. Electr. Insul.*, **EI-18**, 619.
Tallman, D. E., Spinks, G., Dominis, A. & Wallace, G. G. (2002) *J. Sol. St. Electrochem.*, **6**, 73.
Tanaka, J. & Tanaka, M. (1986) in *Handbook of Conducting Polymers* (Skotheim, T. A., ed.), New York, Marcel Dekker, **2**, 1269.
Tang, C. W. (1986) *Appl. Phys. Lett.*, **48**, 183.
Tang, C. W. & VanSlyke, S. A. (1987) *Appl. Phys. Lett.*, **51**, 913.
Tarascon, J. M., Gozdz, A. S., Schmutz, C., Shokoohi, F. & Warren, P. R. (1996) *Solid State Ionics*, **86–8**, 49.
Taylor, D. M., ed. (1999) *Electrostatics*, Inst. Phys. Conf. Ser., **163**, Bristol, Institute of Physics Press.
Taylor, D. M. & Secker, P. E. (1994) *Industrial Electrostatics: Fundamentals and Measurements*, New York, RSP, Wiley.
Taylor, P. L. & Heinonen, O. (2002) *A Quantum Approach to Condensed Matter Physics*, Cambridge, Cambridge University Press.
Tessler, N. (1999) *Adv. Mater.*, **11**, 363.
Theophilou, N., Aznar, R., Munardi, A., Sledz, J., Shue, F. & Naarman, H. (1986) *Synth. Metals*, **16**, 337.
Toussaint, J. M. & Brédas, J.-L. (1995) *Synth. Metals*, **69**, 637.
Touwslager, F. J., Willard, N. P. & de Leeuw, D. M. (2002) *Appl. Phys. Lett.*, **81**, 4556.
Tretiak, S., Saxena, A., Martin, R. L. & Bishop, A. R. (2003) *Proc. Nat. Acad. Sci. USA*, **100**, 2185.
Tsubokawa, N., Tsuchida, M. & Nakazawa, Y. (2001) *Sensors and Actuators, B*, **79**, 92.
Tsukamoto, J. (1992) *Adv. in Phys.*, **41**, 509.
Tsukamoto, J., Takahashi, A. & Kawasaki, K. (1990) *Japan J. Appl. Phys.*, **29**, 125.
Tzamalis, G., Zaidi, N. A., Homes, C. C. & Monkman, A. P. (2001) *J. Phys.: Condens. Matter*, **13**, 6297.
Ubbelohde, A. R. (1952) *An Introduction to Modern Thermodynamical Principles*, Oxford, Oxford University Press.
Ueno, N., Sugita, K., Seki, K. & Inokuchi, H. (1986) *Phys. Rev., B*, **34**, 6386.
Uhlir, A. (1955) *Bell Systems Tech. J.*, 105.
Ulański, J. (1990) *Synth. Metals*, **39**, 13.
Valdes, L. B. (1954) *Proc. Inst. Radio Engrs.*, **42**, 420.
van der Pauw, L. J. (1961) *Philips Res. Repts.*, **16**, 187.
van Hal, P. A., Wienk, M. M., Kroon, J. M., Verhees, W. J. H., Slooff, L. H., van Gennip, W. J. H., Jonkheijm, P. & Janssen, R. A. J. (2003) *Adv. Mater.*, **15**, 118.
van Turnhout, J. (1975) *Thermally Stimulated Discharge of Polymer Electrets*, Amsterdam, Elsevier.
Varlot, K., Martin, J. M., Quet, C. & Kihn, Y. (1997) *Ultramicroscopy*, **68**, 123.
Varlot, K., Martin, J. M. & Quet, C. (1998) *J. of Microscopy Pt.2*, **191**, 187.
Varlot, K., Martin, J. M. & Quet, C. (2001) *Micron*, **32**, 371.
Vercelli, B., Zecchin, S., Comisso, N., Zotti, G., Berlin, A., Dalcanale, E. & Groenendaal, L. (2002) *Chem. Mater.*, **14**, 4768.
Veres, J., Ogier, S. D., Leeming, S. W., Cupertino, D. C. & Khaffaf, S. M. (2003) *Adv. Func. Mater.*, **13**, 199.
Vermier, I. E., Kim, N. Y. & Laibinis, P. E. (1999) *Appl. Phys. Lett.*, **74**, 3860.
Voet, A. (1980) *Rubber & Chem. Techn.*, **54**, 42.
Voet, A., Whitten, W. N. & Cook, F. R. (1965) *Kolloid Z.*, **201**, 39.
Vogel, A. I., Cresswell, W. T., Jeffrey, G. H. & Leicester, J. (1952) *J. Chem. Soc.*, 514.

Volkenstein, M. V. (1963) *Configurational Statistics of Polymeric Chains*, New York, Interscience.
von Hippel, A. R., ed. (1954) *Dielectric Materials and Applications*, New York, Wiley.
von Hippel, A. R. & Maurer, R. J. (1941) *Phys. Rev.*, **59**, 820.
von Malm, N., Steiger, J., Heil, H., Schmechel, R. & von Seggern, H. (2002) *J. Appl. Phys.*, **92**, 7564.
Wagner, K. W. (1914) *Arch. Elektrotechn.*, **2**, 371.
Wakihara, M. & Yamamoto, O. (1998) *Lithium Ion Batteries*, Tokyo, Wiley-VCH.
Walker, A. B., Kambili, A. & Martin, S. J. (2002) *J. Phys.: Condens. Matter*, **14**, 9825.
Waltman, R. J. & Bargon, J. (1986) *Can. J. Chem.*, **64**, 76.
Wang, C., Kilitziraki, M., Pålsson, L.-O., Bryce, M. R., Monkman, A. P. & Samuel, I. D. W. (2001) *Adv. Func. Mater.*, **11**, 47.
Wang, F. E. & Chiu, Y.-N. (1976) *Chem. Phys.*, **12**, 225.
Warman, J. M., Gelinck, G. H. & de Haas, M. P. (2002) *J. Phys.: Condens. Matter*, **14**, 9935.
Weaver, M. S., Michalski, L. A., Rajan, K., Rothman, M. A., Silvernail, J. A, Brown, J. J., Burrows, P. E., Graff, G. L., Gross, M. E., Martin, P. M., Hall, M., Mast, E., Bonham, C, Bennett, W. & Zumhoff, M. (2002) *Appl. Phys. Lett.*, **81**, 2929.
Wegewijs, B. R., Dicker, G., Piris, J., García, A. A., de Haas, M. P. & Warman, J. M. (2000) *Chem. Phys. Lett.*, **332**, 79.
Wegner, G. (1969) *Z. Naturforsch.*, **24b**, 824.
Weibull, W. (1951) *J. Appl. Mech.*, **18**, 293.
Weiser, G. (1992) *Phys. Rev., B*, **45**, 14076.
Weiser, G. & Horvath, A. (1998) *Chem. Phys.*, **227**, 153.
Wessling, B. (1998) *Synth. Metals*, **93**, 143.
Wesson, L. G. (1948) *Tables of Electric Dipole Moments*, Cambridge, MA, Technology Press.
White, D. & Bott, D. C. (1984) *Polym. Commun.*, **25**, 98.
Whitehead, S. (1951) *Dielectric Breakdown of Solids*, Oxford, Oxford University Press.
Willbourn, A. H. (1958) *Trans. Faraday Soc.*, **54**, 717.
Williams, G. & Watts, D. C. (1970) *Trans. Faraday Soc.*, **66**, 80.
Williams, J. M., Ferraro, J. R., Thorn, R. J., Carlson, K. D., Geiser, U., Wang, H. H., Kini, A. M. & Whangbo, M.-H. (1992) *Organic Superconductors (Including Fullerenes)*, Englewood Cliffs, Prentice-Hall.
Williams, M. L., Landel, R. F. & Ferry, J. D. (1955) *J. Amer. Chem. Soc.*, **77**, 3701.
Wohlgenannt, M., Tandon, K., Mazumdar, S., Ramasesha, S. & Vardeny, Z. V. (2001) *Nature*, **409**, 494.
Wong, J. Y., Langer, R. & Ingber, D. E. (1994) *Proc. Nat. Acad. Sci. USA*, **91**, 3201.
Works, C. N. (1947) *J. Appl. Phys.*, **18**, 605.
Wudl, F., Kobayashi, M. & Heeger, A. J. (1984) *J. Org. Chem.*, **49**, 3382.
Xia, R., Heliotis, G., Hou, Y. & Bradley, D. D. C. (2003) *Org. Electron.*, **4**, 165.
Yamada, Y., Saito, M., Shinozaki, T. & Takeichi, M. (1999) *U. S. Patent*, 5,965,064.
Yang, Y. & Heeger, A. J. (1994) *Nature*, **372**, 344.
Yau, S.-T., Barisci, J. N. & Spinks, G. M. (1999) *Appl. Phys. Lett.*, **74**, 667.
Yesodha, S. K., Pillai, C. K. S. & Tsutsumi, N. (2004) *Prog. Polym. Sci.*, **29**, 45.
Young, R. J. (1984) in *Polydiacetylenes: Synthesis, Structure, and Electronic Properties* (Bloor, D. & Chance, R. R., eds.), Dordrecht, Martinus Nijhoff, p. 335.
Yu, L., Chan, W. K., Peng, Z. & Gharavi, A. (1996) *Acc. Chem. Res.*, **29**, 13.
Zhang, F., Jonforsen, M. D., Johansson, D. M., Andersson, M. R. & Inganäs, O. (2003) *Synth. Metals*, **138**, 555.

Zhang, L. G., Fan, Y., Ma, B. L., Xu, X. Y., Kong, X. G., Shen, D. Z. & Li, Y. J. (2002) *Thin Sol. Films*, **419**, 194.
Zhokhavets, U., Goldhahn, R., Gobsch, G. & Schliefke, W. (2003) *Synth. Metals*, **138**, 491.
Zhou, M. & Heinze, J. (1999) *Electrochim. Acta*, **44**, 1733.
Zilker, S. J. (2000) *Chem. Phys. Chem.*, **1**, 72.
Zilker, S. J. (2002) *Chem. Phys. Chem.*, **3**, 333.
Zimm, B. H. (1956) *J. Chem. Phys.*, **24**, 269.
Zisman, W. A. (1932) *Rev. Sci. Instrum.*, **3**, 367.
Zojer, E., Knupfer, M., Shuai, Z., Brédas, J. L., Fink, J. & Leising, G. (2000) *J. Phys.: Condens. Matter*, **12**, 1753.

Index